VOLUME ONE HUNDRED AND TWENTY EIGHT

ADVANCES IN PROTEIN CHEMISTRY AND STRUCTURAL BIOLOGY

Membrane Proteins

VOLUME ONE HUNDRED AND TWENTY EIGHT

ADVANCES IN PROTEIN CHEMISTRY AND STRUCTURAL BIOLOGY

Membrane Proteins

Edited by

ROSSEN DONEV
MicroPharm Ltd,
United Kingdom

ELSEVIER

ACADEMIC PRESS
An imprint of Elsevier

Academic Press is an imprint of Elsevier
50 Hampshire Street, 5th Floor, Cambridge, MA 02139, United States
525 B Street, Suite 1650, San Diego, CA 92101, United States
The Boulevard, Langford Lane, Kidlington, Oxford OX5 1GB, United Kingdom
125 London Wall, London, EC2Y 5AS, United Kingdom

First edition 2022

ISBN: 978-0-323-98895-7
ISSN: 1876-1623

For information on all Academic Press publications
visit our website at https://www.elsevier.com/books-and-journals

Publisher: Zoe Kruze
Acquisitions Editor: Leticia Lima
Developmental Editor: Naiza Ermin Mendoza
Production Project Manager: Abdulla Sait
Cover Designer: Matthew Limbert

Typeset by STRAIVE, India

Contents

Contributors *xi*

1. Architects of their own environment: How membrane proteins shape the Gram-negative cell envelope **1**

Johannes Thoma and Björn M. Burmann

1. Introduction 2
2. Protein insertion and translocation by the SecYEG complex 3
3. Outer membrane β-barrel folding at the BAM and TAM complexes 6
4. Lipoprotein targeting and sorting by the Lol system 8
5. Synthesis and transport of lipopolysaccharide to the outer membrane 10
6. The blind spot: Phospholipid synthesis, insertion and trafficking 13
7. Shaping the cell envelope: Membrane attachment and detachment 15
8. Membrane protein turnover: AAA+ proteases and beyond 18
9. Sensing membrane defects and maintenance of membrane integrity 20
10. Conclusion 22

References 22

2. Modulation of host cellular responses by gram-negative bacterial porins **35**

Arpita Sharma, Shashi Prakash Yadav, Dwipjyoti Sarma, and Arunika Mukhopadhaya

1. Introduction 36
2. Structure of gram-negative bacterial porins 38
3. Functions of the gram-negative bacterial porins 40
4. Important uses of gram-negative bacterial porins 59
5. Conclusion 63

Acknowledgment 63

References 64

3. Pore formation by pore forming membrane proteins towards infections **79**

Achinta Sannigrahi and Krishnananda Chattopadhyay

1. Introduction 80
2. PFT mechanism of action and structural classification 81
3. α-PFTs 82

4. β-PFTs 86
5. Conformational switch driven membrane pore formation 92
6. PFT effects and cellular defense mechanisms 98
7. Calcium-dependent membrane repair mechanisms 100
8. Therapeutic development 101
9. Conclusion 103
Reference 103

4. Transmembrane β-barrel proteins of bacteria: From structure to function **113**

Simen Hermansen, Dirk Linke, and Jack C. Leo

1. Introduction 114
2. Structure of β-barrel proteins 124
3. Folding and membrane integration of β-barrels 127
4. Functions of transmembrane β-barrels 135
5. Evolution of β-barrels 145
6. Medical relevance of bacterial β-barrel proteins 146
7. Conclusions and future perspectives 150
Disclaimer 150
References 151

5. Ribosome-membrane crosstalk: Co-translational targeting pathways of proteins across membranes in prokaryotes and eukaryotes **163**

Krishnamoorthi Srinivasan, Aneek Banerjee, Priya Baid, Ankit Dhur, and Jayati Sengupta

1. Introduction 164
2. Membrane translocation via a conserved channel mechanism 169
3. Molecular machinery of bacterial protein transport 171
4. Co-translational protein transport in eukaryotes: 175
5. Concluding remarks 188
Acknowledgments 190
Conflict of interest statement 190
Author contributions 190
References 190

6. Mechanisms of long-distance allosteric couplings in proton-binding membrane transporters **199**

Ana-Nicoleta Bondar

1. Introduction 200
2. Proton sensing and proton transfer in human physiology and disease 200

3. Titratable protein sidechains in protein environments 201
4. Protonation-coupled changes in membrane protein structure and dynamics 203
5. Conclusions 228
Acknowledgments 229
References 229

7. Structures and functions of the membrane-damaging pore-forming proteins 241
Anish Kumar Mondal and Kausik Chattopadhyay

1. Introduction 242
2. α-PFPs 245
3. β-PFPs 253
4. Membrane-attack complex and perforin family 264
5. Gasdermins 268
6. PFPs from multicellular eukaryotes 270
7. Conclusion 273
Acknowledgments 275
References 275

8. Ganglioside binding domains in proteins: Physiological and pathological mechanisms 289
Fodil Azzaz, Nouara Yahi, Coralie Di Scala, Henri Chahinian, and Jacques Fantini

1. Structural features of gangliosides 290
2. Membrane localization of gangliosides 292
3. Molecular interactions: The "NH trick" 294
4. Flat surfaces and chalices: Same gangliosides, two topologies 300
5. Ganglioside binding domains (GBDs) 301
6. Helical GBDs 301
7. Loop shaped GBDs 303
8. Large flat GBDs 305
9. Gangliosides and synapses 306
10. Gangliosides and bacterial toxins 308
11. Gangliosides and virus infection 311
12. Gangliosides and neurodegenerative diseases 313
13. Conclusion and perspectives 315
References 317

9. Development of enhanced conformational sampling methods to probe the activation landscape of GPCRs **325**

Ravinder Abrol, Erik Serrano, and Luis Jaimes Santiago

1. Introduction 326
2. Molecular dynamics based methods 332
3. Discrete conformational sampling methods 343
4. Future of GPCR conformational sampling is hybrid approaches 347

Acknowledgments 348

References 348

10. Molecular determinants of GPCR pharmacogenetics: Deconstructing the population variants in β_2-adrenergic receptor **361**

Manali Joshi, Siddhanta V. Nikte, and Durba Sengupta

1. Introduction 362
2. Computational tools to analyze SNPs in GPCRs 379
3. An overview of the pharmacogenomics in β_2-AR 383
4. Conclusions 390

Acknowledgments 391

References 391

11. The role of membranes in function and dysfunction of intrinsically disordered amyloidogenic proteins **397**

Bani Kumar Pathak, Sandip Dey, Sukanya Mozumder, and Jayati Sengupta

1. Introduction 398
2. Human disease-related, widely known IDPs 400
3. Association of IDPs with biological membranes 404
4. Implications of IDP-membrane interactions in physiological functions 410
5. Pathological implications of IDP-membrane interactions 412
6. Outlook 419

Acknowledgments 420

Conflict of interest statement 420

Author contributions 420

References 421

12. Dendritic spine membrane proteome and its alterations in autistic spectrum disorder **435**

Ana Sofía Vallés and Francisco J. Barrantes

1. Introduction 436
2. Molecular architecture of the dendritic spine 439
3. Cholinergic neurotransmission and nicotinic receptors in dendritic spines 443
4. Spine alterations in neurological and neuropsychiatric disease 452
5. Spine dysfunction in autism spectrum disorder 453
6. Dysfunctional neurotransmission in autistic spectrum disorder 454
7. Concluding remarks 458

References 459

Contributors

Ravinder Abrol
Department of Chemistry and Biochemistry, California State University, Northridge, CA, United States

Fodil Azzaz
INSERM UMR_S 1072; Aix-Marseille Université, Marseille, France

Priya Baid
Division of Structural Biology and Bioinformatics, CSIR-Indian Institute of Chemical Biology, Kolkata, India

Aneek Banerjee
Division of Structural Biology and Bioinformatics, CSIR-Indian Institute of Chemical Biology, Kolkata; Academy of Scientific and Innovative Research (AcSIR), Ghaziabad, India

Francisco J. Barrantes
Instituto de Investigaciones Biomédicas (BIOMED), UCA-CONICET, Buenos Aires, Argentina

Ana-Nicoleta Bondar
University of Bucharest, Faculty of Physics, Măgurele, Romania; Forschungszentrum Jülich, Institute of Computational Biomedicine, Jülich, Germany

Björn M. Burmann
Wallenberg Centre for Molecular and Translational Medicine; Department of Chemistry and Molecular Biology, University of Gothenburg, Göteborg, Sweden

Henri Chahinian
INSERM UMR_S 1072; Aix-Marseille Université, Marseille, France

Kausik Chattopadhyay
Department of Biological Sciences, Indian Institute of Science Education and Research Mohali, Mohali, Punjab, India

Krishnananda Chattopadhyay
Structural Biology and Bioinformatics Division, CSIR-Indian Institute of Chemical Biology, Kolkata, West Bengal, India

Sandip Dey
Structural Biology and Bioinformatics Division, CSIR-Indian Institute of Chemical Biology, Jadavpur, Kolkata, India

Ankit Dhur
Division of Structural Biology and Bioinformatics, CSIR-Indian Institute of Chemical Biology, Kolkata, India

Coralie Di Scala
Neuroscience Center-HiLIFE, University of Helsinki, Helsinki, Finland

Jacques Fantini
INSERM UMR_S 1072; Aix-Marseille Université, Marseille, France

Simen Hermansen
Section for Genetics and Evolutionary Biology, Department of Biosciences, University of Oslo, Oslo, Norway

Manali Joshi
Bioinformatics Centre, Savitribai Phule Pune University, Pune, India

Jack C. Leo
Antimicrobial resistance, Omics and Microbiota Group, Department of Biosciences, Nottingham Trent University, Nottingham, United Kingdom

Dirk Linke
Section for Genetics and Evolutionary Biology, Department of Biosciences, University of Oslo, Oslo, Norway

Anish Kumar Mondal
Department of Biological Sciences, Indian Institute of Science Education and Research Mohali, Mohali, Punjab, India

Sukanya Mozumder
Structural Biology and Bioinformatics Division, CSIR-Indian Institute of Chemical Biology, Jadavpur, Kolkata; Academy of Scientific and Innovative Research (AcSIR), Ghaziabad, India

Arunika Mukhopadhaya
Department of Biological Sciences, Indian Institute of Science Education and Research Mohali, Mohali, Punjab, India

Siddhanta V. Nikte
CSIR-National Chemical Laboratory, Dr. Homi Bhabha Road, Pune; Academy of Scientific and Innovative Research (AcSIR), Ghaziabad, India

Bani Kumar Pathak
Structural Biology and Bioinformatics Division, CSIR-Indian Institute of Chemical Biology, Jadavpur, Kolkata, India

Achinta Sannigrahi
Department of Chemical Engineering, Indian Institute of Science, Bengaluru, Karnataka, India

Luis Jaimes Santiago
Department of Chemistry and Biochemistry, California State University, Northridge, CA, United States

Dwipjyoti Sarma
Department of Biological Sciences, Indian Institute of Science Education and Research Mohali, Mohali, Punjab, India

Durba Sengupta
CSIR-National Chemical Laboratory, Dr. Homi Bhabha Road, Pune; Academy of Scientific and Innovative Research (AcSIR), Ghaziabad, India

Jayati Sengupta
Division of Structural Biology and Bioinformatics, CSIR-Indian Institute of Chemical Biology, Kolkata; Academy of Scientific and Innovative Research (AcSIR), Ghaziabad; Structural Biology and Bioinformatics Division, CSIR-Indian Institute of Chemical Biology, Jadavpur, Kolkata, India

Erik Serrano
Department of Chemistry and Biochemistry, California State University, Northridge, CA, United States

Arpita Sharma
Department of Biological Sciences, Indian Institute of Science Education and Research Mohali, Mohali, Punjab, India

Krishnamoorthi Srinivasan
Division of Structural Biology and Bioinformatics, CSIR-Indian Institute of Chemical Biology, Kolkata; Academy of Scientific and Innovative Research (AcSIR), Ghaziabad, India

Johannes Thoma
Wallenberg Centre for Molecular and Translational Medicine; Department of Chemistry and Molecular Biology, University of Gothenburg, Göteborg, Sweden

Ana Sofía Vallés
Instituto de Investigaciones Bioquímicas de Bahía Blanca (UNS-CONICET), Bahía Blanca, Argentina

Shashi Prakash Yadav
Department of Biological Sciences, Indian Institute of Science Education and Research Mohali, Mohali, Punjab, India

Nouara Yahi
INSERM UMR_S 1072; Aix-Marseille Université, Marseille, France

CHAPTER ONE

Architects of their own environment: How membrane proteins shape the Gram-negative cell envelope

Johannes Thoma[a,b,*] and Björn M. Burmann[a,b]
[a]Wallenberg Centre for Molecular and Translational Medicine, University of Gothenburg, Göteborg, Sweden
[b]Department of Chemistry and Molecular Biology, University of Gothenburg, Göteborg, Sweden
*Corresponding author: e-mail address: johannes.thoma@gu.se

Contents

1. Introduction 2
2. Protein insertion and translocation by the SecYEG complex 3
3. Outer membrane β-barrel folding at the BAM and TAM complexes 6
4. Lipoprotein targeting and sorting by the Lol system 8
5. Synthesis and transport of lipopolysaccharide to the outer membrane 10
6. The blind spot: Phospholipid synthesis, insertion and trafficking 13
7. Shaping the cell envelope: Membrane attachment and detachment 15
8. Membrane protein turnover: AAA+ proteases and beyond 18
9. Sensing membrane defects and maintenance of membrane integrity 20
10. Conclusion 22
References 22

Abstract

Gram-negative bacteria are surrounded by a complex multilayered cell envelope, consisting of an inner and an outer membrane, and separated by the aqueous periplasm, which contains a thin peptidoglycan cell wall. These bacteria employ an arsenal of highly specialized membrane protein machineries to ensure the correct assembly and maintenance of the membranes forming the cell envelope. Here, we review the diverse protein systems, which perform these functions in *Escherichia coli*, such as the folding and insertion of membrane proteins, the transport of lipoproteins and lipopolysaccharide within the cell envelope, the targeting of phospholipids, and the regulation of mistargeted envelope components. Some of these protein machineries have been known for a long time, yet still hold surprises. Others have only recently been described and some are still missing pieces or yet remain to be discovered.

Advances in Protein Chemistry and Structural Biology, Volume 128
ISSN 1876-1623
https://doi.org/10.1016/bs.apcsb.2021.10.001

1. Introduction

One decisive event in the early days of evolution is marked by the formation of membrane vesicles, spontaneously formed and bound together by the hydrophobic effect, which allowed for the spatial separation of individual reaction compartments and thus can be seen as a major step toward the development of the first protocells (Chen & Walde, 2010). However, restricted to the selective permeability of amphiphile membranes, such protocells suffered from severe limitations to molecular transport, which could be overcome only through the incorporation of transmembrane enzymes, allowing for the specific uptake and release of reactants (Monnard & Deamer, 2002). Fast forward several billion years, the lipids and proteins forming biological membranes have co-evolved to form one highly complex functional unit, wherein lipids provide the structural framework, which membrane proteins equip with manifold functions (Hanczyc & Monnard, 2017). Importantly, proteins aren't merely functional parts of biological membranes, regulating the directed transport of molecules, stimuli, and information across the bilayer, but are critically involved in membrane biogenesis. Membrane proteins are responsible for every step from lipid synthesis to membrane assembly and even eventual disassembly. Membrane proteins insert other membrane proteins, aid as well as monitor their folding, and regulate their eventual degradation. Membrane proteins, more precisely a subset of membrane proteins, can therefore be seen as the architects of biological membranes, ensuring their structural and functional integrity. This chapter is dedicated to this group of membrane proteins. Using the Gram-negative model organism *E. coli* as an example, we discuss the diverse membrane protein machineries, which ensure and maintain the functional integrity of the membranes forming their cell envelope.

Gram-negative bacteria are surrounded by two highly complex, yet structurally vastly different membranes. The inner membrane (IM), similar to the majority of cellular membranes, is assembled from phospholipids, namely, Phophatidylethanolamine (PE), Phophatidylglycerol (PG), and cardiolipin (CL), in a ratio of roughly 70:20:10 (Dowhan & Bogdanov, 2002). While this lipid composition is also found in the inner leaflet of the outer membrane (OM), its outer leaflet is formed almost exclusively by lipopolysaccharide (LPS). Much like the differing lipid composition, the membrane proteins of bacterial membranes are very different. Inner membrane proteins (IMPs) are characterized by α-helical transmembrane segments, whereas outer membrane proteins (OMPs) are predominantly defined by transmembrane β-strands (Silhavy, Kahne, & Walker, 2010). Due to these differences, Gram-negative bacteria maintain highly specialized cellular machineries to ensure the assembly and homeostasis of these membranes. Here, we summarize what is known about these machineries and review recent findings covering the structural and mechanistic details underlying their function. We begin with the machineries responsible for the folding of transmembrane proteins at the IM and OM and the machinery for lipoprotein targeting and insertion. We cover the biogenesis and targeting of LPS, as well as the transport of phospholipids, and finally look at the systems responsible for ensuring membrane integrity and the machinery responsible for providing structural support to the cell envelope.

2. Protein insertion and translocation by the SecYEG complex

In principle, membrane proteins can fold into lipid bilayers in an unassisted process solely relying on their inherent biophysical properties—the hydrophobic core of the bilayer presents a substantial energetic barrier, which all hydrophilic elements of a protein need to overcome during the process of folding, but the final folded membrane-inserted state of a protein often is a sufficiently deep energetic minimum to funnel the protein directly into (Bowie, 2005; Fleming, 2014; White & Wimley, 1999). Nevertheless, despite being frequently observed in biophysical experiments *in vitro*, the unassisted folding of membrane proteins into model membranes is a highly inefficient process (Harris & Booth, 2012). *In vivo*, the folding process is streamlined by means of the SecYEG translocon residing in the inner membrane (IM), which works in concert with several accessory proteins modulating its function as either a protein translocase or membrane protein foldase (Fig. 1) (Cranford-Smith & Huber, 2018).

Proteins targeted to the cell envelope are distinguished by an amino-terminal signal sequence, which exits the ribosome as the first part of the nascent protein chain during biosynthesis. Thereby, the hydrophobicity of the signal peptide is indicative for the targeting of the nascent protein to the so-called translocon complex *via* either the co- or to the post-translational pathway (Hegde & Bernstein, 2006).

The co-translational pathway is characterized by the involvement of the ribonucleoprotein signal recognition particle (SRP) (Akopian, Shen, Zhang, & Shan, 2013). SRP recognizes and subsequently binds to the signal

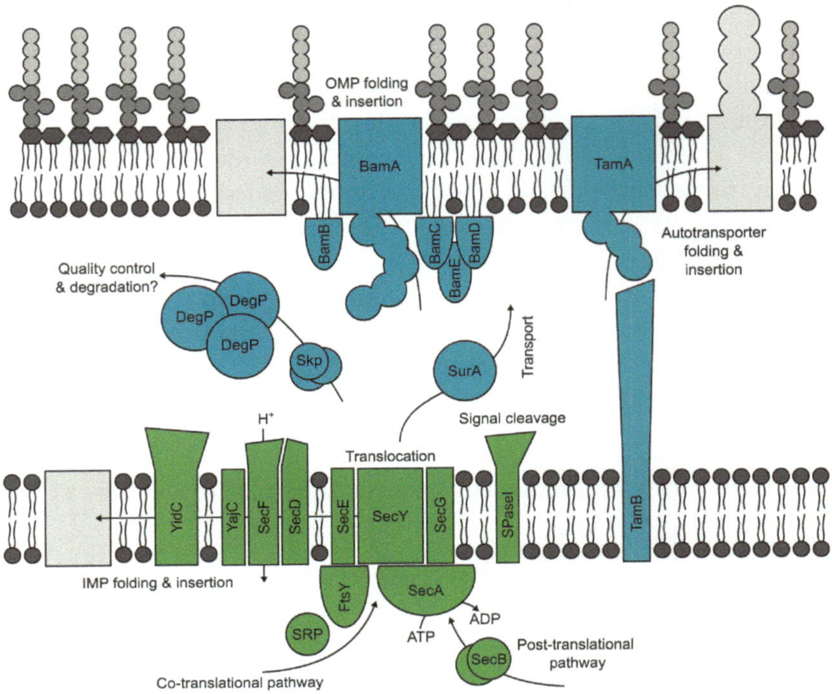

Fig. 1 Transport, folding and insertion pathways of transmembrane proteins. Transmembrane proteins are co- or post-translationally targeted to the SecYEG translocon complex, which inserts inner membrane proteins laterally and secretes outer membrane proteins into the periplasm. Outer membrane proteins are transported to outer membrane and inserted by the BAM complex, whereas autotransporters are inserted by the TAM module.

peptide causing a temporary stalling of the protein synthesis, which is released once the SRP-ribosome complex engaged the membrane associated SRP receptor FtsY, leading to the transfer of the signal-peptide to the SecYEG translocon complex (Elvekrog & Walter, 2015; Saraogi & Shan, 2014). Once directed to the translocon, the SRP–FtsY complex dissociates and the amino-terminal signal peptide interacts directly with the SecYEG complex to facilitate the co-translational insertion of the nascent membrane protein (Draycheva, Lee, & Wintermeyer, 2018).

In the alternative post-translational pathway, proteins are fully synthesized within the cytosol and received by the soluble holdase chaperones SecB and/or Trigger Factor (TF) (De Geyter et al., 2020; Sala, Bordes, & Genevaux, 2014). Together these chaperones protect nascent proteins from misfolding and aggregation by maintaining them in an unfolded, highly flexible yet completely folding competent state characteristic for holdases (Burmann & Hiller, 2012; Burmann, Wang, & Hiller, 2013; Huang, Rossi, Saio, & Kalodimos, 2016; Morgado, Burmann, Sharpe, Mazur, & Hiller, 2017). Thereby, proteins are subsequently targeted and inserted into the translocon complex in an ATP-dependent process driven by the auxiliary SecA protein (Collinson, 2017; Cranford-Smith & Huber, 2018; Gelis et al., 2007). Recently, SecA has been implied to also be involved in a conjoined so-called uncoupled co-translational pathway, directing nascent proteins to SecYEG without a direct interaction between the ribosome and the translocon (Huber et al., 2017). The SecYEG complex is not only responsible for folding and insertion of proteins into the IM, but also for translocating periplasmic as well as OMPs across the and into the periplasm (Oswald, Njenga, Natriashvili, Sarmah, & Koch, 2021). SecYEG thus has an important sorting function, whereby hydrophobic α-helical transmembrane segments of IMPs are detected and inserted into the membrane, whereas the less hydrophobic segments of periplasmic proteins or OMPs are shuttled through the translocon and into the periplasm (White & von Heijne, 2008). Proteins designated to the IM are almost exclusively targeted to the co-translational pathway, whereas the majority of secreted proteins are targeted to the post-translational pathway (Smets, Loos, Karamanou, & Economou, 2019).

SecYEG itself forms a heterotrimeric transmembrane complex residing in the IM and comprises SecY as the main translocon channel supplemented by the accessory proteins SecE and SecG (Komar et al., 2016). Thereby, 10 transmembrane α-helices of the main component SecY, together with three helices contributed by SecE and two by SecG form the core segment—a protein conducting channel (Breyton, Haase, Rapoport, Kühlbrandt, & Collinson, 2002; Tanaka et al., 2015). This complex can be complemented by SecA, harboring two nucleotide binding domains that can bind to the cytoplasmic side of SecYEG in either a monomeric or dimeric form. However, the details of these different quaternary structures are still debated and remain under investigation (Banerjee, Lindenthal, & Oliver, 2017; Gouridis et al., 2013). In the cellular context, SecYEG has been observed to often occur within a much larger protein complex, termed the holo-translocon (Komar et al., 2016), where it is complemented by four additional proteins, the trimeric accessory complex SecDF–YajC (Tsukazaki et al., 2011) as well as the membrane protein insertase termed YidC (Kuhn & Kiefer, 2017).

The translocation or membrane insertion process is initiated by binding of the amino-terminal signal sequence to the central region of the translocon. The signal peptides of IMPs, mostly equivalent with their first transmembrane helix, can be released into the IM through a lateral gate of the translocon, sequentially followed by the succeeding transmembrane segments. In contrast, in case of substrates secreted into the periplasm the signal peptides remain bound, resulting in a hairpin-like arrangement of the nascent protein chain within SecYEG during translocation (Papanikou, Karamanou, & Economou, 2007; Park et al., 2014). Whereas during membrane insertion the nascent protein is directly fed into the translocon by the ribosome, translocation is energetically driven either by the ATPase activity of SecA or by the accessory proteins SecD and SecF coupled to the proton-motive force (PMF) across the IM (Economou, 1999; Ricci & Silhavy, 2012; Schiebel, Driessen, Hartl, & Wickner, 1991; Tsukazaki et al., 2011). Moreover, translocon activity seems to be dependent on the presence of cardiolipin (Corey et al., 2018). Mechanistically, when driven by SecA, translocation has been proposed to occur *via* a "Brownian ratchet" mechanism, whereby proteins can freely move into the translocon, but are blocked from escaping in the reverse direction, and as a consequence can only proceed in one direction (Allen et al., 2016). In contrast, SecDF was proposed to act as a motor protein enhancing the rate of translocation of unfolded polypeptides by exerting a pulling force from the periplasmic side of the IM (Tsukazaki, 2018). Translocation into the periplasm is finalized by cleavage of the amino-terminal signal peptide by the membrane bound signal peptidase SPaseI (Paetzel, Karla, Strynadka, & Dalbey, 2002). The folding and membrane insertion of IMPs by the SecYEG translocon is aided by the membrane protein insertase YidC, which can also insert a subset of IMPs in an Sec-independent manner (Welte et al., 2012). YidC is thought to form a hydrophilic groove inside the IM, which facilitates the translocation and flipping of individual α-helices and thereby guiding IMPs towards their correct topology (Nagamori, Smirnova, & Kaback, 2004; Serdiuk et al., 2016).

3. Outer membrane β-barrel folding at the BAM and TAM complexes

Much like the Sec translocon at the inner membrane, the β-barrel assembly machinery (BAM) complex facilitates membrane protein folding into the bacterial OM (Fig. 1) (Ricci & Silhavy, 2019). Despite their phenomenological similarity, mechanistically the two machineries are vastly different. As an additional mechanistic constraint, the BAM machinery cannot rely on cellular energy sources such as ATP or the PMF, both of which are absent in the periplasm, but rather needs to function as the last link in an energetically downwards directed transport chain, mainly driven by entropy (Schiffrin, Brockwell, & Radford, 2017). Following biosynthesis, translocation by SecYEG, and cleavage of the signal peptide by SPaseI, nascent OMPs are received by soluble periplasmic holdase chaperones, primarily SurA (Sala et al., 2014; Saraogi & Shan, 2014). SurA, structurally highly related to the cytosolic Trigger Factor, is responsible for the bulk transport of OMPs to the OM and maintains its clients in a dynamic unfolded state throughout their passage across the aqueous periplasm, while preserving their folding-competence (Mas, Thoma, & Hiller, 2019). A secondary transport pathway has been hypothesized, formed by the periplasmic chaperone Skp and the protease/chaperone DegP (Sklar, Wu, Kahne, & Silhavy, 2007). However, in contrast to a deletion of *surA*, a deletion of the *skp* gene has only mild effects on OM integrity (Sklar et al., 2007). Moreover, Skp has been found to tightly bind to and consequentially suppress the membrane integration of unfolded clients (Burmann et al., 2013; Thoma, Burmann, Hiller, & Müller, 2015), indicating a role for Skp in protein quality control together with DegP, rather than in a redundant transport pathway.

For the majority of OMPs, folding and membrane insertion occurs through interaction with the hetero-pentameric BAM complex (Ricci & Silhavy, 2019). This complex comprises the central component BamA, a 16-stranded β-barrel with five periplasmic protein transport associated (POTRA) domains, as well as the four accessory lipoproteins BamB–E, which assemble onto BamA at the inner surface of the OM (Bakelar, Buchan, & Noinaj, 2016). To date, the roles of the individual accessory proteins as well as the molecular details by which BAM inserts incoming OMPs remain elusive. A stepwise insertion mechanism, based on β-strand augmentation along the seam of the BamA β-barrel seems plausible (Tomasek & Kahne, 2021). Such a mechanism is corroborated by the recent experimental observation of trapped intermediates between BamA and incoming OMPs (Doyle & Bernstein, 2019; Tomasek et al., 2020), which were also found in the mitochondrial BAM analog Sam50 (Höhr et al., 2018). Moreover, peptidomimetic antibiotics interfering with the β-barrel seam have been demonstrated to suppress BamA activity, further supporting the proposed functional model of β-strand augmentation (Kaur et al., 2021; Luther et al., 2019). Importantly, it could be shown that OMPs can efficiently fold into lipid bilayers from a SurA-stabilized state in a self-guided process as long as a few transmembrane β-hairpins are inserted in the membrane to form a folding template for the incoming polypeptide (Thoma et al., 2015). It is therefore conceivable that BAM is only required for the initial steps of membrane insertion, which is further corroborated by the observation that BAM can locally disturb and thin lipid bilayers, potentially providing the structural framework for further facilitating membrane protein insertion (Horne, Brockwell, & Radford, 2020; Hussain & Bernstein, 2018; Iadanza et al., 2020; Schiffrin et al., 2017). Recently, it has also been proposed, that BAM complexes in combination with SurA might form a periplasm-spanning super-complex in direct contact with the holo-translocon in the IM (Alvira et al., 2020). While the structural basis for such a complex remains unclear, from a cellular perspective, the spatial clustering of translocation and membrane insertion seems intriguing and in fact membrane regions of locally high insertion activity have already been observed (Gunasinghe et al., 2018; Rassam et al., 2015).

While the BAM inserts the majority of OMPs into the OM, with the translocation and assembly module (TAM), comprising TamA in the OM and TamB in the IM, a homologous system exists for the insertion of a subgroup of OMPs, so-called autotransporters, which are distinguished by extended extracellular passenger-domains, which protrude from a central transmembrane β-barrel (Albenne & Ieva, 2017; Ranava, Caumont-Sarcos, Albenne, & Ieva, 2018; Selkrig et al., 2012). For the particular case of autotransporters, the formation of a hybrid barrel with TamA during membrane insertion was initially proposed, much like the later proposed mechanism for β-barrel insertion by the BAM complex (Gruss et al., 2013). This represents an intriguing mechanism, since only such a hybrid barrel would form a pore sufficiently large for translocation of the passenger domain into to extracellular space (Gruss et al., 2013).

4. Lipoprotein targeting and sorting by the Lol system

In addition to transmembrane proteins, a second important component of bacterial membranes are lipoproteins and the genome of *E. coli* contains ~100 different lipoproteins (Juncker et al., 2003). These membrane-attached proteins are anchored to the membrane through a lipid moiety and are commonly found in the outer leaflet of the IM as well as the inner leaflet of the OM and therefore residing almost exclusively in the periplasm (Narita & Tokuda, 2010; Zückert, 2014). Following their synthesis in the bacterial cytosol, lipoproteins are shuttled into the periplasm by the Sec translocon. The membrane targeting signal sequence of lipoproteins is distinguished by a motif of four amino acids, the so-called lipobox, with the consensus sequence LAGC (Inouye, Wang, & Sezikawa, 1977). Proteins containing the lipobox motif are not processed by SPaseI but instead recognized by the lipoprotein diacylglyceryl transferase Lgt, which transfers the diacylglyceryl moiety of PG to the cysteine of the lipobox forming a thioether linkage (Fig. 2) (Sankaran & Wu, 1994). Subsequently, the signal peptide is cleaved by the lipoprotein specific signal peptidase Lsp (Vogeley et al., 2016). The third and final modification step is executed by the acyltransferase Lnt, which transfers a fatty acid chain from (typically)

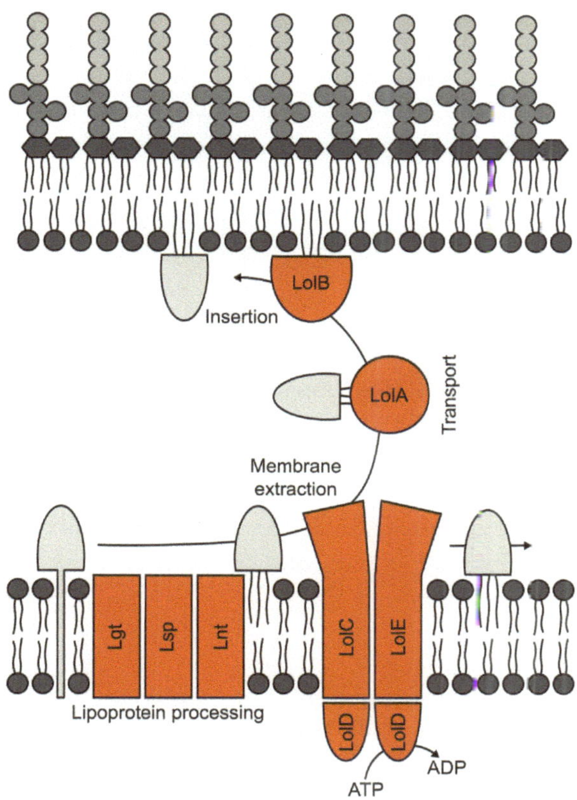

Fig. 2 Processing, sorting and transport of lipoproteins. Following translocation, lipoproteins are lipidated by the Lgt/Lsp/Lnt system. The Lol system is responsible for sorting, extracting and transporting outer membrane lipoproteins to the outer membrane.

PE to the amino-terminal cysteine of the nascent lipoprotein (Wiktor et al., 2017). The localization of lipoproteins to the IM and the OM, respectively, is generally dictated by the amino acid following the amino-terminal cysteine, where aspartic acid is indicative of IM-bound lipoproteins and all other amino acids result in direction to the OM (Yamaguchi, Yu, & Inouye, 1988). OM-bound lipoproteins are recognized and extracted from the IM by the ABC transporter complex LolCDE, transported across the periplasm by the soluble LolA protein and finally deposited in the OM by the lipoprotein LolB (Grabowicz, 2019). The recently solved structure of the LolCDE complex revealed a heterodimeric arrangement of the α-helical transmembrane bundles of LolC and LolE, with one cytosolic nucleotide binding domain of LolD attached to each (Tang et al., 2021). During membrane extraction, binding of a lipoprotein to LolCDE appears to induce a large structural transition, which lifts the lipid moieties out of the membrane plane, allowing engagement and binding of the periplasmic LolA to the periplasmic domain of LolC—a motion similarly proposed for mechanotransmission by the efflux pump MacB (Crow, Greene, Kaplan, & Koronakis, 2017; Kaplan, Greene, Crow, & Koronakis, 2018). The lipoprotein is then handed off to LolA upon ATP binding, the hydrolysis of which results in the release of the lipoprotein-bound LolA and allows LolCDE to retransition to the relaxed state (Tang, Chang, Zhang, et al., 2021). Little is known about the underlying molecular mechanism by which LolB facilitates the release of lipoproteins from LolA and their final anchoring into the OM (Grabowicz, 2019; Hayashi et al., 2014). Interestingly, the lipidation and OM-localization of LolB itself only seems to be important for targeting lipoproteins to the correct membrane but not required for the insertion mechanism, since LolB has been shown to facilitate lipoprotein deposition into both envelope membranes when present in a free soluble periplasmic form (Tsukahara, Mukaiyama, Okuda, Narita, & Tokuda, 2009).

5. Synthesis and transport of lipopolysaccharide to the outer membrane

Arguably the most complex molecule found in the membranes of the Gram-negative cell envelope is lipopolysaccharide (LPS). LPS consists of Lipid A, which has a diglucosamine phosphate headgroup and six alkyl chains, a core-oligo-saccharide subdivided into a conserved inner and a variable outer core, and an extended O-antigen polysaccharide (David, Yethon, & Whitfield, 1998; Raetz & Whitfield, 2002; Reyes, González, Jiménez, Herrera, & Andrade, 2012). The individual components of LPS are synthesized by more than 100 different enzymes at the cytosolic surface of the IM, assembled to form the full LPS molecule at the periplasmic surface of the IM, subsequently transported to the OM and finally flipped into its outer leaflet (Fig. 3) (Polissi & Sperandeo, 2014). In one synthetic pathway a group of Lpx* proteins conducts the assembly of the Lipid A moiety (Okuda, Sherman, Silhavy, Ruiz, & Kahne, 2016; Raetz & Whitfield, 2002). Next, the individual (between 8 and 15) sugar units of the core oligosaccharide are sequentially transferred to Lipid A by enzymes mostly of the Waa* group (Mamat, Skurnik, & Bengoechea, 2011; Valvano, 2014). Moreover, several additional enzymes are involved in the eventual

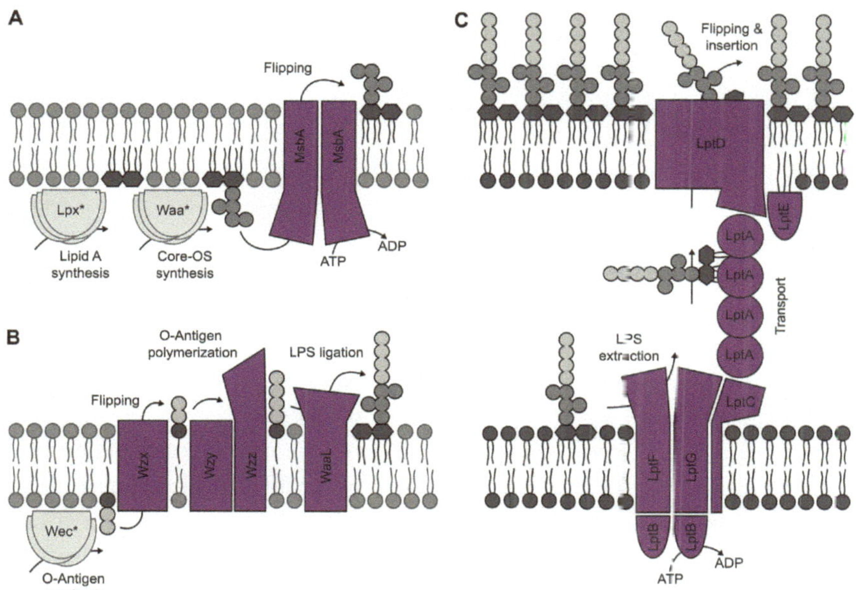

Fig. 3 Synthesis and transport of lipopolysaccharide. (A) Lipid A is synthesized by a group of Lpx* enzymes. Following addition of the core-oligosaccharide by a group of Waa* enzymes, the lipopolysaccharide core is flipped across the inner membrane by MsbA. (B) O-Antigen is synthesized by a group of Wec* enzymes, flipped across the inner membrane, polymerized within the periplasm and ligated to the pre-assembled lipopolysaccharide core. (C) Fully assembled lipopolysaccharide is extracted from the inner membrane, transported to and inserted into the outer membrane by the Lpt system.

subsequent modification of the phosphate groups of Lipid A (Valvano, 2014). The latter plays a crucial role in the context of antimicrobial resistance to polymyxin antibiotics (Tran et al., 2005). Once the core oligosaccharide is assembled, Lipid A is flipped across the IM by the homodimeric flippase MsbA in an energy-dependent manner (Voss & Stephen Trent, 2018). Each monomer of the ABC transporter MsbA constists of a transmembrane bundle of six α-helices and a nucleotide binding domain exposed at the cytosolic side of the IM (Ward, Reyes, Yu, Roth, & Chang, 2007). Thereby MsbA supposedly functions by a so-called trap and flip mechanism, during which the nascent Lipid A enters the nucleotide free MsbA and is subsequently lifted into the outer leaflet upon ATP binding with the headgroup still facing the cytosol, before being flipped and finally translocated upon ATP hydrolysis (Mi et al., 2017). It has also been proposed that this transport process is energized not only through ATP hydrolysis but coupled to the PMF across the IM (Singh et al., 2016). The repeating saccharide moiety of the O-antigen is synthesized linked to an undecaprenyl diphosphate lipid anchor in a separate pathway at the inner surface of the IM by several Wec* enzymes (Rai & Mitchell, 2020). Once assembled this O-antigen-linked lipid is flipped into the outer layer of the IM by the flippase Wzx (Whitfield, Williams, & Kelly, 2020). Multiple units of the O-antigen are then linked within the periplasmic space by the O-antigen polymerase Wzy and its co-polymerase Wzz (Whitfield et al., 2020). In the final biosynthetic step the O-antigen is linked to the preassembled Lipid A by the O-antigen ligase WaaL to form the full LPS (Ruan, Monjarás Feria, Hamad, & Valvano, 2018).

An intricate envelope spanning machinery, the multiprotein LptA–F complex, is responsible for shuttling newly synthesized LPS from the outer leaflet of the IM across the periplasm and into the outer leaflet of the OM (Bishop, 2019; Sperandeo, Martorana, & Polissi, 2017). This system comprises the $LptB_2FGC$ complex in the IM and the LptDE complex in the OM, bridged by the periplasmic LptA. $LptB_2FG$ forms an ABC transporter assembled from a heterodimeric transmembrane helix bundle of LptF and LptG, each linked to a cytosolic nucleotide binding domain formed by LptB (Dong, Tang, Zhang, & Dong, 2017). This core complex is complemented by LptC, which contributes one additional transmembrane helix to the complex, linked to a periplasmic domain, which forms the connection to the periplasmic LptA (Li, Orlando, & Liao, 2019; Owens et al., 2019). The LptDE complex in the OM is to a large part formed by the 24-stranded transmembrane β-barrel LptD, which possesses an exposed periplasmic domain, and is complemented by the lipoprotein LptE, which was found to reside inside the lumen of the LptD barrel in the so-far determined crystal structures (Dong et al., 2014; Qiao, Luo, Zhao, Zhang, & Huang, 2014). The component connecting the two membrane bound complexes is LptA, which has structural similarity with the periplasmic domains of LptC and LptD (Botos et al., 2016) and forms a periplasm-spanning bridge between the respective complexes (Sherman et al., 2018). Together the whole complex functions like an escalator, where the individual LPS molecules are one after the other extracted from the outer leaflet of the IM by $LptB_2FGC$, then they slide along the LptA bridge and are subsequently flipped into the outer leaflet of the OM by LptDE (Li, Gu, Dong, Wang, & Dong, 2015; Okuda et al., 2016). Thereby, in the absence of external energy sources inside the periplasm, the entire process is thought to be driven by ATP hydrolysis occurring at the cytosolic side of the IM by $LptB_2FG$, feeding the LPS molecules into the protein bridge (Gu et al., 2015; Okuda, Freinkman, & Kahne, 2012).

6. The blind spot: Phospholipid synthesis, insertion and trafficking

In contrast to the transport and insertion pathway of LPS into the outer leaflet of the OM, so far only very few details are known about phospholipid trafficking in the IM as well as the OM. While the biosynthetic pathway of PE, PG, and CL synthesis, known as the Kennedy pathway, has been described decades ago, only very little structural and mechanistic information is available (Dowhan & Bogdanov, 2021). Lipid synthesis is accomplished through a series of membrane-attached and transmembrane proteins at the cytoplasmic side of the IM (Fig. 4) (Parsons & Rock, 2013). Briefly, starting from phosphatidic acid (PA) as a common precursor, PA is either converted to PE *via* phosphatidylserine (PS) by the synthases PssA and Psd, or to PG *via* phosphatidylglycerophosphate (PGP) by the transferase PgsA and the phosphatases PgpABC (Zhang & Rock, 2008). CL can be formed either from two PG molecules or PG and PE by the cardiolipin synthases ClsA/B/C (Romantsov, Guan, & Wood, 2009). Thus far, it remains unknown how phospholipids are transferred between the inner and outer leaflet of the IM and no dedicated flippases potentially required for transfer have been identified. In fact, the Lipid A flippase MsbA has been suggested to be involved in phospholipid flipping (Doerrler, Gibbons, Christian, & Raetz, 2004) and thus might serve as a universal lipid

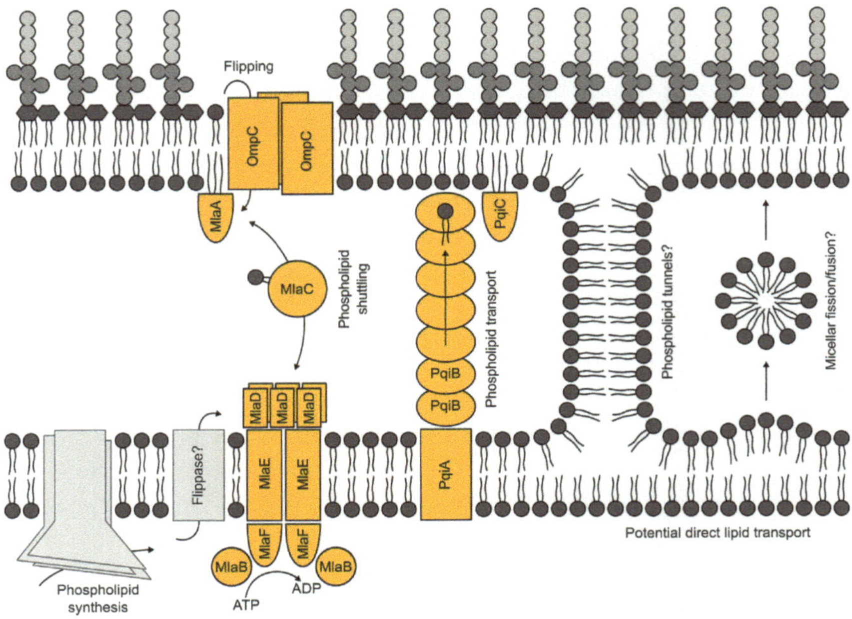

Fig. 4 Synthesis and transport of phospholipids. Phospholipids are synthesized in the Kennedy pathway. Several pathways are suggested to mediate lipid transport between the inner and outer membrane, including the protein-based Mla and Pqi systems and protein-independent lipid transfer *via* lipid tunnels or micellar transport.

flippase, however, MsbA dependent phospholipid flipping remains yet to be observed in the cellular context. *In vitro,* the flipping of phospholipids can be induced by α-helical membrane spanning peptides with a hydrophobic core, it is therefore also conceivable that phospholipid flipping occurs unspecifically along multiple transmembrane proteins (Kol, De Kroon, Rijkers, Killian, & De Kruijff, 2001).

Even less is known about phospholipid transport to the inner leaflet of the OM. Three potential mechanisms have been proposed, awaiting experimental verification: (i) protein-based shuttling either with the help of soluble periplasmic transport proteins or periplasm spanning protein escalator, reminiscent to the Lpt system; (ii) lipid exchange *via* the release of micelles or lipid vesicles from the outer leaflet of the IM and their subsequent fusion with the inner leaflet of the OM and (iii) continuous transport by means of "lipid tunnels," directly connecting the outer leaflet of the IM and their fusion with the inner leaflet of the OM (Powers & Stephen Trent, 2019; Shrivastava & Chng, 2019).

Initially, a complex of the trimeric porin OmpC (and potentially OmpF) with the lipoprotein MlaA has been implicated in playing a role in the maintenance of OM asymmetry, by acting as a flippase to remove phospholipids from the outer leaflet of the OM (Chong, Woo, & Chng 2015). The system is complemented by the soluble periplasmic MlaC and the ABC transporter MlaFEDB residing in the IM. The latter comprises a homodimeric transmembrane bundle of MlaE, each of which is connected to a cytosolic nucleotide binding domain of MlaF and an additionally bound MlaB, and is completed by a periplasmic hexameric assembly of MlaD (Coudray et al., 2020; Tang et al., 2021).

Despite its initially proposed involvement in retrograde phospholipid transport, MlaFEDB has been suggested to be capable of extracting phospholipids from the IM, therefore indicating a potential bi-directional lipid transport by the Mla system (Chi et al., 2020; Hughes et al., 2019; Tang, Chang, Qiao, et al., 2021). Structures of MlaFEDB in different states suggest a ratchet-like motion similar to the mechanism described for other ABC transporters; however, it is thus far unclear whether such motion is involved in the extraction or deposition of phospholipids in the outer leaflet of the IM.

In addition, recently two systems potentially forming hydrophobic protein tunnels between the IM and OM have been described with LetB and PqiB (Ekiert et al., 2017; Isom et al., 2020), which might be complemented by their respective corresponding membrane components LetA and PqiA in the IM, as well as PqiC in the OM (Nakayama & Zhang-Akiyama, 2017). While to date lipid transport along these systems remains hypothetical, phospholipids have been shown to rapidly exchange between the inner leaflet of the OM and the outer leaflet of the IM already four decades ago (Donohue-Rolfe & Schaechter, 1980), thus, the existence of one or more bidirectional transport systems certainly seems highly plausible.

7. Shaping the cell envelope: Membrane attachment and detachment

The three layers of the Gram-negative cell envelope aren't simply stacked upon each other but are directly connected with the aid of several proteins. In fact, two of the most abundant proteins of the proteome of *E. coli*, the murein-lipoprotein Lpp and the transmembrane protein OmpA, are involved in linking the OM to the underlying peptidoglycan layer, which provides structural stability to the cellular envelope (Fig. 5)

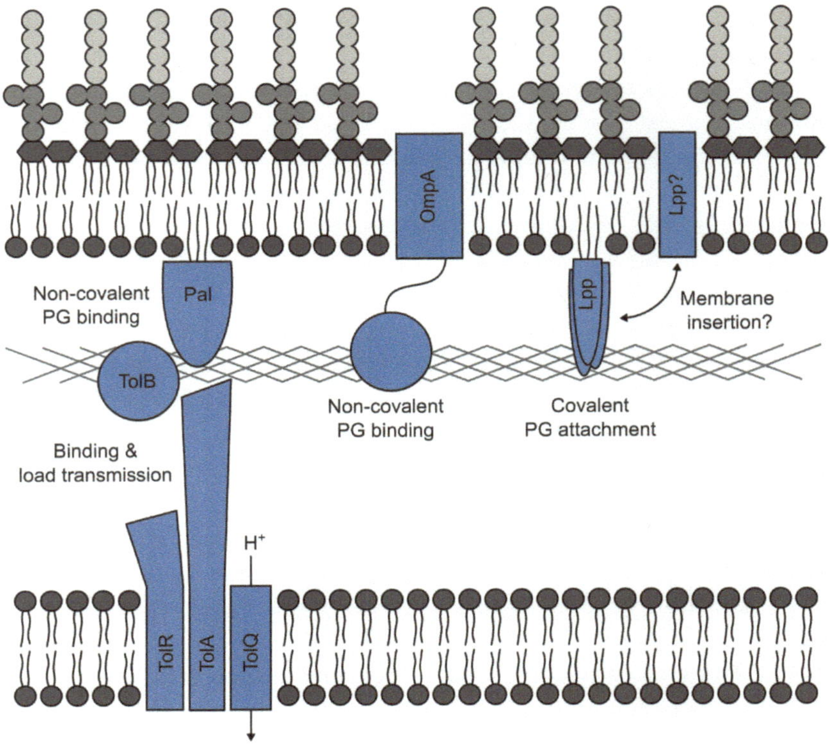

Fig. 5 Anchoring and membrane attachment. The lipoprotein Lpp can be covalently linked to the peptidoglycan, the outer membrane protein OmpA and lipoprotein Pal bind peptidoglycan noncovalently. Pal interacts with the Tol system in the inner membrane.

(Silhavy et al., 2010). The cell envelope of a single *E. coli* cell can contain up to 1 million copies of Lpp (Li, Burkhardt, Gross, & Weissman, 2014). It forms an extended α-helix in a trimeric coiled-coil arrangement and is anchored in the inner leaflet of the OM by the lipid moiety of the triacylated the amino-terminal end (Shu, Liu, Ji, & Lu, 2000). Lpp can be covalently linked to the peptidoglycan through a carboxy-terminal lysine residue by periplasmic transpeptidases (Asmar & Collet, 2018). Thereby Lpp plays a vital role in maintaining the correct distance between OM and peptidoglycan as well as the mechanical properties of the envelope (Mathelié-Guinlet, Asmar, Collet, & Dufrêne, 2020). In fact, increasing the length of Lpp proportionately increases the width of the bacterial periplasm (Asmar et al., 2017). Interestingly, two-thirds of all Lpp molecules are not bound to the peptidoglycan but present in a free form. This free form does not co-localize with the peptidoglycan-bound form inside the periplasm but has been found to span the OM and be surface-accessible—the physiological role of this transmembrane form remains completely miraculous (Cowles, Li, Semmelhack, Cristea, & Silhavy, 2011).

In form of the transmembrane protein OmpA a second major component of the OM, present at ~200,000 copies per cell (Li et al., 2014), is involved in anchoring the OM to the underlying peptidoglycan. According to the current structural model, the monomeric OmpA is a two-domain protein comprising an amino-terminal 8-stranded transmembrane β-barrel domain embedded in the OM and a soluble carboxy-terminal peptidoglycan-binding domain in the periplasm, connected by a flexible linker (Ishida, Garcia-Herrero, & Vogel, 2014; Ortiz-Suarez, Samsudin, Piggot, Bond, & Khalid, 2016; Pautsch & Schulz, 1998). Unlike Lpp, OmpA binds noncovalently and potentially in a flexible manner to the peptidoglycan and it remains to be elucidated to what extend the two proteins work cooperatively to stabilize the OM (Samsudin, Ortiz-Suarez, Piggot, Bond, & Khalid, 2016). While it remains unclear, whether tethering the OM to the underlying peptidoglycan is the sole function of OmpA, the protein has been hypothesized to be involved in absorbing mechanical stress during expansion of the periplasm (Bosshart et al., 2012). Moreover, multiple *in vitro* approaches indicate the possibility of OmpA existing in two distinct structural states, a closed small-pore state as observed in multiple available 3D structures and a large open-pore state, capable of molecular transport (Arora, Rinehart, Szabo, & Tamm, 2000; Negoda, Negoda, & Reusch, 2010; Reusch, 2012); however, for the latter structural and mechanistical details so far remain elusive.

A third protein linking the OM to the underlying peptidoglycan is the peptidoglycan-associated outer membrane lipoprotein Pal, which together with the Tol system connects all three layers of the cell envelope and is also involved in invaginating the OM during cell division (Godlewska, Wiśniewska, Pietras, & Jagusztyn-Krynicka, 2009). The latter consists of the IMPs TolA, TolQ and TolR and is energized through the PMF across the IM (Cascales, Lloubès, & Sturgis, 2001), much like the ortholog TonB/ExbB/ExbD system, which is responsible for the mechanical gating of TonB-dependent outer membrane receptors like FhuA and BtuB (Sverzhinsky et al., 2015) as well as the MotA/MotB of the flagellar motor complex (Kitao & Nishihara, 2017). Furthermore, interaction between Pal and the Tol system appears to depend on the soluble periplasmic protein TolB, which might mediate the interaction between Pal and TolA

(Petiti et al., 2019). In addition, TolR has been implicated in peptidoglycan binding and could thereby also play an important structural role by tethering the IM to the peptidoglycan (Boags, 2019), a role which has also been ascribed to MotB (Minamino, Terahara, Kojima, & Namba, 2018).

The Tol–Pal system has been implicated to play a key role in the rearrangements occurring in the cell envelope during bacterial cell division (Szczepaniak, Press, & Kleanthous, 2020). While the latter is primarily driven by the Min and FtsZ systems, which orchestrate assembly and constriction of the *Z*-ring during division (Adams & Errington, 2009; Eswara & Ramamurthi, 2017), the Tol–Pal system is thought to be a decisive factor for the disassembly of peptidoglycan and invaginating the OM at the site of constriction (Petiti et al., 2019; Szczepaniak et al., 2020). Furthermore, disruptions of the three systems anchoring the OM to the peptidoglycan have been shown to substantially increase the release of outer membrane vesicles (OMVs), thus indicating a weakening of the overall stability of the cellular envelope (Schwechheimer & Kuehn, 2015). OMVs, suggested to be involved in manifold biological processes, including cellular communication, quorum sensing and pathogen-host interactions (Kulp & Kuehn, 2010), are continuously released by Gram-negative bacteria through blebbing from the OM by locally bulging the membrane outwards (Toyofuku, Nomura, & Eberl, 2019). To date it remains unclear whether envelope destabilizations occur as part of a deliberate process to promote the targeted release of OMVs or, *vice versa*, whether OMVs are an unintentional result of envelope instabilities (McBroom, Johnson, Vemulapalli, & Kuehn, 2006; Schwechheimer, Kulp, & Kuehn, 2014). However, the selectivity of OMV compositions and their release in response to specific stimuli suggest thus far unresolved underlying directed processes (Guerrero-Mandujano, Hernández-Cortez, Ibarra, & Castro-Escarpulli, 2017).

8. Membrane protein turnover: AAA+ proteases and beyond

One important factor to maintaining membrane integrity is membrane protein quality control and thus the removal of misfolded and dysfunctional proteins from the membrane (Fig. 6) (Dalbey, Wang, & van Dijl, 2012; Katikaridis, Bohl, & Mogk, 2021). Due to the high energetic cost of protein unfolding, proteases with a disaggregase function are typically ATP driven, and *E. coli* employs several so-called ATPases associated with diverse cellular activities (AAA+) proteases (Mahmoud & Chien, 2018; Olivares, Baker, & Sauer, 2015); however, only the AAA+ metalloprotease

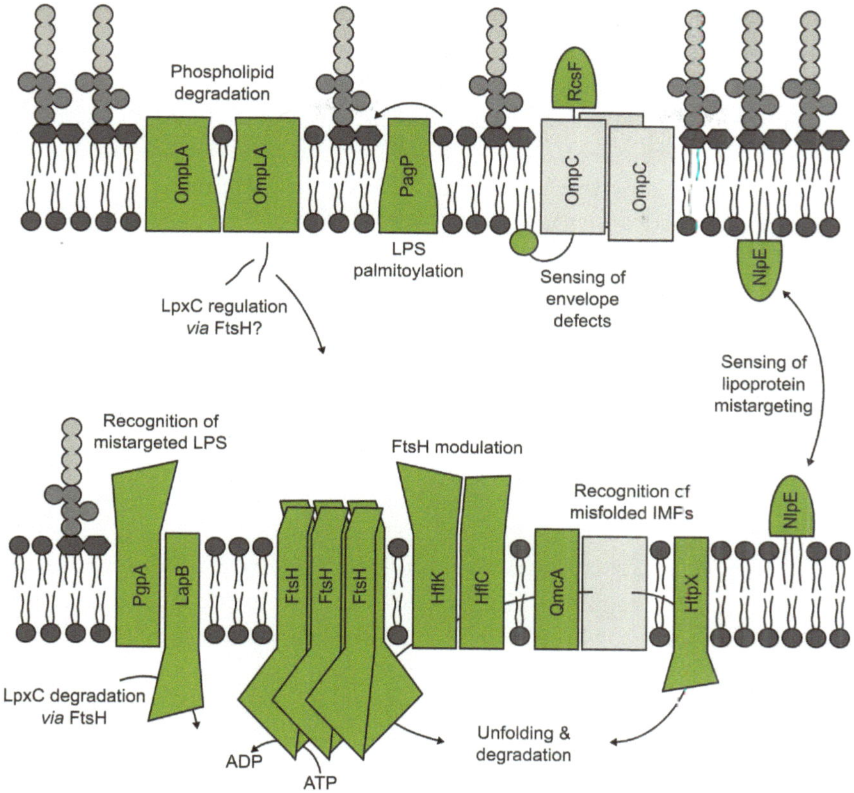

Fig. 6 Maintenance of membrane integrity. The inner membrane proteases FtsH and HtpX degrade misfolded inner membrane proteins. OmpLA and PagP recognize and degrade mistargeted phospholipids in the outer leaflet of the outer membrane. PgpA detects accumulation of lipopolysaccharide in the inner membrane. The lipoprotein NlpE detects lipoprotein targeting defects, RcsF detects defects in envelope assembly.

protease FtsH is membrane-bound residing in the IM (Ito & Akiyama, 2005; Sauer & Baker, 2011). FtsH forms a homo-hexameric complex in the IM, with each monomer contributing two transmembrane α-helices, a small periplasmic domain, and a large cytosolic AAA+ domain (Langklotz, Baumann, & Narberhaus, 2012). However, within the IM the complex is typically complemented by two additional proteins, HflK and HflC, to form the so-called FtsH holo-enzyme (Saikawa, Akiyama, & Ito, 2004). Thereby, HflK and HflC are thought to act in the functional regulation of FtsH (Bandyopadhyay, Parua, Datta, & Parrack, 2010). FtsH itself is proposed to function by mechanically dislocating protein substrates from the membrane in a sequential manner through an extension on its cytoplasmic domains (Dalbey et al., 2012). Recently, cryo-EM structures of FtsH in multiple conformations provided mechanistical insight suggesting substrate unfolding *via* a tilting motion of the cytosolic domain, thereby facilitating lateral substrate entry (Carvalho et al., 2020). Nevertheless, to date the structural details underlying FtsH function remain unresolved. In addition to HflKC, the prohibitin like protein QmcA has been suggested to detect misfolded segments of transmembrane proteins and consequentially target them to FtsH for subsequent degradation (Chiba, Ito, & Akiyama, 2006).

A second protease suggested to be involved in the quality control of membrane proteins within the IM is the heat-shock induced protease HtpX, which unlike FtsH functions in an ATP-independent manner (Shimohata, Chiba, Saikawa, Ito, & Akiyama, 2002). To date no structural data is available for HtpX and little is known about its functional details, however, it has been shown to partially complement the function of FtsH (Chiba et al., 2006).

Whereas the ATP-driven active unfolding and extraction of membrane proteins is possible at the IM, in absence of cellular energy sources an analogous process at the OM seems unlikely, in particular since the extraordinary thermal stability of OMPs suggests generally low energy folded states (Moon, Zaccai, Fleming, Gessmann, & Fleming, 2013) and the same amount of energy would be required in order to release these proteins from the OM. As an alternative route for OMP turnover the involvement of OMVs as molecular waste containers has been proposed (McBroom & Kuehn, 2007). Nevertheless, the typically high reproductive rates of Gram-negative bacteria with generation times as short as 20 min raise the question whether membrane protein turnover in fact constitutes a relevant problem. In fact, it has been shown that OMP biogenesis occurs mostly in the center of cells, resulting in old OMPs being pushed out towards the cellular poles and yielding repository cells upon cell division (Rassam et al., 2015). Interestingly, if the same concept applied to the IM, this would in turn imply more of a post-transcriptional regulatory role of AAA+ proteases such as FtsH, rather than an exigency for protein quality control (Bittner, Arends, & Narberhaus, 2017).

9. Sensing membrane defects and maintenance of membrane integrity

Assembly and homeostasis of bacterial membranes are tightly regulated processes and consequently several membrane proteins are responsible for detecting membrane defects (Fig. 6) (Grabowicz & Silhavy, 2017). Such regulation is not limited to the removal of misfolded or dysfunctional proteins, as is accomplished by AAA+ proteases like FtsH, but also applies to mistargeted phospholipids and LPS. Thereby the transmembrane protein PbgA (YejM) senses LPS in the outer leaflet of the IM and in response downregulates LPS synthesis by triggering the degradation of LpxC by FtsH in order to prevent the cytotoxic accumulation of LPS in the IM (Clairfeuille et al., 2020; Nguyen, Kelly, Qiu, & Misra, 2020). To this end PbgA presumably works in concert with LapA and LapB (YciM) by inhibiting the LapB-induced degradation of LpxC by FtsH in absence of LPS (Fivenson & Bernhardt, 2020; Klein, Kobylak, Lindner, Stupak, & Raina, 2014). Likewise, with OmpLA and PagP the OM contains two integral membrane proteins, which detect impaired membrane asymmetry due to phospholipids present in the outer leaflet of the OM. The phospholipase OmpLA forms a homo-dimer in the OM consisting of two 12-stranded transmembrane β-barrels and degrades phospholipids mislocalized to the outer leaflet (Snijder et al., 1999). In addition, the resulting acyl-chains might act as a secondary messenger downregulating the FtsH dependent degradation of LpxC and consequentially resulting in an increased LPS production (May & Silhavy, 2018). The second enzyme targeting mislocalized phospholipids is the Lipid A palmitoyltransferase PagP, an 8-stranded monomeric β-barrel which transfers palmitate chains from phospholipids to Lipid A (Bishop et al., 2000). Interestingly, PagP has been shown to be upregulated in response to Mg^{2+}-limitation under control of the PhoPQ system (Szczesny, Beloin, & Ghigo, 2018) and might thus counteract LPS destabilization caused by the shortage of divalent ions (Clifton et al., 2015; Rice & Wereszczynski, 2018). Finally, as discussed above, the Mla system might play a role in removing phospholipids from the outer leaflet of the OM.

Much like the systems responding to mistargeted lipids, the lipoprotein NlpE is responsible for sensing perturbations in lipoprotein targeting and counteracts the resulting envelope stress by activating the Cpx stress response pathway (Delhaye, Laloux, & Collet, 2019; May, Lehman, Mitchell, & Grabowicz, 2019). Similarly, the surface exposed lipoprotein RcsF can activate the Rcs envelope stress response (Laloux & Collet, 2017). RcsF is quite a peculiar lipoprotein, where a soluble globular domain is separated from the lipidated amino-terminus by a long unstructured linker, which can be threaded through the lumen of a porin (Konovalova, Mitchell, & Silhavy, 2016). While initially suspected to be oriented with the globular domain inside the periplasm, recent structural work suggests that the globular domain is exposed at the cell surface and the with the acylated amino-terminus embedded in the inner leaflet of the OM (Rodríguez-Alonso et al., 2020). Thereby, the globular domain might be shuttled to the cell surface while the porin folds at the BAM complex, similar to the mechanism described above for passenger domain translocation during autotransporter folding at TamA (Gruss et al., 2013).

10. Conclusion

Despite surrounding one of the best studied organisms on our planet and despite decades of research the membranes of the *E. coli* cell envelope still hold many secrets. Many of the protein machineries responsible for assembling and maintaining these membranes have been found and described, whereas others remain concealed, for example, systems involved in phospholipid transport within the envelope. Some machineries have been described in great detail, and elaborate models delineating the molecular details underlying their function have been established, whereas the structural and functional details of others remain vague. Even thoroughly studied proteins like Lpp and OmpA hold surprises and mysteries, such as the transmembrane state of Lpp or the putative large open-pore state of OmpA, both of which are yet lacking any structural basis. One major obstacle to obtaining deep structural and functional insight into the protein machineries shaping the cell envelope (in particular the ones residing in the OM) is the lack of suitable methods to study these systems in context with their native membrane environment—to date, virtually all structural investigations of membrane proteins heavily rely on the use of membrane mimetics, which can only inadequately resemble the complex asymmetric architecture of native membranes (Thoma & Burmann, 2021). However, methods to study membrane proteins under native conditions, such as in-cell NMR spectroscopy, *in situ* NMR spectroscopy, and cryo-electron tomography are on the rise and will hopefully help gaining a deeper and more complete picture of the ways proteins function within the cell envelope in the near future (Castaño-Díez & Zanetti, 2019; Ladizhansky, 2017; Thoma & Burmann, 2020).

References

Adams, D. W., & Errington, J. (2009). Bacterial cell division: Assembly, maintenance and disassembly of the Z ring. *Nature Reviews Microbiology*, 7(9), 642–653. https://doi.org/10.1038/nrmicro2198.

Akopian, D., Shen, K., Zhang, X., & Shan, S. O. (2013). Signal recognition particle: An essential protein-targeting machine. *Annual Review of Biochemistry*, *82*. https://doi.org/10.1146/annurev-biochem-072711-164732.

Albenne, C., & Ieva, R. (2017). Job contenders: Roles of the β-barrel assembly machinery and the translocation and assembly module in autotransporter secretion. *Molecular Microbiology*, *106*(September), 505–517. https://doi.org/10.1111/mmi.13832.

Allen, W. J., Corey, R. A., Oatley, P., Sessions, R. B., Baldwin, S. A., Radford, S. E., et al. (2016). Two-way communication between SecY and SecA suggests a brownian ratchet mechanism for protein translocation. *eLife*, *5*(MAY2016), 1–23. https://doi.org/10.7554/eLife.15598.

Alvira, S., Watkins, D. W., Troman, L., Allen, W. J., Lorriman, J. S., Degliesposti, G., et al. (2020). Inter-membrane association of the sec and bam translocons for bacterial outer-membrane biogenesis. *eLife*, *9*, 1–24. https://doi.org/10.7554/eLife.60669.

Arora, A., Rinehart, D., Szabo, G., & Tamm, L. K. (2000). Refolded outer membrane protein A of Escherichia coli forms ion channels with two conductance states in planar lipid bilayers. *Journal of Biological Chemistry*, *275*(3), 1594–1600. https://doi.org/10.1074/jbc.275.3.1594.

Asmar, A. T., & Collet, J. F. (2018). Lpp, the Braun lipoprotein, turns 50—Major achievements and remaining issues. *FEMS Microbiology Letters*, *365*(18), 1–8. https://doi.org/10.1093/femsle/fny199.

Asmar, A. T., Ferreira, J. L., Cohen, E. J., Cho, S. H., Beeby, M., Hughes, K. T., et al. (2017). Communication across the bacterial cell envelope depends on the size of the periplasm. *PLoS Biology*, *15*(12), 1–16. https://doi.org/10.1371/journal.pbio.2004303.

Bakelar, J., Buchan, S. K., & Noinaj, N. (2016). The structure of the β-barrel assembly machinery complex. *Science (New York, N.Y.)*, *351*(6269), 180–186. https://doi.org/10.1126/science.aad3460.

Bandyopadhyay, K., Parua, P. K., Datta, A. B., & Parrack, P. (2010). Escherichia coli HflK and HflC can individually inhibit the HflB (FtsH)-mediated proteolysis of λCII in vitro. *Archives of Biochemistry and Biophysics*, *501*(2), 239–243. https://doi.org/10.1016/j.abb.2010.06.030.

Banerjee, T., Lindenthal, C., & Oliver, D. (2017). SecA functions in vivo as a discrete anti-parallel dimer to promote protein transport. *Molecular Microbiology*, *103*(3), 439–451. https://doi.org/10.1111/mmi.13567.

Bishop, R. E. (2019). Lipopolysaccharide transport ratcheted up. *Nature*, *567*, 471–472.

Bishop, R. E., Gibbons, H. S., Guina, T., Trent, M. S., Miller, S. I., & Raetz, C. R. (2000). Transfer of palmitate from phospholipids to lipid A in outer membranes of gram-negative bacteria. *The EMBO Journal*, *19*(19), 5071–5080. https://doi.org/10.1093/emboj/19.19.5071.

Bittner, L. M., Arends, J., & Narberhaus, F. (2017). When, how and why? Regulated proteolysis by the essential FtsH protease in Escherichia coli. *Biological Chemistry*, *398*(5–6), 625–635. https://doi.org/10.1515/hsz-2016-0302.

Boags, A. T. (2019). Binding from both sides: TolR and full-length OmpA bind and maintain the local structure of the E. Coli Cell Wall. *Biophysical Journal*. *116*(3), 203a–204a. https://doi.org/10.1016/j.bpj.2018.11.1125.

Bosshart, P. D., Iordanov, I., Garzon-Coral, C., Demange, P., Engel, A., Milon, A., et al. (2012). The transmembrane protein KpOmpA anchoring the outer membrane of Klebsiella pneumoniae unfolds and refolds in response to tensile load. *Structure*, *20*(1), 121–127. https://doi.org/10.1016/j.str.2011.11.002.

Botos, I., Majdalani, N., Mayclin, S. J., McCarthy, J. G., Lundquist, K., Wojtowicz, D., et al. (2016). Structural and functional characterization of the LPS transporter LptDE from gram-negative pathogens. *Structure*, *24*(6), 965–976. https://doi.org/10.1016/j.str.2016.03.026.

Bowie, J. U. (2005). Solving the membrane protein folding problem. *Nature*, *438*(7068), 581–589. https://doi.org/10.1038/nature04395.

Breyton, C., Haase, W., Rapoport, T. A., Kühlbrandt, W., & Collinson, I. (2002). Three-dimensional structure of the bacterial protein-translocation complex SecYEG. *Nature*, *418*(6898), 662–665. https://doi.org/10.1038/nature00827.

Burmann, B. M., & Hiller, S. (2012). Solution NMR studies of membrane-protein--chaperone complexes. *Chimia, 66*(10), 759–763. https://doi.org/10.2533/chimia.2012.759.

Burmann, B. M., Wang, C., & Hiller, S. (2013). Conformation and dynamics of the periplasmic membrane-protein-chaperone complexes OmpX-Skp and tOmpA-Skp. *Nature Structural & Molecular Biology, 20*(11), 1265–1272. https://doi.org/10.1038/nsmb.2677.

Carvalho, V., Prabudiansyah, I., Kovacik, L., Chami, M., Kieffer, R., Van Der Valk, R., et al. (2020). The cytoplasmic domain of the AAA+ protease FtsH is tilted with respect to the membrane to facilitate substrate entry. *Journal of Biological Chemistry, 296*, 100029. https://doi.org/10.1074/jbc.ra120.014739.

Cascales, E., Lloubès, R., & Sturgis, J. N. (2001). The TolQ-TolR proteins energize TolA and share homologies with the flagellar motor proteins MotA-MotB. *Molecular Microbiology, 42*(3), 795–807. https://doi.org/10.1046/j.1365-2958.2001.02673.x.

Castaño-Díez, D., & Zanetti, G. (2019). In situ structure determination by subtomogram averaging. *Current Opinion in Structural Biology, 58*, 68–75. https://doi.org/10.1016/j.sbi.2019.05.011.

Chen, I. A., & Walde, P. (2010). From self-assembled vesicles to protocells. *Cold Spring Harbor Perspectives in Biology, 2*(7), 1–13. https://doi.org/10.1101/cshperspect.a002170.

Chi, X., Fan, Q., Zhang, Y., Liang, K., Wan, L., Zhou, Q., et al. (2020). Structural mechanism of phospholipids translocation by MlaFEDB complex. *Cell Research, 30*(12), 1127–1135. https://doi.org/10.1038/s41422-020-00404-6.

Chiba, S., Ito, K., & Akiyama, Y. (2006). The Escherichia coli plasma membrane contains two PHB (prohibitin homology) domain protein complexes of opposite orientations. *Molecular Microbiology, 60*(2), 448–457. https://doi.org/10.1111/j.1365-2958.2006.05104.x.

Chong, Z. S., Woo, W. F., & Chng, S. S. (2015). Osmoporin OmpC forms a complex with MlaA to maintain outer membrane lipid asymmetry in Escherichia coli. *Molecular Microbiology, 98*(6), 1133–1146. https://doi.org/10.1111/mmi.13202.

Clairfeuille, T., Buchholz, K. R., Li, Q., Verschueren, E., Liu, P., Sangaraju, D., et al. (2020). Structure of the essential inner membrane lipopolysaccharide–PbgA complex. *Nature, 584*(7821), 479–483. https://doi.org/10.1038/s41586-020-2597-x.

Clifton, L. A., Skoda, M. W. A., Le Brun, A. P., Ciesielski, F., Kuzmenko, I., Holt, S. A., et al. (2015). Effect of divalent cation removal on the structure of gram-negative bacterial outer membrane models. *Langmuir, 31*(1), 404–412. https://doi.org/10.1021/la504407v.

Collinson, I. (2017). SecA-a new twist in the tale. *Journal of Bacteriology, 199*(2). https://doi.org/10.1128/JB.00736-16.

Corey, R. A., Pyle, E., Allen, W. J., Watkins, D. W., Casiraghi, M., Miroux, B., et al. (2018). Specific cardiolipin–SecY interactions are required for proton-motive force stimulation of protein secretion. *Proceedings of the National Academy of Sciences of the United States of America, 115*(31), 7967–7972. https://doi.org/10.1073/pnas.1721536115.

Coudray, N., Isom, G. L., Macrae, M. R., Saiduddin, M. N., Bhabha, G., & Ekiert, D. C. (2020). Structure of bacterial phospholipid transporter MlaFEDB with substrate bound. *eLife, 9*, 1–73. https://doi.org/10.7554/eLife.62518.

Cowles, C. E., Li, Y., Semmelhack, M. F., Cristea, I. M., & Silhavy, T. J. (2011). The free and bound forms of Lpp occupy distinct subcellular locations in Escherichia coli. *Molecular Microbiology, 79*(5), 1168–1181. https://doi.org/10.1111/j.1365-2958.2011.07539.x.

Cranford-Smith, T., & Huber, D. (2018). The way is the goal: How SecA transports proteins across the cytoplasmic membrane in bacteria. *FEMS Microbiology Letters, 365*(11), 1–16. https://doi.org/10.1093/femsle/fny093.

Crow, A., Greene, N. P., Kaplan, E., & Koronakis, V. (2017). Structure and mechanotransmission mechanism of the MacB ABC transporter superfamily. *Proceedings of the National Academy of Sciences of the United States of America, 114*(47), 12572–12577. https://doi.org/10.1073/pnas.1712153114.

Dalbey, R. E., Wang, P., & van Dijl, J. M. (2012). Membrane proteases in the bacterial protein secretion and quality control pathway. *Microbiology and Molecular Biology Reviews, 76*(2), 311–330. https://doi.org/10.1128/mmbr.05019-11.

David, E. H., Yethon, J. A., & Whitfield, C. (1998). Molecular basis for structural diversity in the core regions of the lipopolysaccharides of *Escherichia coli* and *salmonella enterica. Molecular Microbiology, 30*(2), 221–232. https://doi.org/10.1046/j.1365-2958.1998.01063.x.

De Geyter, J., Portaliou, A. G., Srinivasu, B., Krishnamurthy, S., Economou, A., & Karamanou, S. (2020). Trigger factor is a bona fide secretory pathway chaperone that interacts with SecB and the translocase. *EMBO Reports, 21*(6), 1–17. https://doi.org/10.15252/embr.201949054.

Delhaye, A., Laloux, G., & Collet, J. F. (2019). The lipoprotein NlpE is a cpx sensor that serves as a sentinel for protein sorting and folding defects in the Escherichia coli envelope. *Journal of Bacteriology, 209*(10). https://doi.org/10.1128/JB.00611-18.

Doerrler, W. T., Gibbons, H. S., Christian, R., & Raetz, H. (2004). MsbA-dependent translocation of lipids across the inner membrane of Escherichia coli. *Journal of Biological Chemistry, 279*(43), 45102–45109. https://doi.org/10.1074/jbc.M408106200.

Dong, H., Tang, X., Zhang, Z., & Dong, C. (2017). Structural insight into lipopolysaccharide transport from the gram-negative bacterial inner membrane to the outer membrane. *Biochimica et Biophysica Acta - Molecular and Cell Biology of Lipids, 1862*(11), 1461–1467. https://doi.org/10.1016/j.bbalip.2017.08.003.

Dong, H., Xiang, Q., Gu, Y., Wang, Z., Paterson, N. G., Stansfeld, P. J., et al. (2014). Structural basis for outer membrane lipopolysaccharide insertion. *Nature, 511*(7507), 52–56. https://doi.org/10.1038/nature13464.

Donohue-Rolfe, A. M., & Schaechter, M. (1980). Translocation of phospholipids from the inner to the outer membrane of Escherichia coli. *Proceedings of the National Academy of Sciences of the United States of America,* 77(4 I), 1867–1871. https://doi.org/10.1073/pnas.77.4.1867.

Dowhan, W., & Bogdanov, M. (2002). Functional roles of lipids in membranes. In *Biochemistry of lipids, lipoproteins and membranes* (pp. 1–35). Retrieved from https://doi.org/10.1016/B978-0-444-63438-2.00001-8.

Dowhan, W., & Bogdanov, M. (2021). Eugene P. Kennedy's legacy: Defining bacterial phospholipid pathways and function. *Frontiers in Molecular Biosciences, 8*(March), 1–22. https://doi.org/10.3389/fmolb.2021.666203.

Doyle, M. T., & Bernstein, H. D. (2019). Bacterial outer membrane proteins assemble via asymmetric interactions with the BamA β-barrel. *Nature Communications, 10*(1), 1–13. https://doi.org/10.1038/s41467-019-11230-9.

Draycheva, A., Lee, S., & Wintermeyer, W. (2018). Cotranslational protein targeting to the membrane: Nascent-chain transfer in a quaternary complex formed at the translocon. *Scientific Reports, 8*(1), 1–12. https://doi.org/10.1038/s41598-018-28262-8.

Economou, A. (1999). Following the leader: Bacterial protein export through the sec pathway. *Trends in Microbiology,* 7(8), 315–320. https://doi.org/10.1016/S0966-842X(99)01555-3.

Ekiert, D. C., Bhabha, G., Isom, G. L., Greenan, G., Ovchinnikov, S., Henderson, I. R., et al. (2017). Architectures of lipid transport Systems for the Bacterial Outer Membrane. *Cell, 169*(2), 273–285.e17. https://doi.org/10.1016/j.cell.2017.03.019.

Elvekrog, M. M., & Walter, P. (2015). Dynamics of co-translational protein targeting. *Current Opinion in Chemical Biology, 29,* 79–86. https://doi.org/10.1016/j.cbpa.2015.09.016.

Eswara, P. J., & Ramamurthi, K. S. (2017). Bacterial cell division: Nonmodels poised to take the spotlight. *Annual Review of Microbiology*, *71*(June), 393–411. https://doi.org/10.1146/annurev-micro-102215-095657.
Fivenson, E. M., & Bernhardt, T. G. (2020). An essential membrane protein modulates the proteolysis of LpxC to control lipopolysaccharide synthesis in escherichia coli. *MBio*, *11*(3). https://doi.org/10.1128/mBio.00939-20.
Fleming, K. G. (2014). Energetics of membrane protein folding. *Annual Review of Biophysics*, *43*, 233–255. https://doi.org/10.1146/annurev-biophys-051013-022926.
Gelis, I., Bonvin, A. M. J. J., Keramisanou, D., Koukaki, M., Gouridis, G., Karamanou, S., et al. (2007). Structural basis for signal-sequence recognition by the translocase motor SecA as determined by NMR. *Cell*, *131*(4), 756–769. https://doi.org/10.1016/j.cell.2007.09.039.
Godlewska, R., Wiśniewska, K., Pietras, Z., & Jagusztyn-Krynicka, E. K. (2009). Peptidoglycan-associated lipoprotein (pal) of gram-negative bacteria: Function, structure, role in pathogenesis and potential application in immunoprophylaxis: Minireview. *FEMS Microbiology Letters*, *298*(1), 1–11. https://doi.org/10.1111/j.1574-6968.2009.01659.x.
Gouridis, G., Karamanou, S., Sardis, M. F., Schärer, M. A., Capitani, G., & Economou, A. (2013). Quaternary dynamics of the SecA motor drive translocase catalysis. *Molecular Cell*, *52*(5), 655–666. https://doi.org/10.1016/j.molcel.2013.10.036.
Grabowicz, M. (2019). Lipoproteins and their trafficking to the outer membrane. *EcoSal Plus*, *8*(2). https://doi.org/10.1128/ecosalplus.esp-0038-2018.
Grabowicz, M., & Silhavy, T. J. (2017). Envelope stress responses: An interconnected safety net. *Trends in Biochemical Sciences*, *42*(3), 232–242. https://doi.org/10.1016/j.tibs.2016.10.002.
Gruss, F., Zähringer, F., Jakob, R. P., Burmann, B. M., Hiller, S., & Maier, T. (2013). The structural basis of autotransporter translocation by TamA. *Nature Structural & Molecular Biology*, *20*(11), 1318–1320. https://doi.org/10.1038/nsmb.2689.
Gu, Y., Stansfeld, P. J., Zeng, Y., Dong, H., Wang, W., & Dong, C. (2015). Lipopolysaccharide is inserted into the outer membrane through an intramembrane hole, a lumen gate, and the lateral opening of lptd. *Structure*, *23*(3), 496–504. https://doi.org/10.1016/j.str.2015.01.001.
Guerrero-Mandujano, A., Hernández-Cortez, C., Ibarra, J. A., & Castro-Escarpulli, G. (2017). The outer membrane vesicles: Secretion system type zero. *Traffic*, *18*(7), 425–432. https://doi.org/10.1111/tra.12488.
Gunasinghe, S. D., Shiota, T., Stubenrauch, C. J., Schulze, K. E., Webb, C. T., Fulcher, A. J., et al. (2018). The WD40 protein BamB mediates coupling of BAM complexes into assembly precincts in the bacterial outer membrane. *Cell Reports*, 2782–2794. https://doi.org/10.1016/j.celrep.2018.04.093.
Hanczyc, M. M., & Monnard, P. A. (2017). Primordial membranes: More than simple container boundaries. *Current Opinion in Chemical Biology*, *40*, 78–86. https://doi.org/10.1016/j.cbpa.2017.07.009.
Harris, N. J., & Booth, P. J. (2012). Folding and stability of membrane transport proteins in vitro. *Biochimica et Biophysica Acta (BBA) - Biomembranes*, *1818*(4), 1055–1066. https://doi.org/10.1016/j.bbamem.2011.11.006.
Hayashi, Y., Tsurumizu, R., Tsukahara, J., Takeda, K., Narita, S. I., Mori, M., et al. (2014). Roles of the protruding loop of factor b essential for the localization of lipoproteins (LolB) in the anchoring of bacterial triacylated proteins to the outer membrane. *Journal of Biological Chemistry*, *289*(15), 10530–10539. https://doi.org/10.1074/jbc.M113.539270.
Hegde, R. S., & Bernstein, H. D. (2006). The surprising complexity of signal sequences. *Trends in Biochemical Sciences*, *31*(10), 563–571. https://doi.org/10.1016/j.tibs.2006.08.004.

Höhr, A. I. C., Lindau, C., Wirth, C., Qiu, J., Stroud, D. A., Kutik, S., et al. (2018). Membrane protein insertion through a mitochondrial β-barrel gate. *Science, 359*(6373), eaah6834. https://doi.org/10.1126/science.aah6834.

Horne, J. E., Brockwell, D. J., & Radford, S. E. (2020). Role of the lipid bilayer in outer membrane protein folding in gram-negative bacteria. *Journal of Biological Chemistry, 295*(30), 10340–10367. https://doi.org/10.1074/jbc.REV120.011473.

Huang, C., Rossi, P., Saio, T., & Kalodimos, C. G. (2016). Structural basis for the antifolding activity of a molecular chaperone. *Nature, 537*(7619), 202–206. https://doi.org/10.1038/nature18965.

Huber, D., Jamshad, M., Hanmer, R., Schibich, D., Döring, K., Marcomini, I., et al. (2017). SecA cotranslationally interacts with nascent substrate proteins in vivo. *Journal of Bacteriology, 199*(2), 1–14. https://doi.org/10.1128/JB.00622-16.

Hughes, G. W., Hall, S. C. L., Laxton, C. S., Sridhar, P., Mahadi, A. H., Hatton, C., et al. (2019). Evidence for phospholipid export from the bacterial inner membrane by the Mla ABC transport system. *Nature Microbiology, 4*(10), 1692–1705. https://doi.org/10.1038/s41564-019-0481-y.

Hussain, S., & Bernstein, H. D. (2018). The bam complex catalyzes efficient insertion of bacterial outer membrane proteins into membrane vesicles of variable lipid composition. *Journal of Biological Chemistry, 293*(8), 2959–2973. https://doi.org/10.1074/jbc.RA117.000349.

Iadanza, M. G., Schiffrin, B., White, P., Watson, M. A., Horne, J. E., Higgins, A. J., et al. (2020). Distortion of the bilayer and dynamics of the BAM complex in lipid nanodiscs. *Communications Biology, 3*, 1–14. https://doi.org/10.1038/s42003-020-01419-w.

Inouye, S., Wang, S., & Sezikawa, J. (1977). Amino acid sequence for the peptide extension on the prolipoprotein of the Escherichia coli outer membrane. *Proceedings of the National Academy of Sciences of the United States of America, 74*(3), 1004–1008. https://doi.org/10.1073/pnas.74.3.1004.

Ishida, H., Garcia-Herrero, A., & Vogel, H. J. (2014). The periplasmic domain of Escherichia coli outer membrane protein A can undergo a localized temperature dependent structural transition. *Biochimica et Biophysica Acta (BBA) - Biomembranes, 1838*(12), 3014–3024. https://doi.org/10.1016/j.bbamem.2014.08.008.

Isom, G. L., Coudray, N., MacRae, M. R., McManus, C. T., Ekiert, D. C., & Bhabha, G. (2020). LetB structure reveals a tunnel for lipid transport across the bacterial envelope. *Cell, 181*(3), 653–664. e19 https://doi.org/10.1016/j.cell.2020.03.030.

Ito, K., & Akiyama, Y. (2005). Cellular functions, mechanism of action, and regulation of FtsH protease. *Annual Review of Microbiology, 59*, 211–231. https://doi.org/10.1146/annurev.micro.59.030804.121316.

Juncker, A. S., Willenbrock, H., von Heijne, G., Brunak, S., Nielsen, H., & Krogh, A. (2003). Prediction of lipoprotein signal peptides in gram-negative bacteria. *Protein Science, 12*(8), 1652–1662. https://doi.org/10.1110/ps.0303703.

Kaplan, E., Greene, N. P., Crow, A., & Koronakis, V. (2018). Insights into bacterial lipoprotein trafficking from a structure of LolA bound to the LolC periplasmic domain. *Proceedings of the National Academy of Sciences of the United States of America, 115*(31), E7389–E7397. https://doi.org/10.1073/pnas.1806822115.

Katikaridis, P., Bohl, V., & Mogk, A. (2021). Resisting the heat: Bacterial Disaggregases rescue cells from devastating protein aggregation. *Frontiers in Molecular Biosciences, 8*(May), 1–14. https://doi.org/10.3389/fmolb.2021.681439.

Kaur, H., Jakob, R. P., Marzinek, J. K., Green, R., Imai, Y., Bolla, J. R., et al. (2021). The antibiotic darobactin mimics a β-strand to inhibit outer membrane insertase. *Nature, 593*(7857), 125–129. https://doi.org/10.1038/s41586-021-03455-w.

Kitao, A., & Nishihara, Y. (2017). Structure of the MotA/B proton channel. In. *Methods in Molecular Biology, 1593*, 133–145. https://doi.org/10.1007/978-1-4939-6927-2_10.

Klein, G., Kobylak, N., Lindner, B., Stupak, A., & Raina, S. (2014). Assembly of lipopolysaccharide in escherichia coli requires the essential LapB heat shock protein. *Journal of Biological Chemistry, 289*(21), 14829–14853. https://doi.org/10.1074/jbc.M113.539494.

Kol, M. A., De Kroon, A. I. P. M., Rijkers, D. T. S., Killian, J. A., & De Kruijff, B. (2001). Membrane-spanning peptides induce phospholipid flop: A model for phospholipid translocation across the inner membrane of E. coli. *Biochemistry, 40*(35), 10500–10506. https://doi.org/10.1021/bi010627+.

Komar, J., Alvira, S., Schulze, R. J., Martin, R., Lycklama A Nijeholt, J. A., Lee, S. C., et al. (2016). Membrane protein insertion and assembly by the bacterial holo-translocon SecYEG-SecDF-YajC-YidC. *Biochemical Journal, 473*(19), 3341–3354. https://doi.org/10.1042/BCJ20160545.

Konovalova, A., Mitchell, A. M., & Silhavy, T. J. (2016). A lipoprotein/b-barrel complex monitors lipopolysaccharide integrity transducing information across the outer membrane. *eLife, 5*(JUNE2016), 1–17. https://doi.org/10.7554/eLife.15276.

Kuhn, A., & Kiefer, D. (2017). Membrane protein insertase YidC in bacteria and archaea. *Molecular Microbiology, 103*(4), 590–594. https://doi.org/10.1111/mmi.13586.

Kulp, A., & Kuehn, M. J. (2010). Biological functions and biogenesis of secreted bacterial outer membrane vesicles. *Annual Review of Microbiology, 64*, 163–184. https://doi.org/10.1146/annurev.micro.091208.073413.

Ladizhansky, V. (2017). Applications of solid-state NMR to membrane proteins. *Biochimica et Biophysica Acta, Proteins and Proteomics, 1865*(11), 1577–1586. https://doi.org/10.1016/j.bbapap.2017.07.004.

Laloux, G., & Collet, J. F. (2017). Major tom to ground control: How lipoproteins communicate extracytoplasmic stress to the decision center of the cell. *Journal of Bacteriology, 199*(21). https://doi.org/10.1128/JB.00216-17.

Langklotz, S., Baumann, U., & Narberhaus, F. (2012). Structure and function of the bacterial AAA protease FtsH. *Biochimica et Biophysica Acta, Molecular Cell Research, 1823*(1), 40–48. https://doi.org/10.1016/j.bbamcr.2011.08.015.

Li, G. W., Burkhardt, D., Gross, C., & Weissman, J. S. (2014). Quantifying absolute protein synthesis rates reveals principles underlying allocation of cellular resources. *Cell, 157*(3), 624–635. https://doi.org/10.1016/j.cell.2014.02.033.

Li, X., Gu, Y., Dong, H., Wang, W., & Dong, C. (2015). Trapped lipopolysaccharide and LptD intermediates reveal lipopolysaccharide translocation steps across the Escherichia coli outer membrane. *Scientific Reports, 5*(July), 1–8. https://doi.org/10.1038/srep11883.

Li, Y., Orlando, B. J., & Liao, M. (2019). Structural basis of lipopolysaccharide extraction by the LptB 2 FGC complex. *Nature, 567*(7749), 486–490. https://doi.org/10.1038/s41586-019-1025-6.

Luther, A., Urfer, M., Zahn, M., Müller, M., Wang, S. Y., Mondal, M., et al. (2019). Chimeric peptidomimetic antibiotics against gram-negative bacteria. *Nature, 576*(7787), 452–458. https://doi.org/10.1038/s41586-019-1665-6.

Mahmoud, S. A., & Chien, P. (2018). Regulated proteolysis in bacteria. *Annual Review of Biochemistry, 87*(April), 677–696. https://doi.org/10.1146/annurev-biochem-062917-012848.

Mamat, U., Skurnik, M., & Bengoechea, J. A. (2011). Lipopolysaccharide Core oligosaccharide biosynthesis and assembly. *Bacterial Lipopolysaccharides*. https://doi.org/10.1007/978-3-7091-0733-1_8.

Mas, G., Thoma, J., & Hiller, S. (2019). The periplasmic chaperones Skp and SurA. In. *Subcellular Biochemistry, 92*, 169–186. https://doi.org/10.1007/978-3-030-18768-2_6.

Mathelié-Guinlet, M., Asmar, A. T., Collet, J. F., & Dufrêne, Y. F. (2020). Lipoprotein Lpp regulates the mechanical properties of the E. coli cell envelope. *Nature. Communications, 11*(1). https://doi.org/10.1038/s41467-020-15489-1.

May, K. L., Lehman, K. M., Mitchell, A. M., & Grabowicz, M. (2019). A stress response monitoring lipoprotein trafficking to the outer membrane. *MBio*, *10*(3), 1–14. https://doi.org/10.1128/mBio.00618-19.

May, K. L., & Silhavy, T. J. (2018). The Escherichia coli phospholipase PldA regulates outer membrane homeostasis via lipid signaling. *MBio*, *9*(2). https://doi.org/10.1128/mBio.00379-18.

McBroom, A. J., Johnson, A. P., Vemulapalli, S., & Kuehn, M. J. (2006). Outer membrane vesicle production by Escherichia coli is independent of membrane instability. *Journal of Bacteriology*, *188*(15), 5385–5392. https://doi.org/10.1128/JB.00498-06.

McBroom, A. J., & Kuehn, M. J. (2007). Release of outer membrane vesicles by gram-negative bacteria is a novel envelope stress response. *Molecular Microbiology*, *63*(2), 545–558. https://doi.org/10.1111/j.1365-2958.2006.05522.x.

Mi, W., Li, Y., Yoon, S. H., Ernst, R. K., Walz, T., & Liao, M. (2017). Structural basis of MsbA-mediated lipopolysaccharide transport. *Nature*, *549*(7671), 233–237. https://doi.org/10.1038/nature23649.

Minamino, T., Terahara, N., Kojima, S., & Namba, K. (2018). Autonomous control mechanism of stator assembly in the bacterial flagellar motor in response to changes in the environment. *Molecular Microbiology*, *109*(6), 723–734. https://doi.org/10.1111/mmi.14092.

Monnard, P. A., & Deamer, D. W. (2002). Membrane self-assembly processes: Steps toward the first cellular life. *Anatomical Record*, *268*(3), 196–207. https://doi.org/10.1002/ar.10154.

Moon, C. P., Zaccai, N. R., Fleming, P. J., Gessmann, D., & Fleming, K. G. (2013). Membrane protein thermodynamic stability may serve as the energy sink for sorting in the periplasm. *Proceedings of the National Academy of Sciences of the United States of America*, *110*(11), 4285–4290. https://doi.org/10.1073/pnas.1212527110.

Morgado, L., Burmann, B. M., Sharpe, T., Mazur, A., & Hiller, S. (2017). The dynamic dimer structure of the chaperone trigger factor. *Nature Communications*, *8*(1), 1992. https://doi.org/10.1038/s41467-017-02196-7.

Nagamori, S., Smirnova, I. N., & Kaback, H. R. (2004). Role of YidC in folding of polytopic membrane proteins. *Journal of Cell Biology*, *165*(1), 53–62. https://doi.org/10.1083/jcb.200402067.

Nakayama, T., & Zhang-Akiyama, Q. M. (2017). pqiABC and yebST, putative mce operons of Escherichia coli, encode transport pathways and contribute to membrane integrity. *Journal of Bacteriology*, *199*(1). https://doi.org/10.1128/JB.00606-16.

Narita, S.-I., & Tokuda, H. (2010). Biogenesis and membrane targeting of lipoproteins. *EcoSal Plus*, *4*(1). https://doi.org/10.1128/ecosalplus.4.3.7.

Negoda, A., Negoda, E., & Reusch, R. N. (2010). Resolving the native conformation of Escherichia coli OmpA. *FEBS Journal*, *277*(21), 4427–4437. https://doi.org/10.1111/j.1742-4658.2010.07823.x.

Nguyen, D., Kelly, K., Qiu, N., & Misra, R. (2020). YejM controls LpxC levels by regulating protease activity of the FtsH/YciM complex of Escherichia coli. *Journal of Bacteriology*, *202*(18). https://doi.org/10.1128/JB.00303-20.

Okuda, S., Freinkman, E., & Kahne, D. (2012). Cytoplasmic ATP hydrolysis powers transport of lipopolysaccharide across the periplasm in E. coli. *Science*, *338*(6111), 1214–1217. https://doi.org/10.1126/science.1228984.

Okuda, S., Sherman, D. J., Silhavy, T. J., Ruiz, N., & Kahne, D. (2016). Lipopolysaccharide transport and assembly at the outer membrane: The PEZ model. *Nature Reviews Microbiology*, *14*(6), 337–345. https://doi.org/10.1038/nrmicro.2016.25.

Olivares, A. O., Baker, T. A., & Sauer, R. T. (2015). Mechanistic insights into bacterial AAA+ proteases and protein-remodelling machines. *Nature Reviews Microbiology*, *14*(1), 33–44. https://doi.org/10.1038/nrmicro.2015.4.

Ortiz-Suarez, M. L. L., Samsudin, F., Piggot, T. J. J., Bond, P. J. J., & Khalid, S. (2016). Full-length OmpA: Structure, function, and membrane interactions predicted by molecular dynamics simulations. *Biophysical Journal, 111*(8), 1692–1702. https://doi.org/10.1016/j.bpj.2016.09.009.

Oswald, J., Njenga, R., Natriashvili, A., Sarmah, P., & Koch, H. G. (2021). The dynamic SecYEG Translocon. *Frontiers in Molecular Biosciences, 8*(April), 1–30. https://doi.org/10.3389/fmolb.2021.664241.

Owens, T. W., Taylor, R. J., Pahil, K. S., Bertani, B. R., Ruiz, N., Kruse, A. C., et al. (2019). Structural basis of unidirectional export of lipopolysaccharide to the cell surface. *Nature, 567*(7749), 550–553. https://doi.org/10.1038/s41586-019-1039-0.

Paetzel, M., Karla, A., Strynadka, N. C. J., & Dalbey, R. E. (2002). Signal peptidases. *Chemical Reviews, 102*(12), 4549–4579. https://doi.org/10.1021/cr010166y.

Papanikou, E., Karamanou, S., & Economou, A. (2007). Bacterial protein secretion through the translocase nanomachine. *Nature Reviews. Microbiology, 5*(11), 839–851. https://doi.org/10.1038/nrmicro1771.

Park, E., Ménétret, J. F., Gumbart, J. C., Ludtke, S. J., Li, W., Whynot, A., et al. (2014). Structure of the SecY channel during initiation of protein translocation. *Nature, 506*(7486), 102–106. https://doi.org/10.1038/nature12720.

Parsons, J. B., & Rock, C. O. (2013). Bacterial lipids: Metabolism and membrane homeostasis. *Progress in Lipid Research, 52*(3), 249–276. https://doi.org/10.1016/j.plipres.2013.02.002.

Pautsch, A., & Schulz, G. E. (1998). Structure of the outer membrane protein A transmembrane domain. *Nature Structural Biology, 5*(11), 1013–1017. https://doi.org/10.1038/2983.

Petiti, M., Serrano, B., Faure, L., Lloubes, R., Mignot, T., & Duché, D. (2019). Tol energy-driven localization of pal and Anchoring to the peptidoglycan promote outer-membrane constriction. *Journal of Molecular Biology, 431*(17), 3275–3288. https://doi.org/10.1016/j.jmb.2019.05.039.

Polissi, A., & Sperandeo, P. (2014). The lipopolysaccharide export pathway in Escherichia coli: Structure, organization and regulated assembly of the Lpt machinery. *Marine Drugs, 12*(2), 1023–1042. https://doi.org/10.3390/md12021023.

Powers, M. J., & Stephen Trent, M. (2019). Intermembrane transport: Glycerophospholipid homeostasis of the gram-negative cell envelope. *Proceedings of the National Academy of Sciences of the United States of America, 116*(35), 17147–17155. https://doi.org/10.1073/pnas.1902026116.

Qiao, S., Luo, Q., Zhao, Y., Zhang, X. C., & Huang, Y. (2014). Structural basis for lipopolysaccharide insertion in the bacterial outer membrane. *Nature, 511*(7507), 108–111. https://doi.org/10.1038/nature13484.

Raetz, C. R., & Whitfield, C. (2002). Lipopolysaccharide endotoxins. *Annual Review of Biochemistry, 71*, 635–700. https://doi.org/10.1146/annurev.biochem.71.110601.135414.

Rai, A. K., & Mitchell, A. M. (2020). Enterobacterial common antigen: Synthesis and function of an enigmatic molecule. *MBio, 11*(4), 1–19. https://doi.org/10.1128/mBio.01914-20.

Ranava, D., Caumont-Sarcos, A., Albenne, C., & Ieva, R. (2018). Bacterial machineries for the assembly of membrane-embedded β-barrel proteins. *FEMS Microbiology Letters, 365*(10), 1–12. https://doi.org/10.1093/femsle/fny087.

Rassam, P., Copeland, N. A., Birkholz, O., Tóth, C., Chavent, M., Duncan, A. L., et al. (2015). Supramolecular assemblies underpin turnover of outer membrane proteins in bacteria. *Nature, 523*, 333–336. https://doi.org/10.1038/nature14461.

Reusch, R. N. (2012). Insights into the structure and assembly of Escherichia coli outer membrane protein A. *FEBS Journal, 10*(1111). https://doi.org/https://doi.org/10.1111/j.1742-4658.2012.08484.x.

Reyes, R. E., González, C. R., Jiménez, R. C., Herrera, M. O., & Andrade, A. A. (2012). Mechanisms of O-antigen structural variation of bacterial lipopolysaccharide (LPS). In *The complex world of polysaccharides* (pp. 71–98).

Ricci, D. P., & Silhavy, T. J. (2012). The bam machine: A molecular cooper. *Biochimica et Biophysica Acta - Biomembranes, 1818*(4), 1067–1084. https://doi.org/10.1016/j.bbamem.2011.08.020.

Ricci, D. P., & Silhavy, T. J. (2019). Outer membrane protein insertion by the β-barrel assembly machine. *EcoSal Plus, 8*(2), 1–9. https://doi.org/10.1128/ecosalplus.esp-0035-2018.

Rice, A., & Wereszczynski, J. (2018). Atomistic scale effects of lipopolysaccharide modifications on bacterial outer membrane defenses. *Biophysical Journal, 114*(6), 1389–1399. https://doi.org/10.1016/j.bpj.2018.02.006.

Rodríguez-Alonso, R., Létoquart, J., Nguyen, V. S., Louis, G., Calabrese, A. N., Iorga, B. I., et al. (2020). Structural insight into the formation of lipoprotein-β-barrel complexes. *Nature Chemical Biology, 16*(9), 1019–1025. https://doi.org/10.1038/s41589-020-0575-0.

Romantsov, T., Guan, Z., & Wood, J. M. (2009). Cardiolipin and the osmotic stress responses of bacteria. *Biochimica et Biophysica Acta - Biomembranes, 1788*(10), 2092–2100. https://doi.org/10.1016/j.bbamem.2009.06.010.

Ruan, X., Monjarás Feria, J., Hamad, M., & Valvano, M. A. (2018). Escherichia coli and Pseudomonas aeruginosa lipopolysaccharide O-antigen ligases share similar membrane topology and biochemical properties. *Molecular Microbiology, 110*(1), 95–113. https://doi.org/10.1111/mmi.14085.

Saikawa, N., Akiyama, Y., & Ito, K. (2004). FtsH exists as an exceptionally large complex containing HflKC in the plasma membrane of Escherichia coli. *Journal of Structural Biology, 146*(1–2), 123–129. https://doi.org/10.1016/j.jsb.2003.09.020.

Sala, A., Bordes, P., & Genevaux, P. (2014). Multitasking SecB chaperones in bacteria. *Frontiers in Microbiology, 5*(DEC), 1–12. https://doi.org/10.3389/fmicb.2014.00666.

Samsudin, F., Ortiz-Suarez, M. L., Piggot, T. J., Bond, P. J., & Khalid, S (2016). OmpA: A flexible clamp for bacterial Cell Wall attachment. *Structure, 24*(12), 2227–2235. https://doi.org/10.1016/j.str.2016.10.009.

Sankaran, K., & Wu, H. C. (1994). Lipid modification of bacterial prolipoprotein. Transfer of diacylglyceryl moiety from phosphatidylglycerol. *Journal of Biological Chemistry, 269*(31), 19701–19706. https://doi.org/10.1016/s0021-9258(17)32077-x.

Saraogi, I., & Shan, S.-o. (2014). Co-translational protein targeting to the bacterial membrane. *Biochimica et Biophysica Acta, Molecular Cell Research, 1843*(8), 1433–1441. https://doi.org/10.1016/j.bbamcr.2013.10.013.

Sauer, R. T., & Baker, T. A. (2011). AAA+ proteases: ATP-fueled machines of protein destruction. *Annual Review of Biochemistry, 80*, 587–612. https://doi.org/10.1146/annurev-biochem-060408-172623.

Schiebel, E., Driessen, A. J. M., Hartl, F. U., & Wickner, W. (1991). ΔμH+ and ATP function at different steps of the catalytic cycle of preprotein translocase. *Cell, 64*(5), 927–939. https://doi.org/10.1016/0092-8674(91)90317-R.

Schiffrin, B., Brockwell, D. J., & Radford, S. E. (2017). Outer membrane protein folding from an energy landscape perspective. *BMC Biology, 15*(1), 1–16. https://doi.org/10.1186/s12915-017-0464-5.

Schiffrin, B., Calabrese, A. N., Higgins, A. J., Humes, J. R., Ashcroft, A. E., Kalli, A. C., et al. (2017). Effects of periplasmic chaperones and membrane thickness on BamA-catalyzed outer-membrane protein folding. *Journal of Molecular Biology, 429*(23), 3776–3792. https://doi.org/10.1016/j.jmb.2017.09.008.

Schwechheimer, C., & Kuehn, M. J. (2015). Outer-membrane vesicles from gram-negative bacteria: Biogenesis and functions. *Nature Reviews Microbiology, 13*(10), 605–619. https://doi.org/10.1038/nrmicro3525.

Schwechheimer, C., Kulp, A., & Kuehn, M. J. (2014). Modulation of bacterial outer membrane vesicle production by envelope structure and content. *BMC Microbiology*, *14*(1), 324. https://doi.org/10.1186/s12866-014-0324-1.

Selkrig, J., Mosbahi, K., Webb, C. T., Belousoff, M. J., Perry, A. J., Wells, T. J., et al. (2012). Discovery of an archetypal protein transport system in bacterial outer membranes. *Nature Structural & Molecular Biology*, *19*(5), 506–510. https://doi.org/10.1038/nsmb.2261.

Serdiuk, T., Balasubramaniam, D., Sugihara, J., Mari, S. A., Kaback, H. R., & Müller, D. J. (2016). YidC assists the stepwise and stochastic folding of membrane proteins. *Nature Chemical Biology*, *12*(11), 911–917. https://doi.org/10.1038/nchembio.2169.

Sherman, D. J., Xie, R., Taylor, R. J., George, A. H., Okuda, S., Foster, P. J., et al. (2018). Lipopolysaccharide is transported to the cell surface by a membrane-Tomembrane protein bridge. *Science*, *359*(6377). https://doi.org/10.1126/science.aar1886.

Shimohata, N., Chiba, S., Saikawa, N., Ito, K., & Akiyama, Y. (2002). The Cpx stress response system of Escherichia coli senses plasma membrane proteins and controls HtpX, a membrane protease with a cytosolic active site. *Genes to Cells*, *7*(7), 653–662. https://doi.org/10.1046/j.1365-2443.2002.00554.x.

Shrivastava, R., & Chng, S. S. (2019). Lipid trafficking across the gram-negative cell envelope. *Journal of Biological Chemistry*, *294*(39), 14175–14184. https://doi.org/10.1074/jbc.AW119.008139.

Shu, W., Liu, J., Ji, H., & Lu, M. (2000). Core structure of the outer membrane lipoprotein from Escherichia coli at 1.9 A resolution. *Journal of Molecular Biology*, *299*(4), 1101–1112. https://doi.org/10.1006/jmbi.2000.3776.

Silhavy, T. J., Kahne, D., & Walker, S. (2010). The bacterial cell envelope. *Cold Spring Harbor Perspectives in Biology*, *2*(5), 1–16. https://doi.org/10.1101/cshperspect.a000414.

Singh, H., Velamakanni, S., Deery, M. J., Howard, J., Wei, S. L., & Van Veen, H. W. (2016). ATP-dependent substrate transport by the ABC transporter MsbA is proton-coupled. *Nature Communications*, 7. https://doi.org/10.1038/ncomms12387.

Sklar, J. G., Wu, T., Kahne, D., & Silhavy, T. J. (2007). Defining the roles of the periplasmic chaperones SurA, Skp, and DegP in Escherichia coli. *Genes & Development*, *21*(19), 2473–2484. https://doi.org/10.1101/gad.1581007.

Smets, D., Loos, M. S., Karamanou, S., & Economou, A. (2019). Protein transport across the bacterial plasma membrane by the sec pathway. *Protein Journal*, *38*(3), 262–273. https://doi.org/10.1007/s10930-019-09841-8.

Snijder, H. J., Ubarretxena-Belandia, I., Blaauw, M., Kalk, K. H., Varheij, H. M., Egmond, M. R., et al. (1999). Structural evidence for dimerization-regulated activation of an integral membrane phospholipase. *Nature*, *401*(6754), 717–721. https://doi.org/10.1038/401717a0.

Sperandeo, P., Martorana, A. M., & Polissi, A. (2017). The lipopolysaccharide transport (Lpt) machinery: A nonconventional transporter for lipopolysaccharide assembly at the outer membrane of gram-negative bacteria. *Journal of Biological Chemistry*, *292*(44), 17981–17990. https://doi.org/10.1074/jbc.R117.802512.

Sverzhinsky, A., Chung, J. W., Deme, J. C., Fabre, L., Levey, K. T., Plesa, M., et al. (2015). Membrane protein complex ExbB4-ExbD1-TonB1 from Escherichia coli demonstrates conformational plasticity. *Journal of Bacteriology*, *197*(11), 1873–1885. https://doi.org/10.1128/JB.00069-15.

Szczepaniak, J., Press, C., & Kleanthous, C. (2020). The multifarious roles of Tol-pal in gram-negative bacteria. *FEMS Microbiology Reviews*, *44*(4), 490–506. https://doi.org/10.1093/femsre/fuaa018.

Szczesny, M., Beloin, C., & Ghigo, J. M. (2018). Increased osmolarity in biofilm triggers RcsB-dependent lipid a palmitoylation in Escherichia coli. *MBio*, *9*(4). https://doi.org/10.1128/mBio.01415-18.

Tanaka, Y., Sugano, Y., Takemoto, M., Mori, T., Furukawa, A., Kusakizako, T., et al. (2015). Crystal structures of SecYEG in Lipidic cubic phase elucidate a precise resting and a peptide-bound state. *Cell Reports*, *13*(8), 1561–1568. https://doi.org/10.1016/j.celrep.2015.10.025.

Tang, X., Chang, S., Qiao, W., Luo, Q., Chen, Y., Jia, Z., et al. (2021). Structural insights into outer membrane asymmetry maintenance in gram-negative bacteria by MlaFEDB. *Nature Structural and Molecular Biology*, *28*(1), 81–91. https://doi.org/10.1038/s41594-020-00532-y.

Tang, X., Chang, S., Zhang, K., Luo, Q., Zhang, Z., Wang, T., et al. (2021). Structural basis for bacterial lipoprotein relocation by the transporter LolCDE. *Nature Structural and Molecular Biology*, *28*(4), 347–355. https://doi.org/10.1038/s41594-021-00573-x.

Thoma, J., & Burmann, B. M. (2020). High-resolution in situ NMR spectroscopy of bacterial envelope proteins in outer membrane vesicles. *Biochemistry*, *59*(17), 1656–1660. https://doi.org/10.1021/acs.biochem.9b01123.

Thoma, J., & Burmann, B. M. (2021). Fake it 'till you make it—The pursuit of suitable membrane mimetics for membrane protein biophysics. *International Journal of Molecular Sciences*, *22*(1), 1–22. https://doi.org/10.3390/ijms22010050.

Thoma, J., Burmann, B. M., Hiller, S., & Müller, D. J. (2015). Impact of holdase chaperones Skp and SurA on the folding of β-barrel outer-membrane proteins. *Nature Structural & Molecular Biology*, *22*(10), 795–802. https://doi.org/10.1038/nsmb.3087.

Tomasek, D., & Kahne, D. (2021). The assembly of β-barrel outer membrane proteins. *Current Opinion in Microbiology*, *60*, 16–23. https://doi.org/10.1016/j.mib.2021.01.009.

Tomasek, D., Rawson, S., Lee, J., Wzorek, J. S., Harrison, S. C., Li, Z., et al. (2020). Structure of a nascent membrane protein as it folds on the BAM complex. *Nature*. (November 2019). https://doi.org/https://doi.org/10.1038/s41586-020-2370-1.

Toyofuku, M., Nomura, N., & Eberl, L. (2019). Types and origins of bacterial membrane vesicles. *Nature Reviews Microbiology*, *17*(1), 13–24. https://doi.org/10.1038/s41579-018-0112-2.

Tran, A. X., Lester, M. E., Stead, C. M., Raetz, C. R. H., Maskell, D. J., McGrath, S. C., et al. (2005). Resistance to the antimicrobial peptide polymyxin requires myristoylation of Escherichia coli and salmonella typhimurium lipid A. *Journal of Biological Chemistry*, *280*(31), 28186–28194. https://doi.org/10.1074/jbc.M505020200.

Tsukahara, J., Mukaiyama, K., Okuda, S., Narita, S. I., & Tokuda, H. (2009). Dissection of LolB function - lipoprotein binding, membrane targeting and incorporation of lipoproteins into lipid bilayers. *FEBS Journal*, *276*(16), 4496–4504. https://doi.org/10.1111/j.1742-4658.2009.07156.x.

Tsukazaki, T. (2018). Structure-based working model of SecDF, a proton-driven bacterial protein translocation factor. *FEMS Microbiology Letters*, *365*(12), 1–9. https://doi.org/10.1093/femsle/fny112.

Tsukazaki, T., Mori, H., Echizen, Y., Ishitani, R., Fukai, S., Tanaka, T., et al. (2011). Structure and function of a membrane component SecDF that enhances protein export. *Nature*, *474*(7350), 235–238. https://doi.org/10.1038/nature09980.

Valvano, M. A. (2014). *Genetics and biosynthesis of lipopolysaccharide. Molecular medical microbiology* (Vols. 1–3) (2nd Edition). Elsevier Ltd. https://doi.org/10.1016/B978-0-12-397169-2.00004-4.

Vogeley, L., El Arnaout, T., Bailey, J., Stansfeld, P. J., Boland, C., & Caffrey, M. (2016). Structural basis of lipoprotein signal peptidase II action and inhibition by the antibiotic globomycin. *Science*, *351*(6275), 876–880. https://doi.org/10.1126/science.aad3747.

Voss, B. J., & Stephen Trent, M. (2018). LPS Transport: Flipping Out over MsbA. *Current Biology*, *28*(1), R30–R33. https://doi.org/10.1016/j.cub.2017.10.067.

Ward, A., Reyes, C. L., Yu, J., Roth, C. B., & Chang, G. (2007). Flexibility in the ABC transporter MsbA: Alternating access with a twist. *Proceedings of the National Academy of Sciences of the United States of America, 104*(48), 19005–19010. https://doi.org/10.1073/pnas.0709388104.

Welte, T., Kudva, R., Kuhn, P., Sturm, L., Braig, D., Müller, M., et al. (2012). Promiscuous targeting of polytopic membrane proteins to SecYEG or YidC by the Escherichia coli signal recognition particle. *Molecular Biology of the Cell, 23*(3), 464–479. https://doi.org/10.1091/mbc.E11-07-0590.

White, S. H., & von Heijne, G. (2008). How Translocons select transmembrane helices. *Annual Review of Biophysics, 37*(1), 23–42. https://doi.org/10.1146/annurev.biophys.37.032807.125904.

White, S. H., & Wimley, W. C. (1999). Membrane protein folding and stability: Physical principles. *Annual Review of Biophysics and Biomolecular Structure, 28*, 319–365. https://doi.org/10.1146/annurev.biophys.28.1.319.

Whitfield, C., Williams, D. M., & Kelly, S. D. (2020). Lipopolysaccharide O-antigens-bacterial glycans made to measure. *Journal of Biological Chemistry, 295*(31), 10593–10609. https://doi.org/10.1074/jbc.REV120.009402.

Wiktor, M., Weichert, D., Howe, N., Huang, C. Y., Olieric, V., Boland, C., et al. (2017). Structural insights into the mechanism of the membrane integral N-acyltransferase step in bacterial lipoprotein synthesis. *Nature Communications, 8*(May), 25–28. https://doi.org/10.1038/ncomms15952.

Yamaguchi, K., Yu, F., & Inouye, M. (1988). A single amino acid determinant of the membrane localization of lipoproteins in E. coli. *Cell, 53*(3), 423–432. https://doi.org/10.1016/0092-8674(88)90162-6.

Zhang, Y. M., & Rock, C. O. (2008). Membrane lipid homeostasis in bacteria. *Nature Reviews Microbiology, 6*(3), 222–233. https://doi.org/10.1038/nrmicro1839.

Zückert, W. R. (2014). Secretion of bacterial lipoproteins: Through the cytoplasmic membrane, the periplasm and beyond. *Biochimica et Biophysica Acta, Molecular Cell Research, 1843*(8), 1509–1516. https://doi.org/10.1016/j.bbamcr.2014.04.022.

CHAPTER TWO

Modulation of host cellular responses by gram-negative bacterial porins

Arpita Sharma†, Shashi Prakash Yadav†, Dwipjyoti Sarma, and Arunika Mukhopadhaya*

Department of Biological Sciences, Indian Institute of Science Education and Research Mohali, Mohali, Punjab, India

*Corresponding author: e-mail address: arunika@iisermohali.ac.in

Contents

1. Introduction 36
2. Structure of gram-negative bacterial porins 38
3. Functions of the gram-negative bacterial porins 40
   3.1 Channel function 40
   3.2 Antimicrobial resistance 42
   3.3 Bacteriophage-binding site 46
   3.4 Complement-binding site 49
   3.5 Bacterial pathogenesis 50
   3.6 Modulation of host immune responses by gram-negative bacterial porins 54
   3.7 Involvement of the gram-negative bacterial porins in cell death responses 57
4. Important uses of gram-negative bacterial porins 59
   4.1 Adjuvant properties of gram-negative bacterial porins 59
   4.2 Vaccine potential of the gram-negative bacterial porins 60
   4.3 Porins as a target for the generation of therapeutics 62
   4.4 Porins as the biomarkers 62
5. Conclusion 63
Acknowledgment 63
References 64

Abstract

The outer membrane of a gram-negative bacteria encapsulates the plasma membrane thereby protecting it from the harsh external environment. This membrane acts as a sieving barrier due to the presence of special membrane-spanning proteins called "porins." These porins are β-barrel channel proteins that allow the passive transport

† Contributed equally.

Advances in Protein Chemistry and Structural Biology, Volume 128
ISSN 1876-1623
https://doi.org/10.1016/bs.apcsb.2021.09.004

of hydrophilic molecules and are impermeable to large and charged molecules. Many porins form trimers in the outer membrane. They are abundantly present on the bacterial surface and therefore play various significant roles in the host–bacteria interactions. These include the roles of porins in the adhesion and virulence mechanisms necessary for the pathogenesis, along with providing resistance to the bacteria against the antimicrobial substances. They also act as the receptors for phage and complement proteins and are involved in modulating the host cellular responses. In addition, the potential use of porins as adjuvants, vaccine candidates, therapeutic targets, and biomarkers is now being exploited. In this review, we focus briefly on the structure of the porins along with their important functions and roles in the host-bacteria interactions.

1. Introduction

Porins, hydrophilic transmembrane channel proteins, are one of the most abundant proteins in the outer membrane of gram-negative bacteria (Fig. 1). The outer membrane can have a porin abundance of up to 100%, e.g., porin P100 of *Thermus thermophilus* forms a complete layer called the s-layer composed of only this porin (Caston, Berenguer, de Pedro, & Carrascosa, 1993). Although bacteria possess many porin-encoding genes, the expression of some porins dominates the expression of others, depending on the surrounding niche. For example, in *Salmonella typhi*, major porins are constitutively expressed (e.g., OmpF and OmpC), while the minor porins (e.g., OmpS1 and OmpS2) are expressed at low levels during the infection scenario (Nikaido, 2003). These observations indicate that the expression of porins is tightly regulated as per the bacterial requirement.

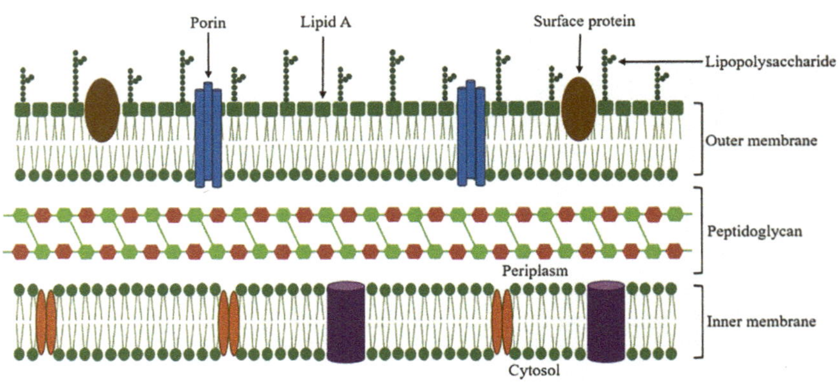

Fig. 1 Gram-negative bacterial envelope showing inner plasma membrane, middle peptidoglycan layer, and outer membrane.

Porins generally allow the passive diffusion of nutrients and waste across the membrane, down the concentration gradient, without any expenditure of ATP. In addition to acting like channels, porins play various other roles. Their general function in prokaryotes is mainly to maintain the osmolarity, transport of salts and nutrients across the membrane, and to contribute towards virulence in some cases. The abundance and localization of the porins in the outer membrane direct their functional significance in various processes.

Porins are also present in the cell wall of gram-positive bacteria and perform similar functions but are less abundant compared to gram-negative bacteria. *Mycobacterium tuberculosis* porin OmpATb is essential for their adaptation in low pH conditions and survival in macrophages (Raynaud et al., 2002). MspA, a major porin of *M. stegmatis* forms a tetramer and has a large pore size of 10 nm (Engelhardt, Heinz, & Niederweis, 2002). Porin tetramers are very unlikely in gram-negative bacteria as they generally form trimers and in some cases dimers or monomers. The pore size of the gram-positive porins is also comparably larger. PorA porin is a major channel in the cell wall of gram-positive bacteria *Corynebacterium glutamicum* (Costa-Riu, Burkovski, Kramer, & Benz, 2003).

In eukaryotes, the mitochondrial outer membrane also possesses porins that help in the transport of metabolites across the mitochondrial outer membrane, regulation of mitochondrial metabolism, and apoptosis (Young, Bay, Hausner, & Court, 2007). Open and closed conformations of some mitochondrial porins are regulated by the voltage across the outer membrane and they show anion selectivity. Thus, they are also known as voltage-gated anion-selective channels (VDACs). The selectivity of VDACs varies within the states as the open state shows anion selectivity and the closed state is more of cation-selective (Camara, Zhou, Wen, Tajkhorshid, & Kwok, 2017; Lemasters & Holmuhamedov, 2006). Apart from transporting solutes and metabolites across the outer mitochondrial membrane, porins also assist in translocating the newly synthesized proteins in the mitochondrial membrane (Zeth, 2010).

Porins are also found in chloroplasts. They behave somewhat like VDACs as they are in open conformation at zero membrane potential, and as the potential rises, they acquire a closed conformation state. They help in the transport of many metabolites as the chloroplast is a center for metabolic reactions like photosynthetic CO_2 fixation, amino acid synthesis and ammonia assimilation (Ulf-IngoFlügge, 2000).

The ubiquitous distribution of porins in almost all the major kingdoms of life and acquisition of more diverse functions shows that they are one of the important biomolecules.

This review focuses briefly on the structure of the gram-negative bacterial porins along with their role in the modulation of the host responses and their potential applications.

2. Structure of gram-negative bacterial porins

The primary sequence of gram-negative bacterial porin consists of homogeneously distributed charged residues in high density. Spectroscopic analyses have shown that they majorly span as β-barrels across the membrane with a very small fraction of α-helix. The first high-resolution X-ray structure of a gram-negative porin was published in 1990 (an outer membrane porin of *Rhodobacter capsulatus*) (Weiss, Wacker, Weckesser, Welte, & Schulz, 1990). The structure was shown to be a trimer containing folds of 16 β-strands arranged in an antiparallel manner with hydrogen bonding between the chains (Fig. 2A and B) (Weiss et al., 1991). After the determination of this trimeric structure, many porins were found to have similar architecture and with very lesser variations. Later, porins with 8–24 β-strands were also discovered (Fairman, Noinaj, & Buchanan, 2011).

Most of the gram-negative porins are known to acquire an oval shape in which the monomer spreads approximately 30–35 Å laterally and approximately 50 Å vertically. The diameter of the pore ranges from 6 Å (for highly specific porins) to 15 Å (for non-specific porins) (Galdiero et al., 2012). Generally, the strands are connected by 8 or 9 long loops which face the extracellular side and 7 or 8 small turns in the periplasmic space. The barrel is closed by the pairing of the first and last β-strand in an antiparallel fashion. Different loops of porins are mainly involved in a specific type of function. For example, loops L1, L2, and L4 are involved in monomer–monomer interactions within the trimer while L5, L6, and L7 are superficially involved in the biological activity of the porin. L3 is involved in the determination of pore permeability. Further, its interaction with peripheral barrel strands might help in the stability of the barrel. L8 folds back into the barrel leading to the formation of a channel opening at the external side. The porin substructures are arranged onto a scaffolding provided by the β-sheet cylinder of the monomer (Galdiero et al., 2012; Schulz, 1993). Porin substructures are divided into three parts:

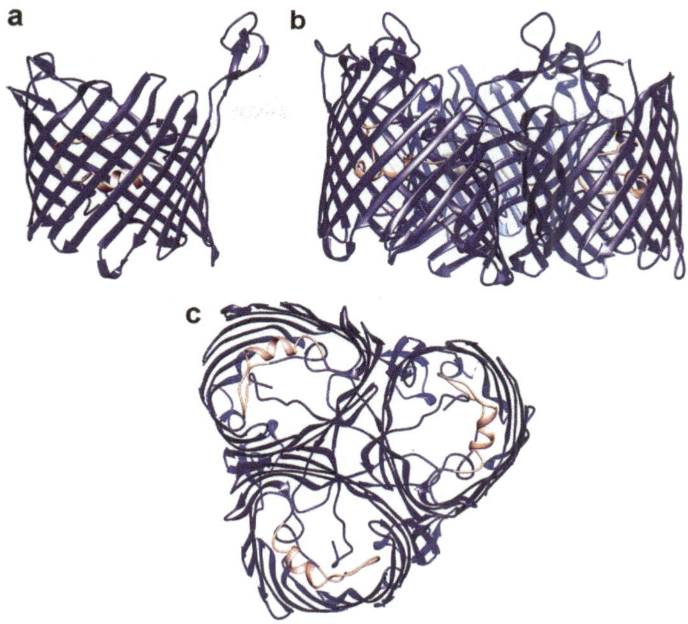

Fig. 2 Structural model of a trimeric porin OmpU (Protein Data Bank ID: 6EHB) of *V. cholerae* visualized using UCSF Chimera (Pettersen et al., 2004). OmpU forms a homotrimer with each monomer containing eight loops. L3 loop forming the eyelet region is shown in light pink. (A) Monomer view from the membrane plane. (B) Trimer view from the membrane plane. (C) Trimer view from the periplasmic side.

a. *External or extracellular mouth and vestibule*: The external vestibule, generally known as the mouth is a cone-like structure with a wide opening towards the outside environment. The long and irregular loops and turns present in this region are highly variable in amino acid composition, length, and conformation between different species and are responsible for hydrophilicity. Further, this vestibule has a high number of charged species (as compared to internal vestibule) which makes a potential gradient.

b. *Constriction zone/Eyelet*: Generally, Loop L3 folds back into the barrel and reduces the free cross-section leading to the formation of an eyelet at the barrel core. Although the potential gradient in the external vestibule allows only prescreened molecules inside, yet the ultimate selection is made by the eyelet. The narrow dimensions, the shape limit, the charge constraint of polar residues, and the chemical nature of the solutes determine the absolute selection of the accepted molecules. Due to the presence of this eyelet, the hydrated ion movement no longer remains free, and nonpolar molecules get filtered.

c. *Internal or periplasmic mouth and vestibule*: This cone region is the least variable region of all the porin structures. They have a very low number of charged species and short looping turns as compared to the external vestibule and provides minimum resistance in the movement of the solutes (Jap & Walian, 1996).

3. Functions of the gram-negative bacterial porins

Gram-negative bacterial porins are known to play various functions (Fig. 3). Some major functions of gram-negative bacterial porins are discussed below.

3.1 Channel function

Based on the ways of diffusion of the solutes and nutrients across the membrane, gram-negative bacterial porins are generally divided into two categories: the first group is called general (non-specific) porins and the second group is specific porins.

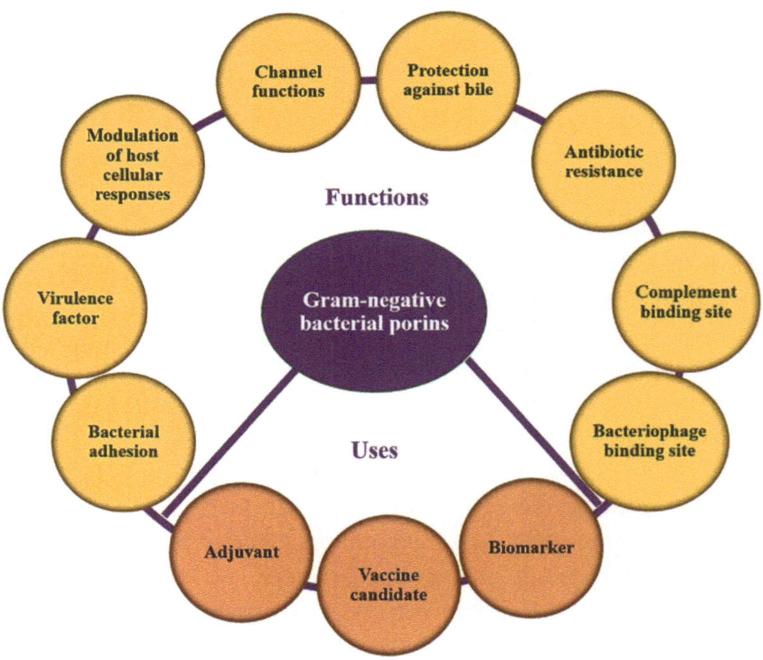

Fig. 3 Functions and usage of gram-negative bacterial porins.

a. *General or non-specific porins*: They are the channels with no particular substrate specificity except for some cations or anions (Koebnik, Locher, & Van Gelder, 2000). Majority of the porins belong to this category. They allow the diffusion of hydrophilic molecules with an exclusion limit of 600 Da and an extreme of 5000 Da. They exclude the entry of larger molecules into the periplasmic space (Benz, 1988). All the gram-negative bacteria are known to contain atleast one general porin in high copy number for the filtration of solutes (Rosenbusch, 1974; Welte, Nestel, Wacker, & Diederichs, 1995). The rate of permeation of solutes depends on the size and molecular weight of the solute and the channel conductance depends linearly on the substrate concentration (Renkin, 1954). These types of porins can also be biased based on the charge. OmpF and OmpC of *E. coli* prefer neutral solutes and cations over anions, whereas, PhoE of *E. coli* and most Neisserial porins prefer anions. Further, the above-mentioned porins of *E. coli* exclude lipophilic solutes (Nikaido, 1994). These porins also show voltage-dependent closing which is regulated by the external field relative to the porin (Welte et al., 1995).

b. *Specific porins*: As the name suggests, specific porins are the channels that have substrate specificity. They accelerate the diffusion of specific solutes required by the cell. They also allow the passage of solutes that are bigger for general porins like large maltodextrins or too slow to penetrate through general porins to serve as a major carbon source (Nikaido, 1994). They contain binding sites with moderate specificity for specific molecules and this binding probably alters the eyelet conformation (Welte et al., 1995). These pores are not voltage-gated (Koebnik et al., 2000) and are usually expressed at low copy numbers and in certain environmental conditions (Welte et al., 1995). Some examples of specific porins include LamB (malto-oligosaccharide-specific maltoporin) and Tsx (nucleoside channel) from *E. coli*, ScrY (sucrose-specific porin) from *S. typhimurium*, OprB (glucose channel), OprD (basic amino acid channel), OprP and OprO (phosphate and pyrophosphate channel, respectively) from *P. aeruginosa* (Ishii, Okajima, & Nakae, 1981; Nikaido, 1994; Schulein, Schmid, & Benzl, 1991; Trias, Rosenberg, & Nikaido, 1988).

Earlier, 16 stranded porins were classified as general porins, while 18 stranded porins were thought to be specific porins, both of which were trimeric. But later on, the discovery of porins having a varying number of β-strands other than 16 or 18 and different quaternary structures made this classification obsolete (Fairman et al., 2011). Also, some non-specific porins were found to bind small and negatively charged substrates.

3.2 Antimicrobial resistance

3.2.1 Protection against bile

Bacteria face many challenges in the environment in which they grow. Generally, natural environments pose different challenges, such as fluctuating temperatures, pH, low availability of nutrients, presence of toxic chemicals, and other competing microbes. Bacteria adapt to these fluctuations by readily fine-tuning their metabolism as per their requirements. The success of most pathogens mainly depends on how rapidly they adapt to environmental changes. Porins are documented to support the bacteria for surviving in these harsh conditions (Benz, 1988). As in the case of *V. cholerae,* strains which majorly express OmpU in the gut are more resistant to bile and anionic detergents than strains that majorly express another porin, OmpT (Provenzano & Klose, 2000). Cation selectivity of OmpU might be a possible explanation for its role in bile resistance as it might impair the passage of anionic bile salts through the pore (Benz, Maier, & Chakraborty, 1997). Similarly, the absence of OmpC porin in *S. typhi* makes it susceptible to bile salts leading to a slow growth rate in sodium deoxycholate-containing medium (Villarreal et al., 2014). AcrAB-TolC efflux system in *E. coli* helps in extruding bile acids from the bacterial cell thus resisting the antimicrobial action of bile (Thanassi, Cheng, & Nikaido, 1997). *E. coli* AcrAD-TolC performs expulsion of deoxycholate and cholic acid making the bacteria resistant to higher concentrations of these bile components (Nishino & Yamaguchi, 2001). Another efflux system of *E. coli* MdtABC-TolC efflux pump extrudes bile salts (Nagakubo, Nishino, Hirata, & Yamaguchi, 2002). In these systems, TolC is a porin helping in the extrusion of molecules across the outer membrane. Sublethal doses of bile salt deoxycholate cause upregulation of operon encoding the components of AcrAB efflux system in *S. typhimurium* (Prouty, Brodsky, Falkow, & Gunn, 2004; Urdaneta & Casadesús, 2018). Further, AcrAB-TolC efflux pump also helps in the adaptation of *S. typhimurium* to bile. *S. typhimurium* SL1344 downregulates the expression of its major porins OmpC and OmpD in presence of bile thus restricting the entry of bile from these channels (Gunn, 2000; Ruiz, Kahne, & Silhavy, 2006).

3.2.2 Antibiotic resistance

Bacteria employ many strategies to counteract antibiotics. Some of them start resisting the uptake of antibiotics by changing the integrity of their outer membrane or by acquiring mutations in the channel proteins. Some porins involved in antibiotic resistance are enlisted in Table 1. *Neisseria gonorrhoeae* becomes resistant to β-lactams and tetracycline by acquiring a mutation in the loop region of porin PorB (Gill et al., 1998). In *N. meningitidis,* a mutation in the PorB-encoding gene affecting a single amino acid in the eyelet region causes resistance to ampicillin (Bartsch et al., 2021). *E. coli* OmpF porin acts as the major entry route of several antibiotics making the bacteria susceptible to them. On the other hand, OmpA and OmpC porins in *E. coli* help in the maintenance of membrane integrity, thus making the bacteria more resistant to the antibiotics (Choi & Lee, 2019). The double mutant of *ompA* and *ompC* is more susceptible to antibiotics than that of single mutants (Choi & Lee, 2019). Researchers have shown that D116A mutation in the loop L3 of *V. cholerae* OmpU (VcOmpU) shows a drastically increased resistance to cefotaxime, ceftazidime, and ceftriaxone (Pagel et al., 2007). VcOmpU is also known to confer resistance against polymyxin B (Mathur & Waldor, 2004). Some of the bacterial species have evolved efflux pumps that expel the antibiotics crossing the membrane to reach the cytoplasm. These are more specifically called tripartite efflux pumps of gram-negative bacteria. Porins also coordinate with other molecules to form efflux pumps which expel the antibacterial drugs to the extracellular environment. Like OprM, an outer membrane porin of *P. aeruginosa,* forms an efflux pump with MexAB to get rid of a class of β-lactams like carbenicillin and ceftazidime (Masuda et al., 2000). Some of the bacterial species came up with active enzymes which can degrade a whole class of antibiotics. The evolution of β-lactamases is such a strategy to combat antibiotics. These enzymes act on β-lactams by degrading them. Porins are reported to aid in β-lactamase mediated antibiotic resistance in some cases. Both porins and β-lactamase have been found in the outer membrane vesicles (OMVs) derived from the β-lactam resistant *E. coli.* Some observations suggest that OMVs from the parent strain are more efficient at degrading some β-lactams like cefoperazone and cefotaxime than OMVs from the *ompC* or *ompF* knockout strains (Kim et al., 2020). This difference might be because of the role of OmpF and OmpC as a route of entry for β-lactams in the OMVs. So, in this manner, porins act as a route for antibiotics to the vesicle lumen, and β-lactamase present in the lumen degrades the entering antibiotics. Thus, these OMVs loaded with porins and β-lactamases clear the antibiotics before they can attack the bacteria.

Table 1 Role of gram-negative bacterial porins in antibiotic resistance.

Porin	Efflux system	Bacterial species	Antibiotic resistance to	References
PorB		*N. gonorrhoeae*	β-Lactams and tetracyclines	Gill et al. (1998)
PorB		*N. meningitidis*	Ampicillin	Bartsch et al. (2021)
OmpF		*E. coli*	β-Lactams	Jaffe, Chabbert, and Semonin (1982)
OmpC		*E. coli*	β-Lactams	Jaffe et al. (1982)
OprD		*P. aeruginosa*	Carbapenems	Ballestero et al. (1996)
OmpC		*Enterobacter aerogenes*	Carbapenems	Sturenburg, Sobottka, Mack, and Laufs (2002)
OmpF		*E. aerogenes*	Carbapenems	Sturenburg et al. (2002)
Omp36		*E. aerogenes*	Imipenem, cefepime, and cefpirome	Thiolas, Bornet, Davin-Regli, Pages, and Bollet (2004)
OmpF		*Serratia marcescens*	β-Lactams	Suh et al. (2010) and Weindorf, Schmidt, and Martin (1998)
OmpC		*S. marcescens*	β-Lactams	Weindorf et al. (1998)
OmpK35		*K. pneumoniae*	Cephalosporins, carbapenems, fluoroquinolones, and chloramphenicol	Chen, Lauderdale, Ho, and Lo (2003), Domenech-Sanchez et al. (2003), and Doumith, Ellington, Livermore, and Woodford (2009)
OmpK36		*K. pneumoniae*	Carbapenems	Doumith et al. (2009) and Song et al. (2009)

OmpF		*S. enterica*	Chloramphenicol	Toro, Lobos, Calderon, Rodriguez, and Mora (1990)
OprA	AmrAB-OprA	*Burkholderia pseudomallei*	Aminoglycosides and macrolides	Moore, DeShazer, Reckseidler, Weissman, and Woods (1999)
OpcM	CeoAB-OpcM	*B. cenocepacia*	Chloramphenicol, fluoroquinolones, and trimethoprim	Nair, Cheung, Griffith, and Burns (2004)
TolC	VexAB-TolC	*V. cholerae*	Erythromycin and novobiocin	Bina, Provenzano, Wang, Bina, and Mekalanos (2006)
OprM	MexJK-OprM	*P. aeruginosa*	Aminoglycosides, clindamycin, erythromycin, and tetracycline	Chuanchuen, Narasaki, and Schweizer (2002)
OprM	MexAB-OprM	*P. aeruginosa*	Carbenicillin and ceftazidime	Masuda et al. (2000)
TolC	AcrAB-TolC	*E. coli*	Cephalosporins, novobiocin, tetracyclines, and fluoroquinolones	Nikaido (2011) and Nikaido and Takatsuka (2009)
TolC	AcrAB-TolC	*H. influenzae*	Novobiocin, rifampin, and erythromycin	Sanchez, Pan, Vinas, and Nikaido (1997)

3.2.3 Resistance to other antimicrobial substances

Antimicrobial peptides (AMPs) are a short chain of amino acids that vary from 10 to 60 residues. They are widely distributed in almost every kingdom of life, from prokaryotes to higher vertebrates such as humans. They are known for their broad spectrum of antimicrobial activity. AMPs employ many strategies like bacterial membrane permeabilization, enzymatic attack on the cell wall, and interference with intracellular components. OmpU of *V. splendidus* makes the bacteria resistant to AMPs like defensins, polymyxin B, protegrin, and bacteria permeability-increasing proteins (BPI) (Duperthuy et al., 2010). VcOmpU is reported to play a key role in its resistance to a human BPI-derived bioactive peptide P2 (Mathur & Waldor, 2004). OmpU is crucial for the basal level expression of a master transcriptional regulator σ^E. Researchers have shown that exposure to a sublethal dose of AMP activates σ^E-mediated periplasmic stress response. Deletion of *ompU* abrogates the σ^E activation and thus the periplasmic stress response against the AMPs (Mathur, Davis, & Waldor, 2007). PorB of *N. meningitidis* is known to act as a channel in expelling AMPs out of the bacterial periplasm (Tzeng & Stephens, 2015).

Apart from producing AMPs to counter the disease-causing pathogens, host cells also produce reactive oxygen species (ROS). ROS is produced at various sites like mitochondria, cytoplasm, peroxisome, plasma membrane, and endoplasmic reticulum. ROS shows a broad spectrum of antimicrobial activity. Among other strategies, bacteria also use porins to protect themselves from oxidative stresses (van der Heijden et al., 2016). OmpU is involved in the regulated transport of manganese, a critical factor in *V. cholerae* ROS resistance (Jiang et al., 2020). Researchers have shown that deletion of OmpU makes the bacterium more susceptible to ROS while supplementing manganese in the medium restores ROS resistance in *ompU* deletion mutants (Jiang et al., 2020). *S. typhimurium* regulates its two major porins OmpA and OmpC to reduce the outer membrane permeability to peroxide under *in vitro* conditions (van der Heijden et al., 2016).

3.3 Bacteriophage-binding site

Bacteriophages are a group of viruses that specifically attack bacteria. For infecting the bacteria, the phage first binds to the bacterial cell surface followed by injection of the phage genetic material into its cytoplasm. Therefore, the attachment of phage to the bacterial cell surface is crucial for both infection and determination of host specificity (Stone, Campbell, Grant, & McAuliffe, 2019). Porins often act as a bacteriophage-binding site (Table 2) and act as receptors for bacteriophages (Mangalea &

Table 2 Bacteriophage-binding site in various porins.

Porin	Bacterial species	Phage	Phage receptor	References
OmpC + LamB	*E. coli* K12	Bp7	gp38	Chen et al. (2020)
OmpF (in absence of LamB)	*E. coli*	λ	J protein	Meyer et al. (2012)
LamB	*E. coli*	λ K10	gpJ	Chatterjee and Rothenberg (2012) and Hancock and Reeves (1976)
OmpC	*E. coli*	T4	LTFs	Yu and Mizushima (1982)
OmpA	*E. coli*	vB_EcoM-ep3	Dpo41	Wang et al. (2019)
OmpF	*E. coli*	K20, Tu1a, and T2		Riede, Degen, and Henning (1985) and Silverman and Benson (1987)
PhoE	*E. coli*	TC45 and its host range derivative TC45 *hr* N3		Chai and Foulds (1978)
OmpX, FhuA	*E. coli*	Various tailed phages		Drexler, Dannull, Hindennach, Mutschler, and Henning (1991) and Parent et al. (2014)
OmpC, if OmpA is absent	*E. coli*	Ox2		Morona and Henning (1984)
OmpA + OmpC	*S. flexneri*	Sf6		Parent et al. (2014)
OprM	*P. aeruginosa*	OMKO1		Chan et al. (2016)
Ail, OmpF	*Y. pestis*	Yep-phi	Tail fiber protein	Zhao et al. (2018)
OmpF	*Y. enterocolitica*	ϕR1-RT, TG1		Leon-Velarde et al. (2016)
OmpF	*Yersinia*	fPS-90 and fPS-2		Salem, Pajunen, Jun, and Skurnik (2021)
OmpK36	*K. pneumoniae*	GH-K3		Cai et al. (2018)
OmpC	*Salmonella*	S16 Gifsy-1/Gifsy-2	LTFs	Ho and Slauch (2001) and Marti et al. (2013)
OmpC, OmpF, and OmpK	*V. alginolyticus*	*Tequatroviruses* and *Schizotequatroviruses*		Skliros et al. (2021)
RopA1	*S. meliloti*	ΦM12 and N3		Crook, Draper, Guillory, and Griffitts (2013)

Duerkop, 2020). Below are some of the examples of gram-negative bacteria and their porins which are involved in phage binding and recognition.

- *E. coli*: It employs many different porins present in its outer membrane for binding to phages. The amino acid segment of 75–110 of PhoE porin contributes to the receptor site for PhoE-specific phages while the receptor site for OmpC-specific phages is somewhat defined by the amino acid segment 140–279 (Tommassen, van der Ley, van Zeijl, & Agterberg, 1985). OmpA also exploits its amino-terminal moiety (constituting the membrane part of the protein and comprising of about 180 AA residues) in the bacteriophage receptor functions (Morona, Kramer, & Henning, 1985).

  Out of these porins, the function of OmpC as a phage receptor has been studied widely. Phages T4, ss4, Tu1b, Hy2, Bp7, Me1, PA-2, and PP01 are known to interact with the bacteria via OmpC (Chen et al., 2020; Datta, Arden, & Henning, 1977; Yu & Mizushima, 1982). In some cases, deletion of *ompC* may not eliminate phage sensitivity completely, indicating the role of OmpC as a co-receptor or the involvement of other receptors (Montag, Schwarz, & Henning, 1989). The tailed bacteriophage T4 is known to adsorb to *E. coli* via two discrete ways, OmpC-independent (like *E. coli* B) or OmpC-dependent (like *E. coli* K12) (Furukawa, Yamada, & Mizushima, 1979; Henning & Jann, 1979; Prehm, Jann, Jann, Schmidt, & Stirm, 1976). Phage Bp7 employs its gp38 receptor for binding reversibly to OmpC of *E. coli* K12 along with its maltoporin LamB (Chen et al., 2020). This maltoporin was also shown to act as a receptor for the bacteriophage λ and K10 (Hancock & Reeves, 1976; Randall-Hazelbauer & Schwartz, 1973). gpJ of phage λ is known to interact with LamB during the attachment to host cell surface. Substitution of this J gene from phage λ with the tail fiber gene from a related phage 434 resulted in the binding of phage to OmpC, which the latter uses for infection. This indicates the importance of the C-terminal of gpJ protein for the host specificity (Chatterjee & Rothenberg, 2012).
- *Shigella*: Studies on phage Sf6 suggests the requirement of porins and LPS of *S. flexneri* for the release of its genome. For the entry, Sf6 utilizes LPS as a primary receptor and either porin OmpA or OmpC as secondary receptors, with OmpA being the preferred one (Parent et al., 2014). Unlike other phages where the phage interacts with a specific site on the porin, Sf6 interacts with the whole surface of OmpA. (Porcek & Parent, 2015).

- *Yersinia*: Yep-phi, a T7-related bacteriophage, binds to the outer membrane proteins Ail and OmpF along with LPS of *Y. pestis* (Zhao et al., 2018). Two myoviridae phages ϕR1-RT and TG1 infect *Y. enterocolitica* via binding to OmpF (Leon-Velarde et al., 2016).
- *Salmonella:* S16, a *Salmonella* phage utilizes OmpC porin along with LPS for its adsorption to host cells using its long tail fibers (LTFs) (Marti et al., 2013). Gifsy-1/Gifsy-2 phages also bind to OmpC on the bacterial surface (Ho & Slauch, 2001).
- *Vibrio*: For *Tequatrovirus* phages and *Schizotequatroviruses* phages, OmpC, OmpF, and OmpK porins of *V. alginolyticus* are involved in phage adsorption (Inoue, Matsuzaki, & Tanaka, 1995; Skliros et al., 2021).

Further, phages can swap from a favored receptor to another receptor as a result of mutations (Nguyen, Molineux, Springman, & Bull, 2012). For example, in case of *E. coli,* phage Ox2 host range mutants can bind to OmpC in absence of OmpA (Morona & Henning, 1984). Similarly, the J protein of phage λ can utilize OmpF for binding when LamB was absent or several mutations occurred in the J protein (Meyer et al., 2012).

3.4 Complement-binding site

Complement is a complex part of the innate immune system that acts as the first line of defense against infectious microbes and altered host cells. The complement system comprises of several plasma proteins that once activated work together in a cascade to form membrane attack complex (MAC) or opsonize pathogens and induce a series of inflammatory responses thereby eliminating the infection and maintaining homeostasis (Merle, Church, Fremeaux-Bacchi, & Roumenina, 2015). Many bacterial porins are known to act as binding sites for the components of complement cascades and play a role in their activation.

Porins of *S. typhimurium*, *S. minnesota*, *K. pneumoniae,* and *Aeromonas hydrophila* are known to bind to complement factor C1q, thereafter triggering the classical or alternative pathway (Albertí et al., 1996; Galdiero, Tufano, Sommese, Folgore, & Tedesco, 1984; Latsch, Mollerfeld, Ringsdorf, & Loos, 1990). OmpK36 of *K. pneumoniae* binds to C1q in its native state in the bacterial outer membrane and activates the complement pathway. Afterward, there is an accumulation of C3b and C5b-9 on the porin (Alberti et al., 1993; Albertí et al., 1995). Strains which have smooth LPS are serum-resistant and bind less C1q compared to serum-sensitive rough-LPS strains (Albertí et al., 1996). *A. hydrophila* porin II activates classical pathway via binding to C1q. Here also, strains with O-antigen LPS bind less C1q than serum-sensitive strains, because of decreased accessibility of the outer membrane proteins (Merino et al., 1998).

As the activation of host complement protein leads to killing of the pathogen, pathogens employ different strategies to escape from the complement attack. One such way is the recruitment of classical pathway inhibitor C4b-binding protein (C4bp) to their surface (Bettoni et al., 2019). Recruitment of C4bp results in decreased activation of complement cascade by hijacking the regulator of the alternative pathway, host factor H (fH), a molecule leading to downregulation of C3b deposition and inactivation of deposited C3b (Lewis et al., 2013). Surface-exposed loop 5 of *N. gonorrhoeae* porin PorB1A directly binds to fH and therefore resists killing (Ram et al., 1998). It also binds to C4bp present in the serum probably through its loop 1. This downregulation of the classical pathway, therefore, helps the bacteria to evade complement deposition on its surface. PorB1B also binds to C4bp by its loops 5 and 7 (Ram et al., 2001). *N. meningitidis* porin B2 leads to inhibition of the alternative pathway which helps in virulence in infant rat models and humans (Lewis et al., 2013; Lewis, Vu, Granoff, & Ram, 2014). PorB3 of *N. meningitidis* along with its surface protein NspA binds to fH and evades its bactericidal activity thereby conferring resistance to the bacteria (Giuntini, Pajon, Ram, & Granoff, 2015). *N. meningitidis* strains lacking PorA bound significantly less C4bp and therefore led to more killing of bacteria (Jarva, Ram, Vogel, Blom, & Meri, 2005). The N-terminal of *E. coli* OmpA makes hydrophobic interactions with C4bp thereby conferring serum resistance to the bacteria (Prasadarao, Blom, Villoutreix, & Linsangan, 2002). Porin D of *P. aeruginosa* binds to vitronectin, a soluble regulatory glycoprotein of MAC, and therefore inhibits the lysis of the bacteria (Paulsson et al., 2015).

3.5 Bacterial pathogenesis

3.5.1 Adhesion

When a pathogen infects a host, one of the preliminary events includes the attachment between the pathogen and the host cells. This attachment is required for colonization and/or internalization of the bacteria. Some of the outer membrane porins were found to play a role in adhesion (Table 3).

Many OmpA and OmpA-like porins present on the outer membrane of various gram-negative bacteria are known to mediate the adhesion of the bacteria to the host. OmpA of *E. coli* K1 causing newborn meningitis (NBM) plays a role in the adhesion of the bacteria to capillary endothelium

Table 3 Gram-negative bacterial porins involved in adhesion.

Porin	Bacterial species	Target	References
OmpA	*E. coli* K1	Capillary endothelium and astrocytes of CNS	Meier, Oelschlaeger, Merkert, Korhonen, and Hacker (1996) and Wu et al. (2009)
OmpA	*E. coli*	Leukocytes and macrophages	Jeannin et al. (2002)
OmpA	*K. pneumoniae*	Leukocytes and macrophages	Jeannin et al. (2002)
OmpA	*M. haemolytica*	Epithelial cells on mucosal surfaces	Ayalew, Confer, Hartson, and Shrestha (2010)
OmpA	*A. actinomycetemcomitans*	Epithelial cells on mucosal surfaces	Kajiya et al. (2011)
P5 protein	*H. influenzae*	Epithelial cells on mucosal surfaces	Thanavala and Lugade (2011)
OmpA	*E. coli*	BMECs	Teng et al. (2006)
OmpA	*Actinobacillus suis*	BMECs	Ojha, Lacouture, Gottschalk, and MacInnes (2010)
OmpA	*N. gonorrhoeae*	Epithelial cells, human cervical carcinoma, and endometrial carcinoma cells	Serino et al. (2007)
OmpA	EHEC, EPEC	HeLa and Caco-2 epithelial cells	Torres, Jeter, Langley, and Matthysse (2005) and Torres and Kaper (2003)
OmpA	*A. baumannii*	Fibronectin	Smani, McConnell, and Pachon (2012)
PmOmpA	*P. multocida*	Fibronectin	Dabo, Confer, and Quijano-Blas (2003)

Continued

Table 3 Gram-negative bacterial porins involved in adhesion.—cont'd

Porin	Bacterial species	Target	References
PIII	*N. gonorrhoeae*	Human cervical and urethral cells	Leuzzi et al. (2013)
OpaA	*Neisseria*	Fibronectin	van Putten, Duensing, and Cole (1998)
Omp33	*A. baumannii*	Fibronectin	Smani, Dominguez-Herrera, and Pachon (2013)
Omp43	*Bartonella henselae*	Fibronectin	Dabo, Confer, Saliki, and Anderson (2006)
Pap31	*Bartonella henselae*	Fibronectin	Dabo et al. (2006)
OprF	*Pseudomonas*	Lung epithelial cells	Azghani, Idell, Bains, and Hancock (2002)
OmpC, OmpF	APEC	BMECs	Hejair et al. (2017)
Porin aha	*A. veronii*	EPC cells	Song et al. (2019)
LamB-like proteins	*Aeromonas* species	ECM	Vazquez-Juarez, Romero, and Ascencio (2004)
CadF	*Clostridium jejuni*	Fibronectin	Konkel, Garvis, Tipton, Anderson, and Cieplak (1997)
OmpU	*V. vulnificus*	Fibronectin, RGD	Goo et al. (2006)
OmpU	*V. mimicus*	EPC cells	Liu et al. (2015)

and astrocytes of the central nervous system (Meier et al., 1996; Wu et al., 2009). OmpA of pathogenic strains of *E. coli* and *K. pneumoniae* mediates adhesion to leukocytes and macrophages (Jeannin et al., 2002). OmpA in ruminant respiratory pathogen *Mannheimia haemolytica,* oral pathogen *Aggregatibacter actinomycetemcomitans,* and OmpA homolog P5 protein in human respiratory pathogen nontypeable *Haemophilus influenzae* are also exploited as adhesins to epithelial cells on mucosal surfaces (Ayalew et al., 2010; Kajiya et al., 2011; Thanavala & Lugade, 2011). Another report from Teng et al. shows the adhesion of purified *E. coli* OmpA to brain microvascular endothelial cells (BMECs). Furthermore, this OmpA porin positively regulates the expression of another adhesin, type 1 fimbriae, the absence of which also decreases binding to the host (Teng et al., 2006). Serino et al. has discovered the role of *N. gonorrhoeae* OmpA in adhesion and invasion of epithelial cells using a deficient isogenic mutant of *ompA* (Serino et al., 2007). OmpA of Enterohemorrhagic *E. coli* (EHEC) was found to have adhesive characteristics in HeLa and Caco-2 cells (Torres & Kaper, 2003). This characteristic of OmpA is also found in Enteropathogenic *E. coli* (EPEC) but not in other strains of *E. coli.* OmpA of *Acinetobacter baumannii* and PmOmpA of *Pasteurella multocida* are known to interact with the host cells via extracellular matrix component fibronectin (Dabo et al., 2003; Smani et al., 2012).

Homologous to the C-terminal domain of OmpA, porin PIII of *N. gonorrhoeae* is employed for adhering to human cervical and urethral cells (Leuzzi et al., 2013). Along with PIII, Neisserial OpaA, Omp43, and Pap31 of *Bartonella henselae* and Omp33 of *A. baumannii* also bind to fibronectin (Dabo et al., 2006; Smani et al., 2013; van Putten et al., 1998).

In *Pseudomonas,* OprF protein binds to lung epithelial cells and colonizes the airway epithelium during pulmonary infection (Azghani et al., 2002). Further, the *oprF* mutant was found to show a decreased adhesion to rat glial and Caco-2 colorectal cells (Fito-Boncompte et al., 2011). Inactivation of *ompC* and *ompF* porin genes decreases the adhesion and colonization of avian pathogenic *E. coli* (APEC) to mouse BMECs (Hejair et al., 2017). Inactivation of porin gene *aha* from *A. veronii,* which infects mammals and aquatic organisms, decreased its adhesion to epithelioma papulosum cyprinid (EPC) cells (Song et al., 2019). Other outer membrane proteins and LamB-like proteins of various *Aeromonas* species adhere to extracellular matrix components or bind to host cells as adhesins (Vazquez-Juarez et al., 2004; Yang et al., 2019).

Pathogenic *Vibrio* species are also known to employ their outer membrane porins for adhesion. *V. vulnificus* OmpU helps in adherence of the bacteria to the host via fibronectin and RGD tripeptide (Goo et al., 2006). In the case of *V. mimicus*, the N-terminal of OmpU was found to be involved in the adherence of the bacteria to the host EPC cells (Liu et al., 2015).

3.5.2 Virulence

Porins not only mediate the adhesion of the bacteria to the host cells, but they can also hijack other machineries of the cell and their absence may lead to lesser virulence. Many porins are documented to play key roles in the pathogenesis of bacteria. Deletion of *ompF* and *ompC* of APEC led to the lesser invasion, colonization, and proliferation of the bacteria (Hejair et al., 2017). Some porins of *Salmonella* are known to inhibit phagocytosis in macrophages by elevating the levels of cyclic adenosine monophosphate (cAMP). High levels of cAMP inactivate RhoA kinase thus suspending the actin remodeling necessary for phagocytosis. Meningococcal porins of the outer membrane of *N. meningitidis* inhibit phagocytosis, degranulation, opsonin receptor expression, and actin polymerization (Bjerknes, Guttormsen, Solberg, & Wetzler, 1994). OmpB and OmpC of *S. flexneri* have a role in the virulence of the bacteria as the double deletion mutant is unable to spread from one epithelial cell to another (Bernardini, Sanna, Fontaine, & Sansonetti, 1993). Deletion of porin *oprF* of *P. aeruginosa* causes impaired secretion of ExoT and ExoS toxins through the type III secretion system (T3SS) and hampers the production of the quorum-sensing-dependent virulence factors pyocyanin, elastase, lectin PA-1L, and exotoxin A. Deletion of another porin *oprC* severely impairs bacterial motility and quorum-sensing systems, as well as lowers the levels of LPS and pyocyanin in *P. aeruginosa*. Microinjection of OmpA of *A. baumannii* in embryos of *Xenopus laevis* leads to impaired embryogenesis.

3.6 Modulation of host immune responses by gram-negative bacterial porins

The immune system helps to prevent or limit the pathogens to keep the body free of infection. Modulation of these host immune responses is one of the evasion strategies that bacteria have evolved for the spread of infection.

3.6.1 Modulation of the host innate immune system by porins

Innate immune cells have special receptor proteins called pattern recognition receptors (PRRs) which are specialized in recognizing conserved patterns in the molecules present on the pathogens. These patterns in the biomolecules of the pathogens are known as pathogen-associated molecular patterns (PAMPs). PRRs can be divided into different groups based on their specificities for different PAMPs. Once a PAMP activates a PRR, an intracellular signaling cascade is initiated leading to the upregulation of cytokine and chemokine expression, important for the induction of inflammatory responses. Several porins are documented in the literature for acting as PAMPs and can mount immune responses in the host. VcOmpU is known to induce the production of proinflammatory mediators, such as IL-6, TNFα, ROS, and NO in macrophages and monocytes (Sakharwade, Sharma, & Mukhopadhaya, 2013). As a PAMP, VcOmpU is recognized by PRR, TLR1/2 heterodimer in monocytes, however, in macrophages, it also employs scavenger receptor CD36 for the production of ROS (Khan, Sharma, & Mukhopadhaya, 2015; Prasad, Dhar, & Mukhopadhaya, 2019). VcOmpU-mediated production of proinflammatory cytokines involves the activation of mitogen-activated protein kinases (MAPKs), p38, and c-Jun N-terminal kinase (JNK) (Prasad et al., 2019). In monocytes, activation of both p38 and JNK is TLR-dependent. However, in macrophages, JNK activation is dependent upon CD36 (Prasad et al., 2019). Similar to VcOmpU, *V. parahaemolyticus* OmpU (VpOmpU) activates monocytes and macrophages in a TLR-dependent manner involving MAPKs p38 and JNK. Further, VpOmpU is recognized by both TLR1/2 and TLR2/6 heterodimer in macrophages; however, only TLR1/2 heterodimer is involved in monocytes (Gulati, Kumar, & Mukhopadhaya, 2019). PorB porin of *N. meningitidis* induces the production of proinflammatory cytokines like TNFα and IL-6 in MAPK-dependent fashion in macrophages (Platt, MacLeod, Massari, Liu, & Wetzler, 2013). Further, PorB upregulates TLR2 as well as co-stimulatory molecules CD40, CD86, and other activation marker CD69 in macrophages (Platt et al., 2013). Porin from *H. influenzae* is also reported to induce the production of proinflammatory cytokines, such as IL-6 and TNFα in monocytes and macrophages in a TLR2-dependent manner (Galdiero et al., 2004). Similar to LPS, the porin fraction of *S. typhimurium* is known to induce the release of TNFα, IL-6, and IL-8 in monocytes. However, the signaling is independent of CD14 and partially dependent on integrin CD11a/CD18 (Galdiero et al., 2001). Porins of *P. multocida, P. haemolytica,* and *S. typhimurium* alone cannot induce

NO release but when incubated with IFNγ lead to production and release of NO in murine peritoneal macrophages (Marcatili et al., 2000).

3.6.2 Modulation of the host adaptive immune system by porins

It is now widely accepted that outer membrane porins of gram-negative bacteria not only have innate immune modulation capacity but they can also potently activate the adaptive immune system.

Studies done on LPS nonresponsive mice strain C3H/HeJ have revealed that Neisserial porins are potent inducers of B cell activation, and cause upregulation of co-stimulatory molecule CD86 on the surface of B lymphocytes (Wetzler, 1996). Other than being a B cell mitogen, it is also reported that *N. meningitidis* PorB has the potential to induce DC maturation. Further, this maturation is marked by the upregulated expression of both MHC I and MHC II as well as CD86. This upregulation is found to be TLR2 and myeloid differentiation factor 88 (MyD88)-dependent (Singleton, Massari, & Wetzler, 2005). PorB also uplifted the ability of DCs to activate T cells. PorA of *N. meningitidis* also activates human monocyte-derived DCs (mo-DCs), causing their maturation. It is marked by the decrease in receptor-mediated endocytosis, enhanced expression of MHCs along with co-stimulatory molecules and, the production of chemokines like IL-8, C-C motif chemokine ligand 5 (Ccl5), macrophage inflammatory protein (MIP)-1α, and MIP-1β (Al-Bader et al., 2004). OmpA of *K. pneumoniae* and *E. coli* O157:H7 have been reported to target DCs and get endocytosed in a TLR-dependent manner. This leads to the production of high IL-12 which eventually causes Th1 polarization. Activation of Tc was also observed with this OmpA (Jeannin et al., 2000; Torres et al., 2006).

Li et al. have reported an *in vivo* mice model where they studied the immunogenicity of OmpC and OmpF porins from pathogenic *E. coli* strain. They observed that both OmpC and OmpF induce high IgG1 and IgG2a titers, indicating the ability of these porins to generate Th2/Th1 mixed responses, and causing opsonophagocytosis of the pathogen (Liu et al., 2012). Porin from *Shigella dysenteriae* type 1 can activate mouse peritoneal B cells via TLR2/6 heterodimer resulting in activation of transcription factor nuclear factor κ light chain enhancer of activated B cell (NF-κB) and further upregulation of co-stimulatory molecules CD86 and CD80. Further, it induces significant production of IgA, in addition to IgG2a and IgM (Ray & Biswas, 2005). Macrophages stimulated with this porin can cause T cell proliferation. Further, *in vivo* studies revealed that immunizing the C57BL/6 mice with this porin cause significant release of IFNγ and no effect in IL-4 production, evidently causing a bias towards Th1 phenotype (Ray & Biswas, 2005). OmpS1 and OmpS2 porins of *S. typhi* upregulate co-stimulatory molecules in macrophages and DCs. Further, OmpS1 induces upregulation of MHC class II in DCs, while OmpS2 upregulates CD40 molecules in both macrophages and DCs (Moreno-Eutimio et al., 2013). A 30 kDa porin of *H. pylori* activates and polarizes human T lymphocytes to either Th1 or Th2 and induces production of IL-6, IL-3, IFNγ, and IL-4 (Tufano et al., 1994).

There has been extensive research to explore the possible role of porins as vaccine candidates or as vaccine adjuvants against various gram-negative bacterial diseases exploiting this potent immunogenic property of porins.

3.7 Involvement of the gram-negative bacterial porins in cell death responses

Many intracellular and extracellular bacteria are known to modulate the cell death pathways of the host to evade the normal host defense responses. One of the most common and well-characterized pathways of host cell death modulated by the bacteria is apoptosis. Many gram-negative bacteria are known to employ their outer membrane porins for the modulation of the host cell death. Several reports are available in the literature which shed light on how porin modulate host cell death.

Neisseria translocate PorB to the mitochondria of the epithelial and phagocytic cells leading to their apoptosis (Muller et al., 2000; Muller et al., 2002). The translocation of PorB to the host mitochondria causes the loss of mitochondrial membrane potential thereby inducing apoptosis which eventually affects the innate immune system (Deo et al., 2018). Another study depicts that PorB enters the mitochondria through TOM complex similar to other mitochondria-targeted proteins, but then it bypasses the normal path of SAM (sorting and assembly machinery) complex. This leads to the accumulation of PorB in the intermembrane space and a fraction of PorB gets integrated into the inner mitochondrial membrane. This forms ATP-regulated pores in the inner mitochondrial membrane which dissipates the membrane potential and damages the cristae structure thereby initiating apoptosis (Kozjak-Pavlovic et al., 2009).

Secretory OMVs of *A. baumannii* get endocytosed into the host cells and deliver OmpA into the host which induces apoptosis after its entry. This OmpA causes changes in mitochondrial morphology along with enhanced levels of ROS, reduced mitochondrial membrane potential, and reduced levels of ATP (Tiku et al., 2021). OmpA in the purified form also causes the release of cytochrome *c* and apoptosis-inducing factor (AIF) from the mitochondria leading to caspase-dependent and AIF-dependent apoptosis in epithelial cells. OmpA at higher concentrations can induce apoptosis in DCs (Choi et al., 2005; Lee et al., 2007). Another porin of *A. baumannii,* Omp33–36 along with induction of caspase-dependent apoptosis can also modulate autophagy (Rumbo et al., 2014).

V. cholerae translocate OmpU porin to the host cell mitochondria and induces loss of mitochondrial membrane potential and release of cytochrome *c* and AIF. However, cytochrome *c* does not induce caspase activation probably due to the low ATP level in the sample and leads to AIF-dependent and caspase-independent cell death (Gupta, Prasad, & Mukhopadhaya, 2015). In yeast cells, *Brucella melitensis* porin Omp2b was found to play a role in the inhibition of Bax-induced cell death (Laloux, Deghelt, de Barsy, Letesson, & De Bolle, 2010). Some reports also reveal the infertility in humans or mice as a consequence of gram-negative bacterial porins. *S. typhimurium* and *P. multocida* are known to cause vaginal infections. Their porins induce spontaneous apoptosis in human spermatozoa eventually threatening their viability (Gorga, Galdiero, Buommino, & Galdiero, 2001). Porin isolated from the outer membrane of *P. aeruginosa,* induced in vitro apoptosis in an epithelial cell line (SVC1) derived from rat seminal vesicle secretory epithelium (Buommino et al., 1999).

Though the above examples are of porins involved in cell death, PorB porin of *N. meningitidis* acts as an inhibitor of apoptosis induced via the intrinsic apoptotic pathway in different cell types (Massari, Gunawardana, Liu, & Wetzler, 2010; Tunbridge et al., 2006). It localizes to the mitochondria after its internalization where it interacts with mitochondrial porin VDAC. The association of PorB with the mitochondria leads to the stabilization of the mitochondrial membrane thereby preventing mitochondrial depolarization and further release of proapoptotic mitochondrial factors (Massari et al., 2010; Massari, Ho, & Wetzler, 2000; Massari, King, Ho, & Wetzler, 2003). This could be due to the similarities in the channel activity and other features of PorB with VDAC which is involved in apoptosis (Rudel et al., 1996). Another gonococcal porin PorB1B is also known to increase the expression of the antiapoptotic factors in the host cells (Binnicker, Williams, & Apicella, 2004). The inhibition of immune cell apoptosis by Neisserial porins could be an integral mechanism of the porin's immunopotentiating ability (Massari et al., 2000). Moreover, the blocking of cell death could be one of the ways used by the bacteria to multiply and adapt to the environment for further infection (Massari, Ram, Macleod, & Wetzler, 2003).

4. Important uses of gram-negative bacterial porins

Gram-negative bacterial porins are involved in major processes of bacterial pathogenesis thereby making them suitable for various applications (Fig. 3). Some of their major uses are explained below.

4.1 Adjuvant properties of gram-negative bacterial porins

Adjuvants are the crucial components in most vaccine preparations used to enhance the immunogenicity of the antigens. Vaccine adjuvants can be broadly classified into two types based on the principle by which they work, viz. "vaccine delivery system" and "immunostimulatory" adjuvants. The immunostimulatory adjuvants are components derived from the pathogen itself, such as PAMPs that enhance the innate immune responses (Perez et al., 2013). The adjuvants, through the engagement of the innate immune players, indirectly affect or modulate the adaptive immune responses. Immunostimulatory adjuvants use the following methods to perform their assigned function: (i) production of chemokines and cytokines to induce inflammation (Goto & Akama, 1982), (ii) increased antigen presentation due to enhancement in the internalization of antigens (Flach et al., 2011), (iii) enhanced activation and maturation of DCs (Krieg, 2002), (iv) activation of the inflammasome (Ivanov, Garanina, Rizvanov, & Khaiboullina, 2020).

Massari et al. have reported the role of Neisserial porins as potential vaccine adjuvants that works through the TLR2-MyD88-NFκB mediated pathway (Massari et al., 2002). Further, they have observed that these porins activate B cells and upregulate CD86 molecules in the APCs to provide the co-stimulatory signal for T cell activation (Wetzler, 1996). *N. meningitidis* PorB and PorA are potent inducers of APC maturation and also activate B cells in a T cell-independent way. Similar to *N. meningitidis* porins, *N. lactamica* (Nlac) PorB acts as a potent adjuvant when used in conjugation with a prototype antigen ovalbumin (OVA) (Liu, Wetzler, & Massari, 2008). The level of Ig produced is greatly enhanced when *N. meningitidis* porin, Class 3 protein, and *N. gonorrhoeae* porins (PorB1A and PorB1B) were used in combination with dextran conjugated anti-IgD antibody (Snapper, Rosas, Kehry, Mond, & Wetzler, 1997). Fusco et al. reported that group B meningococcal polysaccharide (GBMP) is inefficiently immunogenic, but a C-terminal modification of the sialic acid of polysaccharide by N-propionyl group (NPr) can enhance its immunogenicity. Recombinant PorB (rPorB) enabled the vaccine NPr-GBMP (C-terminal modified sialic acid of group B meningococcal polysaccharide by N-propionyl group) to produce high titers of GBMP-specific antibodies in mouse and non-human primates (Fusco, Michon, Tai, & Blake, 1997). Toledo et al. have reported that the purified major porin fractions of *S. typhi* containing both OmpC and OmpF, when co-immunized with less immunogenic OVA or inactivated H1N1 (the pandemic strain of 2009) or typhoid Vi polysaccharides (T-independent antigen), helped in the generation of higher IgG titers against otherwise less immunogenic antigens (Perez-Toledo et al., 2017). FomA from *Fusobacterium nucleatum*, an oral pathogen in humans was studied by Toussi et al. and found that when used with OVA, it enhanced the antibodies generated against OVA. The responses diminished when the same study was performed in TLR2 knockout mice, suggesting the involvement of TLR2 in FomA-mediated modulation of immune responses (Toussi, Liu, & Massari, 2012).

4.2 Vaccine potential of the gram-negative bacterial porins

Vaccine antigens are better if they exist on the outer surface of the bacteria for the proper access to immune cells. As porins are present in the outer membrane and they have potent immunogenic properties to activate the innate and adaptive immune systems of the host, they are important targets for vaccine development. Numerous instances are reported where researchers tried to exploit the immunogenic property of these porins as a potential vaccine candidate.

Tabaraie et al. observed that porins from a smooth strain of *S. typhi* 0-901 and rough mutant strain of *S. typhimurium* Ra-30, when immunized in different groups of mice gave protection against challenge with parent strain as well as heterologous strain (*S. typhimurium* LT2-71 and *S. enteritidis* SH-1269). Further, they observed that administration of antibodies against *S. typhi* 0-901 porin by passive immunization, protected 86.6% of mice against a lethal dose of *S. typhimurium* LT2-71 and 93.3% against *S. enteritidis* SH-1269 (Tabaraie, Sharma, Sharma, Sehgal, & Ganguly, 1994).

The role of OprF porin of *P. aeruginosa* as a potential vaccine candidate to prevent its infection is also well studied. In 1992, Hughes et al. predicted several probable surface-exposed portions in the entire sequence of OprF and synthesized small peptides corresponding to each of these sequences. Further, they immunized the mice with each of these peptides and found three peptides showing desirably high titers of anti-OprF antibodies that could bind to seven immunotype strains of *P. aeruginosa*. They have also observed that the C-terminal portion of the OprF possesses the majority of the B cell epitopes (Hughes, Gilleland, & Gilleland, 1992). Another group immunized mice with glutathione-S-transferase (GST) fused -OprF, -OprI, -OprF-OprI, and -OprI-OprF of *P. aeruginosa* and found that efficient protection was generated in the mice immunized with GST-OprF-OprI fusion protein (von Specht et al., 1995). Bahey-El-Dina et al. used only an N-terminal peptide of OprF in conjunction with BCG (bacille Calmette-Guerin) or alum as adjuvant. They observed a significant generation of IgG1 and IgG2a against OprF and protection against *P. aeruginosa* along with cross-protection against *A. baumannii* (Bahey-El-Din, Mohamed, Sheweita, Haroun, & Zaghloul, 2020).

Zhu et al. generated a potential DNA-based vaccine for Meningococcal infection by injecting plasmid-containing gene for PorB from *N. gonorrhoeae* strain FA1090 (Zhu, Thomas, & Sparling, 2004). PorB and PmpD porins from *Chlamydia trachomatis* serovar D along with recombinant *V. cholerae* ghost was used to generate a dual-specific vaccine against cholera and chlamydial infections (Eko et al., 2011). In another study by Das et al., rabbits were immunized with different outer membrane proteins (43, 42, 30, and 22 kDa) of *V. cholerae* V86 El Tor Inaba strain. They observed that antisera from 43, 42, and 22 kDa protein-immunized rabbits can reduce fluid secretion in the intestine. Further, he observed that this approach was useful even with heterologous strains *V. cholerae* Ogawa and O139 (Das, Chopra, Cantu, & Peterson, 1998). Large yellow croaker, *Pseudosciaena crocea* was immunized with outer membrane proteins OmpW, OmpV, OmpK, OmpU, and TolC of *V. parahaemolyticus* in different combinations—(i) OmpW and OmpV; (ii) OmpK; and (iii) OmpU and TolC. The results showed a relative survival percentage (RSP) of 80–90% in the vaccinated group, while all unvaccinated control fishes died due to vibriosis (Mao, Yu, You, Wei, & Liu, 2007). Another recombinant porin LamB from *V. alginolyticus* was used to immunize zebrafishes in different groups and following vaccination, different groups were challenged with different

Vibrio strains. The RSP with respect to control fishes was found to range from 54.1% to 77.8% in the immunized fishes, suggesting LamB as a potential vaccine candidate for vibriosis (Lun et al., 2014).

4.3 Porins as a target for the generation of therapeutics

Targeting porins for the development of therapeutics can be a good strategy as they are involved in bacterial pathogenesis. In *V. cholerae*, both OmpU and OmpT porins are under the regulation of ToxR regulon (Provenzano & Klose, 2000). During pathogenesis in the gut, there is an increase in the expression of OmpU and a decrease in the expression of OmpT. However, Provenzano et al. have shown that a modified strain expressing OmpT instead of OmpU show retarded growth of the bacterium in presence of bile salts and anionic detergents (Provenzano & Klose, 2000). This modified strain also expressed less of cholera toxin and toxin coregulated pilus which is the major virulence and colonization factor of *V. cholerae* (Provenzano & Klose, 2000). Such observations suggest that the absence of OmpU might lead to lesser colonization and lesser virulence of the bacteria. Therefore, OmpU could be an important therapeutic target. The *ompQ* mutant of *B. bronchiseptica* displayed attenuation of the ability to form a mature biofilm thus making the bacteria less effective at colonizing the respiratory tract of the host (Cattelan et al., 2016).

Deletion of some porins makes the bacteria more resistant to antibiotics. Double deletion of *ompK35* and *ompK36* porins of *K. pneumoniae* leads to an 8- to 16-fold increase in the minimum inhibitory concentration of antibiotics, meropenem, and cefepime (Tsai et al., 2011). These reports suggest that probably these antibiotics enter the bacterium through these porins. Further, deletion of *ompK36* alone or both *ompK35/ompK36* increased the susceptibility of the bacteria to phagocytosis by neutrophils (Tsai et al., 2011). Therefore, according to these reports, porins not only play a role in metabolic fitness but also makes the bacterium more resistant to phagocytes. Therefore, these porins could be good therapeutic targets.

4.4 Porins as the biomarkers

Porins have their roles associated with virulence, colonization, and survival of the bacteria, thus making them one of the important factors present in many highly virulent strains. OmpU porin which is present across all the *Vibrio* species play diverse roles in their pathogenesis and survival. OmpU is used as a biomarker to discriminate between epidemic and non-epidemic or less toxigenic strains as the amino acid sequence of OmpU is more conserved in epidemic strains of the *V. cholerae* (Paauw et al., 2014). Antibody against porins can also be used as a clinical biomarker in some diseases as well. Lander et al. have shown the presence of antibodies against OmpC of *E. coli* in 55% of the patients with Crohn's disease (Landers et al., 2002). Clinicians can use this as one of the serological markers for detecting Crohn's disease.

5. Conclusion

Gram-negative bacterial porins are now being extensively studied with many ongoing aspects, and still a lot to be explored. They are found to play several pivotal roles in the bacterial life cycle and their pathogenesis, ranging from the adhesion to the death of the host cells. Such dynamic behaviors of porins make them indispensable for the bacteria. The discovery of the first high-resolution crystal structure of the outer membrane porin led to the basis of studies on many porin structures and functions. Nevertheless, the structures of many identified bacterial outer membrane porins are unsolved, and the functions of many are not known yet. There is very less knowledge on how the β-barrels of porins are folded and inserted into the outer membrane of the bacteria. The modifications in the expression and/or structure of porins affect the channel properties of the outer membrane along with its pathogenesis. Also, these alterations may have a huge impact on the receptor function of the porins, the sensitivity of the bacteria against antimicrobials, and how the porins modulate the host cellular responses. The discovery of various functions of porins also led to their use as target adjuvants or vaccines and their use in the therapeutics and as biomarkers. With the advent of new and improvised techniques for molecular and structural analyses, the study of porins is now more efficient. Therefore, a better understanding of both the structural and functional aspects of porins will lead to the improvisation and development of novel drug therapies targeting the porins.

Acknowledgment

Protein structural coordinate was retrieved from the protein data bank (PDB; available at rcsb.org). Protein structural models were generated with UCSF Chimera, developed by the Resource for Biocomputing, Visualization, and Informatics at the University of California, San Francisco.

We thank Indian Institute of Science Education and Research Mohali for support.

References

Al-Bader, T., Jolley, K. A., Humphries, H. E., Holloway, J., Heckels, J. E., Semper, A. E., et al. (2004). Activation of human dendritic cells by the PorA protein of Neisseria meningitidis. *Cellular Microbiology*, *6*(7), 651–662. https://doi.org/10.1111/j.1462-5822.2004.00392.x.

Alberti, S., Marques, G., Camprubi, S., Merino, S., Tomas, J. M., Vivanco, F., et al. (1993). C1q binding and activation of the complement classical pathway by Klebsiella pneumoniae outer membrane proteins. *Infection and Immunity*, *61*(3), 852–860. https://doi.org/10.1128/iai.61.3.852-860.1993.

Albertí, S., Marqués, G., Hernández-Allés, S., Rubires, X., Tomás, J. M., Vivanco, F., et al. (1996). Interaction between complement subcomponent C1q and the Klebsiella pneumoniae Porin OmpK36. *Infection and Immunity*, *64*, 4719–4725.

Albertí, S., Rodríquez-Quiñones, F., Schirmer, T., Rummel, G., Tomás, J. M., Rosenbusch, J. P., et al. (1995). A Porin from Klebsiella pneumoniae: Sequence homology, three-dimensional model, and complement binding. *Infection and Immunity*, *63*, 903–910.

Ayalew, S., Confer, A. W., Hartson, S. D., & Shrestha, B. (2010). Immunoproteomic analyses of outer membrane proteins of Mannheimia haemolytica and identification of potential vaccine candidates. *Proteomics*, *10*(11), 2151–2164. https://doi.org/10.1002/pmic.200900557.

Azghani, A. O., Idell, S., Bains, M., & Hancock, R. E. (2002). Pseudomonas aeruginosa outer membrane protein F is an adhesin in bacterial binding to lung epithelial cells in culture. *Microbial Pathogenesis*, *33*(3), 109–114. https://doi.org/10.1006/mpat.2002.0514.

Bahey-El-Din, M., Mohamed, S. A., Sheweita, S. A., Haroun, M., & Zaghloul, T. I. (2020). Recombinant N-terminal outer membrane porin (OprF) of Pseudomonas aeruginosa is a promising vaccine candidate against both P. aeruginosa and some strains of Acinetobacter baumannii. *International Journal of Medical Microbiology*, *310*(3), 151415. https://doi.org/10.1016/j.ijmm.2020.151415.

Ballestero, S., Fernández-Rodríguez, A., Villaverde, R., Escobar, H., Pérez-Díaz, J. C., & Baquero, F. (1996). Carbapenem resistance in Pseudomonas aeruginosa from cystic fibrosis patients. *Journal of Antimicrobial Chemotherapy*, *38*, 39–45.

Bartsch, A., Ives, C. M., Kattner, C., Pein, F., Diehn, M., Tanabe, M., et al. (2021). An antibiotic-resistance conferring mutation in a neisserial porin: Structure, ion flux, and ampicillin binding. *Biochimica et Biophysica Acta (BBA) - Biomembranes*, *1863*(6). https://doi.org/10.1016/j.bbamem.2021.183601.

Benz, R. (1988). Structure and function of porins from gram-negative bacteria. *Annual Review of Microbiology*, *42*, 359–393. https://doi.org/10.1146/annurev.mi.42.100188.002043.

Benz, R., Maier, E., & Chakraborty, T. (1997). Purification of OmpU from vibrio cholerae classical strain 569B: Evidence for the formation of large cation-selective ion-permeable channels by OmpU. *Microbiología*, *13*(3), 321–330.

Bernardini, M. L., Sanna, M. G., Fontaine, A., & Sansonetti, P. J. (1993). OmpC is involved in invasion of epithelial cells by Shigella flexneri. *Infection and Immunity*, *61*(9), 3625–3635. https://doi.org/10.1128/iai.61.9.3625-3635.1993.

Bettoni, S., Shaughnessy, J., Maziarz, K., Ermert, D., Gulati, S., Zheng, B., et al. (2019). C4BP-IgM protein as a therapeutic approach to treat Neisseria gonorrhoeae infections. *JCI Insight*, *4*(23). https://doi.org/10.1172/jci.insight.131886.

Bina, J. E., Provenzano, D., Wang, C., Bina, X. R., & Mekalanos, J. J. (2006). Characterization of the vibrio cholerae vexAB and vexCD efflux systems. *Archives of Microbiology*, *186*(3), 171–181. https://doi.org/10.1007/s00203-006-0133-5.

Binnicker, M. J., Williams, R. D., & Apicella, M. A. (2004). Gonococcal porin IB activates NF-kappaB in human urethral epithelium and increases the expression of host anti-apoptotic factors. *Infection and Immunity*, *72*(11), 6408–6417. https://doi.org/10.1128/IAI.72.11.6408-6417.2004.

Bjerknes, R., Guttormsen, H.-K., Solberg, C. O., & Wetzler, L. M. (1994). Neisserial Porins inhibit human neutrophil actin polymerization, degranulation, opsonin receptor expression, and phagocytosis but prime the neutrophils to increase their oxidative burst. *Infection and Immunity*, *63*, 8.

Buommino, E., Morelli, F., Metafora, S., Rossano, F., Perfetto, B., Baroni, A., et al. (1999). Porin from Pseudomonas aeruginosa induces apoptosis in an epithelial cell line derived from rat seminal vesicles. *Infection and Immunity*, *67*, 4794–4800.

Cai, R., Wu, M., Zhang, H., Zhang, Y., Cheng, M., Guo, Z., et al. (2018). A smooth-type, phage-resistant Klebsiella pneumoniae mutant strain reveals that OmpC is indispensable for infection by phage GH-K3. *Applied and Environmental Microbiology*, *84*(21). https://doi.org/10.1128/AEM.01585-18.

Camara, A. K. S., Zhou, Y., Wen, P. C., Tajkhorshid, E. & Kwok, W. M. (2017). Mitochondrial VDAC1: A key gatekeeper as potential therapeutic target. *Frontiers in Physiology*, *8*, 460. https://doi.org/10.3389/fphys.2017.00460.

Caston, J. R., Berenguer, J., de Pedro, M. A., & Carrascosa, J. L. (1993). S-layer protein from Thermus thermophilus HB8 assembles into porin-like structures. *Molecular Microbiology*, *9*(1), 65–75. https://doi.org/10.1111/j.1365-2958.1993.tb01669.x.

Cattelan, N., Villalba, M. I., Parisi, G., Arnal, L., Serra, D. O., Aguilar, M., et al. (2016). Outer membrane protein OmpQ of Bordetella bronchiseptica is required for mature biofilm formation. *Microbiology (Reading)*, *162*(2), 351–363. https://doi.org/10.1099/mic.0.000224.

Chai, T. J., & Foulds, J. (1978). Two bacteriophages which utilize a new Escherichia coli major outer membrane protein as part of their receptor. *Journal of Bacteriology*, *135*(1), 164–170. https://doi.org/10.1128/jb.135.1.164-170.1978.

Chan, B. K., Sistrom, M., Wertz, J. E., Kortright, K. E., Narayan, D., & Turner, P. E. (2016). Phage selection restores antibiotic sensitivity in MDR Pseudomonas aeruginosa. *Scientific Reports*, *6*, 26717. https://doi.org/10.1038/srep26717.

Chatterjee, S., & Rothenberg, E. (2012). Interaction of bacteriophage l with its E. coli receptor, LamB. *Viruses*, *4*(11), 3162–3178. https://doi.org/10.3390/v4113162.

Chen, F. J., Lauderdale, T. L., Ho, M., & Lo, H. J. (2003). The roles of mutations in gyrA, parC, and ompK35 in fluoroquinolone resistance in Klebsiella pneumoniae. *Microbial Drug Resistance*, *9*(3), 265–271. https://doi.org/10.1089/107662903322286472.

Chen, P., Sun, H., Ren, H., Liu, W., Li, G., & Zhang, C. (2020). LamB, OmpC, and the Core lipopolysaccharide of Escherichia coli K-12 function as receptors of bacteriophage Bp7. *Journal of Virology*, *94*(12). https://doi.org/10.1128/JVI.00325-20.

Choi, C. H., Lee, E. Y., Lee, Y. C., Park, T. I., Kim, H. J., Hyun, S. H., et al. (2005). Outer membrane protein 38 of Acinetobacter baumannii localizes to the mitochondria and induces apoptosis of epithelial cells. *Cellular Microbiology*, *7*(8), 1127–1138. https://doi.org/10.1111/j.1462-5822.2005.00538.x.

Choi, U., & Lee, C.-R. (2019). Distinct roles of outer membrane Porins in antibiotic resistance and membrane integrity in Escherichia coli. *Frontiers in Microbiology*, *10*, 1–9. https://doi.org/10.3389/fmicb.2019.00953.

Chuanchuen, R., Narasaki, C. T., & Schweizer, H. P. (2002). The MexJK efflux pump of Pseudomonas aeruginosa requires OprM for antibiotic efflux but not for efflux of Triclosan. *Journal of Bacteriology*, *184*(18), 5036–5044. https://doi.org/10.1128/jb.184.18.5036-5044.2002.

Costa-Riu, N., Burkovski, A., Kramer, R., & Benz, R. (2003). PorA represents the major cell wall channel of the gram-positive bacterium Corynebacterium glutamicum. *Journal of Bacteriology*, *185*(16), 4779–4786. https://doi.org/10.1128/JB.185.16.4779-4786.2003.

Crook, M. B., Draper, A. L., Guillory, R. J., & Griffitts, J. S. (2013). The Sinorhizobium meliloti essential porin RopA1 is a target for numerous bacteriophages. *Journal of Bacteriology*, *195*(16), 3663–3671. https://doi.org/10.1128/JB.00480-13.

Dabo, S. M., Confer, A. W., & Quijano-Blas, R. A. (2003). Molecular and immunological characterization of Pasteurella multocida serotype a:3 OmpA: Evidence of its role in P. multocida interaction with extracellular matrix molecules. *Microbial Pathogenesis*, *35*(4), 147–157. https://doi.org/10.1016/s0882-4010(03)00098-6.

Dabo, S. M., Confer, A. W., Saliki, J. T., & Anderson, B. E. (2006). Binding of Bartonella henselae to extracellular molecules: Identification of potential adhesins. *Microbial Pathogenesis*, *41*(1), 10–20. https://doi.org/10.1016/j.micpath.2006.04.003.

Das, M., Chopra, A. K., Cantu, J. M., & Peterson, J. W. (1998). Antisera to selected outer membrane proteins of vibrio cholerae protect against challenge with homologous and heterologous strains of V. cholerae. *FEMS Immunology and Medical Microbiology*, *22*(4), 303–308. https://doi.org/10.1111/j.1574-695X.1998.tb01219.x.

Datta, D. B., Arden, B., & Henning, U. (1977). Major proteins of the Escherichia coli outer cell envelope membrane as bacteriophage receptors. *Journal of Bacteriology*, *131*(3), 821–829. https://doi.org/10.1128/jb.131.3.821-829.1977.

Deo, P., Chow, S. H., Hay, I. D., Kleifeld, O., Costin, A., Elgass, K. D., et al. (2018). Outer membrane vesicles from Neisseria gonorrhoeae target PorB to mitochondria and induce apoptosis. *PLoS Pathogens*, *14*(3), e1006945. https://doi.org/10.1371/journal.ppat.1006945.

Domenech-Sanchez, A., Martinez-Martinez, L., Hernandez-Alles, S., del Carmen Conejo, M., Pascual, A., Tomas, J. M., et al. (2003). Role of Klebsiella pneumoniae OmpK35 porin in antimicrobial resistance. *Antimicrobial Agents and Chemotherapy*, *47*(10), 3332–3335. https://doi.org/10.1128/AAC.47.10.3332-3335.2003.

Doumith, M., Ellington, M. J., Livermore, D. M., & Woodford, N. (2009). Molecular mechanisms disrupting porin expression in ertapenem-resistant Klebsiella and Enterobacter spp. clinical isolates from the UK. *Journal of Antimicrobial Chemotherapy*, *63*(4), 659–667. https://doi.org/10.1093/jac/dkp029.

Drexler, K., Dannull, J., Hindennach, I., Mutschler, B., & Henning, U. (1991). Single mutations in a gene for a tail fiber component of an Escherichia coli phage can cause an extension from a protein to a carbohydrate as a receptor. *Journal of Molecular Biology*, *219*(4), 655–663. https://doi.org/10.1016/0022-2836(91)90662-p.

Duperthuy, M., Binesse, J., Le Roux, F., Romestand, B., Caro, A., Got, P., et al. (2010). The major outer membrane protein OmpU of Vibrio splendidus contributes to host antimicrobial peptide resistance and is required for virulence in the oyster Crassostrea gigas. *Environmental Microbiology*, *12*(4), 951–963. https://doi.org/10.1111/j.1462-2920.2009.02138.x.

Eko, F. O., Okenu, D. N., Singh, U. P., He, Q., Black, C., & Igietseme, J. U. (2011). Evaluation of a broadly protective chlamydia-cholera combination vaccine candidate. *Vaccine*, *29*(21), 3802–3810. https://doi.org/10.1016/j.vaccine.2011.03.027.

Engelhardt, H., Heinz, C., & Niederweis, M. (2002). A tetrameric porin limits the cell wall permeability of mycobacterium smegmatis. *The Journal of Biological Chemistry*, *277*(40), 37567–37572. https://doi.org/10.1074/jbc.M206983200.

Fairman, J. W., Noinaj, N., & Buchanan, S. K. (2011). The structural biology of beta-barrel membrane proteins: A summary of recent reports. *Current Opinion in Structural Biology*, *21*(4), 523–531. https://doi.org/10.1016/j.sbi.2011.05.005.

Fito-Boncompte, L., Chapalain, A., Bouffartigues, E., Chaker, H., Lesouhaitier, O., Gicquel, G., et al. (2011). Full virulence of Pseudomonas aeruginosa requires OprF. *Infection and Immunity*, *79*(3), 1176–1186. https://doi.org/10.1128/IAI.00850-10.

Flach, T. L., Ng, G., Hari, A., Desrosiers, M. D., Zhang, P., Ward, S. M., et al. (2011). Alum interaction with dendritic cell membrane lipids is essential for its adjuvanticity. *Nature Medicine*, *17*(4), 479–487. https://doi.org/10.1038/nm.2306.

Furukawa, H., Yamada, H., & Mizushima, S. (1979). Interaction of bacteriophage T4 with reconstituted cell envelopes of Escherichia coli K-12. *Journal of Bacteriology*, *140*(3), 1071–1080. https://doi.org/10.1128/jb.140.3.1071-1080.1979.

Fusco, P. C., Michon, F., Tai, J. Y., & Blake, M. S. (1997). Preclinical evaluation of a novel group B meningococcal conjugate vaccine that ElicitsBactericidal activity in both mice and nonhuman primates. *The Journal of Infectious Diseases*, *175*, 364–372.

Galdiero, M., D'Isanto, M., Vitiello, M., Finamore, E., Peluso, L., & Galdiero, M. (2001). Porins from salmonella enterica serovar typhimurium induce TNF-alpha, IL-6 and IL-8 release by CD14-independent and CD11a/CD18-dependent mechanisms. *Microbiology (Reading)*, *147*(Pt 10), 2697–2704. https://doi.org/10.1099/00221287-147-10-2697.

Galdiero, S., Falanga, A., Cantisani, M., Tarallo, R., Pepa, M. E. D., D'Oriano, V., et al. (2012). Microbe-host interactions: Structure and role of gram-negative bacterial Porins. *Current Protein and Peptide Science*, *13*, 11.

Galdiero, M., Galdiero, M., Finamore, E., Rossano, F., Gambuzza, M., Catania, M. R., et al. (2004). Haemophilus influenzae Porin induces toll-like receptor 2-mediated cytokine production in human monocytes and mouse macrophages. *Infection and Immunity*, *72*(2), 1204–1209. https://doi.org/10.1128/iai.72.2.1204-1209.2004.

Galdiero, F., Tufano, M. A., Sommese, L., Folgore, A., & Tedesco, F. (1984). Activation of complement system by porins extracted from salmonella typhimurium. *Infection and Immunity*, *46*(2), 559–563. https://doi.org/10.1128/iai.46.2.559-563.1984.

Gill, M. J., Simjee, S., Al-hattawi, K., Robertson, B. D., Easmon, C. S., & Ison, C. A. (1998). Gonococcal resistance to b-lactams and tetracycline involves mutation in loop 3 of the Porin encoded at the penB locus. *Antimicrobial Agents and Chemotherapy*, *42*, 5.

Giuntini, S., Pajon, R., Ram, S., & Granoff, D. M. (2015). Binding of complement factor H to PorB3 and NspA enhances resistance of Neisseria meningitidis to anti-factor H binding protein bactericidal activity. *Infection and Immunity*, *83*(4), 1536–1545. https://doi.org/10.1128/IAI.02984-14.

Goo, S. Y., Lee, H. J., Kim, W. H., Han, K. L., Park, D. K., Lee, H. J., et al. (2006). Identification of OmpU of Vibrio vulnificus as a fibronectin-binding protein and its role in bacterial pathogenesis. *Infection and Immunity*, *74*(10), 5586–5594. https://doi.org/10.1128/IAI.00171-06.

Gorga, F., Galdiero, M., Buommino, E., & Galdiero, E. (2001). Porins and lipopolysaccharide induce apoptosis in human spermatozoa. *Clinical and Diagnostic Laboratory Immunology*, *8*(1), 206–208. https://doi.org/10.1128/CDLI.8.1.206-208.2001.

Goto, N., & Akama, K. (1982). Histopathological studies of reactions in mice injected with aluminum-adsorbed tetanus toxoid. *Microbiology and Immunology*, *26*(12), 1121–1132. https://doi.org/10.1111/j.1348-0421.1982.tb00261.x.

Gulati, A., Kumar, R., & Mukhopadhaya, A. (2019). Differential recognition of Vibrio parahaemolyticus OmpU by toll-like receptors in monocytes and macrophages for the induction of Proinflammatory responses. *Infection and Immunity*, *87*(5). https://doi.org/10.1128/iai.00809-18.

Gunn, J. S. (2000). Mechanisms of bacterial resistance and response to bile. *Microbes and Infection*, *2*(8), 907–913. https://doi.org/10.1016/s1286-4579(00)00392-0.

Gupta, S., Prasad, G. V., & Mukhopadhaya, A. (2015). Vibrio cholerae Porin OmpU induces caspase-independent programmed cell death upon translocation to the host cell mitochondria. *The Journal of Biological Chemistry*, *290*(52), 31051–31068. https://doi.org/10.1074/jbc.M115.670182.

Hancock, R. E., & Reeves, P. (1976). Lipopolysaccharide-deficient, bacteriophage-resistant mutants of Escherichia coli K-12. *Journal of Bacteriology*, *127*(1), 98–108. https://doi.org/10.1128/jb.127.1.98-108.1976.

Hejair, H. M. A., Zhu, Y., Ma, J., Zhang, Y., Pan, Z., Zhang, W., et al. (2017). Functional role of ompF and ompC porins in pathogenesis of avian pathogenic Escherichia coli. *Microbial Pathogenesis*, *107*, 29–37. https://doi.org/10.1016/j.micpath.2017.02.033.

Henning, U., & Jann, K. (1979). Two-component nature of bacteriophage T4 receptor activity in Escherichia coli K-12. *Journal of Bacteriology*, *137*(1), 664–666. https://doi.org/10.1128/jb.137.1.664-666.1979.

Ho, T. D., & Slauch, J. M. (2001). OmpC is the receptor for Gifsy-1 and Gifsy-2 bacteriophages of salmonella. *Journal of Bacteriology*, *183*(4), 1495–1498. https://doi.org/10.1128/JB.183.4.1495-1498.2001.

Hughes, E. E., Gilleland, L. B., & Gilleland, H. E., Jr. (1992). Synthetic peptides representing epitopes of outer membrane protein F of Pseudomonas aeruginosa that elicit antibodies reactive with whole cells of heterologous Immunotype strains of P. aeruginosa. *Infection and Immunity*, *60*, 3497–3503.

Inoue, T., Matsuzaki, S., & Tanaka, S. (1995). A 26-kDa outer membrane protein, OmpK, common to vibrio species is the receptor for a broad-host-range vibriophage, KVP40. *FEMS Microbiology Letters*, *125*(1), 101–105. https://doi.org/10.1111/j.1574-6968.1995.tb07342.x.

Ishii, J. N., Okajima, Y., & Nakae, T. (1981). Characterization of lamB protein from the outer membrane of Escherichia coli that forms diffusion pores selective for maltose-maltodextrins. *FEBS Letters*, *134*(2), 217–220. https://doi.org/10.1016/0014-5793(81)80605-9.

Ivanov, K., Garanina, E., Rizvanov, A., & Khaiboullina, S. (2020). Inflammasomes as targets for adjuvants. *Pathogens*, *9*(4). https://doi.org/10.3390/pathogens9040252.

Jaffe, A., Chabbert, Y. A., & Semonin, O. (1982). Role of Porin proteins OmpF and OmpC in the permeation of β-lactams. *Antimicrobial Agents and Chemotherapy*, *22*, 942–948.

Jap, B. K., & Walian, P. J. (1996). Structure and functional mechanism of porins. *Physiological Reviews*, *76*(4), 1073–1088. https://doi.org/10.1152/physrev.1996.76.4.1073.

Jarva, H., Ram, S., Vogel, U., Blom, A. M., & Meri, S. (2005). Binding of the complement inhibitor C4bp to serogroup B Neisseria meningitidis. *Journal of Immunology*, *174*(10), 6299–6307. https://doi.org/10.4049/jimmunol.174.10.6299.

Jeannin, P., Magistrelli, G., Goetsch, L., Haeuw, J. F., Thieblemont, N., Bonnefoy, J. Y., et al. (2002). Outer membrane protein a (OmpA): A new pathogen-associated molecular pattern that interacts with antigen presenting cells-impact on vaccine strategies. *Vaccine*, *20*(Suppl 4), A23–A27. https://doi.org/10.1016/s0264-410x(02)00383-3.

Jeannin, P., Renno, T., Goetsch, L., Miconnet, I., Aubry, J. P., Delneste, Y., et al. (2000). OmpA targets dendritic cells, induces their maturation and delivers antigen into the MHC class I presentation pathway. *Nature Immunology*, *1*(6), 502–509. https://doi.org/10.1038/82751.

Jiang, H. H., Zhou, Y., Liu, M., Larios-Valencia, J., Lee, Z., Wang, H., et al. (2020). Vibrio cholerae virulence activator ToxR regulates manganese transport and resistance to reactive oxygen species. *Infection and Immunity*, *88*(3). https://doi.org/10.1128/IAI.00944-19.

Kajiya, M., Komatsuzawa, H., Papantonakis, A., Seki, M., Makihira, S., Ouhara, K., et al. (2011). Aggregatibacter actinomycetemcomitans Omp29 is associated with bacterial entry to gingival epithelial cells by F-actin rearrangement. *PLoS One*, *6*(4), e18287. https://doi.org/10.1371/journal.pone.0018287.

Khan, J., Sharma, P. K., & Mukhopadhaya, A. (2015). Vibrio cholerae porin OmpU mediates M1-polarization of macrophages/monocytes via TLR1/TLR2 activation. *Immunobiology*, *220*(11), 1199–1209. https://doi.org/10.1016/j.imbio.2015.06.009.

Kim, S. W., Lee, J. S., Park, S. B., Lee, A. R., Jung, J. W., Chun, J. H., et al. (2020). The importance of Porins and β-lactamase in outer membrane vesicles on the hydrolysis of β-lactam antibiotics. *International Journal of Molecular Sciences*, *21*(8). https://doi.org/10.3390/ijms21082822.

Koebnik, R., Locher, K. P., & Van Gelder, P. (2000). Structure and function of bacterial outer membrane proteins:Barrels in a nutshell. *Molecular Microbiology*, *37*(2), 239–253. https://doi.org/10.1046/j.1365-2958.2000.01983.x.

Konkel, M. E., Garvis, S. G., Tipton, S. L., Anderson, D. E., Jr., & Cieplak, W., Jr. (1997). Identification and molecular cloning of a gene encoding a fibronectin-binding protein (CadF) from campylobacter jejuni. *Molecular Microbiology*, *24*(5), 953–963. https://doi.org/10.1046/j.1365-2958.1997.4031771.x.

Kozjak-Pavlovic, V., Dian-Lothrop, E. A., Meinecke, M., Kepp, O., Ross, K., Rajalingam, K., et al. (2009). Bacterial porin disrupts mitochondrial membrane potential and sensitizes host cells to apoptosis. *PLoS Pathogens*, *5*(10), e1000629. https://doi.org/10.1371/journal.ppat.1000629.

Krieg, A. M. (2002). CpG motifs in bacterial DNA and their immune effects. *Annual Review of Immunology*, *20*, 709–760. https://doi.org/10.1146/annurev.immunol.20.100301.064842.

Laloux, G., Deghelt, M., de Barsy, M., Letesson, J. J., & De Bolle, X. (2010). Identification of the essential Brucella melitensis porin Omp2b as a suppressor of Bax-induced cell death in yeast in a genome-wide screening. *PLoS One*, *5*(10), e13274. https://doi.org/10.1371/journal.pone.0013274.

Landers, C. J., Cohavy, O., Misra, R., Yang, H., Lin, Y. C., Braun, J., et al. (2002). Selected loss of tolerance evidenced by Crohn's disease-associated immune responses to auto- and microbial antigens. *Gastroenterology*, *123*(3), 689–699. https://doi.org/10.1053/gast.2002.35379.

Latsch, M., Mollerfeld, J., Ringsdorf, H., & Loos, M. (1990). Studies on the interaction of C1q, a subcomponent of the first component of complement, with porins from Salmonella Minnesota incorporated into artificial membranes. *FEBS Letters*, *276*(1–2), 201–204. https://doi.org/10.1016/0014-5793(90)80542-q.

Lee, J. S., Lee, J. C., Lee, C. M., Jung, I. D., Jeong, Y. I., Seong, E. Y., et al. (2007). Outer membrane protein a of Acinetobacter baumannii induces differentiation of CD4+ T cells toward a Th1 polarizing phenotype through the activation of dendritic cells. *Biochemical Pharmacology*, *74*(1), 86–97. https://doi.org/10.1016/j.bcp.2007.02.012.

Lemasters, J. J., & Holmuhamedov, E. (2006). Voltage-dependent anion channel (VDAC) as mitochondrial governator--thinking outside the box. *Biochimica et Biophysica Acta*, *1762*(2), 181–190. https://doi.org/10.1016/j.bbadis.2005.10.006.

Leon-Velarde, C. G., Happonen, L., Pajunen, M., Leskinen, K., Kropinski, A. M., Mattinen, L., et al. (2016). Yersinia enterocolitica-specific infection by bacteriophages TG1 and varphiR1-RT is dependent on temperature-regulated expression of the phage host receptor OmpF. *Applied and Environmental Microbiology*, *82*(17), 5340–5353. https://doi.org/10.1128/AEM.01594-16.

Leuzzi, R., Nesta, B., Monaci, E., Cartocci, E., Serino, L., Soriani, M., et al. (2013). Neisseria gonorrhoeae PIII has a role on NG1873 outer membrane localization and is involved in bacterial adhesion to human cervical and urethral epithelial cells. *BMC Microbiology*, *13*, 251. https://doi.org/10.1186/1471-2180-13-251.

Lewis, L. A., Vu, D. M., Granoff, D. M., & Ram, S. (2014). Inhibition of the alternative pathway of nonhuman infant complement by porin B2 contributes to virulence of Neisseria meningitidis in the infant rat model. *Infection and Immunity*, *82*(6), 2574–2584. https://doi.org/10.1128/IAI.01517-14.

Lewis, L. A., Vu, D. M., Vasudhev, S., Shaughnessy, J., Granoff, D. M., & Ram, S. (2013). Factor H-dependent alternative pathway inhibition mediated by porin B contributes to virulence of Neisseria meningitidis. *MBio*, *4*(5). https://doi.org/10.1128/mBio.00339-13. e00339-00313.

Liu, C., Chen, Z., Tan, C., Liu, W., Xu, Z., Zhou, R., et al. (2012). Immunogenic characterization of outer membrane porins OmpC and OmpF of porcine extraintestinal pathogenic Escherichia coli. *FEMS Microbiology Letters*, *337*(2), 104–111. https://doi.org/10.1111/1574-6968.12013.

Liu, X., Gao, H., Xiao, N., Liu, Y., Li, J., & Li, L. (2015). Outer membrane protein U (OmpU) mediates adhesion of Vibrio mimicus to host cells via two novel N-terminal motifs. *PLoS One*, *10*(3), e0119026. https://doi.org/10.1371/journal.pone.0119026.

Liu, X., Wetzler, L. M., & Massari, P. (2008). The PorB porin from commensal Neisseria lactamica induces Th1 and Th2 immune responses to ovalbumin in mice and is a potential immune adjuvant. *Vaccine*, *26*(6), 786–796. https://doi.org/10.1016/j.vaccine.2007.11.080.

Lun, J., Xia, C., Yuan, C., Zhang, Y., Zhong, M., Huang, T., et al. (2014). The outer membrane protein, LamB (maltoporin), is a versatile vaccine candidate among the vibrio species. *Vaccine*, *32*(7), 809–815. https://doi.org/10.1016/j.vaccine.2013.12.035.

Mangalea, M. R., & Duerkop, B. A. (2020). Fitness trade-offs resulting from bacteriophage resistance potentiate synergistic antibacterial strategies. *Infection and Immunity*, *88*(7). https://doi.org/10.1128/IAI.00926-19.

Mao, Z., Yu, L., You, Z., Wei, Y., & Liu, Y. (2007). Cloning, expression and immunogenicty analysis of five outer membrane proteins of Vibrio parahaemolyticus zj2003. *Fish & Shellfish Immunology*, *23*(3), 567–575. https://doi.org/10.1016/j.fsi.2007.01.004.

Marcatili, A., D'Isanto, M., Galdierob, M., Pagninib, U., Palombab, E., Vitiello, M., et al. (2000). Role of Pasteurella multocida, Pasteurella haemolytica and Salmonella typhimurium porins on inducible nitric oxide release by murine macrophages. *Research in Microbiology*, *151*(3), 217–228.

Marti, R., Zurfluh, K., Hagens, S., Pianezzi, J., Klumpp, J., & Loessner, M. J. (2013). Long tail fibres of the novel broad-host-range T-even bacteriophage S16 specifically recognize salmonella OmpC. *Molecular Microbiology*, *87*(4), 818–834. https://doi.org/10.1111/mmi.12134.

Massari, P., Gunawardana, J., Liu, X., & Wetzler, L. M. (2010). Meningococcal porin PorB prevents cellular apoptosis in a toll-like receptor 2- and NF-kappaB-independent manner. *Infection and Immunity*, *78*(3), 994–1003. https://doi.org/10.1128/IAI.00156-09.

Massari, P., Henneke, P., Ho, Y., Latz, E., Golenbock, D. T., & Wetzler, L. M. (2002). Cutting edge: Immune stimulation by Neisserial Porins is toll-like receptor 2 and MyD88 dependent. *The Journal of Immunology*, *168*(4), 1533–1537. https://doi.org/10.4049/jimmunol.168.4.1533.

Massari, P., Ho, Y., & Wetzler, L. M. (2000). Neisseria meningitidis porin PorB interacts with mitochondria and protects cells from apoptosis. *Proceedings of the National Academy of Sciences of the United States of America*, *97*(16), 9070–9075. https://doi.org/10.1073/pnas.97.16.9070.

Massari, P., King, C. A., Ho, A. Y., & Wetzler, L. M. (2003). Neisserial PorB is translocated to the mitochondria of HeLa cells infected with Neisseria meningitidis and protects cells from apoptosis. *Cellular Microbiology, 5*(2), 99–109. https://doi.org/10.1046/j.1462-5822.2003.00257.x.

Massari, P., Ram, S., Macleod, H., & Wetzler, L. M. (2003). The role of porins in neisserial pathogenesis and immunity. *Trends in Microbiology, 11*(2), 87–93. https://doi.org/10.1016/s0966-842x(02)00037-9.

Masuda, N., Sakagawa, E., Ohya, S., Gotoh, N., Tsujimoto, H., & Nishino, T. (2000). Substrate specificities of MexAB-OprM, MexCD-OprJ, and MexXY-OprM efflux pumps in Pseudomonas aeruginosa. *Antimicrobial Agents and Chemotherapy, 44*(12), 3322–3327. https://doi.org/10.1128/aac.44.12.3322-3327.2000

Mathur, J., Davis, B. M., & Waldor, M. K. (2007). Antimicrobial peptides activate the vibrio cholerae? Eregulon through an OmpU-dependent signalling pathway. *Molecular Microbiology, 63*(3). https://doi.org/10.1111/j.1365-2958.2006.05544.x.

Mathur, J., & Waldor, M. K. (2004). The vibrio cholerae ToxR-regulated Porin OmpU confers resistance to antimicrobial peptides. *Infection and Immunity, 72*(6), 3577–3583. https://doi.org/10.1128/iai.72.6.3577-3583.2004.

Meier, C., Oelschlaeger, T. A., Merkert, H., Korhonen, T. K., & Hacker, J. (1996). Ability of Escherichia coli isolates that cause meningitis in newborns to invade epithelial and endothelial cells. *Infection and Immunity, 64*(7), 2391–2399. https://doi.org/10.1128/iai.64.7.2391-2399.1996.

Merino, S., Nogueras, M. M., Aguilar, A., Rubires, X., Alberti, S., Benedi, V. J., et al. (1998). Activation of the complement classical pathway (C1q binding) by mesophilic Aeromonas hydrophila outer membrane protein. *Infection and Immunity, 66*(8), 3825–3831. https://doi.org/10.1128/IAI.66.8.3825-3831.1998.

Merle, N. S., Church, S. E., Fremeaux-Bacchi, V., & Roumenina, L. T. (2015). Complement system. Part I. molecular mechanisms of activation and regulation. *Frontiers in Immunology, 6*, 262. https://doi.org/10.3389/fimmu.2015.00262.

Meyer, J. R., Dobias, D. T., Weitz, J. S., Barrick, J. E., Quick, R. T., & Lenski, R. E. (2012). Repeatability and contingency in the evolution of a key innovation in phage lambda. *Science, 335*(6067), 428–432. https://doi.org/10.1126/science.1214449.

Montag, D., Schwarz, H., & Henning, U. (1989). A component of the side tail fiber of Escherichia coli bacteriophage lambda can functionally replace the receptor-recognizing part of a long tail fiber protein of the unrelated bacteriophage T4. *Journal of Bacteriology, 171*(8), 4378–4384. https://doi.org/10.1128/jb.171.8.4378-4384.1989.

Moore, R. A., DeShazer, D., Reckseidler, S., Weissman, A., & Woods, D. E. (1999). Efflux-mediated aminoglycoside and macrolide resistance in Burkholderia pseudomallei. *Antimicrobial Agents and Chemotherapy, 43*(3), 465–470. https://doi.org/10.1128/AAC.43.3.465.

Moreno-Eutimio, M. A., Tenorio-Calvo, A., Pastelin-Palacios, R., Perez-Shibayama, C., Gil-Cruz, C., Lopez-Santiago, R., et al. (2013). Salmonella Typhi OmpS1 and OmpS2 porins are potent protective immunogens with adjuvant properties. *Immunology, 139*(4), 459–471. https://doi.org/10.1111/imm.12093.

Morona, R., & Henning, U. (1984). Host range mutants of bacteriophage Ox2 can use two different outer membrane proteins of Escherichia coli K-12 as receptors. *Journal of Bacteriology, 159*(2), 579–582. https://doi.org/10.1128/jb.159.2.579-582.1984.

Morona, R., Kramer, C., & Henning, U. (1985). Bacteriophage receptor area of outer membrane protein OmpA of Escherichia coli K-12. *Journal of Bacteriology, 164*(2), 539–543. https://doi.org/10.1128/jb.164.2.539-543.1985.

Muller, A., Gunther, D., Brinkmann, V., Hurwitz, R., Meyer, T. F., & Rudel, T. (2000). Targeting of the pro-apoptotic VDAC-like porin (PorB) of Neisseria gonorrhoeae to mitochondria of infected cells. *The EMBO Journal, 19*(20), 5332–5343. https://doi.org/10.1093/emboj/19.20.5332.

Muller, A., Rassow, J., Grimm, J., Machuy, N., Meyer, T. F., & Rudel, T. (2002). VDAC and the bacterial porin PorB of Neisseria gonorrhoeae share mitochondrial import pathways. *The EMBO Journal, 21*(8), 1916–1929. https://doi.org/10.1093/emboj/21.8.1916.

Nagakubo, S., Nishino, K., Hirata, T., & Yamaguchi, A. (2002). The putative response regulator BaeR stimulates multidrug resistance of Escherichia coli via a novel multidrug exporter system, MdtABC. *Journal of Bacteriology, 184*(15), 4161–4167. https://doi.org/10.1128/JB.184.15.4161-4167.2002.

Nair, B. M., Cheung, K. J., Griffith, A., & Burns, J. L. (2004). Salicylate induces an antibiotic efflux pump in Burkholderia cepacia complex genomovar III (B. cenocepacia). *Journal of Clinical Investigation, 113*(3), 464–473. https://doi.org/10.1172/jci200419710.

Nguyen, A. H., Molineux, I. J., Springman, R., & Bull, J. J. (2012). Multiple genetic pathways to similar fitness limits during viral adaptation to a new host. *Evolution, 66*(2), 363–374. https://doi.org/10.1111/j.1558-5646.2011.01433.x.

Nikaido, H. (1994). Porins and specific diffusion channels in bacterial outer membranes. *Journal of Biological Chemistry, 269*(6), 3905–3908. https://doi.org/10.1016/s0021-9258(17)41716-9.

Nikaido, H. (2003). Molecular basis of bacterial outer membrane permeability revisited. *Microbiology and Molecular Biology Reviews, 67*(4), 593–656. https://doi.org/10.1128/MMBR.67.4.593-656.2003.

Nikaido, H. (2011). Structure and mechanism of RND-type multidrug efflux pumps. *Advances in Enzymology and Related Areas of Molecular Biology, 77*, 1–60. https://doi.org/10.1002/9780470920541.ch1.

Nikaido, H., & Takatsuka, Y. (2009). Mechanisms of RND multidrug efflux pumps. *Biochimica et Biophysica Acta, 1794*(5), 769–781. https://doi.org/10.1016/j.bbapap.2008.10.004.

Nishino, K., & Yamaguchi, A. (2001). Analysis of a complete library of putative drug transporter genes in Escherichia coli. *Journal of Bacteriology, 183*(20), 5803–5812. https://doi.org/10.1128/JB.183.20.5803-5812.2001.

Ojha, S., Lacouture, S., Gottschalk, M., & MacInnes, J. I. (2010). Characterization of colonization-deficient mutants of Actinobacillus suis. *Veterinary Microbiology, 140*(1–2), 122–130. https://doi.org/10.1016/j.vetmic.2009.07.014.

Paauw, A., Trip, H., Niemcewicz, M., Sellek, R., Heng, J. M., Mars-Groenendijk, R. H., et al. (2014). OmpU as a biomarker for rapid discrimination between toxigenic and epidemic vibrio cholerae O1/O139 and non-epidemic vibrio cholerae in a modified MALDI-TOF MS assay. *BMC Microbiology, 14*, 158.

Pagel, M., Simonet, V., Li, J., Lallemand, M., Lauman, B., & Delcour, A. H. (2007). Phenotypic characterization of pore mutants of the vibrio cholerae porin OmpU. *Journal of Bacteriology, 189*(23), 8593–8600. https://doi.org/10.1128/JB.01163-07.

Parent, K. N., Erb, M. L., Cardone, G., Nguyen, K., Gilcrease, E. B., Porcek, N. B., et al. (2014). OmpA and OmpC are critical host factors for bacteriophage Sf6 entry in Shigella. *Molecular Microbiology, 92*(1), 47–60. https://doi.org/10.1111/mmi.12536.

Paulsson, M., Singh, B., Al-Jubair, T., Su, Y. C., Hoiby, N., & Riesbeck, K. (2015). Identification of outer membrane Porin D as a vitronectin-binding factor in cystic fibrosis clinical isolates of Pseudomonas aeruginosa. *Journal of Cystic Fibrosis, 14*(5), 600–607. https://doi.org/10.1016/j.jcf.2015.05.005.

Perez, O., Romeu, B., Cabrera, O., Gonzalez, E., Batista-Duharte, A., Labrada, A., et al. (2013). Adjuvants are key factors for the development of future vaccines: Lessons from the Finlay adjuvant platform. *Frontiers in Immunology, 4*, 407. https://doi.org/10.3389/fimmu.2013.00407.

Perez-Toledo, M., Valero-Pacheco, N., Pastelin-Palacios, R., Gil-Cruz, C., Perez-Shibayama, C., Moreno-Eutimio, M. A., et al. (2017). Salmonella Typhi Porins OmpC and OmpF are potent adjuvants for T-dependent and T-independent antigens. *Frontiers in Immunology*, *8*, 230. https://doi.org/10.3389/fimmu.2017.00230.

Pettersen, E. F., Goddard, T. D., Huang, C. C., Couch, G. S., Greenblatt, D. M., Meng, E. C., et al. (2004). UCSF chimera–a visualization system for exploratory research and analysis. *Journal of Computational Chemistry*, *25*(13), 1605–1612. https://doi.org/10.1002/jcc.20084.

Platt, A., MacLeod, H., Massari, P., Liu, X., & Wetzler, L. (2013). In vivo and in vitro characterization of the immune stimulating activity of the Neisserial porin PorB. *PLoS One*, *8*(12), e82171. https://doi.org/10.1371/journal.pone.0082171.

Porcek, N. B., & Parent, K. N. (2015). Key residues of S. flexneri OmpA mediate infection by bacteriophage Sf6. *Journal of Molecular Biology*, *427*(10), 1964–1976. https://doi.org/10.1016/j.jmb.2015.03.012.

Prasad, G., Dhar, V., & Mukhopadhaya, A. (2019). Vibrio cholerae OmpU mediates CD36-dependent reactive oxygen species generation triggering an additional pathway of MAPK activation in macrophages. *Journal of Immunology*, *202*(8), 2431–2450. https://doi.org/10.4049/jimmunol.1800389.

Prasadarao, N. V., Blom, A. M., Villoutreix, B. O., & Linsangan, L. C. (2002). A novel interaction of outer membrane protein a with C4b binding protein mediates serum resistance of Escherichia coli K1. *Journal of Immunology*, *169*(11), 6352–6360. https://doi.org/10.4049/jimmunol.169.11.6352.

Prehm, P., Jann, B., Jann, K., Schmidt, G., & Stirm, S. (1976). On a bacteriophage T3 and T4 receptor region within the cell wall lipopolysaccharide of Escherichia coli B. *Journal of Molecular Biology*, *101*(2), 277–281. https://doi.org/10.1016/0022-2836(76)90377-6.

Prouty, A. M., Brodsky, I. E., Falkow, S., & Gunn, J. S. (2004). Bile-salt-mediated induction of antimicrobial and bile resistance in salmonella typhimurium. *Microbiology (Reading)*, *150*(Pt 4), 775–783. https://doi.org/10.1099/mic.0.26769-0.

Provenzano, D., & Klose, K. E. (2000). Altered expression of the ToxR-regulated porins OmpU and OmpT diminishes vibrio cholerae bile resistance, virulence factor expression, and intestinal colonization. *Proceedings of the National Academy of Sciences of the United States of America*, *97*(18), 10220–10224. https://doi.org/10.1073/pnas.170219997.

Ram, S., Cullinane, M., Blom, A. M., Gulati, S., McQuillen, D. P., Monks, B. G., et al. (2001). Binding of C4b-binding protein to porin: A molecular mechanism of serum resistance of Neisseria gonorrhoeae. *The Journal of Experimental Medicine*, *193*(3), 281–295. https://doi.org/10.1084/jem.193.3.281.

Ram, S., McQuillen, D. P., Gulati, S., Elkins, C., Pangburn, M. K., & Rice, P. A. (1998). Binding of complement factor H to loop 5 of porin protein IA: A molecular mechanism of serum resistance of nonsialylated Neisseria gonorrhoeae. *The Journal of Experimental Medicine*, *188*(4), 671–680. https://doi.org/10.1084/jem.188.4.671.

Randall-Hazelbauer, L., & Schwartz, M. (1973). Isolation of the bacteriophage lambda receptor from Escherichia coli. *Journal of Bacteriology*, *116*(3), 1436–1446. https://doi.org/10.1128/jb.116.3.1436-1446.1973.

Ray, A., & Biswas, T. (2005). Porin of Shigella dysenteriae enhances toll-like receptors 2 and 6 of mouse peritoneal B-2 cells and induces the expression of immunoglobulin M, immunoglobulin G2a and immunoglobulin a. *Immunology*, *114*(1), 94–100. https://doi.org/10.1111/j.1365-2567.2004.02002.x.

Raynaud, C., Papavinasasundaram, K. G., Speight, R. A., Springer, B., Sander, P., Bottger, E. C., et al. (2002). The functions of OmpATb, a pore-forming protein of mycobacterium tuberculosis. *Molecular Microbiology*, *46*(1) 191–201. https://doi.org/10.1046/j.1365-2958.2002.03152.x.

Renkin, E. M. (1954). Filtration, diffusion, and molecular sieving through porous cellulose membranes. *The Journal of General Physiology*, *38*(2), 225–243.
Riede, I., Degen, M., & Henning, U. (1985). The receptor specificity of bacteriophages can be determined by a tail fiber modifying protein. *The EMBO Journal*, *4*(9), 2343–2346.
Rosenbusch, J. P. (1974). Characterization of the major envelope protein from Escherichia coli. Regular arrangement on the peptidoglycan and unusual dodecyl sulfate binding. *The Journal of Biological Chemistry*, *249*(24), 8019–8029.
Rudel, T., Schmid, A., Benz, R., Kolb, H. A., Lang, F., & Meyer, T. F. (1996). Modulation of Neisseria porin (PorB) by cytosolic ATP/GTP of target cells: Parallels between pathogen accommodation and mitochondrial endosymbiosis. *Cell*, *85*(3), 391–402. https://doi.org/10.1016/s0092-8674(00)81117-4.
Ruiz, N., Kahne, D., & Silhavy, T. J. (2006). Advances in understanding bacterial outer-membrane biogenesis. *Nature Reviews. Microbiology*, *4*(1), 57–66. https://doi.org/10.1038/nrmicro1322.
Rumbo, C., Tomas, M., Fernandez Moreira, E., Soares, N. C., Carvajal, M., Santillana, E., et al. (2014). The Acinetobacter baumannii Omp33-36 porin is a virulence factor that induces apoptosis and modulates autophagy in human cells. *Infection and Immunity*, *82*(11), 4666–4680. https://doi.org/10.1128/IAI.02034-14.
Sakharwade, S. C., Sharma, P. K., & Mukhopadhaya, A. (2013). Vibrio cholerae porin OmpU induces pro-inflammatory responses, but down-regulates LPS-mediated effects in RAW 264.7, THP-1 and human PBMCs. *PLoS One*, *8*(9), e76583. https://doi.org/10.1371/journal.pone.0076583.
Salem, M., Pajunen, M. I., Jun, J. W., & Skurnik, M. (2021). T4-like bacteriophages isolated from pig stools infect Yersinia pseudotuberculosis and Yersinia pestis using LPS and OmpF as receptors. *Viruses*, *13*(2). https://doi.org/10.3390/v13020296.
Sanchez, L., Pan, W., Vinas, M., & Nikaido, H. (1997). The acrAB homolog of Haemophilus influenzae codes for a functional multidrug efflux pump. *Journal of Bacteriology*, *179*(21), 6855–6857. https://doi.org/10.1128/jb.179.21.6855-6857.1997.
Schulein, K., Schmid, K., & Benzl, R. (1991). The sugar-specific outer membrane channel ScrY contains functional characteristics of general diffusion pores and substrate-specific porins. *Molecular Microbiology*, *5*(9), 2233–2241. https://doi.org/10.1111/j.1365-2958.1991.tb02153.x.
Schulz, G. E. (1993). Bacterial porins: Structure and function. *Current Opinion in Cell Biology*, *5*, 701–707.
Serino, L., Nesta, B., Leuzzi, R., Fontana, M. R., Monaci, E., Mocca, B. T., et al. (2007). Identification of a new OmpA-like protein in Neisseria gonorrhoeae involved in the binding to human epithelial cells and in vivo colonization. *Molecular Microbiology*, *64*(5), 1391–1403. https://doi.org/10.1111/j.1365-2958.2007.05745.x.
Silverman, J. A., & Benson, S. A. (1987). Bacteriophage K20 requires both the OmpF porin and lipopolysaccharide for receptor function. *Journal of Bacteriology*, *169*(10), 4830–4833. https://doi.org/10.1128/jb.169.10.4830-4833.1987.
Singleton, T. E., Massari, P., & Wetzler, L. M. (2005). Neisserial porin-induced dendritic cell activation is MyD88 and TLR2 dependent. *Journal of Immunology*, *174*(6), 3545–3550. https://doi.org/10.4049/jimmunol.174.6.3545.
Skliros, D., Kalatzis, P. G., Kalloniati, C., Komaitis, F., Papathanasiou, S., Kouri, E. D., et al. (2021). The development of bacteriophage resistance in vibrio alginolyticus depends on a complex metabolic adaptation strategy. *Viruses*, *13*(4). https://doi.org/10.3390/v13040656.
Smani, Y., Dominguez-Herrera, J., & Pachon, J. (2013). Association of the outer membrane protein Omp33 with fitness and virulence of Acinetobacter baumannii. *The Journal of Infectious Diseases*, *208*(10), 1561–1570. https://doi.org/10.1093/infdis/jit386.

Smani, Y., McConnell, M. J., & Pachon, J. (2012). Role of fibronectin in the adhesion of Acinetobacter baumannii to host cells. *PLoS One*, 7(4), e33073. https://doi.org/10.1371/journal.pone.0033073.

Snapper, C. M., Rosas, F. R., Kehry, M. R., Mond, J. J., & Wetzler, L. M. (1997). Neisserial Porins may provide critical second signals to polysaccharide-activated murine B cells for induction of immunoglobulin secretion. *Infection and Immunity*, *65*, 3203–3208.

Song, H. C., Kang, Y. H., Zhang, D. X., Chen, L., Qian, A. D., Shan, X. F., et al. (2019). Great effect of porin(aha) in bacterial adhesion and virulence regulation in Aeromonas veronii. *Microbial Pathogenesis*, *126*, 269–278. https://doi.org/10.1016/j.micpath.2018.11.002.

Song, W., Suh, B., Choi, J. Y., Jeong, S. H., Jeon, E. H., Lee, Y. K., et al. (2009). In vivo selection of carbapenem-resistant Klebsiella pneumoniae by OmpK36 loss during meropenem treatment. *Diagnostic Microbiology and Infectious Disease*, *65*(4), 447–449. https://doi.org/10.1016/j.diagmicrobio.2009.08.010.

Stone, E., Campbell, K., Grant, I., & McAuliffe, O. (2019). Understanding and exploiting phage-host interactions. *Viruses*, *11*(6). https://doi.org/10.3390/v11060567.

Sturenburg, E., Sobottka, I., Mack, D., & Laufs, R. (2002). Cloning and sequencing of Enterobacter aerogenes OmpC-type osmoporin linked to carbapenem resistance. *International Journal of Medical Microbiology*, *291*(8), 649–654. https://doi.org/10.1078/1438-4221-00175.

Suh, B., Bae, I. K., Kim, J., Jeong, S. H., Yong, D., & Lee, K. (2010). Outbreak of meropenem-resistant Serratia marcescens comediated by chromosomal AmpC beta-lactamase overproduction and outer membrane protein loss. *Antimicrobial Agents and Chemotherapy*, *54*(12), 5057–5061. https://doi.org/10.1128/AAC.00768-10.

Tabaraie, B., Sharma, B. K., Sharma, P. R., Sehgal, R., & Ganguly, N. K. (1994). Evaluation of salmonella Porins as a broad Spectrum vaccine candidate. *Microbiology and Immunology*, *38*.

Teng, C. H., Xie, Y., Shin, S., Di Cello, F., Paul-Satyaseela, M., Cai, M., et al. (2006). Effects of ompA deletion on expression of type 1 fimbriae in Escherichia coli K1 strain RS218 and on the association of E. coli with human brain microvascular endothelial cells. *Infection and Immunity*, *74*(10), 5609–5616. https://doi.org/10.1128/IAI.00321-06.

Thanassi, D. G., Cheng, L. W., & Nikaido, H. (1997). Active efflux of bile salts by Escherichia coli. *Journal of Bacteriology*, *179*(8), 2512–2518. https://doi.org/10.1128/jb.179.8.2512-2518.1997.

Thanavala, Y., & Lugade, A. A. (2011). Role of nontypeable Haemophilus influenzae in otitis media and chronic obstructive pulmonary disease. *Advances in Oto-Rhino-Laryngology*, *72*, 170–175. https://doi.org/10.1159/000324785.

Thiolas, A., Bornet, C., Davin-Regli, A., Pages, J. M., & Bollet, C. (2004). Resistance to imipenem, cefepime, and cefpirome associated with mutation in Omp36 osmoporin of Enterobacter aerogenes. *Biochemical and Biophysical Research Communications*, *317*(3), 851–856. https://doi.org/10.1016/j.bbrc.2004.03.130.

Tiku, V., Kofoed, E. M., Yan, D., Kang, J., Xu, M., Reichelt, M., et al. (2021). Outer membrane vesicles containing OmpA induce mitochondrial fragmentation to promote pathogenesis of Acinetobacter baumannii. *Scientific Reports*, *11*(1), 618. https://doi.org/10.1038/s41598-020-79966-9.

Tommassen, J., van der Ley, P., van Zeijl, M., & Agterberg, M. (1985). Localization of functional domains in E. coli K-12 outer membrane porins. *The EMBO Journal*, *4*(6), 1583–1587.

Toro, C. S., Lobos, S. R., Calderon, I., Rodriguez, M., & Mora, G. C. (1990). Clinical isolate of a porinless salmonella typhi resistant to high levels of chloramphenicol. *Antimicrobial Agents and Chemotherapy*, *34*(9), 1715–1719. https://doi.org/10.1128/AAC.34.9.1715.

Torres, A. G., Jeter, C., Langley, W., & Matthysse, A. G. (2005). Differential binding of Escherichia coli O157:H7 to alfalfa, human epithelial cells, and plastic is mediated by a variety of surface structures. *Applied and Environmental Microbiology*, *71*(12), 8008–8015. https://doi.org/10.1128/AEM.71.12.8008-8015.2005.

Torres, A. G., & Kaper, J. B. (2003). Multiple elements controlling adherence of enterohemorrhagic Escherichia coli O157:H7 to HeLa cells. *Infection and Immunity*, *71*(9), 4985–4995. https://doi.org/10.1128/IAI.71.9.4985-4995.2003.

Torres, A. G., Li, Y., Tutt, C. B., Xin, L., Eaves-Pyles, T., & Soong, L. (2006). Outer membrane protein a of Escherichia coli O157:H7 stimulates dendritic cell activation. *Infection and Immunity*, *74*(5), 2676–2685. https://doi.org/10.1128/IAI.74.5.2676-2685.2006.

Toussi, D. N., Liu, X., & Massari, P. (2012). The FomA porin from fusobacterium nucleatum is a toll-like receptor 2 agonist with immune adjuvant activity. *Clinical and Vaccine Immunology*, *19*(7), 1093–1101. https://doi.org/10.1128/CVI.00236-12.

Trias, J., Rosenberg, E. Y., & Nikaido, H. (1988). Specificity of the glucose channel formed by protein D1 of Pseudomonas aeruginosa. *Biochimica et Biophysica Acta*, *938*(3), 493–496. https://doi.org/10.1016/0005-2736(88)90148-4.

Tsai, Y.-K., Fung, C.-P., Lin, J.-C., Chen, J.-H., Chang, F.-Y., Chen, T.-L., et al. (2011). Klebsiella pneumoniaeOuter membrane Porins OmpK35 and OmpK36 play roles in both antimicrobial resistance and virulence. *Antimicrobial Agents and Chemotherapy*, *55*(4), 1485–1493. https://doi.org/10.1128/aac.01275-10.

Tufano, M. A., Rossano, F., Catalanotti, P., Liguori, G., Capasso, C., Ceccarelli, M. T., et al. (1994). Immunobiological activities of helicobacter pylori porins. *Infection and Immunity*, *62*(4), 1392–1399. https://doi.org/10.1128/iai.62.4.1392-1399.1994.

Tunbridge, A. J., Stevanin, T. M., Lee, M., Marriott, H. M., Moir, J. W., Read, R. C., et al. (2006). Inhibition of macrophage apoptosis by Neisseria meningitidis requires nitric oxide detoxification mechanisms. *Infection and Immunity*, *74*(1), 729–733. https://doi.org/10.1128/IAI.74.1.729-733.2006.

Tzeng, Y.-L., & Stephens, D. S. (2015). Antimicrobial peptide resistance in Neisseria meningitidis. *Biochimica et Biophysica Acta (BBA) - Biomembranes*, *1848*(11), 3026–3031. https://doi.org/10.1016/j.bbamem.2015.05.006.

Ulf-IngoFlügge. (2000). Transport in and out of plastids: Does the outer envelope membrane control the flow? *Trends in Plant Science*, *5*, 3.

Urdaneta, V., & Casadesús, J. (2018). Adaptation ofSalmonella entericato bile: essential role of AcrAB-mediated efflux. *Environmental Microbiology*, *20*(4), 1405–1418. https://doi.org/10.1111/1462-2920.14047.

van der Heijden, J., Reynolds, L. A., Deng, W., Mills, A., Scholz, R., Imami, K., et al. (2016). Salmonella rapidly regulates membrane permeability to survive oxidative stress. *MBio*, 7(4). https://doi.org/10.1128/mBio.01238-16.

van Putten, J. P., Duensing, T. D., & Cole, R. L. (1998). Entry of OpaA+ gonococci into HEp-2 cells requires concerted action of glycosaminoglycans, fibronectin and integrin receptors. *Molecular Microbiology*, *29*(1), 369–379. https://doi.org/10.1046/j.1365-2958.1998.00951.x.

Vazquez-Juarez, R. C., Romero, M. J., & Ascencio, F. (2004). Adhesive properties of a LamB-like outer-membrane protein and its contribution to Aeromonas veronii adhesion. *Journal of Applied Microbiology*, *96*(4), 700–708. https://doi.org/10.1111/j.1365-2672.2004.02177.x.

Villarreal, J. M., Becerra-Lobato, N., Rebollar-Flores, J. E., Medina-Aparicio, L., Carbajal-Gómez, E., Zavala-García, M. L., et al. (2014). TheSalmonella entericaserovar TyphiltrR-ompR-ompC-ompFgenes are involved in resistance to the bile salt sodium deoxycholate and in bacterial transformation. *Molecular Microbiology*, *92*(5), 1005–1024. https://doi.org/10.1111/mmi.12610.

von Specht, B. U., Knapp, B., Muth, G., Broker, M., Hungerer, K. D., Diehl, K. D., et al. (1995). Protection of immunocompromised mice against lethal infection with Pseudomonas aeruginosa by active or passive immunization with recombinant P. aeruginosa outer membrane protein F and outer membrane protein I fusion proteins. *Infection and Immunity*, *63*(5), 1855–1862. https://doi.org/10.1128/iai.63.5.1855-1862.1995.

Wang, S., Yin, B., Yu, L., Dang, M., Guo, Z., Yan, G., et al. (2019). Overexpression of AmpC promotes bacteriophage lysis of ampicillin-resistant Escherichia coli. *Frontiers in Microbiology*, *10*, 2973. https://doi.org/10.3389/fmicb.2019.02973.

Weindorf, H., Schmidt, H., & Martin, H. H. (1998). Contribution of overproduced chromosomal beta-lactamase and defective outer membrane porins to resistance to extended-spectrum beta-lactam antibiotics in Serratia marcescens. *The Journal of Antimicrobial Chemotherapy*, *41*(2), 189–195. https://doi.org/10.1093/jac/41.2.189.

Weiss, M. S., Abele, U., Weckesser, J., Welte, W., Schiltz, E., & Schulz, G. E. (1991). Molecular architecture and electrostatic properties of a bacterial porin. *Science*, *254*(5038), 1627–1630. https://doi.org/10.1126/science.1721242.

Weiss, M. S., Wacker, T., Weckesser, J., Welte, W., & Schulz, G. E. (1990). The three-dimensional structure of porin from Rhodobacter capsulatus at 3 a resolution. *FEBS Letters*, *267*(2), 268–272. https://doi.org/10.1016/0014-5793(90)80942-c.

Welte, W., Nestel, U., Wacker, T., & Diederichs, K. (1995). Structure and function of the porin channel. *Kidney International*, *48*(4), 930–940. https://doi.org/10.1038/ki.1995.374.

Wetzler, L. W., Ho, Y., & Reiserr, H. (1996). Neisserial Porins induce B lymphocytes to express costimulatory B7-2 molecules and to proliferate. *Journal of Experimental Medicine*, *183*, 1151–1159.

Wu, H. H., Yang, Y. Y., Hsieh, W. S., Lee, C. H., Leu, S. J., & Chen, M. R. (2009). OmpA is the critical component for Escherichia coli invasion-induced astrocyte activation. *Journal of Neuropathology and Experimental Neurology*, *68*(6), 677–690. https://doi.org/10.1097/NEN.0b013e3181a77d1e.

Yang, B., Zhang, D., Wu, T., Zhang, Z., Raza, S. H. A., Schreurs, N., et al. (2019). Maltoporin (LamB protein) contributes to the virulence and adhesion of Aeromonas veronii TH0426. *Journal of Fish Diseases*, *42*(3), 379–389. https://doi.org/10.1111/jfd.12941.

Young, M. J., Bay, D. C., Hausner, G., & Court, D. A. (2007). The evolutionary history of mitochondrial porins. *BMC Evolutionary Biology*, 7(1). https://doi.org/10.1186/1471-2148-7-31.

Yu, F., & Mizushima, S. (1982). Roles of lipopolysaccharide and outer membrane protein OmpC of Escherichia coli K-12 in the receptor function for bacteriophage T4. *Journal of Bacteriology*, *151*(2), 718–722. https://doi.org/10.1128/jb.151.2.718-722.1982.

Zeth, K. (2010). Structure and evolution of mitochondrial outer membrane proteins of β-barrel topology. *Biochimica et Biophysica Acta (BBA)*, *1797*(6–7), 1292–1299. https://doi.org/10.1016/j.bbabio.2010.04.019.

Zhao, X., Cui, Y., Yan, Y., Du, Z., Tan, Y., Yang, H., et al. (2018). Correction for Zhao et al., "outer membrane proteins ail and OmpF of Yersinia pestis are involved in the adsorption of T7-related bacteriophage yep-phi". *Journal of Virology*, *92*(6). https://doi.org/10.1128/JVI.01966-17.

Zhu, W., Thomas, C. E., & Sparling, P. F. (2004). DNA immunization of mice with a plasmid encoding Neisseria gonorrhea PorB protein by intramuscular injection and epidermal particle bombardment. *Vaccine*, *22*(5–6), 660–669. https://doi.org/10.1016/j.vaccine.2003.08.036.

CHAPTER THREE

Pore formation by pore forming membrane proteins towards infections

Achinta Sannigrahi[a,*] and Krishnananda Chattopadhyay[b,*]

[a]Department of Chemical Engineering, Indian Institute of Science, Bengaluru, Karnataka, India
[b]Structural Biology and Bioinformatics Division, CSIR-Indian Institute of Chemical Biology, Kolkata, West Bengal, India
*Corresponding authors: e-mail address: achintas@iisc.ac.in; krish@iicb.res.in

Contents

1. Introduction 80
2. PFT mechanism of action and structural classification 81
3. α-PFTs 82
   3.1 The colicin family 82
   3.2 The ClyA family 84
   3.3 The actinoporin family 85
4. β-PFTs 86
   4.1 The hemolysin family 86
   4.2 The aerolysin family 87
   4.3 The cholesterol-dependent cytolysin family 89
5. Conformational switch driven membrane pore formation 92
6. PFT effects and cellular defense mechanisms 98
   6.1 MAPK pathways 99
7. Calcium-dependent membrane repair mechanisms 100
8. Therapeutic development 101
   8.1 Therapeutic interventions towards the action of PFTs 101
9. Conclusion 103
Reference 103

Abstract

Over the last 25 years, the biology of membrane proteins including the PFPs-membranes interactions is seeking attention for the development of successful drug molecules against a number of infectious diseases. Pore forming toxins (PFTs), the largest family of PFPs are considered as a group of virulence factors produced in a large number of pathogenic systems which include streptococcus, pneumonia, *Staphylococcus aureus*, *E. coli*, *Mycobacterium tuberculosis*, group A and B streptococci, Corynebacterium diphtheria and many more. PFTs are generally utilized by the disease causing pathogens to disrupt the host first line of defense i.e. host cell membranes

Advances in Protein Chemistry and Structural Biology, Volume 128
ISSN 1876-1623
https://doi.org/10.1016/bs.apcsb.2021.09.001

through pore formation strategy. Although, pore formation is the principal mode of action of the PFTs but they can have additional adverse effects on the hosts including immune evasion. Recently, structural investigation of different PFTs have imparted the molecular mechanistic insights into how PFTs get transformed from its inactive state to active toxic state. On the basis of their structural entity, PFTs have been classified in different types and their mode of actions alters in terms of pore formation and corresponding cellular toxicity. Although pathogen genome analysis can identify the probable PFTs depending upon their structural diversity, there are so many PFTs which utilize the local environmental conditions to generate their pore forming ability using a novel strategy which is known as "conformational switch" of a protein. This conformational switch is considered as characteristics of the phase shifting proteins which were often utilized by many pathogenic systems to protect them from the invaders through allosteric communication between distant regions of the protein. In this chapter, we discuss the structure function relationships of PFTs and how activity of PFTs varies with the change in the environmental conditions has been explored. Finally, we demonstrate these structural insights to develop therapeutic potential to treat the infections caused by multidrug resistant pathogens.

1. Introduction

The first line of defense of a living cell is the biological membrane, which is the essential barrier protecting the cell. Pore forming proteins (PFP) are important factors behind the host cell membrane modifications, which is needed to establish the process of infection. PFPs achieve this process through the conversion of their soluble forms to the membrane bound forms. As a result, these proteins often adopt different conformations as well as structures with one converting into the other during membrane interactions. During this process, monomeric PFP subunit usually self-associates into higher-order oligomeric species which are typically formed with the association of membrane scaffold. Recently, the membrane interactions, and activities in biological system by the membrane proteins, including the PFPs-lipid bilayer interactions is seeking attention for the development of successful drug molecules against a number of infectious diseases (Alouf, 2001). We now have far better understanding of how soluble proteins can be converted into membrane-inserted form by refolding from one stable topology to another, using mechanisms that differ significantly.

Pore forming toxins (PFT) are the largest class of bacterial toxins and constitute a major class of pore-forming proteins (PFPs) found not only in pathogenic and non-pathogenic bacteria but also in all kingdoms of life. Although PFTs were previously considered as unsophisticated proteins that form pores in membranes, the astonishing complexity of their architecture and dynamics has been revealed in the past decade with great detail (Bischofberger, Iacovache, & Van Der Goot, 2012). During infection, through the pore formation and disruption of epithelial membrane barriers and immune system modifications, PFTs advance pathogenic growth and dissemination but the exact contribution of PFTs to these processes is difficult to ascertain. Like other PFPs, PFTs alter the permeability in plasma membrane of their targets cells, potentially causing cell death. They also contribute to the modulations of cellular functions of the hosts (Alves, de Ávila, Chávez-Olórtegui, & Lobato, 2014; Los, Randis, Aroian, & Ratner, 2013). Detailed analyses have shown distinctive pore formation by different PFTs, which in turn influence the phenotypes and immune responses in target cells. Although plasma membrane perforation by PFTs always enhances the membrane permeability, the extent of permeability varies with respect to the structural changes of the interacting proteins. Depending on the pore architecture in the membrane formed by the toxins, the activity of different ion channels get modulated since, pores can transport calcium, potassium, ATP through the bilayer based on the pore sizes. Therefore, a detailed investigation of pore structure is necessary to understand the cellular response to a specific PFT. In addition, during infection the PFT concentration varies based on the body site. Moreover, PFTs remain toxic even at sublytic concentrations as they can still alter cellular properties. Through these different effects, PFTs help in bacterial colonization and growth in host, partially protecting the pathogenic bacteria from the host immune responses (Alonzo & Torres, 2014; Diabate et al., 2015). In this chapter, we discuss recent work that has revealed the pore architectures and mechanisms of pore formation for each family of PFTs. We also show how receptor recognition modification enables different PFTs within the same family to different hosts may finally lead to new theranostic developments to battle against infections by targeting virulence factors and inhibiting bacterial survival.

2. PFT mechanism of action and structural classification

While PFTs are secreted as soluble molecules in aqueous solution, receptor specific recognition and interaction of these water soluble monomeric protein units with membranes lead to the assembly, conformational changes and pore formations (Honda, Ni, Miwatani, Adachi, & Kim, 1992; Parker & Feil, 2005). PFTs are classified on the basis of the structures of their membrane interacting regions that penetrate the host cell plasma membrane. Since these regions generally consist of either α-helices or β-barrels, the PFTs have two major classes: –αPFTs and β-PFTs. The majority of bacterial PFTs are β-PFTs, which also form the most-studied group (Iacovache, van der Goot, & Pernot, 2008). There exists another class of PFTs which are basically the subclass of β-PFTs. These are called large-pore-forming cholesterol-dependent cytolysins (CDCs) which are produced by Gram-positive and some Gram-negative bacteria (Alouf, 2003; Hotze et al., 2013; Iacovache et al., 2008). Although the toxins consisted of small-pore-forming repeat in toxin (RTX) which are produced by Gram-negative bacteria, make a large group of PFTs, the characteristics features based on their properties and the molecular understanding of their pore formation strategy are unknown (Alouf, 2003). Another essential type of classification, particularly in view of host defenses, is depending on the size of pores formed. Either small (0.5–5 nm) or large (20–100 nm) pores are created by bacterial PFTs and partial overlapping behaviors of the cellular defenses are found for different PFT classes. There are several known receptor molecules for PFTs and these include other membrane proteins (e.g., CCR5 and ADAM10), glycosylphosphatidylinositol (GPI)-anchored proteins, lipids, and cholesterol (Alonzo et al., 2013; Bischofberger, Gonzalez, & van der Goot, 2009; Gonzalez, Bischofberger, Pernot, Van Der Goot, & Freche, 2008). Receptor specificity of a particular PFT are one of the most important reason behind the PFT induced target cell tropism and this kind of specificity can be altered using in-vitro genetic modification. Several reviews have been written that widely cover the biochemical characteristics and in vitro effects of eukaryotic and bacterial PFTs (Aroian & Van Der Goot, 2007; Bischofberger et al., 2009; Geny & Popoff, 2006). It is known that for numerous human-pathogenic bacteria, as well as several economically important bacteria which are pathogenic to animals, PFTs are expressed during infection and consequence the increase in the pathogenic virulence and this is evident in vivo. For a smaller group of bacteria, there are also in vivo data on the mechanisms through which their PFTs contribute to infection. Although effects of PFTs are highly diverse, a number of common themes could be identified, which are discussed in this chapter.

3. α-PFTs

3.1 The colicin family

Colicin family proteins-which are one of the major classes of α-PFTs-are produced by *Escherichia coli* in order to kill related bacterial species through pore formation in bacterial outer membrane (Lakey, van der Goot, & Pattus, 1994). They have an important role in regulating the microbial residents, presumably by destroying competitors or invading the niche occupancy (Cascales et al., 2007). When the structures of colicin PFT were not available, it was believed from the bioinformatic analysis that the colicin family PFT contain hydrophobic α-helices, which are similar to other transmembrane proteins. The first PFT structure was solved for colicin A, showed an "inside-out membrane protein" fold and the characteristic was found to be consistent with a protein that can switch between soluble and transmembrane conformation. Colicin induced pore formation which is associated with the insertion of a hydrophobic helical hairpin produces a voltage gated channel which is nonspecific in nature. This voltage gated channel formation results in the membrane modulation that includes ATP depletion, depolarization, and finally cell death (Ridleya, Johnson, & Lakey, 2010; Van der Goot, Gonzalez-Manas, Lakey, & Pattus, 1991). The exact stoichiometry and structure of the pore complex formed by the colicin family proteins remains unresolved due to the formation of unstable oligomeric complexes which are difficult to characterize by structural analysis (Iacovache, Bischofberger, & van der Goot, 2010). It is commonly believed that, during the binding with target membrane, the amphipathic helices part away from the hydrophobic helical hairpin, and enable the insertion of the hairpin into the target membrane. This type of orientational changes are known to be the "umbrella" structure formation which is evident from the previous literature on colicin A family (Ridleya et al., 2010). This is recently ascertained by electron paramagnetic resonance (EPR) investigations using spinlabeled colicin A (Pulagam & Steinhoff, 2013). However, umbrella structure of colicin A is not sufficient for the generation of channel in the target membrane and therefore, it is believed that some other mechanisms are also involved simultaneously with umbrella structure formation. Electrostatic complementarity between colicin A and negative charged component of the bacterial membrane may promote some other amphipathic helix insertion into the membrane (Slatin, Qiu, Jakes, & Finkelstein, 1994). The insertion of the hydrophobic hairpin may trigger multimerization could have been other reason behind that, as found in case of colicin A (that is, dimers) and colicin Ia (that is, trimers of dimers) (Dunkel, Pulagam, Steinhoff, & Klare, 2015; Greig, Radjainia, & Mitra, 2009). Apart from the colicin family proteins, the translocation motif of diphtheria toxin that translocates the functional catalytic unit across low pH environment of the endosomal membrane also shows similar kind of colicin fold (Choe et al., 1992). *Bacillus thuringiensis* secreted Cry toxins, the insecticidal pore-forming protein also contains a colicin fold, and a similar fold of plausible familiar ancestry with colicins and this kind of fold is observed in the membrane translocators that make the tip of bacterial type III secretion systems (Barta et al., 2012). Interestingly, BAX, BCL2 and BAK family which are generally produced by phages also possess similar colicin like fold (Westphal, Dewson, Czabotar, & Kluck, 2011). These proteins participate in the apoptosis process and regulate the permeabilization of the mitochondrial membrane and this behavior is pretty consistent with the functions of colicin family proteins (García-Sáez, Fuertes, Suckale, & Salgado, 2010).

3.2 The ClyA family

The cytolysin A (ClyA) family is one of the major classes of α-PFTs which include the subsequent toxins like, ClyA (or hemolysin E (HlyE), which is secreted by some specific strains of *E. coli*; by *Shigella flexneri* and *Salmonella enterica* (Hunt, Green, & Artymiuk, 2010); and the B component of hemolysin BL enterotoxin (Hbl) and the *Bacillus cereus* derived toxins i.e. non-hemolytic tripartite enterotoxin (Nhe; which is hemolytic despite its name) (Madegowda, Eswaramoorthy, Burley, & Swaminathan, 2008). Previous reports established ClyA PFTs as virulence factors but their receptor preference and the extent of toxicity is still an open field of investigation (Jeßberger, Dietrich, Bock, Didier, & Märtlbauer, 2014). The structure of ClyA has been solved both in its soluble and transmembrane form which in turn unveiling the exceptional pore-formation process of this family of PFTs, which involves significant conformational changes. ClyA, in solution phase remains as an entirely alpha helical elongated protein which exceptionally contains a short hydrophobic β-hairpin structure which is also known as the β-tongue (Mueller, Grauschopf, Maier, Glockshuber, & Ban, 2009; Wallace et al., 2000).

In presence of membrane that contains cholesterol, the hydrophobic β-hairpin gets detached from the core of the protein and inserts into the bilayer. This incidence gives rise to a large conformational rearrangement of the distal amino –terminal which swings around to bind with the membrane. The length of this N-terminal helical region varies in different members of the PFT family and this is evident from the recently solved structure of NheA, a component of the ripartite Nhe toxin, in which the helix is much shorter (Ganash et al., 2013).

The membrane pore formation event is significantly regulated by the variation of the length of this helix region. It has been suggested that

NheA can form β-barrel pores, although the mechanism is not known. There are reports which show that the oligomerization of ClyA occurs through a sequential mechanism which is also comparable to some common features of actinoporin family proteins (Fahie et al., 2013; Vaidyanathan, Sathyanarayana, Maiti, Visweswariah, & Ayappa, 2014). The amphipathic helices of ClyA concentrates in the core of ring-structure and creates a helical barrel which gets inserted into the membrane through wedge-like mechanism. There exists a contrasting pore formation mechanism of ClyA which showed the formation of soluble prepores which are found to be encapsulated within bacterial outer membrane vesicles (OMVs) of *E. coli*. These OMVs are protected from the ClyA induced pore formation and damage because of the absence of cholesterol in bacterial membrane. The *Staphylococcus aureus* hemolysin family originated *Vibrio cholerae* cytolysin (VCC) protein is also recently proposed to show similar OMV-mediated delivery mechanism of PFTs but it is not known how these OMV-embedded pre-pores would be delivered to the target membrane (Elluri et al., 2014).

3.3 The actinoporin family

Another family of alpha PFTs is actinoporin family which are produced by sea anemone (Hinds, Zhang, Anderluh, Hansen, & Norton, 2002; Kristan, Viero, Dalla Serra, Maček, & Anderluh, 2009). They form part of the venom that these sedentary animals rely on to paralyze and digest their prey (which range from plankton to fish and crustaceans), as well as to defend themselves from predators. X-ray crystal structure have been reported for the soluble form of the actinoporin equinatoxin-II, sticholysin-II and fragceatoxin C (FraC produced by *Actinia fragacea*) (Mechaly et al., 2011). The crystal structure of the soluble form of Equinatoxin from sea anamone Actinia equine has been solved at 1.9 Å resolution (Anderluh et al., 2003; Athanasiadis, Anderluh, Maček, & Turk, 2001; Hinds et al., 2002). It is a single domain protein based on 12-strand beta sandwich fold with hydrophobic core and a pair of alpha helices in which amphipathic N-terminal helices detaches from the core of the protein and inserts into the membrane (Malovrh et al., 2003). This membrane insertion step is lipid dependent, with a preference for target membranes that are enriched in sphingomyelin and/or have phase-separated lipid membranes (Kristan et al., 2009). Interestingly, membrane insertion of actinoporins can subsequently induce lipid mixing, as has recently been shown for sticholysin using differential scanning calorimetry and atomic force microscopy (AFM) (Ros et al., 2013). Oligomerization precedes pore formation without any pore intermediate state. Furthermore, X ray structure of FraC dimer revealed a significant conformational change in the membrane inserted N-terminal helical domain toward the protomer state found in pores which in turn suggest that actinoporin protomers may assemble sequentially into a pore without the formation of the prepore (Rojko et al., 2013; Tanaka, Caaveiro, Morante, González-Mañas, & Tsumoto, 2015).

4. β-PFTs

4.1 The hemolysin family

Staphylococcus aureus, the opportunistic human pathogen, a causative agent of pneumonia and sepsis, produces a battery of virulence factors. The mobile genetic elements which encode these types of virulence factors results in diverse expression combinations of the toxins in different pathogenic strains (Schreiber, Chan, & Shorr, 2011). The battery of *S. aureus* virulence factors particularly consist of numerous PFTs, which may be one component e.g. α-hemolysin (Hla), or two components, like in the case of γ-hemolysin AB (HlgAB), HlgCB, Panton–Valentine leukocidin (PVL), leukocidin ED (LukED), and leukocidin AB (LukAB; also known as LukGH). These distinct kind of PFTs are not redundant functionally and target mostly the immune cells along with other cell types and they also interact with different cell surface proteins (DuMont & Torres, 2014). The hemolysin family PFTs play crucial roles in *S. aureus* pathogenesis in different ways: some PFTs kill several cells of the immune system through an immune evasion mechanism, while other PFTs such as Hla promoting bacterial dissemination by disrupting epithelial barriers. It is to be noted that most of the staphylococcal PFT components have a related fold, which is shared also by the δ-toxins from *Clostridium perfringens*, the necrotic enteritis toxin B (NetB) (Savva et al., 2013), and VCC (De & Olson, 2011). It is known that the single-component toxins of the hemolysin family usually form heptameric pores, in contrast, the bi-component component toxins create octameric pores (Jayasinghe & Bayley, 2005), taking together four alternating copies of each subunit (Yamashita et al., 2011).

One of the structurally well characterized PFT family is the hemolysin family. Some examples of the hemolysin family proteins are Hla (Song et al., 1996), Hlg (Yamashita et al., 2014), HlgB (Olson, Nariya, Yokota, Kamio, & Gouaux, 1999), VCC (Olson & Gouaux, 2005) and NetB which have been crystallized in the soluble and/or the pore configuration. The abundant structural information has established reliable and consistent pore models on the basis of homology, as observed in the case *C. perfringens* derived δ-toxin (Huyet et al., 2013). In aqueous solution, *S. aureus* hemolysins generates a compact structure which is rich in β-strands, in which the pre-stem motif which in late stage coverts to the transmembrane domain; forms a three-stranded β-sheet which is crowded against the center of the protein. After forming oligomeric assembly of a heptameric or octameric pre-pore ring-like structure, the pre-stem domain (which resides to the inner hollow space of the ring) gets detached from the core of the protein, reorganizes to an anti-parallel β-hairpin and, in collaboration with adjacent hairpins, creates a 14-stranded or 16-stranded β-barrel. Since the outside region of this β-barrel is hydrophobic, it gradually inserts into the lipid bilayer. The membrane insertion event and amphipathic β-barrel formation are found to be associated events. It is noteworthy that, although necessary for pore formation, the pre-stem loop is not essential for the oligomerization.

4.2 The aerolysin family

One of the most broad families of PFP is the aerolysin family. Gram-negative Aeromonas spp. derived aerolysin is the first described member of aerolysin family and other bacteria originated aerolysins are the Gram-positive human pathogen *Clostridium septicum* produced α-toxin, Pseudomonas entomophila produced monalysin and several parasporins produced by *B. thuringiensis*. Aerolysins give rise to bacterial dissemination in some pathogens such as *Aeromonas hydrophila*, probably by disrupting the epithelial barriers, which would in few cases result profound wound infections. According to the bioinformatics studies, it has been shown that there exists a core aerolysin motif, which is consisted of a pair of two β-strands that are separated by the pre-stem loop, and the presence of this loop is conserved throughout all kingdoms of life. There are numerous eukaryotic aerolysins like, enterolobin from the seeds of the tree *Enterolobium contortisiliquum*, which is typically found in Brazil; snail *Biomphalaria glabrata* produced biomphalysin; and frog *Bombina maxima* produced βγ-CAT. It is to be noted that several aerolysin family members produced by different animals exert effects on immune systems (Zhao et al., 2014). The aqueous form of aerolysin is a multidomain protein with highly elongated structure which shows contrasting behavior in comparison to *S. aureus* hemolysins (Parker et al., 1994). Aerolysin is dimeric as obtained from crystallographic structure determination using high protein concentrations (Abrami, Fivaz, & van der Goot, 2000); in this form, the pre-stem domain creates a β-strand-containing loop inside the elongated domain (Iacovache et al., 2006), and this structural behavior already has been found in as *C. septicum* α-toxin (Melton, Parker, Rossjohn, Buckley, & Tweten, 2004). Aerolysin is typically produced as a protoxin (Howard & Buckley, 1985) and prevents the premature oligomerization by extending carboxy terminal. After getting cleaved of the C-terminal peptide, oligomerization occurs to get the ring like structure composed of heptameric assembly (Iacovache et al., 2011). Very recently, a near atomic-level understanding of the pore-formation mechanism was established for the aerolysin family using a cumulative approaches including cryo-electron microscopy (cryo-EM),X-ray crystallography, molecular modeling and dynamics. By arresting different stage specific oligomeric and intermediate structures, the structure of the mature pore state and the intermediate pre-pore were solved (Degiacomi et al., 2013). First stage of the oligomerization involves the assembly of the monomers into heptameric prepore formation and subsequently the prepore docks onto the membrane bilayer, and then the pre-stem domains slide through a pocket into the inner-nook of the pre-pore (Fig. 2). During the transition from the prepore structure to the pore structure, there occurs a conformational change which converts amphipathic β-hairpins to transmembrane β-barrel. This conformational alteration is accompanied by a concerted swirling mechanism that stretches the extracellular portion of the pore as the β-barrel forms and gets inserted into the membrane. Thus, the aerolysin pores possess a big, horizontal, disc-like extracellular organization, which is to some extent prominent in presence of membrane. Recently, despite a different pore-formation process, a similar architecture of the final pore was recently found for sea cucumber *Cucumaria echinata* produced hemolytic lectin CEL-III (Unno, Goda, & Hatakeyama, 2014). It is shown that an amphipathic β-barrel can potentially shift within or extrude from the membrane and that is why a number of different mechanisms have developed to stabilize the barrel orientation. In the case of Hla derived by *S. aureus*, the cytosolic tip of the transmembrane β-hairpin contains charged residues that gets attachment from the tip of the cytosol, while on the other hand, the tips of the aerolysin transmembrane β-hairpins are hydrophobic and take on a rivet-like configuration that fends off the motion and attaches the barrel inside bilayer core. Moreover, the lumen of aerolysin β-barrels is significantly dissimilar from that of *S. aureus* Hla, with a very elevated concentration of charged residues, which might support the conduction of anionic moieties and enhance the flexibility of the pore. The primary structural features of the aerolysin family proteins vary from one protein to another. However, crystallographic studies, have shown notable conformational similarities in the aerolysins fold with various primary sequences, e.g. ε-toxin (Popoff, 2011), derived from *C. perfringens*, a hemolytic lectin (LSL) produced by the parasitic mushroom Laetiporus sulphureus *C. perfringens* enterotoxin (CPE) (Briggs et al., 2011; Kitadokoro et al., 2011), the Cnidaria produced hydralysin toxins (Sher et al., 2005) and the lysenin toxin produced by the earthworm *Eisenia fetida* (De Colibus et al., 2012). Further studies on the structure and membrane interactions of these proteins are needed to establish the global architecture and membrane insertion mechanism of aerolysin family proteins.

4.3 The cholesterol-dependent cytolysin family

The cholesterol-dependent cytolysin (CDC) family which is a large toxin family is mostly derived from Gram-positive bacteria like, Clostridium, Bacillus, Listeria genera and Streptococcus (Hotze & Tweten, 2012). Very recently, it has been reported that some gram negative bacteria also can produce CDC toxins (Hotze et al., 2013). PFTs influence the infection process in diverse ways depending on the pathogens. There are several examples like: listeriolysin O (LLO) drives the Listeria spp. release once the bacterium gets entry into the target cell through phagocytosis; perfringolysin O (PFO) has a crucial role during the gangrene development which is frequently associated with *C. perfringens* infections; and in Streptococcus pneumonia produced pneumolysin (Ply) is involved for tissue evasion, and also leads to activation of the complement cascade and inflammation. Structural studies have shown that CDCs share the similar type of fold like membrane attack component/ perforin (MACPF) domains, which have primarily been illustrated in the mammalian immune system (Rosado et al., 2007). There are several MACPF motif of PFTs in non-mammalian eukaryotes which are fungi derived pleurotolysin (Rosado et al., 2007), and in apicomplexan parasites evolved perforin-like proteins e.g. *Plasmodium falciparum* and *Toxoplasma gondii* (Deligianni et al., 2013; Roiko & Carruthers, 2009) which are involved in parasite release from the parasitophorous vacuole. In eukaryotes, the MACPF fold also present in bacteria (e.g., *Photorhabdus luminescens* derived Plu-MACPF, BSAP-1 from *Bacteroides fragilis* and BT3439 from *Bacteroides thetaiotaomicron*) (Xu et al., 2010). The exact role of these bacterial MACPF proteins in infection is to be completely revealed.

There exists the crystal structures of several CDCs like *Streptococcus suis* produced suilysin (SLY), *Listeria monocytogenes* derived LLO, *Streptococcus mitis* produced lectinolysin (LLY), *Streptococcus intermedius* derived intermedilysin (ILY), *Bacillus anthracis* produced anthrolysin O (ALO), *Streptococcus pyogenes* derived streptolysin O (SLO).However, due to the stoichiometry of the complexes, there is no report available on the crystallized structure of CDC pore configuration, in which up to 50 monomers pull together into very large ring-like structures. Nevertheless, these structures are acquiescent to investigation by AFM (Tilley, Orlova, Gilbert, Andrew, & Saibil, 2005) and electron microscopy (EM) (Czajkowsky, Hotze, Shao, & Tweten, 2004), imparting information on the pore architecture. The archetypical toxin of this family, PFO, is an extended, multidomain β-sheet-rich protein (Rossjohn, Feil, McKinstry, Tweten, & Parker, 1997). As conflicting to the single β-hairpin exhibited by aerolysin and hemolysin-like toxins, each of CDC molecules bears two amphipathic β-hairpins to the final β-barrel. Surprisingly, in the soluble form, these domains are arranged as two pairs of short α-helices which flank a central β-sandwich. During oligomerization and membrane interactions, they undergo a conformational change as observed in prion-like proteins where α-helix-to-β-strand transition occurs (Shatursky et al., 1999; Shepard et al., 1998). This kind of conversion is also found in the sea cucumber hemolytic lectin CEL-III. Subsequently, the two β-hairpins organize into a faintly curved β-sheet that connects with the neighboring β-sheets in the oligomer to shape a giant β-barrel consisted of 80–200 β-strands, on the basis of the pore stoichiometry. The CDC Oligomerization is found to be occurred through the sequential accumulation of monomers or multimers (Ramachandran, Tweten, & Johnson, 2004); moreover, in some cases there does not occur oligomerization completely, resulting the formation of partial arc like structures, lower stoichiometry open pores that have been thought to be active pores (Sonnen, Plitzko, & Gilbert, 2014) (Fig. 1). Very recently this behavior was validated for SLY using a combination of modeling techniques, AFM, cryo-EM, which suggested a sequential mechanism of oligomerization that creates not only pores but also arc-like entities that are kinetically trapped at the pre-pore structure. Both incomplete arcs and closed rings appear to allocate the capability to generate active pores through lipid exclusion (Leung et al., 2014). The β-strands alignment in correct form is necessary for pore assembly, particularly in case of a large complex. It was observed using disulfide scanning that in the pre-pore complex the β-hairpins which are partially

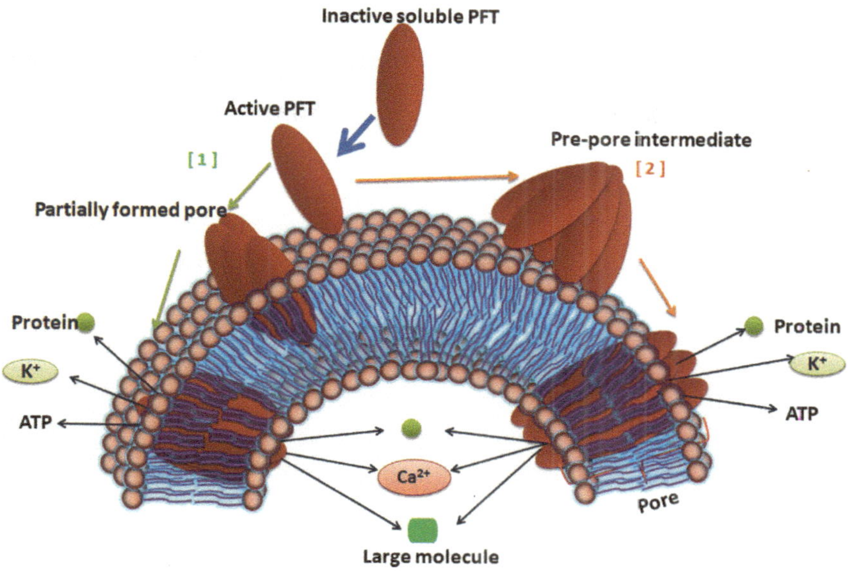

Fig. 1 Mechanism of pore formation. In general the α-PFT monomer first binds with the host cell membrane to generate active PFT conformation which gradually forms the complete pore structure through some partially formed pore structures (mechanism (Alouf, 2001)). On the other hand, the β-PFTs forms prepore conformation before generating the complete pore structure (mechanism (Bischofberger et al., 2012)). In both α-PFT and β-PFT pathways, the final result is the formation of a transmembrane pore with different architecture, stoichiometry, size and conduction features, which promote the influx or efflux of ions, small molecules and proteins through the host membrane, and trigger various secondary responses involved in the repair of the host membrane.

unfolded are significantly dynamic, while in the β-barrel conformation, the pore state is locked, exerting a ~20° tilt in the organization of the β-strands with regard to the membrane. Recently, a study showed that the energy required for the pre-pore to-pore transition is supplied through the formation of a salt-bridge interactions between adjoining monomers at the final stage of oligomerization (Wade et al., 2015). This consequences coordinated disruption of a decisive interface inside each monomer. AFM studies first showed that CDCs undergo an elongation at the membrane surface during pre-pore to pore transition (Czajkowsky et al., 2004). Combining different outcomes using molecular modeling study and EM analysis of CDCs, a general mechanism for CDC pore formation has been established which suggested the simple rotation of the core domain of the toxin during the collapse from the pre-pore to the pore state, and this kind of swirling mechanism reported for aerolysin collapse earlier. The mammalian perforin also showed very similar pore architecture. However, many other toxins have a pore-forming domain or subunit, which is used as a translocon to deliver a catalytic subunit with toxic activity into a target cell. As opposed to the effectors of bacterial secretion systems, bacterial toxins with pore-forming domains are autonomous, as a secretion system apparatus is not required for target cell entry. Although we do not review translocation domains of toxins in this chapter, it is worth mentioning two recent examples of studies that have revealed translocation pore architectures at atomic resolution. First, the transmembrane conformation of the translocation subunit (also known as the protective antigen (PA) of the anthrax toxin was shown to form heptameric or octameric pores (Jiang, Pentelute, Collier, & Zhou, 2015). The most unusual feature of the anthrax pore is the extremely long 105 Å β-barrel structure. The lumen of the pore, which translocates the catalytic subunit, contains a structure known as a Φ-clamp (Krantz et al., 2005) that ratchets the unfolded protein through the lumen while ensuring that the pore remains impermeable to most ions. Second, in *Photorhabdus luminescens*, binding of the TcB and TcC subunits of the Tc toxin complex to the TcA subunit was shown to induce the opening of the translocation channel of the toxin. The channel itself has a hydrophobic and narrow gate region similar to the anthrax Φ-clamp, which probably requires protein unfolding before translocation (Meusch et al., 2014).

5. Conformational switch driven membrane pore formation

Although bioinformatics, computational analysis and genome study have been employed to detect the existence of toxins in pathogen genomes, there exists a special class of toxin which exhibit their toxin nature whenever introduced to external factors. These behaviors rely on the sequence architecture of the proteins. These proteins often use conformational switches which is activated due to the presence of a number of special sequence which can adopt either alpha or beta sheet depending upon the context (Minor & Kim, 1996). These sequences are known as chameleon sequences. Chameleon sequences are not true switches, but rather are sequence fragments that adopt different structures in different proteins. The final structure of the fragment is dictated by nonlocal interactions resulting from the tertiary structure of the protein (Kuznetsov & Rackovsky, 2003; Zhou, Alber, Folkers, Gonnet, & Chelvanayagam, 2000). These sequences are of great interest to the protein switch designers because they enable one to probe the determinants of structural ambiguity. It is well known that some amino acids have a strong propensity to form helices (alanine, leucine, glutamine glutamic acid), others to form strands (valine isoleucine, phenylalanine, tyrosine)and rests some to form irregular conformations (glycine, proline, aspartic acid). Interestingly, chameleon sequences that switch between alpha helix and strand contain more strong helix and/or more strong beta when compared to average alpha helices and normal beta sheets (Alexander, Rozak, Orban, & Bryan, 2005; Dalal & Regan, 2000; He, Yeh, Alexander, Bryan, & Orban, 2005). The adoption of different conformations depending on the packing of the hydrophobic residues creates a binary pattern of the chameleon sequences. Packing of the hydrophobic residues was most easily accommodated by a beta strand with the native binary pattern PHPHPH (where P and H denotes the polar and hydrophobic residues respectively) and an alpha helix with the mutant PHHPPH pattern. More recent works with the Arc repressor has shown that the change in a single methyl group in mutants containing ambiguous binary pattern PHHHPH can lead to conformational switch (Cordes, Burton, Walsh, McKnight, & Sauer, 2000; Cordes, Walsh, McKnight, & Sauer, 1999). It is reported that a slight perturbations in the chameleon sequences by external forces drives the proteins conformational switch. A variety of trigger mechanism have been used including pH shifts, post-translational modification, and ligand binding (Mihara & Takahashi, 1997). Many pathogenic secretory proteins utilize their environment-sensitive conformational switch strategy to promote their infection in hosts. Influenza hemagglutinin is an example of which uses its "spring loaded" metastable conformation in the presence of pH change trigger, which switches this a protein to its fusogenic form (Carr, Chaudhry, & Kim, 1997). *Mycobacterium tuberculosis* is known to modulate host immune responses to facilitate its persistence inside host cells. In this context, TB granuloma is the hallmark of TB infection (Aly et al., 2006; Guirado & Schlesinger, 2013; Kim et al., 2010; Ramakrishnan, 2012; Russell, 2007; Russell, Cardona, Kim, Allain, & Altare, 2009; Saunders & Britton, 2007; Ulrichs & Kaufmann, 2006). Although there is a direct connection between granuloma formation and immune-suppression mediated infection, any detailed understanding on the "complex escape mechanism" of TB is not well-studied. Not only that the presence of toxins are unknown according to the genome analysis of TB. It has been suggested that a small immunogenic protein, MPT63 in TB utilizes its chameleon sequences to switch from native state to alpha helical structure (Ghosh,

Mukherjee, Chattopadhyay, & Ghosh, 2012; Goulding et al., 2002; Manca et al., 1997; Mustafa, 2009; Redchuk et al., 2012; Rolinck-Werninghaus et al., 2003). This alpha helical form undergoes oligomerization to form toxic "oval" shaped morphology at pH 5 (atmosphere inside granulomatous M2 macrophages) (Kundu, Kundu, & Chattopadhyay, 2017; Siromolot, Oliinyk, Kolibo, & Komisarenko, 2016). These toxic oligomers create pores in macrophages and model membrane (Sannigrahi et al., 2018). It is interesting to note that MPT63 oligomerization and oligomer driven membrane pore formation follow different mode of pore formation mechanism which differs from the α-PFTs and β-PFTs. Fig. 2 shows its oligomerization and pore formation process. MPT63 is beta sheet protein but in undergo transition from beta sheet to alpha helix which leads to the oligomerization at low pH even in absence of membrane. When it comes in contact with membrane, the oligomers come closer and form a five unit oligomeric assembly on the membrane resulting the membrane rupture, toxicity consequences cell death. Most interestingly, N-terminal chameleon stretch of MPT63 (19–33 AA) remains conserved in all alpha and beta PFTs (Sannigrahi et al., 2019). The architecture of this domain is responsible for the structural distortion of the most of the pore forming toxins.

There is a phenomenon named as "shape shifting of proteins" which is associated with a number of cellular processes (Benach & Hunt, 2004; Huntington, 2006; Knoverek, Amarasinghe, & Bowman, 2018; Lawrence et al., 2008; Shimaoka et al., 2003). Some pathogenic systems utilize the shape shifting behavior of proteins for the transportation of the specific molecules to the target destinations. One of the best examples of shape shifting is the shape shifting protein channels. Due to specific ion transport process, the ion channels are getting considerable attention (Blobel, 1980; De Keyzer, Van Der Does, & Driessen, 2003; Hamman, Chen, Johnson, & Johnson, 1997; Rapoport, Jungnickel, & Kutay, 1996; Schiebel, Driessen, Hartl, & Wickner, 1991). This protein ion channel often changes their shapes to allow the newly made proteins to pass through membranes before the protein fold into their functional state. This behavior was also observed in bacterial secretion systems. The crystal structure of the Sec YEβ channel presented by Van den Berg et al. provides initial atomic level glimpses of the shape shifting channels (Van den Berg et al., 2004). Previous reports suggested that the channel pore is located at the interface between several heterotrimer e.g. well-characterized translocase derived from *E. coli* strain (Beckmann et al., 2001; Duong, 2003; Yahr & Wickner, 2000). This heterotrimer pore formation process was supported by other

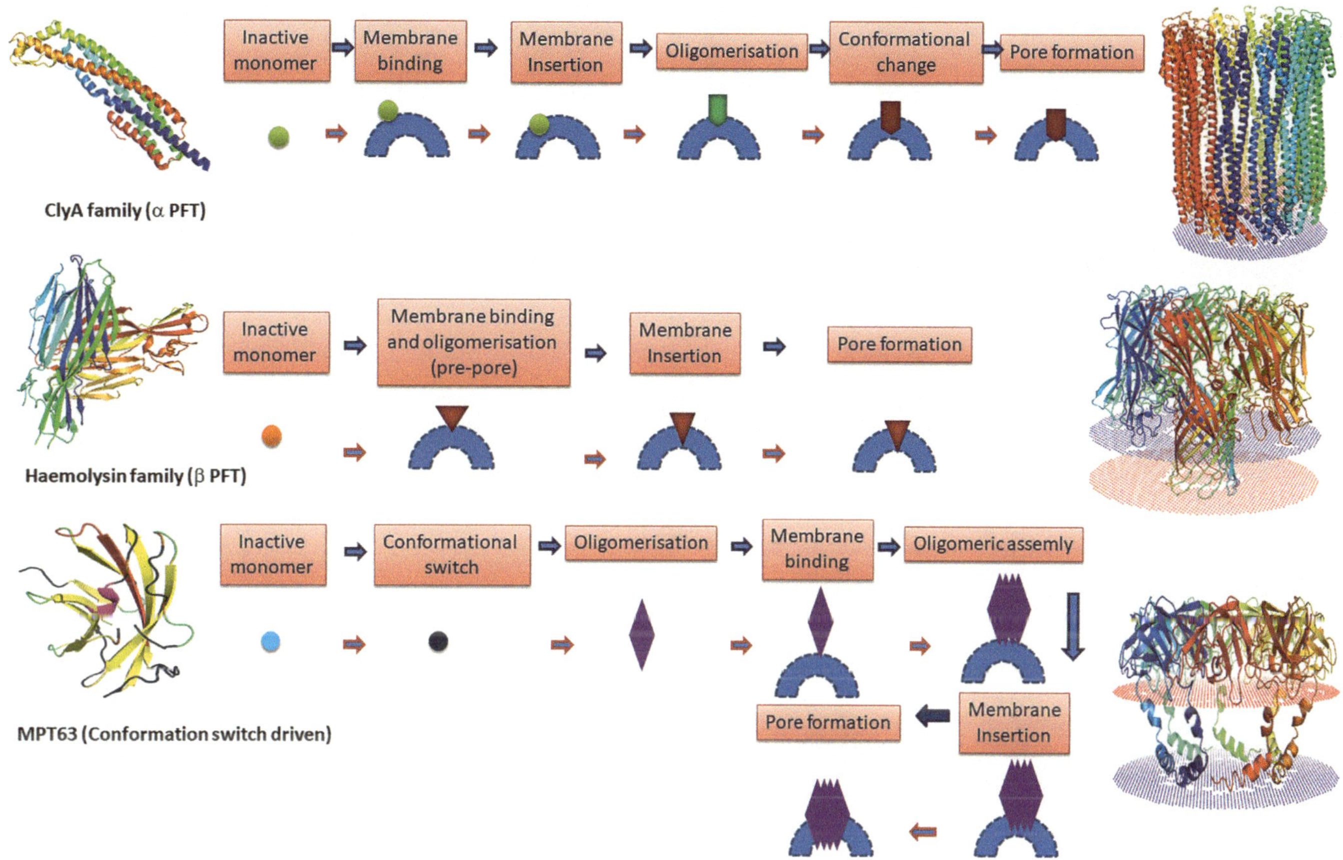

Fig. 2 See figure legend on next page.

biophysical studies which indicated that the several heterotrimers may fuse to create larger channel with respect to the variation of the environmental conditions. All these processes are strongly related with the conformational turnover of the interacting protein species. In some cases, the function of the protein channel is found to be regulated by its known protein partners. For example, SecA protein uses energy from ATP hydrolysis to transport the pre-protein segment through the channel (De Keyzer et al., 2003; Schiebel et al., 1991). On the other hand, the distinguishing between self and non-self is necessary for the adaptive immune system to defend against invading pathogens while avoiding auto immunity. Type-III CRISPR cas systems employ guide RNA to recognize complementary RNA targets which trigger the degradation of both the invader's transcripts and their template DNA (Wang et al., 2019). Wang et al. studied the interactions of various RNA substrates with type-III-A-ribo-nucleoprotein complex and found that Cas-10-the DNase-effector of the complex undergoes rapid conformational fluctuations while target the foreign RNA (Wang et al., 2019). In contrast, this complex remained locked in the static conformation when it was introduced to self RNA. Thus, the conformational fluctuations of the shape shifting protein species effectively protects from the invaders (Wang et al., 2019). The function of a protein is determined on depending upon the different structures which the protein can adopt. NMR spectroscopy which has very often been employed to detect the conformational heterogeneity of the protein through limited ability to characterize the excited states of the protein has made it challenging to explore the connection between the protein functions with its conformational heterogeneity (Wüthrich & Wagner, 1978). The capability of a protein to shift between different ensemble of different structures can give rise to the communication between distant regions of a protein which is known as allostery (Hilser, 2010). The allostery

Fig. 2 The membrane pore formation mechanism of different kind of PFTs. a ClyA family proteins have been mentioned here as an example of α-PFTs.The solved monomeric soluble structure (PDB ID: 1QOY) and the dodecameric structure (PDB ID: 6MRT) of ClyA are shown in the figure along with the mechanism of pore formation. There is no pre-pore structure in case of ClyA. b The Hemolysin family proteins are given as an example of β-PFTs. The solved monomeric soluble structure of α-Hemolysin (PDB ID: 2QK7) and the oligomeric structure (PDB ID: 7AHL) are shown in the figure along with the mechanism of pore formation. The pre-pore formation is one of the key mechanistic steps during membrane pore formation by β-PFT family proteins. c The mechanism of pore formation by MPT63 (PDB ID: 1LMI) by its conformational switch strategy and the predicted membrane-associated structure (model) has been shown. Here oligomeric assembly is the new mechanistic pathway of pore formation.

communication of a protein occurs through two different pathways. One of these is a classic concerted structural change which suggests that effector binding site and ligand binding site each move between two alternative structures in correlated fashion. Ligand binding stabilizes the one structure of the ligand binding sites and thus stabilize the particular structure of the effector binding site. The second pathway is based on the conformational entropy which suggests the correlation of the ligand binding site with different structural states (disordered/ordered states) that can bind the effector. Ligand binding can stabilize the ordered state (Feher, Durrant, Van Wart, & Amaro, 2014; Greener & Sternberg, 2018).

The study of crystal structure of many proteins presents multiple pockets where small molecule binds easily and shows the path of allostery development. Although, this pathway is not applicable for every case, some proteins do not contain any particular drug binding site. It is interesting to note that protein-shape shifting may populate in the excited state through the modulation of the specific pockets which cannot be resolved through the crystal structure analysis. These specific pockets are called cryptic pockets (Greener & Sternberg, 2018). Those cryptic pockets in protein are allosteric in nature and they can easily overlap with the allosteric network. In this way, they generate their potential value towards drug sites. Small molecule binds to those active sites by allosteric way either through activation or inhibition of the process resulting the modulation of the relative populations of different protein structures.

Despite the allosteric effect, cryptic pockets can be considered as the drug target for therapeutic developments. Generally, these cryptic pockets are formed during the protein-conformational switch mediated protein-protein interactions and this interactions leads to the formation of membrane pore forming oligomers. To develop effective drug molecules, one can target these pockets to block the protein-protein interactions. It is the better way to develop effective molecules which can be used as potential competitive inhibitors. There are a little number of methods available to find out protein's important functional sites with respect to small molecule binding for drug delivery and those active sites can not be identified in many cases by normal crystal structure solving method (Knoverek et al., 2018).

A screening method called tethering has been developed to target a specific site on a protein, such as a cryptic site (Erlanson et al., 2000). To achieve this specificity, the method requires the protein to contain a single cysteine residue near the site of interest. Satisfying this requirement often requires one or more mutations to the protein sequence. The protein is incubated with a library of chemical fragments that are capable of forming a disulfide bond with the cysteine. This disulfide tether localizes the compounds to the site of interest. Importantly, a reducing agent is also introduced along with the fragment library. This reducing agent ensures that fragments bind reversibly. Fragments that have strong noncovalent interactions with the target site will tend to stay bound even when the disulfide tether is broken by the reducing agent, while compounds that only form weak interactions with the protein will tend to dissociate. As a result, one can identify tight binding fragments by mass spectrometry. Tethering has successfully identified a number of novel compounds and pockets (Arkin et al., 2003; Ostrem, Peters, Sos, Wells, & Shokat, 2013). Computer simulations provide an alternative approach to discover cryptic sites (Vajda, Beglov, Wakefield, Egbert, & Whitty, 2018). Other approaches attempt to infer cryptic pockets from available crystal structures (Cimermancic et al., 2016) or use alternate simulation strategies, such as the Rosetta software, to identify excited states with cryptic pockets (Johnson & Karanicolas, 2013). The MSM (Markov state models) also captured correlations between the structure of the cryptic pocket and that of the active site, consistent with the allosteric coupling between these sites. Moreover, the model predicted a multitude of new pockets with allosteric coupling to the active site. To target experimentally verified cryptic pockets, a method called Boltzmann docking has been developed to use MSMs to account for the conformational heterogeneity of the target protein (Hart, Ho, Dutta, Gross, & Bowman, 2016). In Boltzmann docking, a library of compounds is docked against a representative structure from every state of an MSM where a pocket of interest is open. Then compounds are ranked based on their population-weighted average docking score. This approach balances the desire for high-affinity interactions against the cost of stabilizing higher-energy excited states.

6. PFT effects and cellular defense mechanisms

The search for mechanisms of action shared by PFTs has involved predominantly in vitro studies on simplified target systems (lipid bilayers, cultured cells, and primary cells) and studies on the in vivo model involving the nematode *Caenorhabditis elegans* and the *Bacillus thuringiensis* crystal toxin PFT Cry5B. These studies have led to an understanding of the molecular requirements for attack, oligomerization, and pore formation. In addition, a number of important host defense and cell death pathways and membrane repair mechanisms have been identified. Nonetheless, several caveats deserve consideration in interpreting these studies, some of which extend to studies discussed in the section on in vivo PFT effects. First, because these studies use purified toxin, cells may be exposed to artificially high doses of toxin. Such doses may not be physiologically relevant and hence could result in responses that are not reflective of those seen during an infection. Second, purified PFTs may behave differently in isolation compared to their behavior in the presence of their pathogen or additional virulence factors. Third, PFTs can affect expression of other genes in their source bacteria (Tseng et al., 2009) and hence may play roles that extend beyond their cytotoxic properties.

6.1 MAPK pathways

Using *C. elegans* and Cry5B, the first functional molecular PFT defense pathways were identified, involving p38 mitogen-activated protein kinase (MAPK) and the N-terminal (JNK)- like MAPK KGB-1 (Huffman et al., 2004). p38 MAPK was shown to be important in mammalian cells in defense against the *Aeromonas hydrophila* PFT aerolysin and *S. aureus* alpha-toxin (Husmann et al., 2006). p38 activation is seen in vivo with *B. thuringiensis* Cry toxins and in vitro with numerous PFTs, including *S. pneumoniae* pneumolysin (PLY), *S. aureus* alpha-toxin, group A streptococcus (GAS) streptolysin O (SLO), *Bacillus anthracis* anthrolysin O (ALO) (Ratner et al., 2006), *Gardnerella vaginalis* vaginolysin (VLY), *A. hydrophila* aerolysin, *Listeria monocytogenes* listeriolysin O (LLO), and *Lactobacillus iners* inerolysin (ILY). p38 was further shown to be activated by and required for defense against Cry toxin in vivo in the lepidopteran *Manduca sexta* and the dipteran *Aedes aegypti* (Ratner et al., 2006). The activation of p38 in response to PFT is thus evolutionarily strongly conserved. Activity of JNK and extracellular signal regulated kinase (ERK) MAPKs was also identified on several occasions in vitro (*M. tuberculosis* 6-kDa early secretory antigenic target [ESAT-6] (Ganguly et al., 2007), *S. pneumoniae* PLY (Stassen et al., 2003), GAS SLO, *A. hydrophila* aerolysin, and L. *monocytogenes* LLO and in vivo, in *C. elegans* (with *B. thuringiensis* Cry5B) (Kao et al., 2011). Activator protein 1 (AP-1; Fos/Jun), functioning downstream of JNK, is involved in *C. elegans* in defenses against PFTs that form small pores (Cry5B) as well as against PFTs that form large pores (SLO). A role for AP-1 in defense against SLO was confirmed in vitro in mammalian cells. LLO and *B. anthracis* protective antigen (PA; a component of anthrax toxin and a PFT can activate ERK, and both p38 and ERK can function to restore potassium homeostasis in cells damaged by PFTs). Thus, the p38, JNK, and perhaps ERK MAPK pathways are arguably the main mediators of physiological PFT defense pathways.

7. Calcium-dependent membrane repair mechanisms

PFT membrane pores often result in an influx of calcium. Ca^{2+} influx is a known trigger of apoptosis, a PFT response that has been observed in various cell types (Aroian & van der Goot, 2007), and it can affect the vesicle trafficking machinery. GAS SLO-induced calcium influx triggers the exocytosis of lysosomes and extracellular release of the lysosomal enzyme acid sphingomyelinase. Acid sphingomyelinase was found to subsequently induce endocytosis, which contributed to membrane repair (Tam et al., 2010). During endocytosis, PFT pores are taken up into the cells, ubiquitinated, and then, through activity of the ESCRT machinery, targeted to lysosomes for degradation (Corrotte, Fernandes, Tam, & Andrews, 2012). *S. aureus* alpha-toxin also enters cells via endocytosis, is transported via late endosomes, and then disappears from the cells. Alpha-toxin multimers, however, were not broken down in acidic compartments but were expelled from cells via exosome-like vesicles called toxosomes (Husmann et al., 2009). A recent study showed that vesicle trafficking pathways also protect cells against PFTs in vivo. Intoxication of *C. elegans* by Cry5B and *V. cholerae* VCC was found to trigger increased rates of endocytosis in intestinal cells. Loss of either of the two key Rab proteins (RAB-5 and RAB-11), master regulators of early endosome and recycling endosome functions, resulted in significant decreases in Cry5B-induced endocytosis in intestinal cells. Loss of RAB-5 and RAB-11 furthermore resulted in strong hypersensitivity of *C. elegans* to Cry5B, and both were required to restore the integrity of the plasma membranes of intestinal cells following Cry5B attack. This demonstrates a correlation between RAB-5, RAB-11, PFT-induced endocytosis, restoration of plasma membrane integrity, and survival of the whole organism. RAB-11 was additionally found to be required for PFT-induced expulsion of microvilli from the enterocyte cell surface, which is hypothesized to be part of the membrane repair mechanism and is also observed in vitro with other PFTs, including *Vibrio parahaemolyticus* thermostable direct hemolysin and GAS SLO (Los et al., 2011).

8. Therapeutic development

8.1 Therapeutic interventions towards the action of PFTs

Since PFTs are the governing factors behind different life threatening diseases, it is utterly important to design and develop therapeutics to negotiate the pore forming activity of the PFTs and to use as new therapeutic agents.

A number of strategies can be employed to target the PFTs and to reduce their toxic effects. These are:

1. Design of small molecules or engineer antibodies which can effectively bind to the specific site of a specific PFT inhibiting membrane pore formation (e.g., oroxylin A, for the inhibition of alpha hemolysin) (Dong et al., 2013; Qiu et al., 2013).
2. The use of engineered synthetic receptor molecules to impair the receptor mediated binding, pore formation and prevention of the PFT recruitment to the membrane (moraviroe, leukocidin ED, CC-chemolysine receptor type 5, CCR5) (Alonzo et al., 2013; Inoshima et al., 2011).
3. Another antimicrobial strategy that targets PFTs is the use of PFTs with reduced toxicity as recombinant toxoid vaccine that triggers an immune response. The toxin structure and pore architecture can be advantageous to enable rational designing of site directed mutagenesis that reduce host membrane binding and cytotoxic efficiency while introducing minimal immunogenic modification. This is shown recently for the variants Net B and ε-toxin (Bokori-Brown et al., 2014; Fernandes da Costa et al., 2014).
4. One more promising strategy is the use of nanoparticles coated with RBC cell membranes which function as reservoir for PFTs which enables the nontoxic use of wild type PFTs and preserves antigen presentation (Hu, Fang, Luk, & Zhang, 2013; Hu & Zhang, 2014).
5. In addition to the above, it is demonstrated for MPT63 protein that nanoparticle conjugation with WT MPT63 results the immunogenic response in cell line (Fig. 3). In contrast, protein engineering using SDM strategy, the help of conformational switch has been employed for retaining the immunogenicity of the protein even after the conjugation with nanoparticles (Sannigrahi et al., 2018). Although the detailed study on this context is absent, this protein engineering technique would be one of the most promising approaches towards the design and development of potential vaccine candidate.

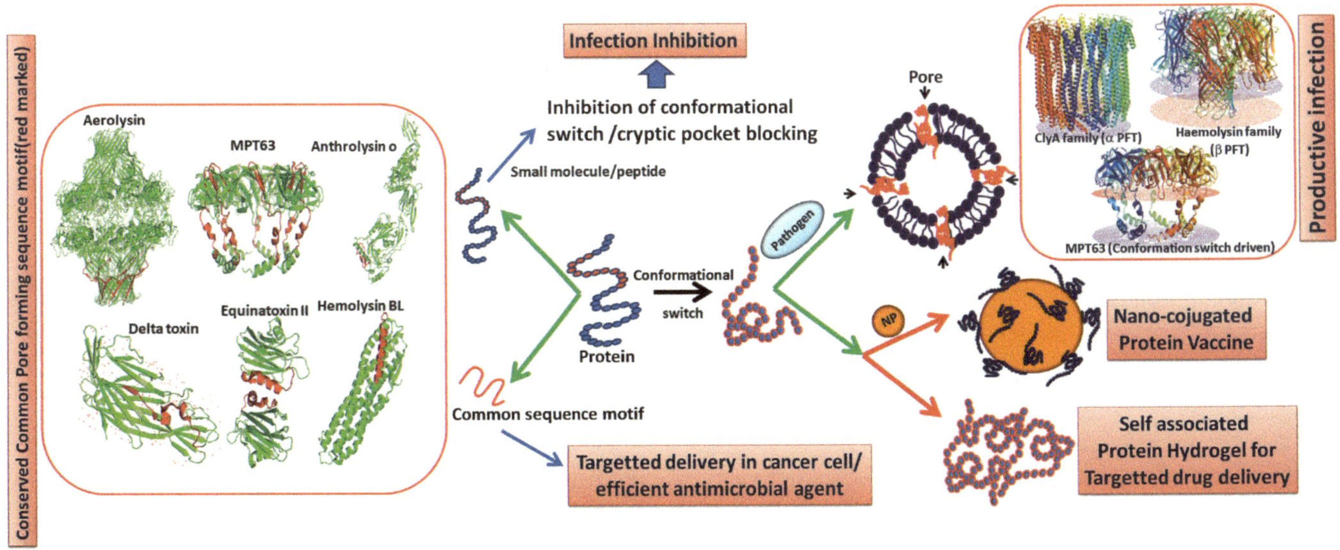

Fig. 3 The schematic representation shows the conformational switch of a protein. There is common sequence motif in different membrane pore forming proteins (PFTs). Pathogen utilizes this conformational switch strategy to evade the host cell membranes. The common motif would be useful for designing appropriate drug candidates, antibody/small molecule/peptide blocker to inhibit various pathogenic infections. The membrane associated structures were obtained using OPM server (http:/opm.phar.umich.edu/).

6. Most interestingly, the conserved domain of all alpha-PFTs and beta PFTs has been identified from the study of MPT63 protein. Therefore the sequence responsible for pore formation should be targeted and a number mutations and design of inhibitory small molecules/nanoparticles or small peptides may be the promising approaches towards the efficient reduction in the activity of PFTs.
7. Finally, it can be suggested that adverse nature of the PFTs can be utilized as antimicrobial agents (Fig. 3) and also can be exploited to promote malignant cell death through pore formation strategy (Hu et al., 2013; Walther et al., 2012; Yang et al., 2006).

9. Conclusion

Pore-forming proteins (PFPs) exist in virtually all domains of life, and by disrupting cellular membranes, depending on the pore size, they cause ion dis-balance, small substances, or even protein efflux/influx, influencing cell's signaling routes and fate. Such pore-forming proteins exist from bacteria to viruses and also shape host defense systems, including innate immunity. There exists a nice structural connection between the functional states of different pathogen-derived proteins. Based on the membrane interacting regions, PFTs are classified into three major classes which are α-PFT, β-PFT and CDC. Nevertheless, there exists another type of PFT which utilizes the conformational switch strategy for membrane pore formation. These types of protein molecules contain specific chameleon sequences which enable them to switch from their inactive to active states during the process of infection. Several viral systems use the fusion domains of their envelope proteins for viral entry. We have shown that there exists a specific ambiguous binary sequence pattern in the pathogen derived protein sequences. These specific domains may remain conserved in multiple pore-forming proteins that utilize environment-sensitive conformational switch strategy. Finally, we have discussed how the conformational switch strategy can be utilized towards novel therapeutics developments.

Reference

Siromolot, A., Oliinyk, O., Kolibo, D., & Komisarenko, S. (2016). Mycobacterium tuberculosis antigens MPT63 and MPT83 increase phagocytic activity of murine peritoneal macrophages. *The Ukrainian Biochemical Journal*, *88*, 62–70.
Abrami, L., Fivaz, M., & van der Goot, F. G. (2000). Adventures of a pore-forming toxin at the target cell surface. *Trends in Microbiology*, *8*, 168–172.

Alexander, P. A., Rozak, D. A., Orban, J., & Bryan, P. N. (2005). Directed evolution of highly homologous proteins with different folds by phage display: Implications for the protein folding code. *Biochemistry, 44*, 14045–14054.

Alonzo, F., III, & Torres, V. J. (2014). The bicomponent pore-forming leucocidins of Staphylococcus aureus. *Microbiology and Molecular Biology Reviews, 78*, 199–230.

Alonzo, F., III, et al. (2013). CCR5 is a receptor for Staphylococcus aureus leukotoxin ED. *Nature, 493*, 51–55.

Alouf, J. (2003). Molecular features of the cytolytic pore-forming bacterial protein toxins. *Folia Microbiologica, 48*, 5.

Alouf, J. E. (2001). Pore-forming bacterial protein toxins: An overview. *Pore-Forming Toxins, 257*, 1–14.

Alves, G. G., de Ávila, R. A. M., Chávez-Olórtegui, C. D., & Lobato, F. C. F. (2014). Clostridium perfringens epsilon toxin: The third most potent bacterial toxin known. *Anaerobe, 30*, 102–107.

Aly, S., et al. (2006). Oxygen status of lung granulomas in Mycobacterium tuberculosis-infected mice. *The Journal of Pathology: A Journal of the Pathological Society of Great Britain and Ireland, 210*, 298–305.

Anderluh, G., et al. (2003). Pore formation by equinatoxin II, a eukaryotic protein toxin, occurs by induction of nonlamellar lipid structures. *Journal of Biological Chemistry, 278*, 45216–45223.

Arkin, M. R., et al. (2003). Binding of small molecules to an adaptive protein–protein interface. *Proceedings of the National Academy of Sciences of the United States of America, 100*, 1603–1608.

Aroian, R., & Van Der Goot, F. (2007). Pore-forming toxins and cellular non-immune defenses (CNIDs). *Current Opinion in Microbiology, 10*, 57–61.

Athanasiadis, A., Anderluh, G., Maček, P., & Turk, D. (2001). Crystal structure of the soluble form of equinatoxin II, a pore-forming toxin from the sea anemone Actinia equina. *Structure, 9*, 341–346.

Barta, M. L., et al. (2012). The structures of coiled-coil domains from type III secretion system translocators reveal homology to pore-forming toxins. *Journal of Molecular Biology, 417*, 395–405.

Beckmann, R., et al. (2001). Architecture of the protein-conducting channel associated with the translating 80S ribosome. *Cell, 107*, 361–372.

Benach, J., & Hunt, J. F. (2004). Cell biology: Shape-shifting protein channel. *Nature, 427*, 24.

Bischofberger, M., Gonzalez, M. R., & van der Goot, F. G. (2009). Membrane injury by pore-forming proteins. *Current Opinion in Cell Biology, 21*, 589–595.

Bischofberger, M., Iacovache, I., & Van Der Goot, F. G. (2012). Pathogenic pore-forming proteins: Function and host response. *Cell Host & Microbe, 12*, 266–275.

Blobel, G. (1980). Intracellular protein topogenesis. *Proceedings of the National Academy of Sciences of the United States of America, 77*, 1496–1500.

Bokori-Brown, M., et al. (2014). Clostridium perfringens epsilon toxin mutant Y30A-Y196A as a recombinant vaccine candidate against enterotoxemia. *Vaccine, 32*, 2682–2687.

Briggs, D. C., et al. (2011). Structure of the food-poisoning Clostridium perfringens enterotoxin reveals similarity to the aerolysin-like pore-forming toxins. *Journal of Molecular Biology, 413*, 138–149.

Carr, C. M., Chaudhry, C., & Kim, P. S. (1997). Influenza hemagglutinin is spring-loaded by a metastable native conformation. *Proceedings of the National Academy of Sciences of the United States of America, 94*, 14306–14313.

Cascales, E., et al. (2007). Colicin biology. *Microbiology and Molecular Biology Reviews, 71*, 158–229.

Choe, S., et al. (1992). The crystal structure of diphtheria toxin. *Nature*, *357*, 216–222.
Cimermancic, P., et al. (2016). CryptoSite: Expanding the druggable proteome by characterization and prediction of cryptic binding sites. *Journal of Molecular Biology*, *428*, 709–719.
Cordes, M. H., Burton, R. E., Walsh, N. P., McKnight, C. J , & Sauer, R. T. (2000). An evolutionary bridge to a new protein fold. *Nature Structural & Molecular Biology*, *7*, 1129.
Cordes, M. H., Walsh, N. P., McKnight, C. J., & Sauer, R. T. (1999). Evolution of a protein fold in vitro. *Science*, *284*, 325–327.
Corrotte, M., Fernandes, M. C., Tam, C., & Andrews, N. W. (2012). Toxin pores endocytosed during plasma membrane repair traffic into the lumen of MVBs for degradation. *Traffic (Copenhagen, Denmark)*, ***13***, 483–494.
Czajkowsky, D. M., Hotze, E. M., Shao, Z., & Tweten, R. K (2004). Vertical collapse of a cytolysin prepore moves its transmembrane beta-hairpins to the membrane. *The EMBO Journal*, *23*, 3206–3215.
Fernandes da Costa, S., et al. (2014). Identification of a key residue for oligomerisation and pore-formation of *Clostridium perfringens* NetB. *Toxins*, *6*, 1049–1061.
Dalal, S., & Regan, L. (2000). Understanding the sequence determinants of conformational switching using protein design. *Protein Science*, *9*, 1651–1659.
De Colibus, L., et al. (2012). Structures of lysenin reveal a shared evolutionary origin for pore-forming proteins and its mode of sphingomyelin recognition. *Structure (London, England : 1993)*, *20*, 1498–1507.
De Keyzer, J., Van Der Does, C., & Driessen, A. (2003). The bacterial translocase: A dynamic protein channel complex. *Cellular and Molecular Life Sciences CMLS*, *60*, 2034–2052.
De, S., & Olson, R. (2011). Crystal structure of the Vibrio cholerae cytolysin heptamer reveals common features among disparate pore-forming toxins. *Proceedings of the National Academy of Sciences of the United States of America*, *108*, 7385–7390.
Degiacomi, M. T., et al. (2013). Molecular assembly of the aerolysin pore reveals a swirling membrane-insertion mechanism. *Nature Chemical Biology*, *9*, 623–629.
Deligianni, E., et al. (2013). A perforin-like protein mediates disruption of the erythrocyte membrane during egress of Plasmodium berghei male gametocytes. *Cellular Microbiology*, *15*, 1438–1455.
Diabate, M., et al. (2015). Escherichia coli α-hemolysin counteracts the anti-virulence innate immune response triggered by the rho GTPase activating toxin CNF1 during bacteremia. *PLoS Pathogens*, *11*, e1004732.
Dong, J., et al. (2013). Oroxylin A inhibits hemolysis via hindering the self-assembly of α-hemolysin heptameric transmembrane pore. *PLoS Computational Biology*, *9*, e1002869.
DuMont, A. L., & Torres, V. J. (2014). Cell targeting by the Staphylococcus aureus pore-forming toxins: It's not just about lipids. *Trends in Microbiology*, *22*, 21–27.
Dunkel, S., Pulagam, L., Steinhoff, H.-J., & Klare, J. (2015). In vivo EPR on spin labeled colicin a reveals an oligomeric assembly of the pore-forming domain in E. coli membranes. *Physical Chemistry Chemical Physics*, *17*, 4875–4878.
Duong, F. (2003). Binding, activation and dissociation of the dimeric SecA ATPase at the dimeric SecYEG translocase. *The EMBO Journal*, *22*, 4375–4384.
Elluri, S., et al. (2014). Outer membrane vesicles mediate transport of biologically active Vibrio cholerae cytolysin (VCC) from V. cholerae strains. *PLoS One*, *9*(e106731).
Erlanson, D. A., et al. (2000). Site-directed ligand discovery. *Proceedings of the National Academy of Sciences of the United States of America*, *97*, 9367–9372.
Fahie, M., et al. (2013). A non-classical assembly pathway of Escherichia coli pore-forming toxin cytolysin A. *Journal of Biological Chemistry*, *288*, 31042–31051.
Feher, V. A., Durrant, J. D., Van Wart, A. T., & Amaro, R. E. (2014). Computational approaches to mapping allosteric pathways. *Current Opinion in Structural Biology*, *25*, 98–103.

Ganash, M., et al. (2013). Structure of the NheA component of the Nhe toxin from Bacillus cereus: Implications for function. *PLoS One*, *8*, e74748.

Ganguly, N., et al. (2007). Mycobacterium tuberculosis 6-kDa early secreted antigenic target (ESAT-6) protein downregulates lipopolysaccharide induced c-myc expression by modulating the extracellular signal regulated kinases 1/2. *BMC Immunology*, *8*, 24.

García-Sáez, A. J., Fuertes, G., Suckale, J., & Salgado, J. (2010). Permeabilization of the outer mitochondrial membrane by Bcl-2 proteins. *Proteins Membrane Binding and Pore Formation*, *677*, 91–105.

Geny, B., & Popoff, M. R. (2006). Bacterial protein toxins and lipids: Pore formation or toxin entry into cells. *Biology of the Cell*, *98*, 667–678.

Ghosh, R., Mukherjee, M., Chattopadhyay, K., & Ghosh, S. (2012). Unusual optical resolution of all four tryptophan residues in MPT63 protein by phosphorescence spectroscopy: Assignment and significance. *The Journal of Physical Chemistry B*, *116*, 12489–12500.

Gonzalez, M., Bischofberger, M., Pernot, L., Van Der Goot, F., & Freche, B. (2008). Bacterial pore-forming toxins: The (w) hole story? *Cellular and Molecular Life Sciences*, *65*, 493–507.

Goulding, C. W., et al. (2002). Crystal structure of a major secreted protein of Mycobacterium tuberculosis—MPT63 at 1.5-Å resolution. *Protein Science*, *11*, 2887–2893.

Greener, J. G., & Sternberg, M. J. (2018). Structure-based prediction of protein allostery. *Current Opinion in Structural Biology*, *50*, 1–8.

Greig, S. L., Radjainia, M., & Mitra, A. K. (2009). Oligomeric structure of colicin IA channel in lipid bilayer membranes. *Journal of Biological Chemistry*, *284*, 16126–16134.

Guirado, E., & Schlesinger, L. (2013). Modeling the Mycobacterium tuberculosis granuloma—The critical battlefield in host immunity and disease. *Frontiers in Immunology*, *4*, 98.

Hamman, B. D., Chen, J.-C., Johnson, E. E., & Johnson, A. E. (1997). The aqueous pore through the translocon has a diameter of 40–60 Å during cotranslational protein translocation at the ER membrane. *Cell*, *89*, 535–544.

Hart, K. M., Ho, C. M., Dutta, S., Gross, M. L., & Bowman, G. R. (2016). Modelling proteins' hidden conformations to predict antibiotic resistance. *Nature Communications*, *7*, 12965.

He, Y., Yeh, D. C., Alexander, P., Bryan, P. N., & Orban, J. (2005). Solution NMR structures of IgG binding domains with artificially evolved high levels of sequence identity but different folds. *Biochemistry*, *44*, 14055–14061.

Hilser, V. J. (2010). An ensemble view of allostery. *Science*, *327*, 653–654.

Hinds, M. G., Zhang, W., Anderluh, G., Hansen, P. E., & Norton, R. S. (2002). Solution structure of the eukaryotic pore-forming cytolysin equinatoxin II: Implications for pore formation. *Journal of Molecular Biology*, *315*, 1219–1229.

Honda, T., Ni, Y., Miwatani, T., Adachi, T., & Kim, J. (1992). The thermostable direct hemolysin of Vibrio parahaemolyticus is a pore-forming toxin. *Canadian Journal of Microbiology*, *38*, 1175–1180.

Hotze, E. M., & Tweten, R. K. (2012). Membrane assembly of the cholesterol-dependent cytolysin pore complex. *Biochimica et Biophysica Acta*, *1818*, 1028–1038.

Hotze, E. M., et al. (2013). Identification and characterization of the first cholesterol-dependent cytolysins from Gram-negative bacteria. *Infection and Immunity*, *81*, 216–225.

Howard, S. P., & Buckley, J. T. (1985). Activation of the hole-forming toxin aerolysin by extracellular processing. *Journal of Bacteriology*, *163*, 336–340.

Hu, C.-M. J., Fang, R. H., Luk, B. T., & Zhang, L. (2013). Nanoparticle-detained toxins for safe and effective vaccination. *Nature Nanotechnology*, *8*, 933.

Hu, C.-M. J., & Zhang, L. (2014). Nanotoxoid vaccines. *Nano Today*, *9*, 401–404.

Huffman, D. L., et al. (2004). Mitogen-activated protein kinase pathways defend against bacterial pore-forming toxins. *Proceedings of the National Academy of Sciences of the United States of America, 101*, 10995–11000.
Hunt, S., Green, J., & Artymiuk, P. J. (2010). Hemolysin E (HlyE, ClyA, SheA) and related toxins. *Proteins Membrane Binding and Pore Formation, 677*, 116–126.
Huntington, J. A. (2006). Shape-shifting serpins–Advantages of a mobile mechanism. *Trends in Biochemical Sciences, 31*, 427–435.
Husmann, M., et al. (2006). Differential role of p38 mitogen activated protein kinase for cellular recovery from attack by pore-forming S. aureus alpha-toxin or streptolysin O. *Biochemical and Biophysical Research Communications, 344*, 1128–1134.
Husmann, M., et al. (2009). Elimination of a bacterial pore-forming toxin by sequential endocytosis and exocytosis. *FEBS Letters, 583*, 337–344.
Huyet, J., et al. (2013). Structural insights into Clostridium perfringens delta toxin pore formation. *PLoS One, 8*, e66673.
Iacovache, I., Bischofberger, M., & van der Goot, F. G. (2010). Structure and assembly of pore-forming proteins. *Current Opinion in Structural Biology, 20*, 241–246.
Iacovache, I., van der Goot, F. G., & Pernot, L. (2008). Pore formation: An ancient yet complex form of attack. *Biochimica et Biophysica Acta (BBA)-Biomembranes, 1778*, 1611–1623.
Iacovache, I., et al. (2006). A rivet model for channel formation by aerolysin-like pore-forming toxins. *The EMBO Journal, 25*, 457–466.
Iacovache, I., et al. (2011). Dual chaperone role of the C-terminal propeptide in folding and oligomerization of the pore-forming toxin aerolysin. *PLoS Pathogens, 7*, e1002135.
Inoshima, I., et al. (2011). A Staphylococcus aureus pore-forming toxin subverts the activity of ADAM10 to cause lethal infection in mice. *Nature Medicine, 17*, 1310.
Jayasinghe, L., & Bayley, H. (2005). The leukocidin pore: Evidence for an octamer with four LukF subunits and four LukS subunits alternating around a central axis. *Protein Science, 14*, 2550–2561.
Jeßberger, N., Dietrich, R., Bock, S., Didier, A., & Märtlbauer, E. (2014). Bacillus cereus enterotoxins act as major virulence factors and exhibit distinct cytotoxicity to different human cell lines. *Toxicon, 77*, 49–57.
Jiang, J., Pentelute, B. L., Collier, R. J., & Zhou, Z. H. (2015). Atomic structure of anthrax protective antigen pore elucidates toxin translocation. *Nature, 521*, 545–549.
Johnson, D. K., & Karanicolas, J. (2013). Druggable protein interaction sites are more predisposed to surface pocket formation than the rest of the protein surface. *PLoS Computational Biology, 9*, e1002951.
Kao, C. Y., et al. (2011). Global functional analyses of cellular responses to pore-forming toxins. *PLoS Pathogens, 7*, e1001314.
Kim, M. J., et al. (2010). Caseation of human tuberculosis granulomas correlates with elevated host lipid metabolism. *EMBO Molecular Medicine, 2*, 258–274.
Kitadokoro, K., et al. (2011). Crystal structure of Clostridium perfringens enterotoxin displays features of beta-pore-forming toxins. *The Journal of Biological Chemistry, 286*, 19549–19555.
Knoverek, C. R., Amarasinghe, G. K., & Bowman, G. R. (2018). Advanced methods for accessing protein shape-shifting present new therapeutic opportunities. *Trends in Biochemical Sciences, 44*, 351–364.
Krantz, B. A., et al. (2005). A phenylalanine clamp catalyzes protein translocation through the anthrax toxin pore. *Science, 309*, 777–781.
Kristan, K.Č., Viero, G., Dalla Serra, M., Maček, P., & Anderluh, G. (2009). Molecular mechanism of pore formation by actinoporins. *Toxicon, 54*, 1125–1134.
Kundu, A., Kundu, S., & Chattopadhyay, K. (2017). The presence of non-native helical structure in the unfolding of a beta-sheet protein MPT63. *Protein Science, 26*, 536–549.

Kuznetsov, I. B., & Rackovsky, S. (2003). On the properties and sequence context of structurally ambivalent fragments in proteins. *Protein Science, 12*, 2420–2433.
Lakey, J. H., van der Goot, F. G., & Pattus, F. (1994). All in the family: The toxic activity of pore-forming colicins. *Toxicology, 87*, 85–108.
Lawrence, S. H., et al. (2008). Shape shifting leads to small-molecule allosteric drug discovery. *Chemistry & Biology, 15*, 586–596.
Leung, C., et al. (2014). Stepwise visualization of membrane pore formation by suilysin, a bacterial cholesterol-dependent cytolysin. *eLife, 3*, e04247.
Los, F. C., Randis, T. M., Aroian, R. V., & Ratner, A. J. (2013). Role of pore-forming toxins in bacterial infectious diseases. *Microbiology and Molecular Biology Reviews, 77*, 173–207.
Los, F. C., et al. (2011). RAB-5- and RAB-11-dependent vesicle-trafficking pathways are required for plasma membrane repair after attack by bacterial pore-forming toxin. *Cell Host & Microbe, 9*, 147–157.
Madegowda, M., Eswaramoorthy, S., Burley, S. K., & Swaminathan, S. (2008). X-ray crystal structure of the B component of hemolysin BL from Bacillus cereus. *Proteins: Structure, Function, and Bioinformatics, 71*, 534–540.
Malovrh, P., et al. (2003). A novel mechanism of pore formation membrane penetration by the N-terminal amphipathic region of equinatoxin. *Journal of Biological Chemistry, 278*, 22678–22685.
Manca, C., et al. (1997). Molecular cloning, purification, and serological characterization of MPT63, a novel antigen secreted by Mycobacterium tuberculosis. *Infection and Immunity, 65*, 16–23.
Mechaly, A. E., et al. (2011). Structural insights into the oligomerization and architecture of eukaryotic membrane pore-forming toxins. *Structure, 19*, 181–191.
Melton, J. A., Parker, M. W., Rossjohn, J., Buckley, J. T., & Tweten, R. K. (2004). The identification and structure of the membrane-spanning domain of the Clostridium septicum alpha toxin. *The Journal of Biological Chemistry, 279*, 14315–14322.
Meusch, D., et al. (2014). Mechanism of Tc toxin action revealed in molecular detail. *Nature, 508*, 61–65.
Mihara, H., & Takahashi, Y. (1997). Engineering peptides and proteins that undergo α-to-β transitions. *Current Opinion in Structural Biology*, 7, 501–508.
Minor, D. L., & Kim, P. S. (1996). Context-dependent secondary structure formation of a designed protein sequence. *Nature, 380*, 730.
Mueller, M., Grauschopf, U., Maier, T., Glockshuber, R., & Ban, N. (2009). The structure of a cytolytic α-helical toxin pore reveals its assembly mechanism. *Nature, 459*, 726–730.
Mustafa, A. (2009). Th1 cell reactivity and HLA-DR binding prediction for promiscuous recognition of MPT63 (Rv1926c), a major secreted protein of Mycobacterium tuberculosis. *Scandinavian Journal of Immunology, 69*, 213–222.
Olson, R., & Gouaux, E. (2005). Crystal structure of the Vibrio cholerae cytolysin (VCC) pro-toxin and its assembly into a heptameric transmembrane pore. *Journal of Molecular Biology, 350*, 997–1016.
Olson, R., Nariya, H., Yokota, K., Kamio, Y., & Gouaux, E. (1999). Crystal structure of staphylococcal LukF delineates conformational changes accompanying formation of a transmembrane channel. *Nature Structural Biology, 6*, 134–140.
Ostrem, J. M., Peters, U., Sos, M. L., Wells, J. A., & Shokat, K. M. (2013). K-Ras (G12C) inhibitors allosterically control GTP affinity and effector interactions. *Nature, 503*, 548.
Parker, M. W., & Feil, S. C. (2005). Pore-forming protein toxins: From structure to function. *Progress in Biophysics and Molecular Biology, 88*, 91–142.
Parker, M. W., et al. (1994). Structure of the Aeromonas toxin proaerolysin in its water-soluble and membrane-channel states. *Nature, 367*, 292–295.

Popoff, M. R. (2011). Epsilon toxin: A fascinating pore-forming toxin. *The FEBS Journal, 278*, 4602–4615.

Pulagam, L. P., & Steinhoff, H.-J. (2013). Acidic pH-induced membrane insertion of colicin a into E. coli natural lipids probed by site-directed spin labeling. *Journal of Molecular Biology, 425*, 1782–1794.

Qiu, J., et al. (2013). Molecular modeling reveals the novel inhibition mechanism and binding mode of three natural compounds to staphylococcal α-hemolysin. *PLoS One, 8*, e80197.

Ramachandran, R., Tweten, R. K., & Johnson, A. E. (2004). Membrane-dependent conformational changes initiate cholesterol-dependent cytolysin oligomerization and intersubunit beta-strand alignment. *Nature Structural & Molecular Biology, 11*, 697–705.

Ramakrishnan, L. (2012). Revisiting the role of the granuloma in tuberculosis. *Nature Reviews Immunology, 12*, 352.

Rapoport, T. A., Jungnickel, B., & Kutay, U. (1996). Protein transport across the eukaryotic endoplasmic reticulum and bacterial inner membranes. *Annual Review of Biochemistry, 65*, 271–303.

Ratner, A. J., et al. (2006). Epithelial cells are sensitive detectors of bacterial pore-forming toxins. *The Journal of Biological Chemistry, 281*, 12994–12998.

Redchuk, T., et al. (2012). Expression of Mycobacterium tuberculosis proteins MPT63 and MPT83 as a fusion: Purification, refolding and immunological characterization. *Journal of Applied Biomedicine, 10*, 169–176.

Ridleya, H., Johnson, C. L., & Lakey, J. H. (2010). Interfacial interactions of pore-forming colicins. *Proteins Membrane Binding and Pore Formation, 677*, 81–90.

Roiko, M. S., & Carruthers, V. B. (2009). New roles for perforins and proteases in apicomplexan egress. *Cellular Microbiology, 11*, 1444–1452.

Rojko, N., et al. (2013). Membrane damage by an α-helical pore-forming protein, equinatoxin II, proceeds through a succession of ordered steps. *Journal of Biological Chemistry, 288*, 23704–23715.

Rolinck-Werninghaus, C., et al. (2003). The potential of recombinant antigens ESAT-6, MPT63 and mig for specific discrimination of Mycobacterium tuberculosis and M. avium infection. *European Journal of Pediatrics, 162*, 534–536.

Ros, U., et al. (2013). The sticholysin family of pore-forming toxins induces the mixing of lipids in membrane domains. *Biochimica et Biophysica Acta (BBA)-Biomembranes, 1828*, 2757–2762.

Rosado, C. J., et al. (2007). A common fold mediates vertebrate defense and bacterial attack. *Science, 317*, 1548–1551.

Rossjohn, J., Feil, S. C., McKinstry, W. J., Tweten, R. K., & Parker, M. W. (1997). Structure of a cholesterol-binding, thiol-activated cytolysin and a model of its membrane form. *Cell, 89*, 685–692.

Russell, D. G. (2007). Who puts the tubercle in tuberculosis? *Nature Reviews Microbiology, 5*, 39.

Russell, D. G., Cardona, P.-J., Kim, M.-J., Allain, S., & Altare, F. (2009). Foamy macrophages and the progression of the human tuberculosis granuloma. *Nature Immunology, 10*, 943.

Sannigrahi, A., et al. (2018). Nanoparticle induced conformational switch between α-Helix and β-sheet attenuates immunogenic response of MPT63. *Langmuir, 34*, 8807–8817.

Sannigrahi, A., et al. (2019). Conformational switch driven membrane pore formation by Mycobacterium secretory protein MPT63 induces macrophage cell death. *ACS Chemical Biology*.

Saunders, B. M., & Britton, W. J. (2007). Life and death in the granuloma: Immunopathology of tuberculosis. *Immunology and Cell Biology, 85*, 103–111.

Savva, C. G., et al. (2013). Molecular architecture and functional analysis of NetB, a pore-forming toxin from Clostridium perfringens. *Journal of Biological Chemistry, 288*, 3512–3522.
Schiebel, E., Driessen, A. J., Hartl, F.-U., & Wickner, W. (1991). $\Delta\mu H+$ and ATP function at different steps of the catalytic cycle of preprotein translocase. *Cell, 64*, 927–939.
Schreiber, M. P., Chan, C. M., & Shorr, A. F. (2011). Bacteremia in Staphylococcus aureus pneumonia: Outcomes and epidemiology. *Journal of Critical Care, 26*, 395–401.
Shatursky, O., et al. (1999). The mechanism of membrane insertion for a cholesterol-dependent cytolysin: A novel paradigm for pore-forming toxins. *Cell, 99*, 293–299.
Shepard, L. A., et al. (1998). Identification of a membrane-spanning domain of the thiol-activated pore-forming toxin Clostridium perfringens perfringolysin O: An alpha-helical to beta-sheet transition identified by fluorescence spectroscopy. *Biochemistry, 37*, 14563–14574.
Sher, D., et al. (2005). Hydralysins, a new category of beta-pore-forming toxins in cnidaria. *The Journal of Biological Chemistry, 280*, 22847–22855.
Shimaoka, M., et al. (2003). Structures of the αL I domain and its complex with ICAM-1 reveal a shape-shifting pathway for integrin regulation. *Cell, 112*, 99–111.
Slatin, S. L., Qiu, X.-Q., Jakes, K. S., & Finkelstein, A. (1994). Identification of a translocated protein segment in a voltage-dependent channel. *Nature, 371*, 158–161.
Song, L., et al. (1996). Structure of staphylococcal α-hemolysin, a heptameric transmembrane pore. *Science, 274*, 1859–1865.
Sonnen, A. F., Plitzko, J. M., & Gilbert, R. J. (2014). Incomplete pneumolysin oligomers form membrane pores. *Open Biology, 4*, 140044.
Stassen, M., et al. (2003). The streptococcal exotoxin streptolysin O activates mast cells to produce tumor necrosis factor alpha by p38 mitogen-activated protein kinase- and protein kinase C-dependent pathways. *Infection and Immunity, 71*, 6171–6177.
Tam, C., et al. (2010). Exocytosis of acid sphingomyelinase by wounded cells promotes endocytosis and plasma membrane repair. *The Journal of Cell Biology, 189*, 1027–1038.
Tanaka, K., Caaveiro, J. M., Morante, K., González-Mañas, J. M., & Tsumoto, K. (2015). Structural basis for self-assembly of a cytolytic pore lined by protein and lipid. *Nature Communications, 6*, 6337.
Tilley, S. J., Orlova, E. V., Gilbert, R. J., Andrew, P. W., & Saibil, H. R. (2005). Structural basis of pore formation by the bacterial toxin pneumolysin. *Cell, 121*, 247–256.
Tseng, C. W., et al. (2009). Staphylococcus aureus Panton-Valentine leukocidin contributes to inflammation and muscle tissue injury. *PLoS One, 4*, e6387.
Ulrichs, T., & Kaufmann, S. H. (2006). New insights into the function of granulomas in human tuberculosis. *The Journal of Pathology: A Journal of the Pathological Society of Great Britain and Ireland, 208*, 261–269.
Unno, H., Goda, S., & Hatakeyama, T. (2014). Hemolytic lectin CEL-III heptamerizes via a large structural transition from α-helices to a β-barrel during the transmembrane pore formation process. *The Journal of Biological Chemistry, 289*, 12805–12812.
Vaidyanathan, M., Sathyanarayana, P., Maiti, P. K., Visweswariah, S. S., & Ayappa, K. (2014). Lysis dynamics and membrane oligomerization pathways for Cytolysin A (ClyA) pore-forming toxin. *RSC Advances, 4*, 4930–4942.
Vajda, S., Beglov, D., Wakefield, A. E., Egbert, M., & Whitty, A. (2018). Cryptic binding sites on proteins: Definition, detection, and druggability. *Current Opinion in Chemical Biology, 44*, 1–8.
Van den Berg, B., et al. (2004). X-ray structure of a protein-conducting channel. *Nature, 427*, 36.
Van der Goot, F., Gonzalez-Manas, J., Lakey, J., & Pattus, F. (1991). A'molten-globule'membrane-insertion intermediate of the pore-forming domain of colicin a. *Nature, 354*, 408–410.

Wade, K. R., et al. (2015). An intermolecular electrostatic interaction controls the prepore-to-pore transition in a cholesterol-dependent cytolysin. *Proceedings of the National Academy of Sciences of the United States of America, 112*, 2204–2209.
Wallace, A. J., et al. (2000). E. coli hemolysin E (HlyE, ClyA, SheA): X-ray crystal structure of the toxin and observation of membrane pores by electron microscopy. *Cell, 100*, 265–276.
Walther, W., et al. (2012). Novel Clostridium perfringens enterotoxin suicide gene therapy for selective treatment of claudin-3-and-4-overexpressing tumors. *Gene Therapy, 19*, 494.
Wang, L., et al. (2019). Dynamics of Cas10 govern discrimination between self and non-self in type III CRISPR-Cas immunity. *Molecular Cell, 73*, 278–290.e274.
Westphal, D., Dewson, G., Czabotar, P. E., & Kluck, R. M. (2011). Molecular biology of Bax and Bak activation and action. *Biochimica et Biophysica Acta (BBA)-Molecular Cell Research, 1813*, 521–531.
Wüthrich, K., & Wagner, G. (1978). Internal motion in globular proteins. *Trends in Biochemical Sciences, 3*, 227–230.
Xu, Q., et al. (2010). Structure of a membrane-attack complex/perforin (MACPF) family protein from the human gut symbiont Bacteroides thetaiotaomicron. *Acta Crystallographica. Section F, Structural Biology and Crystallization Communications, 66*, 1297–1305.
Yahr, T. L., & Wickner, W. T. (2000). Evaluating the oligomeric state of SecYEG in preprotein translocase. *The EMBO Journal, 19*, 4393–4401.
Yamashita, K., et al. (2011). Crystal structure of the octameric pore of staphylococcal γ-hemolysin reveals the β-barrel pore formation mechanism by two components. *Proceedings of the National Academy of Sciences of the United States of America, 108*, 17314–17319.
Yamashita, D., et al. (2014). Molecular basis of transmembrane beta-barrel formation of staphylococcal pore-forming toxins. *Nature Communications, 5*, 1–9.
Yang, W. S., et al. (2006). Suicide cancer gene therapy using pore-forming toxin, streptolysin O. *Molecular Cancer Therapeutics, 5*, 1610–1619.
Zhao, F., et al. (2014). Comprehensive transcriptome profiling and functional analysis of the frog (Bombina maxima) immune system. *DNA Research: An International Journal for Rapid Publication of Reports on Genes and Genomes, 21*, 1–13.
Zhou, X., Alber, F., Folkers, G., Gonnet, G. H., & Chelvanayagam, G. (2000). An analysis of the helix-to-strand transition between peptides with identical sequence. *Proteins: Structure, Function, and Bioinformatics, 41*, 248–256.

CHAPTER FOUR

Transmembrane β-barrel proteins of bacteria: From structure to function

Simen Hermansen[a], Dirk Linke[a], and Jack C. Leo[b,*]
[a]Section for Genetics and Evolutionary Biology, Department of Biosciences, University of Oslo, Oslo, Norway
[b]Antimicrobial resistance, Omics and Microbiota Group, Department of Biosciences, Nottingham Trent University, Nottingham, United Kingdom
*Corresponding author: e-mail address: jack.leo@ntu.ac.uk

Contents

1. Introduction 114
2. Structure of β-barrel proteins 124
   2.1 General structural features of transmembrane β-barrels 124
   2.2 Residue patterns and structural motifs 126
3. Folding and membrane integration of β-barrels 127
   3.1 In vitro folding 127
   3.2 In vivo biogenesis 132
4. Functions of transmembrane β-barrels 135
   4.1 Porins 136
   4.2 TonB-dependent transporters 137
   4.3 Efflux systems 138
   4.4 Protein secretion 139
   4.5 Enzymes 142
   4.6 Adhesins 143
   4.7 Other functions 144
5. Evolution of β-barrels 145
6. Medical relevance of bacterial β-barrel proteins 146
   6.1 β-barrel proteins and antibiotic resistance 147
   6.2 β-barrel proteins as targets of antimicrobials 148
   6.3 β-barrel proteins as antibody and vaccine targets 149
7. Conclusions and future perspectives 150
Disclaimer 150
References 151

Abstract

The outer membrane of Gram-negative bacteria is a specialized organelle conferring protection to the cell against various environmental stresses and resistance to many harmful compounds. The outer membrane has a number of unique features, including

Advances in Protein Chemistry and Structural Biology, Volume 128
ISSN 1876-1623
https://doi.org/10.1016/bs.apcsb.2021.07.002

an asymmetric lipid bilayer, the presence of lipopolysaccharides and an individual proteome. The vast majority of the integral transmembrane proteins in the outer membrane belongs to the family of β-barrel proteins. These evolutionarily related proteins share a cylindrical, anti-parallel β-sheet core fold spanning the outer membrane. The loops and accessory domains attached to the β-barrel allow for a remarkable versatility in function for these proteins, ranging from diffusion pores and transporters to enzymes and adhesins. We summarize the current knowledge on β-barrel structure and folding and give an overview of their functions, evolution, and potential as drug targets.

1. Introduction

Transmembrane proteins fall into two major classes. The transmembrane domains of the more abundant, and therefore more familiar, class consist of hydrophobic α-helices. These proteins are found in all domains of life, and most transmembrane proteins of both prokaryotes and eukaryotes belong to this class.

In addition, a second major class of transmembrane proteins exists. The transmembrane domain of these proteins is not a continuous hydrophobic stretch as in α-helical membrane proteins but is rather a folded domain that relies on its three-dimensional structure to present a hydrophobic surface to the membrane environment. These are the β-barrel proteins, which have a transmembrane domain formed by an antiparallel β-sheet that closes in on itself, forming a cylinder (Fig. 1). The outside of this cylinder is hydrophobic and faces the lipidic membrane. By contrast, the inside of the cylinder, the lumen of β-barrel, is often hydrophilic and can form an aqueous channel through the membrane or host other structures within it.

Transmembrane β-barrel proteins are found in the outer membranes of Gram-negative bacteria, where they constitute almost all the integral membrane proteins. In addition, β-barrel proteins can be found in the outer membranes of mitochondria and plastids in eukaryotic cells. These organelles are remnants of endosymbiotic Gram-negative bacteria, so the β-barrels present in their outer membranes are evolutionary relics from the ancestral bacterial cells (Archibald, 2015). Furthermore, many pore-forming toxins form β-barrel structures (Dal Peraro & van der Goot, 2016). Here, we focus on β-barrel proteins present in Gram-negative bacterial outer membranes.

The size of bacterial β-barrels can vary. The number of strands of classical β-barrels ranges from 8 to 26 in most Gram-negative bacteria, though larger barrels are found, e.g. in the Bacteroidetes, containing 36 β-strands (Fig. 1). β-barrels can further oligomerize, either to form multimers of individual

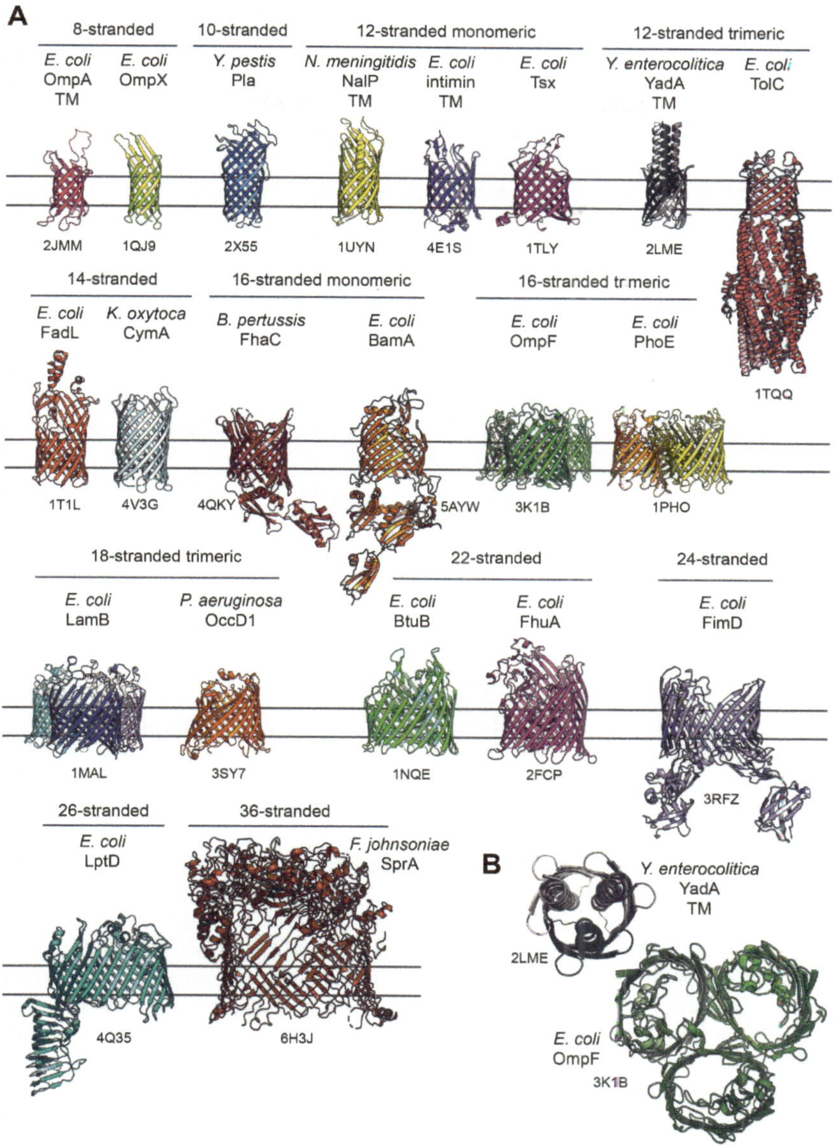

Fig. 1 Examples of classical β-barrel structures. (A) Representative β-barrel structures shown in cartoon representation and viewed from the side. The protein name and source species are given. 8-stranded β-barrels are represented by OmpA and OmpX of *Escherichia coli*, 10-stranded β-barrels by plasminogen activator Pla from *Yersinia pestis*, monomeric 12-stranded β-barrels by NalP from *Neisseria meningitidis* and intimin from *E. coli* (exemplifying type 5a and type 5e secretion systems), and the nucleoside transporter Tsx from *E. coli*. Trimeric 12-stranded β-barrels, where three protomers form a single β-barrel, are exemplified by the *Yersinia* adhesin YadA and *E. coli* TolC; the

(Continued)

β-barrels or a single β-barrel formed by several protomers. An example of the former would be the trimeric porins OmpF and OmpC, whereas the latter are exemplified by the trimeric single β-barrels of TolC and the trimeric autotransporter adhesins such as the *Yersinia* adhesin YadA (Fig. 1).

In bacteria, β-barrel proteins are found in the outer membrane (OM). This membrane is distinct from the inner (plasma)membrane and has several notable characteristics. First, the OM is asymmetric: the inner leaflet of the bilayer consists of phospholipids, but the outer leaflet contains almost exclusively lipopolysaccharides, a type of glycolipid found only in Gram-negative bacteria. Second, the OM is not a semipermeable membrane similar to the inner membrane; rather, it is freely permeable to small hydrophilic molecules but represents a significant diffusion barrier to molecules larger than ~600 Da and many hydrophobic compounds (Vergalli et al., 2020). Third, the lipid:protein ratio of the OM is lower than that of the inner membrane, with the consequence that it is much denser and can therefore be separated from the inner membrane by density gradient centrifugation (Dunstan, Hay, & Lithgow, 2017). The OM is thus crowded by protein with only a few lipid molecules separating the transmembrane proteins, which restricts lateral diffusion of outer membrane proteins and makes the OM quite rigid (Horne et al., 2020; Rassam et al., 2015; Rojas et al., 2018).

Fig. 1—Cont'd protomers are coloured in different shades. 14-stranded β-barrels are represented by the fatty acid transporter FadL from *E. coli* and the cyclodextrin transporter CymA from *Klebsiella oxytoca*. Monomeric 16-stranded β-barrels are represented by FhaC, the two-partner (type 5b) secretion protein for the filamentous haemagglutinin from *Bordetella pertussis*, and the membrane insertase BamA of *E. coli*. Trimeric porins (with the three monomers coloured in different shades) are represented by OmpF and PhoE from *E. coli*. LamB from *E. coli* represents trimeric 18-stranded porins, whereas OccD1 from *Pseudomonas aeruginosa* represents a monomeric 18-stranded β-barrel. 22-stranded TonB-dependent transporters are represented by the vitamin B_{12} importer BtuB and the ferrichrome transporter FhuA, both from *E. coli*. FimD, the usher protein of type 1 fimbriae, is an example of a 24-stranded β-barrel, whereas the lipopolysaccharide exporter LptD exemplifies 26-stranded β-barrels. The largest known classical β-barrel is SprA, the type 9 secretion protein from *Flavobacterium johnsoniae*. The approximate position of the outer membrane is shown with grey lines, with the extracellular side being on top. TM denotes that only the transmembrane domain of the protein is depicted, but the protein would contain addition domains in its native state. The Protein Data Bank identifier for each structure is noted below the protein. (B) Examples of trimeric β-barrels, viewed from the extracellular side to better highlight the differences between proteins that form a single, trimeric β-barrel (exemplified by YadA) and a trimer of three separate β-barrels (exemplified by OmpF). The structures in the figure were produced using PyMOL (Schroedinger). This figure was inspired by Horne, Brockwell, and Radford (2020).

Almost all the transmembrane proteins in the OM are β-barrels. In *E. coli*, only a handful of α-helical proteins are present in the OM (Table 1). An estimated 2–3% of the genes of Gram-negative bacteria code for β-barrel proteins (Wimley, 2003). *E. coli* K-12, the workhorse of molecular biology and the model organism for OM structure and physiology, encodes ~80 putative β-barrel proteins (Table 1). Surprisingly, only two of these are essential for cell viability: BamA and LptD. The former is part of the β-barrel insertion machinery (BAM) and required for insertion of β-barrel proteins into the OM (see Section 3.2) and the latter is needed for lipopolysaccharide transport to the outer leaflet (Botos, Noinaj, & Buchanan, 2017). One reason for the low number of essential β-barrel proteins is the functional redundancy between proteins. For example, the general porins OmpF and OmpC have similar properties and can compensate for each other under many growth conditions (Masi, Winterhalter, & Pagès, 2019).

β-barrels vary in function as well as abundance. OmpA is the most abundant, with over 100,000 copies per cell, followed by the porins OmpF and OmpC and the adhesin OmpX. By contrast, some β-barrels have very low copy numbers or are not expressed at all under standard conditions (Soufi, Krug, Harst, & Macek, 2015). Transmembrane β-barrel proteins play a crucial role in interactions with the environment, and they are involved in processes such as nutrient acquisition, adhesion and biofilm formation, and drug efflux as well as protein secretion (Table 1). In addition, as these proteins are exposed to the surface, they are targets for the immune system and receptors for bacteriophages and bacteriocins.

In addition to the canonical β-barrel proteins described above, it has become clear in recent years that another class of multimeric β-barrels exist. These are associated with protein secretion systems and include the curli secretion pore CsgG and the secretins of type II and type III secretion systems and type IV pili (Goyal et al., 2014; Weaver et al., 2020; Worrall et al., 2016; Yan, Yin, Xu, Zhu, & Li, 2017). While these are β-barrels in the broad sense, they bear several differences to "classical" β-barrels: first, they are much larger than standard β-barrels, with 36 (CsgG) or 56-60 β-strands (secretin). Second, they are not dependent on the BAM for outer membrane insertion but rely on a different mechanism (Dunstan et al., 2015). Similarly, β-barrel toxins are formed by multiple subunits; while most are oligomeric producing 14 or 16-stranded β-barrels, the cholesterol-dependent cytolysins can contain up to 50 subunits and form enormous β-barrels composed of between 50 and 200 β-strands, depending on the toxin (Dal Peraro & van der Goot, 2016).

Table 1 Integral outer membrane proteins of *E. coli* K-12.

Class	Name[a] (UniProt)	UniProt ID	Function (UniProt)	Alternative functions
Classical β-barrels				
8-Stranded				
	OmpA	P0A910	Anchor between peptidoglycan and outer membrane	
	OmpW	P0A915	Uncharacterized	Receptor for colicin S4
	OmpX	P0A917	Uncharacterized	
	PagP [CrcA]	P37001	Lipid A palmitoyltransferase	
10-Stranded				
	OmpP	P34210	Protease	Receptor for bacteriophage Ox2
	OmpT	P09169	Protease	
12-Stranded				
	NanC [YjhA]	P69856	Transport of N-acetylneuraminic acid	
	OmpL [YshA]	P76773	Diffusion pore	
	PldA [OMPLA]	P0A921	Phospholipase A1	
	Tsx [NupA]	P0A927	Transport of nucleosides	
Autotransporters (12-stranded)				
	EaeH [FdeC]	P36943	Type Ve secretion system	Cell adhesion (pseudogene in K-12)
	Flu [Antigen 43]	P39180	Type Va secretion system	Autoaggregation/colony formation
	YaiT	P77199	Type Va secretion system	Uncharacterized (pseudogene in K-12)
	YaiU	P75700	Type Va secretion system	Uncharacterized (pseudogene in K-12)
	YcgI	P76000	Type Va secretion system	Uncharacterized

YcgV	P76017	Type Va secretion system	Uncharacterized (pseudogene in K-12)
YchO	P39165	Type Ve secretion system	Invasin-like
YdbA	P33666	Type Va secretion system	Uncharacterized (pseudogene in K-12)
YdeU	P77286	Type Va secretion system	Uncharacterized
YeeJ	P76347	Type Ve secretion system	Invasin-like
YejO	P33924	Type Va secretion system	Uncharacterized
YfaL	P45508	Type Va secretion system	Uncharacterized
YhjY	P37663	Type Va secretion system	Uncharacterized
YpjA	P52143	Type Va secretion system	Uncharacterized
YuaO [YcbB]	Q9JMS5	Type Va secretion system	Uncharacterized
YuaQ [YchA]	Q9JMS3	Type Va secretion system	Uncharacterized
12-Stranded, trimeric			
CusC	P77211	Copper efflux system component	
MdtP	P32714	Efflux pump component	
MdtQ [YohG]	P33369	Efflux pump component	
TolC	P02930	Efflux pump component	
14-Stranded			
FadL	P10384	Transport of long-chain fatty acids	Receptor for the bacteriophage T2
OmpG	P76045	Diffusion pore	
16-Stranded			
BamA [YaeT]	P0A940	Outer membrane protein insertion	
BscC	P37650	Cellulose export	

Continued

Table 1 Integral outer membrane proteins of *E. coli* K-12.—cont'd

Class	Name[a] (UniProt)	UniProt ID	Function (UniProt)	Alternative functions
	NmpC	P21420	Diffusion pore	Cryptic prophage gene
	OmpC	P06996	Diffusion pore	Receptor for colicin E5 in absence of OmpF
	OmpF	P02931	Diffusion pore	Receptor for the bacteriophage T2, receptor for colicin E5
	OmpN	P77747	Diffusion pore	
	PgaA [YcdS]	P69434	Export of poly-beta-1,6-N-acetyl-D-glucosamine	
	PhoE	P02932	Transport of phosphate and phosphorylated compounds	
	TamA	P0ADE4	Outer mebrane protein insertion (autotransporters)	
18-Stranded				
	BglH [YieC]	P26218	Transport of sugars	
	LamB [Maltoporin]	P02943	Transport of maltose	Receptor for bacteriophage λ
22-Stranded				
	BtuB	P06129	Transport of vitamin B12	Receptor for bacteriophages BF23 and C1, and for colicin A and E
	CirA	P17315	Transport of iron	Receptor for colicins IA and IB
	FecA	P13036	Transport of Fe^{3+} dicitrate	
	FepA	P05825	Transport of iron siderophore (ferrienterobactin)	Receptor for colicins B and D

FhuA	P06971	Transport of iron siderophore (ferrichrome)	Transports the antibiotic albomycin, and acts as a receptor for colicin M, microcin J25 and bacteriophages T1, T5, phi80 and UC-1
FhuE	P16869	Transport of iron siderophore (coprogen, ferrioxamine B, and rhodotorulic acid)	
Fiu [YbiL]	P75780	Transport of iron siderophore (catecholate)	Receptor for microcins M, H47 and E492
YddB	P31827	Uncharacterized TonB-dependent receptor	Probably involved in iron transport
YncD	P76115	Uncharacterized TonB-dependent receptor	
24-Stranded			
EcpC	P77802	Fimbrial usher protein	
ElfC	P75857	Fimbrial usher protein	
FimD	P30130	Fimbrial usher protein	
HtrE	P33129	Fimbrial usher protein	
SfmD	P77468	Fimbrial usher protein	
YbgQ	P75750	Fimbrial usher protein	
YdeT	P76137	Fimbrial usher protein	
YehB	P33341	Fimbrial usher protein	
YfcU	P77196	Fimbrial usher protein	
YhcD	P45420	Fimbrial usher protein	
YqiG	P76655	Fimbrial usher protein	
YraJ	P42915	Fimbrial usher protein	

Continued

Table 1 Integral outer membrane proteins of *E. coli* K-12.—cont'd

Class	Name[a] (UniProt)	UniProt ID	Function (UniProt)	Alternative functions
26-Stranded				
	LptD [OstA]	P31554	LPS assembly	Confers organic solvent tolerance
Unknown number of strands				
	ChiP	P75733 [YbfM]	Transporter of chito-sugars	
	GfcD	P75882	Uncharacterized	Possible role in capsule biosynthesis
	MipA	P0A908	Scaffold protein for peptidoglycan synthesis complex	Role in antibiotic resistance
	NfrA	P31600	Uncharacterized	Allows N4 phage attachment
	UidC	Q47706	Transport of glucuronide	
	YaiO	Q47534	Uncharacterized	
	YdiY	P76206	Uncharacterized	
	YfaZ	P76471	Uncharacterized	
	YfeN	P45564	Uncharacterized	
	YgjJ	P42591	Uncharacterized	
	YiaT	P37681	Uncharacterized	
	YjbH	P32689	Uncharacterized	
Pseudogenes and gene fragments				
	YoeA	P76358	Fragment of possible siderophore receptor	
	YddL	P77519	Fragment of possible porin	
	YedS	P76335	Diffusion pore	Role in resistance to carbapenems

Non-classical β-barrels		
CsgG	P0AEA2	Curli assembly/transport
GspD	P45758	Type II secretion system secretin
FlgH	P0A6S0	Flagellar L-ring protein
α-helical membrane pores		
GfcE	P0A932	Putative polysaccharide export protein
HlyE	P77335	Hemolysin E
Wza	P0A930	involved in exopolysaccharide biosynthesis/ transport

[a]Names derived from UniProt. Alternative names are given in square brackets.
This is based on the dataset provided by BOMP (Berven, Flikka, Jensen, & Eidhammer, 2004) following manual curation. Additional *E. coli* outer membrane proteins missing from this dataset were searched for in UniProt with the terms "*Escherichia coli* (strain K12) [83333]," "outer membrane" AND reviewed: yes. Overlaps with the BOMP dataset were included automatically; others were checked manually using BLAST or HHPred (Söding, Biegert, & Lupas, 2005). Information about function and strand number were extracted from UniProt where available, otherwise the strand number was estimated based on homology searches.

In the following, we will discuss the structure of "classical" bacterial β-barrel proteins in more detail and describe how the proteins fold both in vitro and in vivo. In addition, we give an overview of the possible functions of β-barrels as well as briefly cover the evolution and medical relevance of these proteins.

2. Structure of β-barrel proteins

2.1 General structural features of transmembrane β-barrels

The majority of integral outer membrane proteins share a common antiparallel β-barrel conformation (Koebnik, Locher, & Van Gelder, 2000). β-barrels are β-sheets, where the N- and C-terminal β-strands of the sheet bind together to close the sheet into a cylindrical structure. Neighbouring β-strands are connected to each other by hydrogen bonding between the amide hydrogens and carbonyl oxygens of the protein backbone. All classified membrane-integral β-barrels in bacteria are anti-parallel where neighbouring strands are alternately connected on each side of the OM by long loops on the extracellular side and by shorter turns on the periplasmic side (Kleinschmidt, 2015) (Fig. 2A).

Two parameters, the strand number (n) and the shear (S) can describe the tertiary architecture of β-barrels even though the barrels themselves might vary substantially in their primary structure and function. The strand number designates the number of β-strands that make up the barrel and is directly related to the size of the β-barrel (Fig. 1). The shear indicates how tilted the β-strands are relative to the lateral axis of the β-barrel (Murzin, Lesk, & Chothia, 1994). One can calculate S by tracing hydrogen-bonded α-carbons, facing the same direction, around the barrel from strand to strand, until the starting strand has been reached. S is then defined as the offset in residues between the final and starting residue (McLachlan, 1979) (Fig. 2B and C). If S is positive the β-barrel is said to be right-handed. β-barrels with negative shears (left-handed) have not been observed in nature (Schulz, 2002).

The strand number of β-barrels from Gram-negative bacteria is always even and typically ranges between 8 and 26 strands (Fig. 1 and Table 1). The even strand number means that the C-terminal strand of the protein is always pointed toward the periplasm, and both the C- and N-termini of the β-barrel domain reside in the periplasm. A β-barrel with $S = 0$ would have β-strands that are completely perpendicular relative to the plane of the OM.

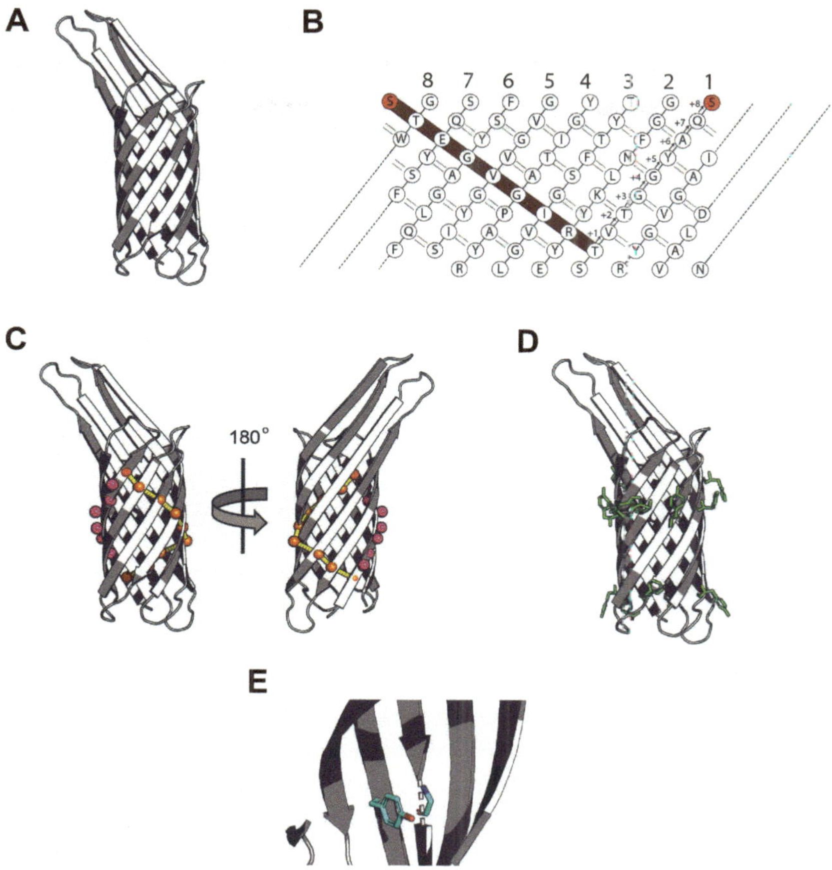

Fig. 2 β-barrel structure. (A) The structure of β-barrels is represented by the 8-stranded *E. coli* protein OmpX (PDB ID: 1QJ9). The extracellular side is toward the top. (B) Schematic of how the shear number (*S*) is calculated. The shear number of Ompx is found by crossing from the starting residue (in red) to the residue in the neighbouring β-strand along the direction of the hydrogen bonds (dotted double lines; path highlighted in brown) until arriving at the original strand. The offset in residues (demonstrated by the small arrows) corresponds to *S*. In the case of OmpX, $S=8$. (C) The shear number demonstrated on the structure of OmpX. The starting residue is highlighted as a red sphere, neighbouring residues are shown as orange spheres and the pathway is highlighted in yellow. To demonstrate the offset, the residues in the β-strand with the starting residue are shown in pink. (D) The aromatic girdle of OmpX. Aromatic residues (Phe, Tyr, Trp) are shown in stick representation and coloured green. (E) The mortise and tenon motif of OmpX, where the lumenal Tyr80 side chain folds over Gly112. The structures in the figure were prepared using PyMOL (Schroedinger). (For interpretation of the references to colour in this figure legend, the reader is referred to the web version of this article.)

Most OM β-barrels have a shear typically between n and $2n$. The optimal relationship between the shear and strand number for these barrels have been estimated to be $S = n + 4$ (Murzin et al., 1994). It has been suggested that the shear of β-barrels is required to reduce steric hindrance between the inward facing residues and that the right-handed twist improves stability (Sansom & Kerr, 1995).

2.2 Residue patterns and structural motifs

2.2.1 The transmembrane β-strands

The individual β-strands form as a result of alternating hydrophobic and hydrophilic residues in a dyad repeat pattern (Wimley, 2002). The hydrophobic residues are partitioned into the hydrophobic environment of the membrane, while the soluble residues point in the opposite direction, into the barrel lumen. β-barrels therefore have a highly hydrophobic lipid facing outer surface and a more polar inner surface that lines the barrel lumen. The structural role of the outward facing hydrophobic residues is reflected in their conservation pattern, where the outward facing hydrophobic residues are more frequently substituted compared to the inward facing residues. The hydrophobic residues interact with the hydrophobic core of the membrane. The major selective constraint on these residues lies in their hydrophobicity so they may be substituted with alternative hydrophobic residues without jeopardizing the structural integrity of the barrel (Stapleton, Whitehead, & Nanda, 2015). The inward facing residues are generally more conserved as they have important functional roles such as interacting with substrates passing through the barrel lumen and forming salt bridges important for the structural integrity of the β-barrel (Rath, Sharpe, & Hiller, 2020).

Aromatic residues have several important roles in transmembrane β-barrel domains. Aromatic residues are enriched at the membrane-water interfaces, where they form what is generally referred to as the "aromatic girdle" (Fig. 2D). The aromatic girdle interacts with the lipid head groups and functions as an interfacial anchor that holds the β-barrel within the membrane bilayer (Hong, Park, Flores Jiménez, Rinehart, & Tamm, 2007).

Inward facing tyrosine residues often form a structural motif with a glycine on a neighbouring strand that improves the structural stability of β-barrel. This interaction is referred to as the "mortise and tenon" motif and is highly conserved in integral β-barrels (Michalik et al., 2017). Due to the lack of any side chain on the glycine, the bulky hydroxyphenyl group of tyrosine can occupy the resulting space (Fig. 2E). This interaction then drives the annealing of neighbouring β-strands during folding (Leyton et al., 2014).

2.2.2 Loops and turns

The difference in size between the extracellular loops and periplasmic turns makes it possible to determine the correct orientation of β-barrels in the OM. The periplasmic turns consist of 5–6 residues on average and their main function is to connect the transmembrane strands (Franklin & Slusky, 2018). It is possible that the turns have a key role in folding of β-barrels as they might promote formation of β-hairpins (Muñoz, Thompson, Hofrichter, & Eaton, 1997).

The extracellular loops are often much longer than the periplasmic turns and have several important functions specific to individual β-barrels (see Section 4). In certain cases, the loops might be folded in the barrel lumen where the loops either restricts the barrel diameter or completely obstruct the barrel. Controlling the effective diameter of the β-barrel is important to control ion permeation and selectivity (see Section 5).

Several structural studies on β-barrels indicate that the extracellular loops are amenable to modifications without jeopardizing the structure of the β-barrel (Koebnik, 1999). This makes the loops of β-barrels exciting possible targets for protein engineering, aiming at biotechnological applications such as surface display (Slusky, 2017).

3. Folding and membrane integration of β-barrels

3.1 In vitro folding

The study of structure, function and folding of β-barrels has been facilitated by techniques that makes it possible to fold denatured membrane proteins in vitro. Folding of integral β-barrels can be completed in vitro by transferring the denatured protein to a suitable hydrophobic environment that, to a certain extent, mimics the outer membrane. β-barrels folded in vitro retain their biological functions, they are suitable for biochemical experiments and can be used for structure determination (Popot, 2014).

3.1.1 Inclusion bodies and denaturants

Folding experiments of β-barrels require a decent amount of protein to be purified. This is usually achieved by heterologous expression in *E. coli*. The expression is typically done without the signal peptide for the protein to aggregate and accumulate in inclusion bodies. This is done out of convenience, as the yield of protein recovered from inclusion bodies is higher compared to purifying protein from the outer membrane (Popot, 2014).

A denaturant is then required in order to solubilize the aggregated protein from collected inclusion bodies (Surrey, Schmid, & Jähnig, 1996). The denaturant, generally urea or guanidinium hydrochloride, stabilizes the unfolded protein in a conformation amenable for folding (Popot, 2014). Urea buffers might be preferred as they are compatible with anion exchange columns; however, prolonged exposure of the protein to urea should be minimized, as it might cause extensive carbamylation, which could affect the folding efficiency and structure of the protein (Stark, Stein, & Moore, 1960). Guanidium hydrochloride is a stronger denaturant than urea and can therefore be used at lower concentrations to solubilize inclusion bodies (Popot, 2014). The choice of denaturant also depends on the refolding conditions: guanidium hydrochloride precipitates with certain zwitterionic detergents like the zwitterionic sulfobetaines that are frequently used to fold β-barrels in vitro (Hjelmeland, 1990).

3.1.2 Folding media: Lipids, detergent and amphipols

β-barrels do not require additional energy to fold in vitro. The proteins fold spontaneously, though not always efficiently, when the denaturant is diluted in the presence of a suitable hydrophobic environment. Lipid bilayers, detergent micelles and amphipols are commonly utilized to fold β-barrels (Popot, 2014).

The mechanical and electrostatic properties of the folding medium determine its utility for in vitro folding of β-barrels. The ideal folding medium will ultimately depend on the specific protein, but certain principles are shared between good folding media. One general requirement for all types of folding media is that they should be basic. Below pH 8, folding yields are generally reduced, and when the pH drops below the isoelectric point of the protein, several β-barrels lose their capacity for in vitro folding (Burgess, Dao, Stanley, & Fleming, 2008).

Folding into lipid bilayers requires the bilayer to be sufficiently flexible and fluid. These characteristics depend largely on the elastic stress between the leaflets of the bilayer. The fluidity and flexibility of the bilayer can be improved by heating; for example, OmpX folded into phosphatidylcholine (PC) bilayers most efficiently around 70 °C (Maurya, Chaturvedi, & Mahalakshmi, 2013). Lipids with shorter acyl chains are better for folding β-barrels as these lipids also result in more flexible bilayers (Burgess et al., 2008). Bilayers that exhibit a higher degree of positive curvature across the membrane normal are more efficient for folding membrane proteins (Booth & Curnow, 2009). The tendency for positive, negative, or neutral bilayer curvature largely depends on the headgroup of the lipid. Negatively charged lipid head groups reduce the rate of folding (Patel, Behrens-Kneip, Holst, & Kleinschmidt, 2009). Short chain lipids ($\sim C_{10}$) with PC headgroups are commonly used for folding β-barrels. PC is zwitterionic and induces a positive curvature on the bilayer that reduces elastic stress.

Folding media containing detergent micelles are the most utilized for refolding β-barrels (Popot, 2014). As with folding media based on lipid bilayers, certain requirements must be met by the detergent for it to be suitable for folding. The detergent must be used above its critical micellar concentration, the head group of the detergent should not carry a strong electrical charge, and shorter acyl chains are preferable for folding (Kleinschmidt, den Blaauwen, Driessen, & Tamm, 1999). Mild uncharged and zwitterionic detergents are preferred for folding β-barrels. The ideal detergent for folding a specific β-barrel must be determined experimentally, as there is no common detergent that can fold all β-barrels while preserving structure and function (Linke, 2014; Orwick-Rydmark, Arnold, & Linke, 2016).

Synthetic amphipathic polymers or "amphipols" are a less utilized method for folding β-barrels (Popot, 2014). Amphipols seem to improve certain experimental techniques that are based on mass spectrometry (Watkinson et al., 2015). The most common amphipol used for folding β-barrels is A8-35 (Kleinschmidt & Popot, 2014).

3.1.3 Electrophoretic analysis of folding: Heat modifiability assays

Folded β-barrels resist denaturation by SDS at room temperature due the extensive hydrogen bonding network of the β-barrel domain (Noinaj, Kuszak, & Buchanan, 2015). For β-barrels to denature in SDS they need to be heated, usually to 80 °C or more. The heating typically results in a marked change in the electrophoretic mobility when the protein is separated by SDS-PAGE. The change in electrophoretic mobility between heated and unheated protein is referred to as "heat modifiability" and corresponds to a breakdown of the β-barrel structure (Nakamura & Mizushima, 1976). SDS binds the protein upon denaturation, which inhibits the protein from refolding. The unfolded protein has an electrophoretic mobility that accurately corresponds to its molecular weight. Smaller β-barrels tend to have a decreased electrophoretic mobility when folded and therefore migrate with a higher apparent molecular weight than the unfolded protein, while larger folded β-barrels might have an increased electrophoretic mobility, therefore appearing to have a lower molecular weight than the unfolded protein (Arnold, Poynor, Nussberger, Lupas, & Linke, 2007). The apparent molecular weight of the folded protein depends on the density of the gel: heat modifiability is more pronounced in low density gels (Heller, 1978).

Heat modifiability has become a reliable method for studying in vitro folding kinetics of β-barrels (Kleinschmidt, 2015). This is done by initiating folding of the β-barrels in vitro and quenching the folding in samples collected at different time points by addition of SDS. The SDS binds to the unfolded protein and prevents it from folding. The folded protein will resist denaturation at room temperature, which results in a stable mixture of the folded and unfolded protein at a specific timepoint in the folding reaction. The SDS-quenched sample can then be analyzed by electrophoresis to separate the folded and unfolded protein. Gel densitometry can then be employed to calculate the fraction of folded protein in the sample (Noinaj et al., 2015). By collecting several samples at different time points of the folding reaction it is possible to track the transition from unfolded to folded protein. Heat-modifiability can also be used to study structural stability of β-barrels by sampling the amount of folded protein in different concentrations of denaturant or at different temperatures. Heat modifiability is compatible with protein folding into detergent micelles, lipid bilayers and amphipols (Arnold et al., 2007; Pocanschi, Popot, & Kleinschmidt, 2013; Rath, Sharpe, Kohl, & Hiller, 2019).

3.1.4 Spectrophotometric analysis of folding: Fluorescence and absorbance

Spectrophotometric analysis of β-barrels folding into lipid bilayers has provided valuable information about the kinetics of folding and membrane insertion (Pocanschi et al., 2013). Most of the fluorescence emitted from β-barrels derives from tryptophan residues which are commonly located in the aromatic girdle. The other aromatic residues have little contribution to the total fluorescence. When tryptophan is excited at 290 nm it fluoresces; however, the emission maximum depends on the surrounding environment of the tryptophan residue. A tryptophan buried within a hydrophobic bilayer will display a blue-shifted emission peak (~330 nm) relative to the emission peak of solvent exposed tryptophans (~350 nm). The observed peak fluorescence can then be used to deduce the folded state of the protein. Folded protein will have buried tryptophan residues with a blue-shifted emission peak (Surrey & Jähnig, 1995).

Tryptophan fluorescence can be blocked by using fluorescence quenchers. Fluorescence quenching techniques are powerful as they allow the accurate determination of the degree of solvent exposure of specific tryptophan residues (Kleinschmidt & Tamm, 1999). This is typically done with single tryptophan mutants, where all tryptophan residues in the protein are replaced with phenylalanine residues except a specific tryptophan of interest. Fluorescence quenching with lipids containing brominated acyl chains have been utilized to track individual tryptophan residues in OmpA as the β-barrel domain formed within a lipid bilayer. Since the position of the fluorescence quencher is known it is possible to determine the location of tryptophan residues within the lipid bilayer based on the degree of observed fluorescence quenching (Kleinschmidt & Tamm, 1999).

Spin-labeled cysteines can be used for short-range fluorescence quenching of neighbouring tryptophan residues. This method has been used to track formation of the β-strands of OmpA folding into lipid bilayer (Kleinschmidt, Bulieris, Qu, Dogterom, & den Blaauwen, 2011). By adding spin-labeled cysteine to the neighbouring strand of a tryptophan it is possible to identify when the strands associate by detecting quenching of the tryptophan fluorescence. Data collected from site-specific fluorescence experiments have been used to derive a scheme for the mechanism of in vitro folding of β-barrels into lipid bilayers (Kleinschmidt et al., 2011): the transition from denatured protein to a fully formed β-barrel is initiated as the protein adsorbs to the bilayer. It appears that the protein initially lies flat on the surface of the bilayer in a "molten disc" conformation (Kleinschmidt, Wiener, & Tamm, 1999). The β-strands then begin to associate, penetrating into the lipid bilayer simultaneously as they form (Kleinschmidt et al., 2011).

Circular dichroism (CD) spectroscopy is a technique based on the absorbance by chiral molecules of circularly polarized light (Kelly, Jess, & Price, 2005). The technique allows for secondary structure determination of proteins and is frequently used to examine the folding and stability of transmembrane β-barrel proteins. CD spectroscopy first demonstrated that the change in electrophoretic mobility between heated and unheated β-barrels in SDS correspond to a breakdown of the β-barrel (Nakamura & Mizushima, 1976). CD spectra of β-barrels dissolved in water revealed that some β-barrels can form partially folded soluble conformations containing a mixture of α-helical and β-stranded content. This partially folded conformation may eventually aggregate or adhere to a folding medium where the correctly folded β-barrel can form (Surrey & Jähnig, 1995). The reason why β-barrels can remain partially soluble in water is explained by the amphipathic characteristic of their β-strands that contain a mixture of polar and hydrophobic residues.

3.2 In vivo biogenesis

Elucidation of how transmembrane β-barrels fold in vivo has lagged behind the characterization of in vitro folding. Folding and membrane insertion of β-barrels into the OM is a complex process requiring a high level of coordination (Fig. 3). While many of the steps leading up to OM insertion were known, how β-barrels folded into the OM remained shrouded in mystery until the discovery of the BAM (Voulhoux, Bos, Geurtsen, Mols, & Tommassen, 2003). This spurred a frenetic research effort in the field, and the past two decades have uncovered the central players in β-barrel biogenesis. Despite the advances made, there are still many open questions, particularly regarding the functioning of the BAM.

3.2.1 Biosynthesis and export to the periplasm

During or after translation by the ribosome, the unfolded β-barrels are recognized based on a cleavable, N-terminal signal peptide that is an integral part of the pre-protein. Translocation though the cytosolic membrane is dependent on the general secretion (Sec) pathway and more specifically, the SecYEG translocon. Around 96% of proteins targeted for the bacterial cell envelope utilize this pathway (Tsirigotaki, De Geyter, Šoštaric, Economou, & Karamanou, 2017). Unfolded proteins with the appropriate signal peptide can be targeted toward SecYEG by a post-translational pathway or by a co-translational pathway (Tsirigotaki et al., 2017). The co-translational pathway is preferred for integral inner membrane proteins, while secreted protein prefer the post-translational pathway (Hegde & Bernstein, 2006). The co-translational pathway involves binding of the Signal Recognition Particle (SRP) to the signal peptide on nascent protein when emerging from the ribosome. The post-translational pathway relies mainly on the chaperone SecA. SecA can bind the nascent protein on the ribosome or the free, fully translated pre-protein in the cytosol. SecA then interacts with SecY and uses ATP hydrolysis and the proton motive force to

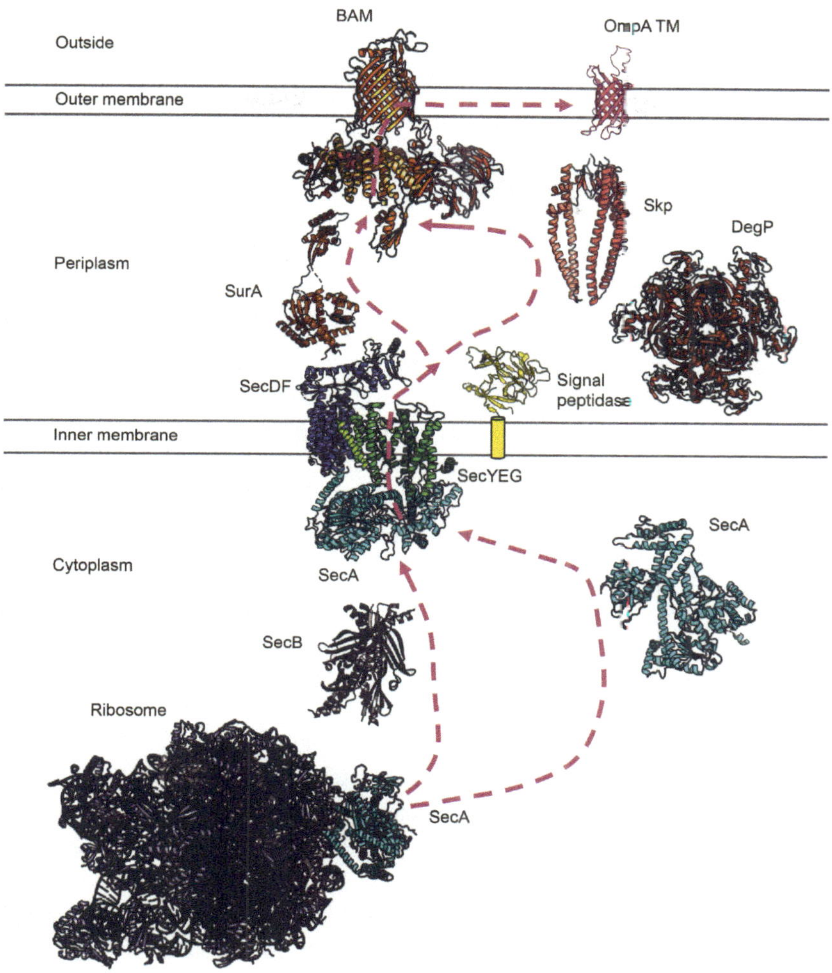

Fig. 3 Biogenesis of transmembrane β-barrels. Like all proteins, β-barrels are synthesized on ribosomes in the cytosol. The nascent β-barrel interacts with chaperones like SecA and SecB. The chaperones deliver the β-barrel to the Sec translocon in the inner membrane, where SecA uses its ATPase activity to thread the β-barrel through the translocon into the periplasm. Signal peptidase removes the signal peptide before periplasmic chaperones (SurA or Skp/DegP) deliver the β-barrel to the BAM for outer membrane insertion. The mature β-barrel is then released into the outer membrane. Recent evidence suggests the Sec, BAM and chaperones form a supercomplex spanning the periplasm. The direction of β-barrel movement is shown by magenta arrows. The PDB IDs used to make the figure are 1TF5 (SecA), 1QYN (SecB), 6S0K (ribosome + SecA), 3DIN (SecYEG + SecA), 3AQP (SecDF), 1KN9 (signal peptidase), 1M5Y (SurA), 1SG2 (Skp), 1KY9 (DegP), 5AYW (BAM), 2JMM (OmpA). (For interpretation of the references to colour in this figure legend, the reader is referred to the web version of this article.)

"push" the protein through the channel (Chatzi, Sardis, Economou, & Karamanou, 2014). The pathway is aided by other chaperones, particularly SecB, that help to maintain the unfolded state of the pre-protein to ensure it remains compatible with the SecYEG translocase (Chatzi et al., 2014).

3.2.2 Transport through the periplasm

After translocation through the cytoplasmic membrane is complete, the pre-protein is still tethered to the membrane by the signal peptide. Proteolytic cleavage by signal peptidase I releases the unfolded β-barrel into the periplasmic space (Paetzel, Karla, Strynadka, & Dalbey, 2002). Aggregation of the β-barrels in the periplasm is prevented by the action of multiple chaperones that recognize and bind unfolded β-barrels and preserve them in state that is competent for outer membrane insertion. β-barrel proteins can take several somewhat redundant pathways through the periplasm (Hagan, Silhavy, & Kahne, 2011; Hussain, Peterson, & Bernstein, 2020). Most β-barrels appear to be transported by the chaperone SurA (Dalbey & Kuhn, 2012). SurA is not essential as there are several other chaperones available in the periplasm that can complement SurA (Plummer & Fleming, 2016). These include DegP and Skp, two chaperones connected to the σ^E stress response (Mogensen & Otzen, 2005). The stress response is triggered upon accumulation of unfolded protein in the periplasm and is also activated at elevated temperatures (Duguay & Silhavy, 2004). At these temperatures, DegP can assemble into a cage-like structure that surrounds misfolded proteins, and acts as a protease (Krojer et al., 2008).

3.2.3 Outer membrane insertion

In Gram-negative bacteria β-barrels are folded and inserted into the OM by the action of the BAM. In *E. coli*, the BAM complex consists of a 16-stranded β-barrel (BamA) and four associated lipoproteins (BamB, BamC, BamD and BamE) (Knowles, Scott-Tucker, Overduin, & Henderson, 2009). The exact composition of the BAM varies between bacterial species, however (Webb, Heinz, & Lithgow, 2012). A BamA homologue is always present though, and homologues of BamA are also found in the OM of mitochondria and chloroplasts (Noinaj et al., 2013).

The function of the BAM is essential for diderm prokaryotes, but the details of the mechanisms of how the complex inserts and folds β-barrels into the OM are still under investigation. Over the past few years, increasing evidence points toward an export complex spanning the periplasm, containing components of the Sec, the BAM and periplasmic chaperones, suggesting that translocation through the inner membrane, transit across the periplasm and outer membrane insertion are highly coordinated (Alvira et al., 2020; Wang et al., 2016).

The BAM recognizes client proteins by a conserved signal, termed the β-signal (Paramasivam, Habeck, & Linke, 2012; Robert et al., 2006). This signal is usually found in the C-terminal strand of β-barrel proteins, where alternating aromatic/hydrophobic and hydrophilic residues at the C-terminus of the strand constitute the signal, which is to some extent species-specific (Paramasivam et al., 2012). However, in BamA itself, the β-signal is located on β-strand 14 rather than 16 (Hagan, Wzorek, & Kahne, 2015). The β-signal is probably recognized by BamD, the other essential component of the BAM, before handing over the client protein to BamA (Hagan et al., 2015). Very recent data indicate that there may be a second, β-signal-independent pathway for β-barrel assembly that requires SurA (Wang, Peterson, & Bernstein, 2021).

Currently, several alternative models for the mechanism of in vivo membrane insertion of β-barrel proteins by the BAM are debated. It seems that the folding mechanism of the BAM is linked to conformational changes in the N- and C-terminal strands of BamA (Wu, Stephenson, Gichaba, & Noinaj, 2020). The terminal strands are short, with reduced hydrogen bonding, and do not fit within the dimensions of the OM. The strands cause local perturbations in the OM and thereby create a point of entry for unfolded β-barrels (Doerner & Sousa, 2017). According to the "budding/template model," the terminal strands can open laterally for entry of β-barrel proteins to the OM and act as a template for unfolded β-barrels to form hydrogen bonds (Lee et al., 2019; Tomasek et al., 2020).

4. Functions of transmembrane β-barrels

The β-barrel scaffold is a versatile one that allows the proteins with this domain to perform a variety of functions (Table 1). The cylindrical shape of β-barrels forms a pore that traverses the outer membrane and it is intuitively clear that such a structure is suited for transport. Indeed, many β-barrel proteins are involved in the transport of nutrients into the cell, or the secretion or excretion of a variety of substances out of the cell. However, transport is not the only possible function; in addition, these proteins often act as adhesins or enzymes. In multi-domain proteins, the β-barrel domain is usually either an outer membrane anchor or involved in secretion, or both, whereas the other functionalities are present in the other, water-soluble domains.

Examples would be OmpA, whose C-terminal periplasmic domain binds to peptidoglycan (Park et al., 2011), or autotransporters, where the β-barrel domain is involved in secreting the effector domains to the outside of the cell and in many cases also anchoring them to the bacterial surface (Meuskens, Saragliadis, Leo, & Linke, 2019). In the following, we will be limiting our discussion to functions residing within the β-barrel domain itself. In proteins with functions other than transport, the activity is generally mediated by the extracellular loops of the protein or a periplasmic or extracellular extension. It is debatable when an extended loop becomes a domain in its own right [e.g., the adhesin BabA from *Helicobacter pylori* (Coppens et al., 2018)]. However, as can be seen below, also residues within the transmembrane β-strands themselves can be important for activity.

4.1 Porins

The main transport proteins in the OM are the porins, which form the aqueous channels that allow the diffusion of small molecules into (and out of) the periplasm. An often-quoted size limit for diffusion through the OM is 600 Da; however, the shape and electrostatic properties of the molecule determine its ability to diffuse through the porin channels (Schulz, 2004).

The porins can be classed as either general porins or specific porins (Nikaido, 2003). In *E. coli*, the general porins are OmpF, OmpC and PhoE. These are relatively non-selective channel proteins; however, Omp F and OmpC are more selective for cations, whereas PhoE prefers anions (Masi et al., 2019). They are homotrimers where each monomer forms its own 16-stranded β-barrel (Fig. 1). Diffusion through the porins is governed by the "constriction zone" approximately halfway through the channel. This is caused by the extracellular loop 3, which folds into the pore and generates a partial occlusion. This loop, together with charged residues in the β-barrel wall, creates a transverse electric field across the pore that dictates the electrochemical nature of molecules that can diffuse easily through the OM (Vergalli et al., 2020). In addition to these, *E. coli* also encodes for two cryptic general porins, i.e. proteins whose expression levels are generally undetectable. The first of these genes, *nmpC*, is a pseudogene in most wild-type *E. coli* strains, but this can become activated and complement *ompC* and *ompF* when these genes deleted (Blasband, Marcotte, & Schnaitman, 1986; Pugsley & Schnaitman, 1978). The second is *ompN*, which is not expressed under standard growth conditions but is upregulated due to antimicrobial stress (Dam, Pagès, & Masi, 2017; Prilipov, Phale, Koebnik, Widmer, & Rosenbusch, 1998).

By contrast, the specific porins are larger, containing 18 β-strands. As their name suggests, these proteins have much narrower substrate specificity, usually transporting only a few structurally similar molecules. Examples are the maltose transporter LamB from *E. coli* the sucrose transporter ScrY from *Salmonella enterica*, and the Occ family of transporters from *P. aeruginosa* (Eren et al., 2012; Forst, Wacker, Welte, & Diederichs, 1998; Schirmer, Keller, Wang, & Rosenbusch, 1995). Like the general porins, LamB and ScrY are homotrimeric, but the Occ proteins crystallize as monomers. The substrate specificity of these channels is determined by a region analogous to the constriction zone in the general porins. In LamB, for example, a helically patterned ring of aromatic residues composes a "greasy slide" that provides the specificity for glucosyl residues (Schirmer et al., 1995) In addition, there are several other specific transporters with smaller β-barrel domains. These include the nucleoside transporter Tsx (12 β-strands), the fatty acid transporter FadL (14 β-strands) of *E. coli* (van den Berg, Black, Clemons, & Rapoport, 2004; Ye & van den Berg, 2004). An intriguing recent discovery is the 14-stranded ligand-gated CymA porin of *Klebsiella oxytoca*, which is specific for bulky cyclodextrin molecules (van den Berg, Bhamidimarri, Prajapati, Kleinekathöfer, & Winterhalter, 2015).

4.2 TonB-dependent transporters

Transport via porins is energy-independent and requires that any compounds entering the cell do so moving down a concentration gradient. However, many substances that are scarce in the environment need to be taken up by the cell against a concentration gradient. For this, active transport is needed. The periplasm does not contain ATP, nor does the leaky nature of the OM allow for maintaining ion gradients, so these usual energy sources of the cell are not available at the OM. Gram-negative bacteria have solved this problem by employing an energy transduction system from the inner membrane across the periplasm. This is mediated by the inner membrane TonB/ExbB/ExbD complex, where ExbB and ExbD form a proton channel and the water-soluble domain of TonB spans the entire periplasm (Celia, Noinaj, & Buchanan, 2020). ExbB/ExbD use the proton-motive force to energize a poorly understood conformational change in TonB that allows the energy to be utilized at the OM.

TonB interacts with a family of 22-stranded β-barrel proteins termed TonB-dependent transporters (Noinaj, Guillier, Barnard, & Buchanan, 2010). These proteins are characterized by a plug domain that occludes the pore of the β-barrel and is involved in binding to the transport substrate. The plug domain is preceded by a short motif called the TonB box, which interacts with TonB when the substrate is bound to the transporter. The interaction causes a conformational change in the plug domain that allows the transported substate to enter the periplasm (Bolam & van den Berg, 2018). In *E. coli*, TonB-dependent transporters are used to import iron-siderophore complexes and vitamin B12 (Noinaj et al., 2010). In other bacteria, heme, host iron-binding proteins, thiamine, carbohydrates, copper and nickel are taken up by TonB-dependent transporters into the cell (Schauer, Rodionov, & de Reuse, 2008). TonB-dependent transporters are particularly abundant in the Bacteroidetes, where many are found in conjunction with surface-exposed lipoproteins and involved in the import of complex carbohydrates into the cell, exemplified by the SusC/SusD pair (Pollet, Martin, & Koropatkin, 2021).

4.3 Efflux systems

Drug efflux is a major contributor to the increase of antibiotic resistance. Active efflux pumps belonging to the root-nodulation-cell division, ATP-binding cassette, major facilitator superfamily, small multidrug resistance or multidrug and toxin extrusion classes are responsible for removing antibiotics and other harmful compounds, including detergents, bile salts, disinfectants, heavy metals and solvents, from the bacterial cytoplasm (Venter, Mowla, Ohene-Agyei, & Ma, 2015). This leads to reduced concentrations in the cell. Many of these pumps are coupled to transport through the OM. In these cases, the export channel is formed by a tripartite system consisting of the efflux pump in the inner membrane, a periplasmic adapter protein (also referred to as a membrane fusion protein), and an outer membrane channel protein (Neuberger, Du, & Luisi, 2018).

In *E. coli*, the OM channel is usually formed by the trimeric β-barrel protein TolC. The transmembrane domain of TolC is a 12-stranded β-barrel, with each protomer contributing four β-strands to the structure. However, the barrel extends into the periplasm as an α-helical barrel, the lumen of which forms a periplasmic tunnel leading to the OM (Koronakis, Eswaran, & Hughes, 2004) (Fig. 1). By itself, the periplasmic entry to the tunnel is occluded, presumably to prevent leakage across the OM, as the β-barrel domain is constitutively open (Koronakis, Sharff, Koronakis, Luisi, & Hughes, 2000). Upon interaction with the efflux pump, the periplasmic opening expands, allowing substrates translocated by the pump to enter the tunnel and to diffuse out of the cell via the β-barrel domain (Du et al., 2014). TolC has three orthologues in *E. coli* (Table 1), one of which, CusC, is involved in the efflux of toxic copper and silver ions (Kulathila, Kulathila, Indic, & van den Berg, 2011). The best studied homologue outside *E. coli* is OprM from *P. aeruginosa*, which contributes to the very high intrinsic resistance of this organism (Phan, Picard, & Broutin, 2015). Interestingly, both CusC and OprM are lipoproteins, with an extended N-terminal domain looping back along to the α-barrel of these proteins to the OM. The function of this additional anchoring in the OM is not clear.

4.4 Protein secretion

Protein secretion into the extracellular medium is necessary to interact with the environment, e.g. for nutrient acquisition, adhesion and biofilm formation, and virulence. Gram-negative bacteria have evolved a variety of secretion systems to export proteins to the cell surface or into the medium (Christie, 2019). Some of these are large, multi-subunit protein complexes spanning the entire cell envelope; others have an OM secretion component and rely on the Sec or Tat pathways for translocation of the substrate proteins into the periplasm. As mentioned earlier, some of the larger complexes have multimeric β-barrel components (type 2, 3 and 8 secretion systems), which are assembled by a different pathway to classical β-barrels (Dunstan et al., 2015). However, four secretion systems rely on classical β-barrels for substrate secretion: type 1, type 5, type 9 secretion systems and the chaperone-usher pathway.

Type 1 secretion systems are closely related to efflux pumps and share the same general architecture of an inner membrane ATP-binding cassette protein, a periplasmic fusion protein and an OM secretion pore (Kanonenberg, Spitz, Erenburg, Beer, & Schmitt, 2018). The OM component is TolC or one of its homologues (Thomas, Holland, & Schmitt, 2014); thus, TolC is involved in exporting both small molecules and proteins. The prototypical substrate for type 1 secretion systems is the haemolysin HlyA of *E. coli*, but also adhesins, enzymes and other toxins can be transported by type 1 secretion systems (Kanonenberg et al., 2018).

Type 5 secretion systems also contain a β-barrel secretion unit. In most subclasses this is part of the same polypeptide chain as the secreted portion or passenger (Fan, Chauhan, Udatha, Leo, & Linke, 2016). The exception is the type 5b or two-partner secretion systems, where the transporter and passenger are separate polypeptides. Due to passenger and translocator being part of the same polypeptide, it was thought that these proteins could "secrete themselves," and no other factors were needed; hence the name "autotransporter" (Klauser, Pohlner, & Meyer, 1993). This turned out to be a misconception, and now it is clear that the BAM plays a crucial role in autotransporter biogenesis (Doyle & Bernstein, 2021; Ryoo et al., 2020). However, the term "autotransporter" is still applicable, as the information for passenger secretion as well as the energy for secretion is provided by the autotransporter protein itself (Leo, Grin, & Linke, 2012). The translocator domain is either a 12-stranded β-barrel (types 5a, 5c and 5e), or a 16-stranded β-barrel related to BamA (types 5b, 5d) (Fan et al., 2016). For all other classes the β-barrel is monomeric, but in type Vc secretion systems, both the β-barrel and the passenger are obligate homotrimers, with each monomer contributing four β-strands to the β-barrel (Meng, Surana, St Geme, Joseph, & Waksman, 2006; Shahid et al., 2012). The translocator forms the secretion pore, probably together with BamA, and at least in subclasses 5a, 5c and 5e, the passenger is secreted via a hairpin intermediate (Chauhan et al., 2019; Junker, Besingi, & Clark, 2009; Oberhettinger, Leo, Linke, Autenrieth, & Schütz, 2015). Folding of the passenger at the extracellular surface provides most of the free energy for secretion (Drobnak, Braselmann, & Clark, 2015; Leo et al., 2016), though other properties such as charge distribution may contribute (Kang'ethe & Bernstein, 2013). Once secreted, autotransporter passengers can perform a variety of functions, many of which are linked to virulence (Meuskens et al., 2019).

Pili and fimbriae are long filamentous structures formed from up to thousands of subunits that function as adhesins or form a mesh-like capsule around bacterial cells (Lukaszczyk, Pradhan, & Remaut, 2019). Apart from the contractile type IV pili and type V pili from the class Bacteroidia, these structures are assembled by the chaperone-usher pathway in Gram-negative bacteria. The usher protein is a large (24-stranded) β-barrel (Phan et al., 2011). This acts not only as the secretion pore for the pilin subunits but also as the assembly point where the pilins are non-covalently linked to each other by the donor-strand exchange mechanism (Geibel & Waksman, 2014).

This exchange requires the participation of the additional periplasmic domains of the usher, and the process leads to extrusion of the pilins to the surface of the bacterium through the usher protein.

The type 9 secretion system is restricted to the Bacteroidetes but is widespread in this phylum and involved in secretion of virulence factors and gliding motility. The system is a multiprotein complex spanning the cell envelope containing at least 15 proteins (Gorasia, Veith, & Reynolds, 2020). The outer membrane translocon, SprA/Sov of the type 9 secretion system is the largest known classical β-barrel with 36 β-strands and extended, folded loop domains covering the extracellular side of the protein (Lauber, Deme, Lea, & Berks, 2018). The lumen of SprA is large enough to accommodate a folded substrate protein. The β-barrel of SprA contains a lateral opening toward the extracellular surface of the membrane; this opening is occluded by a second β-barrel protein, PorV, a loop of which extends through the opening into the lumen of the SprA barrel. In a second structure of SprA, PorV is missing, and SprA is associated with a periplasmic Plug protein that stops the translocon pore on the periplasmic side (Lauber et al., 2018). As PorV and the Plug do not bind to SprA simultaneously, a model was put forward where PorV connects to a substrate protein in the pore and triggers a change in SprA releasing the PorV-substrate complex through the lateral opening. The Plug protein then binds to SprA to prevent leakage of periplasmic contents until PorV again docks with SprA.

In addition to the systems above, some other secretion systems that have not yet been accepted into the official scheme employ β-barrel proteins. Recently, a TonB-dependent secretion system was demonstrated in *Myxococcus xanthus* (Gómez-Santos, Glatter, Koebnik, Świątek-Połatyńska, & Søgaard-Andersen, 2019). In this system, secretion of the PopC protease requires the TonB-dependent transporter Oar. Using a heterologous system, PopC secretion was shown to require TonB. In *P. aeruginosa*, the amyloid protein FapC is not exported by a CsgG-like pore. Rather, it is secreted by FapF, a trimer of 12-stranded β-barrels that have an N-terminal periplasmic coiled coil extension (Rouse et al., 2017). This extension may help in recruiting FapC monomers for secretion (Rouse et al., 2018). Additionally, the porin OmpF may be involved in protein export in *E. coli*. Evidence suggests that the small secreted protein YebF is transported across the OM by OmpF with assistance from OmpC and OmpX (Prehna et al., 2012). The details of this pathway remain to be elucidated; however, it was suggested that YebF might be secreted as a colicin "in reverse" and be threaded through the pore of OmpF.

4.5 Enzymes

While it is easy to see how the structure of β-barrels lends itself to the transport of solutes and even macromolecules, it is less obvious how other functions, such as enzymatic activity, might be incorporated. Nevertheless, the β-barrel fold is versatile enough to also accommodate enzymatic activity and there are several examples of β-barrel proteins that are enzymes. Autotransporter passengers often encompass proteases or lipases (Meuskens et al., 2019). In general, these activities are not provided by the β-barrel domain itself but are found as separate domains in the soluble passenger region. However, some autotransporter β-barrels do undergo an autoproteolytic reaction to release the passenger domain into the extracellular space; an example would be EspP of *E. coli*, where an aspartate residue in the lumen of the β-barrel catalyzes autoproteolysis of the linker segment leading to passenger release (Anderson, Dautin, Barnard, & Bernstein, 2007; Barnard, Dautin, Lukacik, Bernstein, & Buchanan, 2007).

A class of proteases consisting of only a β-barrel domain are the omptins, represented by OmpT from *E. coli*, PgtE from *S. enterica* and the plasminogen activator Pla from *Yersinia pestis* (Haiko, Suomalainen, Ojala, Laehteenmaeki, & Korhonen, 2009). These are small, 10-stranded β-barrel proteins with aspartic protease activity. This activity is located in the long β-strands of the barrel extending into the extracellular medium, where the catalytic His-Asp dyad activates a water molecule that attacks substrate molecules. A second Asp-Asp dyad contributes by stabilizing the attacking water (Eren, Murphy, Goguen, & van den Berg, 2010; Vandeputte-Rutten et al., 2001). In addition, omptins require a bound lipopolysaccharide molecule for activity (Eren & van den Berg, 2012; Kramer et al., 2002; Kukkonen et al., 2004). Pla is involved in fibrinolysis through the generation of plasmin, whereas PgtE activates matrix metalloprotease 9 and aids in serum resistance by cleaving complement components and inactivating antimicrobial peptides. By contrast, OmpT appears to be mostly a housekeeping protease (Haiko et al., 2009).

β-Barrels can also have phospholipase activity. PldA (previously OMPLA for outer membrane phospholipase A) is a 12-stranded β-barrel that dimerizes to form a functional calcium-dependent phospholipase. The active site for phospholipid degradation is formed between the faces of β-barrel, i.e. in the hydrophobic membrane environment (Snijder et al., 1999). The active site is close to the outer leaflet side of the β-barrel, consistent with PldA's role in degrading wayward phospholipids and thus removing them from the outer leaflet (May & Silhavy, 2018).

Several β-barrel enzymes are involved in the modification of the lipidic moiety of lipopolysaccharide, lipid A. These include the palmitoyl transferase PagP in *E. coli*, which transfers a palmitate group from phospholipids to lipid A, thus generating hepta-acylated lipid A, rather than the usual hexa-acylated version (Bishop, 2005). This has been suggested to both improve membrane integrity when phospholipids are present in the membrane, to increase resistance against cationic antimicrobial peptides, and to reduce the inflammatory potential of lipid A. Unusually for a β-barrel protein, the binding site for the palmitate group appears to be within the lumen of the barrel (Ahn et al., 2004). Another β-barrel proteins involved in lipid A modifications are PagL from *S. enterica* and *P. aeruginosa*, which deacylates hexa-acylated lipidA, leading to antagonism of Toll-like receptor 4 signaling and a reduced innate immune response (Rutten et al., 2006).

4.6 Adhesins

An important role for β-barrel proteins is in bacterial adhesion. β-barrels are involved in the secretion of adhesive domains and structures, exemplified by type 5 secretion systems and chaperone-usher pili (see Section 4.4). In addition, β-barrel domains themselves can be adhesive, though these are not as common as other, more extended adhesin structures because β-barrels are often masked by other cell-surface structures such as lipopolysaccharide O-antigens and are therefore not exposed to bind to receptors on cells or in tissues (Biedzka-Sarek, Jarva, Hyytiäinen, Meri, & Skurnik, 2008; Tran, Doyle, & Morona, 2013; van der Ley, de Graaff, & Tommassen, 1986). Examples of adhesive β-barrels are members of the OmpX family. These are small, 8-stranded β-barrels that mediate bacterial adhesion and invasion, as well as defence against host responses in some cases (Koebnik et al., 2000). The prototype of this family is the attachment and invasion locus protein Ail from the pathogenic Yersiniae (Kolodziejek, Hovde, & Minnich, 2012). Ail plays a role in binding to host cells and allowing delivery of type 3 secretion system effectors into host cells by binding to fibronectin and laminin. Ail can also mediate invasion of host cells and further protect the bacteria by conferring serum resistance through binding complement regulatory factors, a trait it shares with its *Salmonella* homologue Rck (Ho et al., 2011; Ho, Jarva, & Meri, 2010).

In addition to binding to host cells, Ail of *Y. pestis* mediates binding to itself. This results in autoaggregation of the bacteria, which can protect against various stressors and lead to biofilm formation (Trunk, Khalil, & Leo, 2018). Other β-barrels mediating autoaggregation include the heat-resistant agglutinin Hra1 from *E. coli*, which is also 8-stranded β-barrel protein (Glaubman et al., 2016). Hra1 is the prototype of a family of β-barrels including Hra2, Hek, Hes and Tia proteins that mediate autoaggregation, and binding to and invasion of host cells (Bondì et al., 2017; Fagan & Smith, 2007; Mancini et al., 2011; Montero et al., 2017).

4.7 Other functions

An important function of β-barrel proteins is maintaining the integrity of the OM. OmpA plays an important role in this by anchoring the OM to the peptidoglycan cell wall (Park et al., 2011). *E. coli* mutants lacking both OmpA and Braun's lipoprotein Lpp, an OM lipoprotein that binds covalently to peptidoglycan, display a rounded cell shape and increased OM vesicle production, demonstrating the importance of these proteins in maintaining the cell shape and OM integrity (Choi & Lee, 2019; Samsudin, Boags, Piggot, & Khalid, 2017; Sonntag, Schwarz, Hirota, & Henning, 1978).

As mentioned in Section 3.2, the central proteins involved in the biogenesis of barrel proteins is itself a β-barrels. BamA is an essential protein and cells depleted of BamA eventually die (Wu et al., 2005). Outer membrane stress is monitored by the complexation of the lipoprotein RcsF with OmpA, OmpC or OmpF (Dekoninck et al., 2020; Konovalova, Perlman, Cowles, & Silhavy, 2014). Under normal conditions, the N-terminus of RcsF is sequestered by these proteins; however, when the OM is perturbed, the N-terminus of RcsF is released into the periplasm, where it can interact with IgaA, resulting in a signaling cascade to initiate the periplasmic stress response (Cho et al., 2014). Complex formation between RcsF and the porins is mediated by BamA (Rodríguez-Alonso et al., 2020). In addition, OmpF and OmpC play an important role in maintaining OM integrity. This is through interaction with MlaA, an α-helical OM lipoprotein that inserts into the OM and is involved in the retrograde transport of phospholipids from the OM outer leaflet to the inner leaflet (Abellon-Ruiz et al., 2017). The lipopolysaccharide transporter LptD is another essential protein in *E. coli* that is also needed for the maintenance of lipid asymmetry in the OM (Botos et al., 2017).

Finally, β-barrels can function as the receptors for bacteriophages and bacteriocins. Though these are not "functions" desirable for the cell, binding to a surface receptor is a critical step in the bacteriophage life cycle. A classic example of a β-barrel phage receptor is the LamB porin, which is the receptor for coliphage λ (Chatterjee & Rothenberg, 2012). Other *E. coli* β-barrel proteins act as receptors for different phages (Hantke, 2020). Bacteriocins are antimicrobial peptides or proteins that target bacteria closely related to the producer organism that lack an associated immunity protein; thus, they are mediators of inter-strain competition (Simons, Alhanout, & Duval, 2020). To exert their toxic function, bacteriocins must be imported into the cell. In Gram-negative bacteria, this requires crossing the OM. Colicins (bacteriocins produced by and targeting *E. coli*) use β-barrel proteins as their receptor and/or OM tranlocators, either porins or TonB-dependent importers (Kim, Tarr, & Penfold, 2014). Another example of inter-strain competition is contact-dependent inhibition systems. These are two-partner (type 5b) secretion systems which export a bacteriotoxin targeting closely related strains lacking an immunity protein (Ikryannikova, Kurbatov, Gorokhovets, Zamyatnin, & Andrey, 2020). The receptor for the toxin is a β-barrel protein; for example, BamA, OmpF and OmpC serve as contact-dependent toxin receptors in *E. coli*.

5. Evolution of β-barrels

It remains unclear whether diderm bacteria evolved from a monoderm ancestor by adding an additional membrane layer to their surface, or whether today's Gram-positive bacteria have evolved through secondary loss of the outer membrane from a diderm ancestor (Megrian, Taib, Witwinowski, Beloin, & Gribaldo, 2020). This discussion is further complicated by the discovery of diderm lineages that might have arisen independently from the prototypical Gram-negative species, the Proteobacteria (Taib et al., 2020).

In any event, the evolution of a second, outer membrane must have had benefits in terms of protection, possibly against antibiotic substances (Gupta, 2011), but comes with additional problems in solute and nutrient transport, ion homeostasis, and protein secretion. The outer membrane does not need to be sealed completely from in- and efflux of ions, as it is not involved in maintaining chemical gradients for energy conservation. β-barrel proteins are typically leaky proteins in terms of ion conductance, and even the smallest exemplars such as the 8-stranded β-barrel OmpX show some conductance (Arnold et al., 2007). While this poses no problem in the outer membrane, such pores could probably not have evolved in the cytoplasmic membrane where it is essential to maintain a proton (or other ion) gradient. The evolution of two different families of membrane proteins, α-helical and β-barrel-based, that populate two different membranes also enables some aspects of protein sorting (to the correct membrane) that are based on sequence features (see Section 3.2).

For Proteobacteria and related bacterial phyla, there is ample evidence that at least most β-barrel proteins in the outer membrane have evolved from one ancestral peptide or protein. This evidence stems from bioinformatics investigations showing that the evolutionary unit of β-barrel diversification is a β-hairpin, and that this hairpin can be traced to a common ancestor of all β-barrel proteins using remote homology detection algorithms (Remmert, Biegert, Linke, Lupas, & Söding, 2010). This principle of homology has been widely used to develop tools that identify β-barrel protein genes in genomes and to predict their strand numbers or their structure (Remmert, Linke, Lupas, & Söding, 2009; Tsaousis, Hamodrakas, & Bagos, 2017). Experimental evidence that larger β-barrel proteins can be relatively easily evolved by duplicating one or several β-hairpins in their sequence further supports the hypothesis of a common ancestor of β-barrels (Arnold et al., 2007). A common ancestry is also suggested by the fact that all β-barrels seem to be inserted into the OM by the same, highly conserved machinery, the BAM (Webb et al., 2012), and that this is based on the homologous C-terminal β-signal (Paramasivam et al., 2012) (see Section 3.2). This suggests an evolutionary link between all substrates, and points to an evolutionarily conserved insertion mechanism for all β-barrel proteins (Leo & Linke, 2018).

As a by-product of the ever-increasing amount of genomic data for Proteobacteria, it has been shown that among the β-barrel proteins, the 16-stranded and 22-stranded β-barrels are over-represented compared to other strand numbers (Paramasivam et al., 2012). This probably reflects the function of these two sub-classes of β-barrels, where the 16-stranded barrels are typically porins, while the large family of 22-stranded barrels are the TonB-dependent transporters (see Section 4).

Evolution of β-barrels is a continuing process. Especially the medicinal use of antibiotics has introduced new evolutionary pressures on outer membrane and its capacity to keep harmful substances away from the machineries that fulfil core cellular functions, as discussed in the next section.

6. Medical relevance of bacterial β-barrel proteins

As surface exposed proteins, transmembrane β-barrels of Gram-negative bacteria are central to the bacterium-host interface. β-barrels are required for many functions of bacteria, both commensal and pathogenic, within the host environment. These include adhesion, immune evasion and/or modulation, adhesion and biofilm formation, virulence factor secretion, and scavenging for nutrients. In addition, β-barrels are the sites of permeation into the cells for many antibiotics, and changes in their structure or expression can result in reduced permeability to antibiotics, contributing to antibiotic resistance. However, inhibition of β-barrel function could be a novel approach for the development of antimicrobial compounds. The extracellular loops of β-barrels are exposed to the immune system and therefore potential vaccine targets. In this section we give a brief overview of the roles that β-barrels play in medical treatment of bacterial infections.

6.1 β-barrel proteins and antibiotic resistance

β-barrels, particularly the general porins like OmpF and OmpC, are the channels by which many antibiotics gain entrance into the cell. The importance of OmpF and OmpC in antibiotic uptake in *E. coli* can be contrasted with the much higher levels of resistance in organisms like *P. aeruginosa* and *Acinetobacter baumannii*. The lack of large-channel, 16-stranded general porins contributes to the very low permeability of the OM in these latter species and the consequently higher intrinsic resistance to antibiotics (Zgurskaya, López, & Gnanakaran, 2015). In *E. coli*, OmpF is a particularly important diffusion pore for many antibiotics, including carbapenems (Choi & Lee, 2019; Vergalli et al., 2020). Many drug-resistant strains of *E. coli* and its close relatives like *Klebsiella pneumoniae* or *Enterobacter cloacae* have either reduced porin expression or have lost some porin genes entirely, leading to a more impermeable OM. Another alternative to reduce antibiotic influx is to accumulate point mutations within the porins, especially in loop 3 and the constriction zone (see Section 4.1). Changes in these regions alters the electrochemical properties of the constriction zone and consequently impacts on the types of molecules that can diffuse through the pore (Vergalli et al., 2020). Reduced OM permeation of antibiotics in combination with efficient drug efflux systems (see Section 4.3) leads to a highly effective permeability barrier to antibiotics and a rise in resistance (Li, Plésiat, & Nikaido, 2015). Porin loss comes with a fitness cost for most strains; however, up-regulation of alternative porins with lower permeability for antibiotics compensates for this cost (Knopp & Andersson, 2015).

6.2 β-barrel proteins as targets of antimicrobials

Although most classical antibiotics utilize porins to gain entry into the cells, none actually target β-barrels themselves. The lack of such molecules is probably because, in *E. coli*, only two β-barrel proteins are essential and therefore compounds that do affect specific β-barrel functions would probably be missed in drug screens. However, it is somewhat surprising that no antibiotics were found to target β-barrel or lipopolysaccharide biogenesis, the two processes that have an essential β-barrel component. Only approximately 10 years ago was a compound found that targets LptD from *P. aeruginosa* (Srinivas et al., 2010). This compound, L27-11, a peptidomimetic antibiotic, interferes with lipopolysaccharide transport and leads to loss of OM integrity. However, L27-11 does not target the β-barrel domain of LptD but rather its periplasmic domain (Andolina et al., 2018). This is reminiscent of the so-called "pilicides," compounds that prevent the formation of pili by interfering with the chaperone-usher complex formation (Pinkner et al., 2006). The pilicides prevent the chaperone-pilin complex from docking on the periplasmic N-terminal domain of the usher, thus inhibiting pilus biogenesis and acting as anti-virulence drugs. Another example of an anti-virulence compound is the cyclic hexapeptide AOA-2 that targets the extracellular loops of OmpA and was shown to reduce adhesion to host cells and biofilm formation in *E. coli*, *P. aeruginosa* and *A. baumannii* (Vila-Farrés et al., 2017).

A more recent discovery was a naturally occurring antibiotic that targets BamA. This is darobactin, a peptide antibiotic isolated from a strain of *Photorhabdus*, a symbiotic bacterium of entomopathogenic nematodes (Imai et al., 2019). Darobactin was demonstrated to be a β-signal analogue that outcompetes native β-barrels for binding to BamA, leading to depletion of β-barrels from the OM (Kaur et al., 2021). As the interaction is not heavily dependent on the amino acid side chains at the lateral gate binding site, and the ones that are involved are needed for proper BamA function, it has been suggested that resistance to darobactin is unlikely to occur. Some other compounds have also been found to inhibit BamA: the molecule MRL-494 inhibits β-barrel activity by binding to BamA (Hart et al., 2019), and IMB-H4 is a compound that inhibits the BamA-BamD interaction based on a yeast two-hybrid screen and displays antibacterial activity (Li et al., 2020). In another study screening for autotransporter inhibitors, the compound VUF15259 was found as a hit (Steenhuis et al., 2019); however, this turned out to be an anti-BamA compound rather than affecting the autotransporter directly.

In the battle against the rise of antimicrobial resistance, inhibiting drug efflux is a promising avenue of research. Indole derivatives targeting TolC were shown to resensitize bacteria to several classes of antibiotics, presumably by blocking the efflux channel through the OM (Zeng et al., 2010). Other approaches include using bacteriophages or bacteriocins as alternatives to antibiotics as potential therapeutics (Royer, Morais, & da Fonseca Batistão, 2021; Telhig, Said, Zirah, Fliss, & Rebuffat, 2020). As β-barrel proteins are often the primary receptors for phage and bacteriocins, studying the interactions between these factors is of central importance for any therapeutic applications.

6.3 β-barrel proteins as antibody and vaccine targets

The extracellular loops of β-barrels are necessarily surface-exposed and therefore potential targets for the adaptive immune system. Despite this, β-barrels have not played a major role in vaccine development till date. This is probably largely due to the additional effort involved in purifying recombinant β-barrels, which requires the use of detergents or other solubilizing agents to keep these largely hydrophobic proteins in solution. However, over the past decade or so, outer membrane vesicles, which contain β-barrels and other surface structures derived from the outer membrane, have become attractive for vaccine development and indeed are already components of some vaccines (Micoli & MacLennan, 2020). Recently, interest has picked up for using some β-barrels as immunogens in vaccine formulations. For example, LptD has been found to be a promising vaccine candidate against several *Vibrio* species (Zha, Li, Li, Ye, & Pan, 2016). The other essential β-barrel, BamA, has shown promise as a vaccine candidate against some bacterial species, but seems inefficient against others despite inducing a strong antibody response (Su, Wan, Mohamed, & Nathan, 2010; Wedege et al., 2013).

β-barrels are not only used as antigens but also in a different way for vaccine development: as a scaffold to present antigens to the immune system. Some β-barrels, most notably autotransporter β-barrels, can be used to display heterologous protein domains and epitopes on the cell surface for a variety of applications, including vaccine development (van Ulsen, Zinner, Jong, & Luirink, 2018). This technology can be combined with other recent developments, such as protein ligation systems, to produce multivalent vaccines displayed on bacterial surfaces or outer membrane vesicles (Hatlem, Trunk, Linke, & Leo, 2019).

In addition to the polyclonal production of antibodies associated with an adaptive immune response, monoclonal antibodies might be used as antimicrobials in their own right. Recently, bactericidal monoclonal antibodies were discovered that target BamA (Storek et al., 2018). The exact mechanism of how these antibodies inhibit BamA function is unclear, but they target extracellular epitopes distant from the lateral gate. Interestingly, increased membrane fluidity potentiated the bactericidal effect of one of the antibodies, with implications for the mechanism of function of BamA.

7. Conclusions and future perspectives

β-barrel proteins of the Gram-negative bacterial outer membrane are a versatile class of transmembrane proteins. The past 40 years have seen major advances in understanding the structure, function and folding of these proteins. Though the general architecture of β-barrel proteins has been known for 30 years (Weiss et al., 1991), there is still much to learn about these molecules. For example, a recent bioinformatic analysis suggested that bacteria contain OM proteins with multiple transmembrane β-barrel domains (Solan, Pereira, Lupas, Kolodny, & Ben-Tal, 2021). Thus, this class of proteins still contains surprises. As β-barrels play central roles in virulence-related functions as well as biofilm formation and antibiotic resistance, understanding how these proteins fold and integrate into the OM is of critical importance. β-barrels, particularly the essential BamA and LptD, are attractive targets for novel antibiotics—indeed, the very recent discovery of darobactin shows that nature has already taken advantage of the former vulnerability. However, also non-essential β-barrels could be considered as candidates for vaccine development or designing novel antimicrobials. Impairing β-barrel folding or function would have consequences for OM integrity, the ability to produce virulence factors or biofilm, or a reduction in general fitness, which could all have therapeutic effects in an infection setting. Therefore, studying this class of proteins is still of paramount biological and medical interest.

Disclaimer

Parts of this chapter are based on an MSc thesis submitted to and approved by the University of Oslo [Hermansen, S. (2020). *Investigating the effects of loop modifications on the folding of outer membrane protein X*. The University of Oslo, Oslo, Norway].

References

Abellon-Ruiz, J., Kaptan, S. S., Basle, A., Claudi, B., Bumann, D., Kleinekathofer, U., et al. (2017). Structural basis for maintenance of bacterial outer membrane lipid asymmetry. *Nature Microbiology*, *2*(12), 1616–1623.

Ahn, V. E., Lo, E. I., Engel, C. K., Chen, L., Hwang, P. M., Kay, L. E., et al. (2004). A hydrocarbon ruler measures palmitate in the enzymatic acylation of endotoxin. *The EMBO Journal*, *23*(15), 2931–2941.

Alvira, S., Watkins, D. W., Troman, L., Allen, W. J., Lorriman, J. S., Degliesposti, G., et al. (2020). Inter-membrane association of the Sec and BAM translocons for bacterial outer-membrane biogenesis. *eLife*, *9*, e60669.

Anderson, D. E., Dautin, N., Barnard, T. J., & Bernstein, H. D. (2007). Cleavage of a bacterial autotransporter by an evolutionarily convergent autocatalytic mechanism. *The EMBO Journal*, *26*(7), 1942–1952.

Andolina, G., Bencze, L., Zerbe, K., Müller, M., Steinmann, J., Kocherla, H., et al. (2018). A peptidomimetic antibiotic interacts with the periplasmic domain of LptD from *Pseudomonas aeruginosa*. *ACS Chemical Biology*, *13*(3), 666–675.

Archibald, J. (2015). Endosymbiosis and eukaryotic cell evolution. *Current Biology*, *25*(19), R911–R921.

Arnold, T., Poynor, M., Nussberger, S., Lupas, A. N., & Linke, D. (2007). Gene duplication of the eight-stranded β-Barrel OmpX produces a functional pore: A scenario for the evolution of transmembrane β-barrels. *Journal of Molecular Biology*, *366*(4), 1174–1184.

Barnard, T. J., Dautin, N., Lukacik, P., Bernstein, H. D., & Buchanan, S. K. (2007). Autotransporter structure reveals intra-barrel cleavage followed by conformational changes. *Nature Structural & Molecular Biology*, *14*(12), 1214–1220.

Berven, F. S., Flikka, K., Jensen, H. B., & Eidhammer, I. (2004). BOMP: A program to predict integral β-barrel outer membrane proteins encoded within genomes of Gram-negative bacteria. *Nucleic Acids Research*, *32*(Suppl. 2), W394–W399.

Biedzka-Sarek, M., Jarva, H., Hyytiäinen, H., Meri, S., & Skurnik, M. (2008). Characterization of complement factor H binding to *Yersinia enterocolitica* serotype O:3. *Infection and Immunity*, *76*(9), 4100–4109.

Bishop, R. E. (2005). The lipid A palmitoyltransferase PagP: Molecular mechanisms and role in bacterial pathogenesis. *Molecular Microbiology*, *57*(4), 900–912.

Blasband, A. J., Marcotte, W. R., & Schnaitman, C. A. (1986). Structure of the *lC* and *nmpC* outer membrane porin protein genes of lambdoid bacteriophage. *Journal of Biological Chemistry*, *261*(27), 12723–12732.

Bolam, D. N., & van den Berg, B. (2018). TonB-dependent transport by the gut microbiota: Novel aspects of an old problem. *Current Opinion in Structural Biology*, *51*, 35–43.

Bondì, R., Chiani, P., Michelacci, V., Minelli, F., Caprioli, A., & Morabito, S. (2017). The gene *tia*, harbored by the subtilase-encoding pathogenicity island, is involved in the ability of Locus of Enterocyte Effacement-negative Shiga toxin-producing *Escherichia coli* strains to invade monolayers of epithelial cells. *Infection and Immunity*, *85*(12), 613.

Booth, P. J., & Curnow, P. (2009). Folding scene investigation: Membrane proteins. *Current Opinion in Structural Biology*, *19*(1), 8–13.

Botos, I., Noinaj, N., & Buchanan, S. K. (2017). Insertion of proteins and lipopolysaccharide into the bacterial outer membrane. *Philosophical Transactions of the Royal Society of London. Series B, Biological Sciences*, *372*(1726), 20160224.

Burgess, N. K., Dao, T. P., Stanley, A. M., & Fleming, K. G. (2008). β-barrel proteins that reside in the *Escherichia coli* outer membrane *in vivo* demonstrate varied folding behavior *in vitro*. *The Journal of Biological Chemistry*, *283*(39), 26748.

Celia, H., Noinaj, N., & Buchanan, S. K. (2020). Structure and stoichiometry of the Ton molecular motor. *International Journal of Molecular Sciences*, *21*(2), 375.

Chatterjee, S., & Rothenberg, E. (2012). Interaction of bacteriophage λ with its *E. coli* Receptor, LamB. *Viruses*, *4*(11), 3162–3178.

Chatzi, A., Sardis, M., Economou, A., & Karamanou, S. (2014). SecA-mediated targeting and translocation of secretory proteins. *Biochimica et Biophysica Acta, 1843*(8), 1466.
Chauhan, N., Hatlem, D., Orwick-Rydmark, M., Schneider, K., Floetenmeyer, M., van Rossum, B., et al. (2019). Insights into the autotransport process of a trimeric autotransporter, *Yersinia* Adhesin A (YadA). *Molecular Microbiology, 111*(3), 844–862.
Cho, S., Szewczyk, J., Pesavento, C., Zietek, M., Banzhaf, M., Roszczenko, P., et al. (2014). Detecting envelope stress by monitoring β-barrel assembly. *Cell, 159*(7), 1652–1664.
Choi, U., & Lee, C. (2019). Distinct roles of outer membrane porins in antibiotic resistance and membrane integrity in *Escherichia coli. Frontiers in Microbiology, 10*, 953.
Christie, P. (2019). The rich tapestry of bacterial protein translocation systems. *The Protein Journal, 38*(4), 389–408.
Coppens, F., Castaldo, G., Debraekeleer, A., Subedi, S., Moonens, K., Lo, A., et al. (2018). Hop-family *Helicobacter* outer membrane adhesins form a novel class of Type 5-like secretion proteins with an interrupted β-barrel domain. *Molecular Microbiology, 110*(1), 33–46.
Dal Peraro, M., & van der Goot, F. G. (2016). Pore-forming toxins: Ancient, but never really out of fashion. *Nature Reviews. Microbiology, 14*(2), 77–92.
Dalbey, R. E., & Kuhn, A. (2012). Protein traffic in Gram-negative bacteria—How exported and secreted proteins find their way. *FEMS Microbiology Reviews, 36*(6), 1023–1045.
Dam, S., Pagès, J., & Masi, M. (2017). Dual regulation of the small RNA MicC and the quiescent porin OmpN in response to antibiotic stress in *Escherichia coli. Antibiotics (Basel), 6*(4).
Dekoninck, K., Létoquart, J., Laguri, C., Demange, P., Bevernaegie, R., Simorre, J., et al. (2020). Defining the function of OmpA in the Rcs stress response. *eLife, 9*, e60861.
Doerner, P. A., & Sousa, M. C. (2017). Extreme dynamics in the BamA β-barrel seam. *Biochemistry, 56*(24), 3142–3149.
Doyle, M. T., & Bernstein, H. D. (2021). BamA forms a translocation channel for polypeptide export across the bacterial outer membrane. *Molecular Cell, 81*(9), 2000–2012.e3.
Drobnak, I., Braselmann, E., & Clark, P. L. (2015). Multiple driving forces required for efficient secretion of autotransporter virulence proteins. *The Journal of Biological Chemistry, 290*(16), 10104–10116.
Du, D., Wang, Z., James, N. R., Voss, J. E., Klimont, E., Ohene-Agyei, T., et al. (2014). Structure of the AcrAB–TolC multidrug efflux pump. *Nature, 509*(7501), 512–515.
Duguay, A. R., & Silhavy, T. J. (2004). Quality control in the bacterial periplasm. *Biochimica et Biophysica Acta, 1694*(1), 121–134.
Dunstan, R. A., Hay, I. D., & Lithgow, T. (2017). Defining membrane protein localization by isopycnic density gradients. *Methods in Molecular Biology (Clifton, N.J.), 1615*, 81.
Dunstan, R. A., Hay, I. D., Wilksch, J. J., Schittenhelm, R. B., Purcell, A. W., Clark, J., et al. (2015). Assembly of the secretion pores GspD, Wza and CsgG into bacterial outer membranes does not require the Omp85 proteins BamA or TamA. *Molecular Microbiology, 97*(4), 616–629.
Eren, E., Murphy, M., Goguen, J., & van den Berg, B. (2010). An active site water network in the plasminogen activator Pla from *Yersinia pestis. Structure, 18*(7), 809–818.
Eren, E., & van den Berg, B. (2012). Structural basis for activation of an integral membrane protease by lipopolysaccharide. *The Journal of Biological Chemistry, 287*(28), 23971–23976.
Eren, E., Vijayaraghavan, J., Liu, J., Cheneke, B. R., Touw, D. S., Lepore, B. W., et al. (2012). Substrate specificity within a family of outer membrane carboxylate channels. *PLoS Biology, 10*(1), e1001242.
Fagan, R. P., & Smith, S. G. J. (2007). The Hek outer membrane protein of *Escherichia coli* is an auto-aggregating adhesin and invasin. *FEMS Microbiology Letters, 269*(2), 248.
Fan, E., Chauhan, N., Udatha, D. B., Leo, J. C., & Linke, D. (2016). Type V secretion systems in bacteria. *Microbiology Spectrum, 4*(1), (VMBF-2015).

Forst, D., Wacker, T., Welte, W., & Diederichs, K. (1998). Structure of the sucrose-specific porin ScrY from *Salmonella* typhimurium and its complex with sucrose. *Nature Structural Biology*, *5*(1), 37–46.
Franklin, M. W., & Slusky, J. S. G. (2018). Tight turns of outer membrane proteins: An analysis of sequence, structure, and hydrogen bonding. *Journal of Molecular Biology*, *430*(18), 3251–3265.
Geibel, S., & Waksman, G. (2014). The molecular dissection of the chaperone-usher pathway. *Biochimica et Biophysica Acta (BBA) - Molecular Cell Research*, *1843*(8), 1559.
Glaubman, J., Hofmann, J., Bonney, M. E., Park, S., Thomas, J. M., Kokona, B., et al. (2016). Self-association motifs in the enteroaggregative *Escherichia coli* heat-resistant agglutinin 1. *Microbiology (Reading, England)*, *162*(7), 1091–1102.
Gómez-Santos, N., Glatter, T., Koebnik, R., Świątek-Połatyńska, M. A., & Søgaard-Andersen, L. (2019). A TonB-dependent transporter is required for secretion of protease PopC across the bacterial outer membrane. *Nature Communications*, *10*(1), 1360
Gorasia, D. G., Veith, P. D., & Reynolds, E. C. (2020). The type IX secretion system: Advances in structure, function and organisation. *Microorganisms (Basel)*, *8*(8).
Goyal, P., Krasteva, P. V., Van Gerven, N., Gubellini, F., Van den Broeck, I., Troupiotis-Tsaïlaki, A., et al. (2014). Structural and mechanistic insights into the bacterial amyloid secretion channel CsgG. *Nature*, *516*(7530), 250–253.
Gupta, R. (2011). Origin of diderm (Gram-negative) bacteria: Antibiotic selection pressure rather than endosymbiosis likely led to the evolution of bacterial cells with two membranes. *Antonie Van Leeuwenhoek*, *100*(2), 171–182.
Hagan, C. L., Silhavy, T. J., & Kahne, D. (2011). β-Barrel membrane protein assembly by the Bam complex. *Annual Review of Biochemistry*, *80*(1), 189–210.
Hagan, C. L., Wzorek, J. S., & Kahne, D. (2015). Inhibition of the β-barrel assembly machine by a peptide that binds BamD. *Proceedings of the National Academy of Sciences of the United States of America*, *112*(7), 2011–2016.
Haiko, J., Suomalainen, M., Ojala, T., Laehteenmaeki, K., & Korhonen, T. K. (2009). Breaking barriers—Attack on innate immune defences by omptin surface proteases of enterobacterial pathogens. *Innate Immunity*, *15*(2), 67–80.
Hantke, K. (2020). Compilation of *Escherichia coli* K-12 outer membrane phage receptors—Their function and some historical remarks. *FEMS Microbiology Letters*, *367*(2).
Hart, E. M., Mitchell, A. M., Konovalova, A., Grabowicz, M., Sheng, J., Han, X., et al. (2019). A small-molecule inhibitor of BamA impervious to efflux and the outer membrane permeability barrier. *Proceedings of the National Academy of Sciences of the United States of America*, *116*(43), 21748–21757.
Hatlem, D., Trunk, T., Linke, D., & Leo, J. C. (2019). Catching a SPY: Using the SpyCatcher-SpyTag and related systems for labeling and localizing bacterial proteins. *International Journal of Molecular Sciences*, *20*(9), 2129.
Hegde, R. S., & Bernstein, H. D. (2006). The surprising complexity of signal sequences. *Trends in Biochemical Sciences*, *31*(10), 563–571.
Heller, K. B. (1978). Apparent molecular weights of a heat-modifiable protein from the outer membrane of *Escherichia coli* in gels with different acrylamide concentrations. *Journal of Bacteriology*, *134*(3), 1181–1183.
Hjelmeland, L. M. (1990). Solubilization of native membrane proteins. *Methods in Enzymology*, *182*, 253–264.
Ho, D. K., Jarva, H., & Meri, S. (2010). Human complement factor H binds to outer membrane protein Rck of *Salmonella*. *The Journal of Immunology*, *185*(3), 1763–1769.
Ho, D. K., Tissari, J., Järvinen, H. M., Blom, A. M., Meri, S., & Jarva, H. (2011). Functional recruitment of human complement inhibitor C4b-binding protein to outer membrane protein Rck of *Salmonella*. *PLoS One*, *6*(11), e27546.

Hong, H., Park, S., Flores Jiménez, R. H., Rinehart, D., & Tamm, L. K. (2007). Role of aromatic side chains in the folding and thermodynamic stability of integral membrane proteins. *Journal of the American Chemical Society*, *129*(26), 8320–8327.
Horne, J. E., Brockwell, D. J., & Radford, S. E. (2020). Role of the lipid bilayer in outer membrane protein folding in Gram-negative bacteria. *The Journal of Biological Chemistry*, *295*(30), 10340–10367.
Hussain, S., Peterson, J. H., & Bernstein, H. D. (2020). Bam complex-mediated assembly of bacterial outer membrane proteins synthesized in an in vitro translation system. *Scientific Reports*, *10*(1), 4557.
Ikryannikova, L. N., Kurbatov, L. K., Gorokhovets, N. V., Zamyatnin, J., & Andrey, A. (2020). Contact-dependent growth inhibition in bacteria: Do not get too close! *International Journal of Molecular Sciences*, *21*(21), 7990.
Imai, Y., Meyer, K. J., Iinishi, A., Favre-Godal, Q., Green, R., Manuse, S., et al. (2019). A new antibiotic selectively kills Gram-negative pathogens. *Nature*, *576*(7787), 459–464.
Junker, M., Besingi, R. N., & Clark, P. L. (2009). Vectorial transport and folding of an autotransporter virulence protein during outer membrane secretion. *Molecular Microbiology*, *71*(5), 1323–1332.
Kang'ethe, W., & Bernstein, H. D. (2013). Charge-dependent secretion of an intrinsically disordered protein via the autotransporter pathway. *Proceedings of the National Academy of Sciences of the United States of America*, *110*(45), E4246–E4255.
Kanonenberg, K., Spitz, O., Erenburg, I. N., Beer, T., & Schmitt, L. (2018). Type I secretion system-it takes three and a substrate. *FEMS Microbiology Letters*, *365*(11), (fny094).
Kaur, H., Jakob, R. P., Marzinek, J. K., Green, R., Imai, Y., Bolla, J. R., et al. (2021). The antibiotic darobactin mimics a β-strand to inhibit outer membrane insertase. *Nature*, *593*, 125–129.
Kelly, S. M., Jess, T. J., & Price, N. C. (2005). How to study proteins by circular dichroism. *Biochimica et Biophysica Acta (BBA) - Proteins and Proteomics*, *1751*(2), 119–139.
Kim, Y. C., Tarr, A. W., & Penfold, C. N. (2014). Colicin import into *E. coli* cells: A model system for insights into the import mechanisms of bacteriocins. *Biochimica et Biophysica Acta (BBA) - Molecular Cell Research*, *1843*(8), 1717–1731.
Klauser, T., Pohlner, J., & Meyer, T. F. (1993). The secretion pathway of IgA protease-type proteins in Gram-negative bacteria. *BioEssays*, *15*(12), 799–805.
Kleinschmidt, J. H. (2015). Folding of β-barrel membrane proteins in lipid bilayers—Unassisted and assisted folding and insertion. *Biochimica et Biophysica Acta (BBA) - Biomembranes*, *1848*(9), 1927.
Kleinschmidt, J. H., Bulieris, P. V., Qu, J., Dogterom, M., & den Blaauwen, T. (2011). Association of neighboring β-strands of outer membrane protein A in lipid bilayers revealed by site-directed fluorescence quenching. *Journal of Molecular Biology*, *407*(2), 316–332.
Kleinschmidt, J. H., den Blaauwen, T., Driessen, A. J. M., & Tamm, L. K. (1999). Outer membrane protein A of *Escherichia coli* inserts and folds into lipid bilayers by a concerted mechanism. *Biochemistry*, *38*(16), 5006–5016.
Kleinschmidt, J. H., & Popot, J. (2014). Folding and stability of integral membrane proteins in amphipols. *Archives of Biochemistry and Biophysics*, *564*, 327–343.
Kleinschmidt, J. H., & Tamm, L. K. (1999). Time-resolved distance determination by tryptophan fluorescence quenching: Probing intermediates in membrane protein folding. *Biochemistry*, *38*(16), 4996–5005.
Kleinschmidt, J. H., Wiener, M. C., & Tamm, L. K. (1999). Outer membrane protein A of *E. coli* folds into detergent micelles, but not in the presence of monomeric detergent. *Protein Science*, *8*(10), 2065–2071.
Knopp, M., & Andersson, D. I. (2015). Amelioration of the fitness costs of antibiotic resistance due to reduced outer membrane permeability by upregulation of alternative porins. *Molecular Biology and Evolution*, *32*(12), 3252–3263.

Knowles, T. J., Scott-Tucker, A., Overduin, M., & Henderson, I. R. (2009). Membrane protein architects: The role of the BAM complex in outer membrane protein assembly. *Nature Reviews Microbiology*, 7(3), 206–214. https://doi.org/10.1038/nrmicro2069.

Koebnik, R. (1999). Structural and functional roles of the surface-exposed loops of the β-barrel membrane protein OmpA from *Escherichia coli*. *Journal of Bacteriology*, *181*(12), 3688–3694.

Koebnik, R., Locher, K. P., & Van Gelder, P. (2000). Structure and function of bacterial outer membrane proteins: Barrels in a nutshell. *Molecular Microbiology*, *37*(2), 239–253.

Kolodziejek, A. M., Hovde, C. J., & Minnich, S. A. (2012). *Yersinia pestis* Ail: Multiple roles of a single protein. *Frontiers in Cellular and Infection Microbiology*, *2*, 103.

Konovalova, A., Perlman, D. H., Cowles, C. E., & Silhavy, T. J. (2014). Transmembrane domain of surface-exposed outer membrane lipoprotein RcsF is threaded through the lumen of β-barrel proteins. *Proceedings of the National Academy of Sciences of the United States of America*, *111*(41), E4350–E4358.

Koronakis, V., Eswaran, J., & Hughes, C. (2004). Structure and function of TolC: The bacterial exit duct for proteins and drugs. *Annual Review of Biochemistry*, *73*(1), 467–489.

Koronakis, V., Sharff, A., Koronakis, E., Luisi, B., & Hughes, C. (2000). Crystal structure of the bacterial membrane protein TolC central to multidrug efflux and protein export. *Nature*, *405*(6789), 914–919.

Kramer, R. A., Brandenburg, K., Vandeputte-Rutten, L., Werkhoven, M., Gros, P., Dekker, N., et al. (2002). Lipopolysaccharide regions involved in the activation of *Escherichia coli* outer membrane protease OmpT. *European Journal of Biochemistry*, *269*(6), 1746–1752.

Krojer, T., Sawa, J., Schäfer, E., Saibil, H. R., Ehrmann, M., & Clausen, T. (2008). Structural basis for the regulated protease and chaperone function of DegP. *Nature*, *453*(7197), 885–890.

Kukkonen, M., Suomalainen, M., Kyllönen, P., Lähteenmäki, K., Lång, H., Virkola, R., et al. (2004). Lack of O-antigen is essential for plasminogen activation by *Yersinia pestis* and *Salmonella enterica*. *Molecular Microbiology*, *51*(1), 215–225.

Kulathila, R., Kulathila, R., Indic, M., & van den Berg, B. (2011). Crystal structure of *Escherichia coli* CusC, the outer membrane component of a heavy metal efflux pump. *PLoS One*, *6*(1), e15610.

Lauber, F., Deme, J. C., Lea, S. M., & Berks, B. C. (2018). Type 9 secretion system structures reveal a new protein transport mechanism. *Nature*, *564*(7734), 77–82.

Lee, J., Tomasek, D., Santos, T. M., May, M. D., Meuskens, I., & Kahne, D. (2019). Formation of a β-barrel membrane protein is catalyzed by the interior surface of the assembly machine protein BamA. *eLife*, *8*, e49787.

Leo, J. C., Grin, I., & Linke, D. (2012). Type V secretion: Mechanism(s) of autotransport through the bacterial outer membrane. *Philosophical Transactions of the Royal Society of London. Series B, Biological Sciences*, *367*(1592), 1088–1101.

Leo, J. C., & Linke, D. (2018). A unified model for BAM function that takes into account type Vc secretion and species differences in BAM composition. *AIMS Microbiology*, *4*(3), 455–468.

Leo, J. C., Oberhettinger, P., Yoshimoto, S., Udatha, D. B., Morth, J. P., Schütz, M., et al. (2016). Secretion of the intimin passenger domain is driven by protein folding. *The Journal of Biological Chemistry*, *291*(38), 20096–20112.

Leyton, D. L., Johnson, M. D., Thapa, R., Huysmans, G. H., Dunstan, R. A., Celik, N., et al. (2014). A mortise–tenon joint in the transmembrane domain modulates autotransporter assembly into bacterial outer membranes. *Nature Communications*, *5*(1), 4239.

Li, X., Plésiat, P., & Nikaido, H. (2015). The challenge of efflux-mediated antibiotic resistance in Gram-negative bacteria. *Clinical Microbiology Reviews*, *28*(2), 337–418.

Li, Y., Zhu, X., Zhang, J., Lin, Y., You, X., Chen, M., et al. (2020). Identification of a compound that inhibits the growth of Gram-negative bacteria by blocking BamA–BamD interaction. *Frontiers in Microbiology*, *11*, 1252.

Linke, D. (2014). Explanatory chapter: Choosing the right detergent. *Methods in Enzymology*, *541*, 141.

Lukaszczyk, M., Pradhan, B., & Remaut, H. (2019). The biosynthesis and structures of bacterial pili. In A. Kuhn (Ed.), *Subcellular Biochemistry: Vol. 92. Bacterial cell walls and membranes*. Cham: Springer.

Mancini, J., Weckselblatt, B., Chung, Y. K., Durante, J. C., Andelman, S., Glaubman, J., et al. (2011). The heat-resistant agglutinin family includes a novel adhesin from enteroaggregative *Escherichia coli* strain 60A. *Journal of Bacteriology*, *193*(18), 4813–4820.

Masi, M., Winterhalter, M., & Pagès, J.-M. (2019). Outer membrane porins. In A. Kuhn (Ed.), *Subcellular Biochemistry: Vol. 92. Bacterial cell walls and membranes*. Cham: Springer.

Maurya, S. R., Chaturvedi, D., & Mahalakshmi, R. (2013). Modulating lipid dynamics and membrane fluidity to drive rapid folding of a transmembrane barrel. *Scientific Reports*, *3*(1), 1989.

May, K. L., & Silhavy, T. J. (2018). The *Escherichia coli* phospholipase PldA regulates outer membrane homeostasis via lipid signaling. *MBio*, *9*(2), 379.

McLachlan, A. D. (1979). Gene duplications in the structural evolution of chymotrypsin. *Journal of Molecular Biology*, *128*(1), 49–79.

Megrian, D., Taib, N., Witwinowski, J., Beloin, C., & Gribaldo, S. (2020). One or two membranes? Diderm Firmicutes challenge the Gram-positive/Gram-negative divide. *Molecular Microbiology*, *113*(3), 659–671.

Meng, G., Surana, N. K., St Geme, I. I. I., Joseph, W., & Waksman, G. (2006). Structure of the outer membrane translocator domain of the Haemophilus influenzae Hia trimeric autotransporter. *The EMBO Journal*, *25*(11), 2297–2304.

Meuskens, I., Saragliadis, A., Leo, J. C., & Linke, D. (2019). Type V secretion systems: An overview of passenger domain functions. *Frontiers in Microbiology*, *10*, 1163.

Michalik, M., Orwick-Rydmark, M., Habeck, M., Alva, V., Arnold, T., & Linke, D. (2017). An evolutionarily conserved glycine-tyrosine motif forms a folding core in outer membrane proteins. *PLoS One*, *12*(8), e0182016.

Micoli, F., & MacLennan, C. A. (2020). Outer membrane vesicle vaccines. *Seminars in Immunology*, *50*, 101433.

Mogensen, J. E., & Otzen, D. E. (2005). Interactions between folding factors and bacterial outer membrane proteins. *Molecular Microbiology*, *57*(2), 326–346.

Montero, D. A., Velasco, J., Del Canto, F., Puente, J. L., Padola, N. L., Rasko, D. A., et al. (2017). Locus of adhesion and autoaggregation (LAA), a pathogenicity island present in emerging Shiga toxin-producing *Escherichia coli* strains. *Scientific Reports*, 7(1), 7011.

Muñoz, V., Thompson, P. A., Hofrichter, J., & Eaton, W. A. (1997). Folding dynamics and mechanism of β-hairpin formation. *Nature*, *390*(6656), 196–199.

Murzin, A. G., Lesk, A. M., & Chothia, C. (1994). Principles determining the structure of β-sheet barrels in proteins I. A theoretical analysis. *Journal of Molecular Biology*, *236*(5), 1369–1381.

Nakamura, K., & Mizushima, S. (1976). Effects of heating in dodecyl sulfate solution on the conformation and electrophoretic mobility of isolated major outer membrane proteins from *Escherichia coli* K-12. *Journal of Biochemistry (Tokyo)*, *80*(6), 1411–1422.

Neuberger, A., Du, D., & Luisi, B. F. (2018). Structure and mechanism of bacterial tripartite efflux pumps. *Research in Microbiology*, *169*(7-8), 401–413.

Nikaido, H. (2003). Molecular basis of bacterial outer membrane permeability revisited. *Microbiology and Molecular Biology Reviews*, *67*(4), 593–656.

Noinaj, N., Guillier, M., Barnard, T. J., & Buchanan, S. K. (2010). TonB-dependent transporters: Regulation, structure, and function. *Annual Review of Microbiology*, *64*(1), 43–60.

Noinaj, N., Kuszak, A. J., & Buchanan, S. K. (2015). Heat modifiability of outer membrane proteins from Gram-negative bacteria. *Methods in Molecular Biology (Clifton, N.J.), 1329*, 51.

Noinaj, N., Kuszak, A. J., Gumbart, J. C., Lukacik, P., Chang, H., Easley, N. C., et al. (2013). Structural insight into the biogenesis of β-barrel membrane proteins. *Nature, 501*(7467), 385–390.

Oberhettinger, P., Leo, J. C., Linke, D., Autenrieth, I. B., & Schütz, M. S. (2015). The inverse autotransporter intimin exports its passenger domain via a hairpin intermediate. *The Journal of Biological Chemistry, 290*(3), 1837–1849.

Orwick-Rydmark, M., Arnold, T., & Linke, D. (2016). The use of detergents to purify membrane proteins. *Current Protocols in Protein Science, 84*, 4.8.1.

Paetzel, M., Karla, A., Strynadka, N. C. J., & Dalbey, R. E. (2002). Signal peptidases. *Chemical Reviews, 102*(12), 4549–4580.

Paramasivam, N., Habeck, M., & Linke, D. (2012). Is the C-terminal insertional signal in Gram-negative bacterial outer membrane proteins species-specific or not? *BMC Genomics, 13*(1), 510.

Park, J. S., Lee, W. C., Yeo, K. J., Ryu, K., Kumarasiri, M., Hesek, D., et al. (2011). Mechanism of anchoring of OmpA protein to the cell wall peptidoglycan of the Gram-negative bacterial outer membrane. *The FASEB Journal, 26*(1), 219–228.

Patel, G. J., Behrens-Kneip, S., Holst, O., & Kleinschmidt, J. H. (2009). The periplasmic chaperone Skp facilitates targeting, insertion, and folding of OmpA into lipid membranes with a negative membrane surface potential. *Biochemistry, 48*(43), 10235–10245.

Phan, G., Picard, M., & Broutin, I. (2015). Focus on the outer membrane factor OprM, the forgotten player from efflux pumps assemblies. *Antibiotics (Basel), 4*(4), 544–566.

Phan, G., Remaut, H., Wang, T., Allen, W., Pirker, K., Lebedev, A., et al. (2011). Crystal structure of the FimD usher bound to its cognate FimC-FimH substrate. *Nature, 474*(7349).

Pinkner, J. S., Remaut, H., Buelens, F., Miller, E., Åberg, V., Pemberton, N., et al. (2006). Rationally designed small compounds inhibit pilus biogenesis in uropathogenic bacteria. *Proceedings of the National Academy of Sciences of the United States of America, 103*(47), 17897–17902.

Plummer, A. M., & Fleming, K. G. (2016). From chaperones to the membrane with a BAM! *Trends in Biochemical Sciences, 41*(10), 872–882.

Pocanschi, C. L., Popot, J., & Kleinschmidt, J. H. (2013). Folding and stability of outer membrane protein A (OmpA) from *Escherichia coli* in an amphipathic polymer, amphipol A8-35. *European Biophysics Journal, 42*(2-3).

Pollet, R. M., Martin, L. M., & Koropatkin, N. M. (2021). TonB-dependent transporters in the Bacteroidetes: Unique domain structures and potential functions. *Molecular Microbiology, 115*(3), 490–501.

Popot, J. (2014). Folding membrane proteins in vitro: A table and some comments. *Archives of Biochemistry and Biophysics, 564*, 314–326.

Prehna, G., Zhang, G., Gong, X., Duszyk, M., Okon, M., McIntosh, L., et al. (2012). A protein export pathway involving *Escherichia coli* porins. *Structure, 20*(7), 1154–1166.

Prilipov, A., Phale, P. S., Koebnik, R., Widmer, C., & Rosenbusch, J. P. (1998). Identification and characterization of two quiescent porin genes, *nmpC* and *ompN*, in *Escherichia coli* BE. *Journal of Bacteriology, 180*(13), 3388–3392.

Pugsley, A. P., & Schnaitman, C. A. (1978). Identification of three genes controlling production of new outer membrane pore proteins in *Escherichia coli* K-12. *Journal of Bacteriology, 135*(3), 1118–1129.

Rassam, P., Copeland, N. A., Birkholz, O., Tóth, C., Chavent, M., Duncan, A. L., et al. (2015). Supramolecular assemblies underpin turnover of outer membrane proteins in bacteria. *Nature, 523*(7560), 333–336.

Rath, P., Sharpe, T., & Hiller, S. (2020). The electrostatic core of the outer membrane protein X from *E. coli*. *Biochimica et Biophysica Acta (BBA) - Biomembranes*, *1862*(1), 183031.

Rath, P., Sharpe, T., Kohl, B., & Hiller, S. (2019). Two-state folding of the outer membrane protein X into a lipid bilayer membrane. *Angewandte Chemie*, *131*(9), 2691–2695.

Remmert, M., Biegert, A., Linke, D., Lupas, A. N., & Söding, J. (2010). Evolution of outer membrane β-barrels from an ancestral ββ hairpin. *Molecular Biology and Evolution*, *27*(6), 1348.

Remmert, M., Linke, D., Lupas, A. N., & Söding, J. (2009). HHomp-prediction and classification of outer membrane proteins. *Nucleic Acids Research*, *37*(Suppl. 2), W446–W451.

Robert, V. M. V., Volokhina, E. B., Senf, F., Bos, M. P., van Gelder, P., & Tommassen, J. P. M. (2006). Assembly factor Omp85 recognizes its outer membrane protein substrates by a species-specific C-terminal motif. *PLoS Biology*, *4*(11), 1984–e377.

Rodríguez-Alonso, R., Létoquart, J., Nguyen, V. S., Louis, G., Calabrese, A. N., Iorga, B. I., et al. (2020). Structural insight into the formation of lipoprotein-β-barrel complexes. *Nature Chemical Biology*, *16*(9), 1019–1025.

Rojas, E. R., Billings, G., Odermatt, P. D., Auer, G. K., Zhu, L., Miguel, A., et al. (2018). The outer membrane is an essential load-bearing element in Gram-negative bacteria. *Nature*, *559*(7715), 617–621.

Rouse, S. L., Hawthorne, W. J., Berry, J., Chorev, D. S., Ionescu, S. A., Lambert, S., et al. (2017). A new class of hybrid secretion system is employed in *Pseudomonas* amyloid biogenesis. *Nature Communications*, *8*(1), 263.

Rouse, S. L., Stylianou, F., Wu, H. Y. G., Berry, J., Sewell, L., Morgan, R. M. L., et al. (2018). The FapF amyloid secretion transporter possesses an atypical asymmetric coiled coil. *Journal of Molecular Biology*, *430*(20), 3863–3871.

Royer, S., Morais, A. P., & da Fonseca Batistão, D. W. (2021). Phage therapy as strategy to face post-antibiotic era: A guide to beginners and experts. *Archives of Microbiology*, *203*(4), 1271–1279.

Rutten, L., Geurtsen, J., Lambert, W., Smolenaers, J. J. M., Bonvin, A. M., de Haan, A., et al. (2006). Crystal structure and catalytic mechanism of the LPS 3-O-deacylase PagL from *Pseudomonas aeruginosa*. *Proceedings of the National Academy of Sciences of the United States of America*, *103*(18), 7071–7076.

Ryoo, D., Rydmark, M. O., Pang, Y. T., Lundquist, K. P., Linke, D., & Gumbart, J. C. (2020). BamA is required for autotransporter secretion. *Biochimica et Biophysica Acta (BBA) - General Subjects*, *1864*(7), 129581.

Samsudin, F., Boags, A., Piggot, T. J., & Khalid, S. (2017). Braun's lipoprotein facilitates OmpA interaction with the *Escherichia coli* cell wall. *Biophysical Journal*, *113*(7), 1496–1504.

Sansom, M. S., & Kerr, I. D. (1995). Transbilayer pores formed by β-barrels: Molecular modeling of pore structures and properties. *Biophysical Journal*, *69*(4), 1334–1343.

Schauer, K., Rodionov, D. A., & de Reuse, H. (2008). New substrates for TonB-dependent transport: Do we only see the 'tip of the iceberg'? *Trends in Biochemical Sciences*, *33*(7), 330–338.

Schirmer, T., Keller, T. A., Wang, Y., & Rosenbusch, J. P. (1995). Structural basis for sugar translocation through maltoporin channels at 3.1 angstrom resolution. *Science*, *267*(5197), 512.

Schulz, G. E. (2002). The structure of bacterial outer membrane proteins. *Biochimica et Biophysica Acta (BBA) - Biomembranes*, *1565*(2), 308–317.

Schulz, G. E. (2004). The structures of general porins. In R. Benz (Ed.), *Bacterial and eukaryotic porins* (pp. 25–40). Weinheim: Wiley-VCH Verlag GmbH & Co. KGaA.

Shahid, S. A., Bardiaux, B., Franks, W. T., Krabben, L., Habeck, M., Van Rossum, B., et al. (2012). Membrane-protein structure determination by solid-state NMR spectroscopy of microcrystals. *Nature Methods*, *9*(12), 1212–1217.

Simons, A., Alhanout, K., & Duval, R. E. (2020). Bacteriocins, antimicrobial peptides from bacterial origin: Overview of their biology and their impact against multidrug-resistant bacteria. *Microorganisms*, *8*(5), 639.

Slusky, J. S. (2017). Outer membrane protein design. *Current Opinion in Structural Biology*, *45*, 45–52.

Snijder, H. J., Ubarretxena-Belandia, I., Blaauw, M., Kalk, K. H., Verhelj, H. M., Egmond, M. R., et al. (1999). Structural evidence for dimerization-regulated activation of an integral membrane phospholipase. *Nature*, *401*(6754), 717–721.

Söding, J., Biegert, A., & Lupas, A. N. (2005). The HHpred interactive server for protein homology detection and structure prediction. *Nucleic Acids Research*, *33*(Web Server), W244–W248.

Solan, R., Pereira, J., Lupas, A. N., Kolodny, R., & Ben-Tal, N. (2021). Gram-negative outer membrane proteins with multiple β-barrel domains. *bioRxiv*. https://doi.org/10.1101/2021.02.17.431615.

Sonntag, I., Schwarz, H., Hirota, Y., & Henning, U. (1978). Cell envelope and shape of *Escherichia coli*: Multiple mutants missing the outer membrane lipoprotein and other major outer membrane proteins. *Journal of Bacteriology*, *136*(1), 280–285.

Soufi, B., Krug, K., Harst, A., & Macek, B. (2015). Characterization of the *E. coli* proteome and its modifications during growth and ethanol stress. *Frontiers in Microbiology*, *6*, 103.

Srinivas, N., Jetter, P., Ueberbacher, B. J., Werneburg, M., Zerbe, K., Steinmann, J., et al. (2010). Peptidomimetic antibiotics target outer-membrane biogenesis in *Pseudomonas aeruginosa*. *Science*, *327*(5968), 1010–1013.

Stapleton, J. A., Whitehead, T. A., & Nanda, V. (2015). Computational redesign of the lipid-facing surface of the outer membrane protein OmpA. *Proceedings of the National Academy of Sciences of the United States of America*, *112*(31), 9632–9637.

Stark, G. R., Stein, W. H., & Moore, S. (1960). Reactions of the cyanate present in aqueous urea with amino acids and proteins. *The Journal of Biological Chemistry*, *235*(11), 3177–3181.

Steenhuis, M., Abdallah, A. M., de Munnik, S. M., Kuhne, S., Sterk, G., van den Berg van Saparoea, B., et al. (2019). Inhibition of autotransporter biogenesis by small molecules. *Molecular Microbiology*, *112*(1), 81–98.

Storek, K. M., Auerbach, M. R., Shi, H., Garcia, N. K., Sun, D., Nickerson, N. N., et al. (2018). Monoclonal antibody targeting the β-barrel assembly machine of *Escherichia coli* is bactericidal. *Proceedings of the National Academy of Sciences of the United States of America*, *115*(14), 3692–3697.

Su, Y., Wan, K., Mohamed, R., & Nathan, S. (2010). Immunization with the recombinant *Burkholderia pseudomallei* outer membrane protein Omp85 induces protective immunity in mice. *Vaccine*, *28*(31), 5005–5011.

Surrey, T., & Jähnig, F. (1995). Kinetics of folding and membrane insertion of a β-barrel membrane protein. *The Journal of Biological Chemistry*, *270*(47).

Surrey, T., Schmid, A., & Jähnig, F. (1996). Folding and membrane insertion of the trimeric β-barrel protein OmpF. *Biochemistry*, *35*(7), 2283–2288.

Taib, N., Megrian, D., Witwinowski, J., Adam, P., Poppleton, D., Borrel, G., et al. (2020). Genome-wide analysis of the Firmicutes illuminates the diderm/monoderm transition. *Nature Ecology & Evolution*, *4*(12), 1661–1672.

Telhig, S., Said, L. B., Zirah, S., Fliss, I., & Rebuffat, S. (2020). Bacteriocins to thwart bacterial resistance in Gram negative bacteria. *Frontiers in Microbiology*, *11*, 586433.

Thomas, S., Holland, I. B., & Schmitt, L. (2014). The type 1 secretion pathway—The hemolysin system and beyond. *Biochimica et Biophysica Acta (BBA) - Molecular Cell Research, 1843*(8), 1629.
Tomasek, D., Rawson, S., Lee, J., Wzorek, J. S., Harrison, S. C., Li, Z., et al. (2020). Structure of a nascent membrane protein as it folds on the BAM complex. *Nature, 80*(1), 473–478.
Tran, E. N. H., Doyle, M. T., & Morona, R. (2013). LPS unmasking of *Shigella flexneri* reveals preferential localisation of tagged outer membrane protease IcsP to septa and new poles. *PLoS One, 8*(7), e70508.
Trunk, T., Khalil, H. S., & Leo, C. J. (2018). Bacterial autoaggregation. *AIMS Microbiology, 4*(1), 140–164.
Tsaousis, G. N., Hamodrakas, S. J., & Bagos, P. G. (2017). Predicting beta barrel transmembrane proteins using HMMs. *Methods in Molecular Biology (Clifton, N.J.), 1552*, 43–61.
Tsirigotaki, A., De Geyter, J., Šoštaric, N., Economou, A., & Karamanou, S. (2017). Protein export through the bacterial Sec pathway. *Nature Reviews Microbiology, 15*(1), 21–36.
van den Berg, B., Bhamidimarri, S. P., Prajapati, J. D., Kleinekathöfer, U., & Winterhalter, M. (2015). Outer-membrane translocation of bulky small molecules by passive diffusion. *Scientific Reports, 5*(1), 11927.
van den Berg, B., Black, P. N., Clemons, W. M., Jr., & Rapoport, T. A. (2004). Crystal structure of the long-chain fatty acid transporter FadL. *Science, 304*(5676), 1506–1509.
van der Ley, P., de Graaff, P., & Tommassen, J. (1986). Shielding of *Escherichia coli* outer membrane proteins as receptors for bacteriophages and colicins by O-antigenic chains of lipopolysaccharide. *Journal of Bacteriology, 168*(1), 450–452.
van Ulsen, P., Zinner, K. M., Jong, W. S. P., & Luirink, J. (2018). On display: Autotransporter secretion and application. *FEMS Microbiology Letters, 365*(18).
Vandeputte-Rutten, L., Kramer, R. A., Kroon, J., Dekker, N., Egmond, M. R., & Gros, P. (2001). Crystal structure of the outer membrane protease OmpT from *Escherichia coli* suggests a novel catalytic site. *The EMBO Journal, 20*(18), 5033–5039.
Venter, H., Mowla, R., Ohene-Agyei, T., & Ma, S. (2015). RND-type drug efflux pumps from Gram-negative bacteria: Molecular mechanism and inhibition. *Frontiers in Microbiology, 6*, 377.
Vergalli, J., Bodrenko, I. V., Masi, M., Moynié, L., Acosta-Gutiérrez, S., Naismith, J. H., et al. (2020). Porins and small-molecule translocation across the outer membrane of Gram-negative bacteria. *Nature Reviews. Microbiology, 18*(3), 164–176.
Vila-Farrés, X., Parra-Millán, R., Sánchez-Encinales, V., Varese, M., Ayerbe-Algaba, R., Bayó, N., et al. (2017). Combating virulence of Gram-negative bacilli by OmpA inhibition. *Scientific Reports*, 7(1), 14683.
Voulhoux, R., Bos, M. P., Geurtsen, J., Mols, M., & Tommassen, J. (2003). Role of a highly conserved bacterial protein in outer membrane protein assembly. *Science, 299*(5604), 262–265.
Wang, X., Peterson, J. H., & Bernstein, H. D. (2021). Bacterial outer membrane proteins are targeted to the Bam complex by two parallel mechanisms. *MBio, 12*(3).
Wang, Y., Wang, R., Jin, F., Liu, Y., Yu, J., Fu, X., et al. (2016). A supercomplex spanning the inner and outer membranes mediates the biogenesis of β-barrel outer membrane proteins in bacteria. *The Journal of Biological Chemistry, 291*(32), 16720–16729.
Watkinson, T. G., Calabrese, A. N., Giusti, F., Zoonens, M., Radford, S. E., & Ashcroft, A. E. (2015). Systematic analysis of the use of amphipathic polymers for studies of outer membrane proteins using mass spectrometry. *International Journal of Mass Spectrometry, 391*, 54–61.
Weaver, S. J., Ortega, D. R., Sazinsky, M. H., Dalia, T. N., Dalia, A. B., & Jensen, G. J. (2020). CryoEM structure of the type IVa pilus secretin required for natural competence in *Vibrio cholerae*. *Nature Communications, 11*(1), 5080.

Webb, C. T., Heinz, E., & Lithgow, T. (2012). Evolution of the β-barrel assembly machinery. *Trends in Microbiology*, *20*(12), 612–620.
Wedege, E., Lie, K., Bolstad, K., Weynants, V. E., Halstensen, A., Herstad, T. K., et al. (2013). Meningococcal Omp85 in detergent-extracted outer membrane vesicle vaccines induces high levels of non-functional antibodies in mice. *Scandinavian Journal of Immunology*, 77(6), 452–459.
Weiss, M. S., Abele, U., Weckesser, J., Welte, W., Schiltz, E., & Schulz, G. E. (1991). Molecular architecture and electrostatic properties of a bacterial porin. *Science*, *254*(5038), 1627–1630.
Wimley, W. C. (2002). Toward genomic identification of β-barrel membrane proteins: Composition and architecture of known structures. *Protein Science*, *11*(2), 301–312.
Wimley, W. C. (2003). The versatile β-barrel membrane protein. *Current Opinion in Structural Biology*, *13*(4), 404–411.
Worrall, L. J., Hong, C., Vuckovic, M., Deng, W., Bergeron, J. R. C., Majewski, D. D., et al. (2016). Near-atomic-resolution cryo-EM analysis of the *Salmonella* T3S injectisome basal body. *Nature*, *540*(7634), 597–601.
Wu, T., Malinverni, J., Ruiz, N., Kim, S., Silhavy, T. J., & Kahne, D. (2005). Identification of a multicomponent complex required for outer membrane biogenesis in *Escherichia coli*. *Cell*, *121*(2), 235–245.
Wu, R., Stephenson, R., Gichaba, A., & Noinaj, N. (2020). The big BAM theory: An open and closed case? *Biochimica et Biophysica Acta (BBA) - Biomembranes*, *1862*(1), 183062.
Yan, Z., Yin, M., Xu, D., Zhu, Y., & Li, X. (2017). Structural insights into the secretin translocation channel in the type II secretion system. *Nature Structural & Molecular Biology*, *24*(2), 177–183.
Ye, J., & van den Berg, B. (2004). Crystal structure of the bacterial nucleoside transporter Tsx. *The EMBO Journal*, *23*(16), 3187–3195.
Zeng, B., Wang, H., Zou, L., Zhang, A., Yang, X., & Guan, Z. (2010). Evaluation and target validation of indole derivatives as inhibitors of the AcrAB-TolC efflux pump. *Bioscience, Biotechnology, and Biochemistry*, *74*(11), 2237–2241.
Zgurskaya, H. I., López, C. A., & Gnanakaran, S. (2015). Permeability barrier of Gram-negative cell envelopes and approaches to bypass it. *ACS Infectious Diseases*, *1*(11), 512–522.
Zha, Z., Li, C., Li, W., Ye, Z., & Pan, J. (2016). LptD is a promising vaccine antigen and potential immunotherapeutic target for protection against *Vibrio* species infection. *Scientific Reports*, *6*(1), 38577.

CHAPTER FIVE

Ribosome-membrane crosstalk: Co-translational targeting pathways of proteins across membranes in prokaryotes and eukaryotes

Krishnamoorthi Srinivasan[a,b], Aneek Banerjee[a,b], Priya Baid[a], Ankit Dhur[a], and Jayati Sengupta[a,b,*]

[a]Division of Structural Biology and Bioinformatics, CSIR-Indian Institute of Chemical Biology, Kolkata, India
[b]Academy of Scientific and Innovative Research (AcSIR), Ghaziabad, India
*Corresponding author: e-mail address: jayati@iicb.res.in

Contents

1. Introduction 164
2. Membrane translocation via a conserved channel mechanism 169
3. Molecular machinery of bacterial protein transport 171
   3.1 SRP-mediated protein targeting 171
   3.2 Non-SRP mediated pathways 174
4. Co-translational protein transport in eukaryotes: 175
   4.1 Cytosolic ribosome 176
   4.2 Mitochondrial ribosome 182
   4.3 Chloroplast ribosome 186
5. Concluding remarks 188
Acknowledgments 190
Conflict of interest statement 190
Author contributions 190
References 190

Abstract

Ribosomes are the molecular machine of living cells designed for decoding mRNA-encoded genetic information into protein. Being sophisticated machinery, both in design and function, the ribosome not only carries out protein synthesis, but also coordinates several other ribosome-associated cellular processes. One such process is the translocation of proteins across or into the membrane depending on their secretory or membrane-associated nature. These proteins comprise a large portion of a cell's proteome and act as key factors for cellular survival as well as several crucial functional pathways. Protein transport to extra- and intra-cytosolic compartments (across the

Advances in Protein Chemistry and Structural Biology, Volume 128
ISSN 1876-1623
https://doi.org/10.1016/bs.apcsb.2021.07.003

eukaryotic endoplasmic reticulum (ER) or across the prokaryotic plasma membrane) or insertion into membranes majorly occurs through an evolutionarily conserved protein-conducting channel called translocon (eukaryotic Sec61 or prokaryotic SecYEG channels). Targeting proteins to the membrane-bound translocon may occur via post-translational or *co*-translational modes and it is often mediated by recognition of an N-terminal signal sequence in the newly synthesizes polypeptide chain. Co-translational translocation is coupled to protein synthesis where the ribosome-nascent chain complex (RNC) itself is targeted to the translocon. Here, in the light of recent advances in structural and functional studies, we discuss our current understanding of the mechanistic models of co-translational translocation, coordinated by the actively translating ribosomes, in prokaryotes and eukaryotes.

1. Introduction

Protein translation is one of the most energetically expensive processes in a cell, and is carried out by universally conserved molecular machine called ribosomes. Ribosomes are ribonucleoproteins, composed of an assembly of RNA (called ribosomal RNA or rRNA) and proteins (r-proteins), which read genetic instructions encoded in the mRNA and stitch together a specific chain of amino acids to synthesize proteins. The core of the ribosome is essentially rRNA, and ribosome is actually a ribozyme (ribonucleic acid enzymes), as the rRNA alone is sufficient to catalyze the peptidyl transfer reaction (Cech, 2000).

The 3D structure of the ribosome core is highly conserved throughout the tree of life and this reflects the central role of ribosome in cellular function. The ribosome consists of two subunits, the large subunit (LSU) (60S for eukaryotes and 50S for bacteria) and the small subunit (SSU) (40S for eukaryotes and 30S for bacteria) (Lake, 1985). While the SSU is involved in mRNA decoding, the LSU is involved in peptide bond formation between amino acids to synthesize proteins, which are released through an exit tunnel present in the LSU.

In the early days it was thought that ribosome structure would be almost fully conserved in all life forms (Moore, Traut, Noller, Pearson, & Delius, 1968), from archaea to bacteria, from eukaryotes to the organelles (mitochondria and chloroplast). However, researches over the years have shown that, despite the overall similarity, these ribosomes are indeed very different (Genuth & Barna, 2018). While the core of the ribosome is effectively conserved, the periphery is often very different (Fig. 1). Eukaryotic cytosolic ribosomes are larger and more complex than their bacterial counterparts; bear longer rRNA (with specific extensions, called expansion segments),

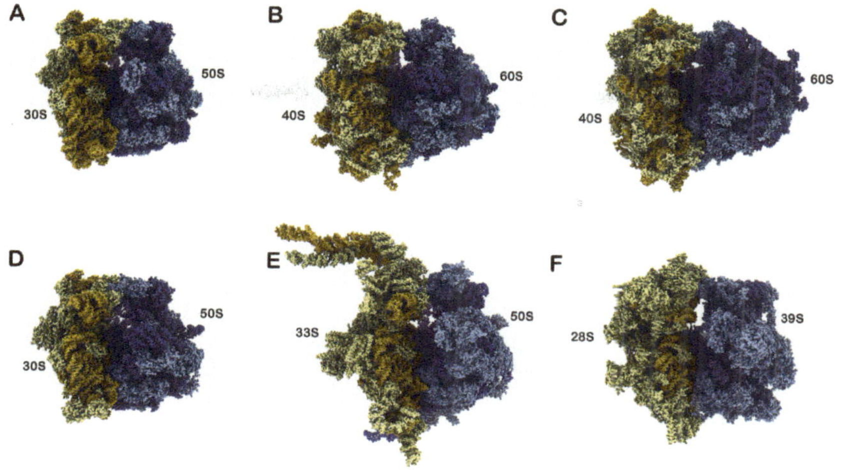

Fig. 1 Seeing is believing: Visualizing the ribosomes from different domains of life. (A) *E. coli* 70S ribosome (PDB ID: 7D6Z). (B) Yeast 80S cytosolic ribosome (PDB ID: 6T4Q). (C) Human 80S cytosolic ribosome (PDB ID: 6EK0). (D) Spinach 70S chloroplast ribosome (PDB ID: 5X8P). (E) Arabidopsis 78S mitochondrial ribosome (PDB ID: 6XYW). (F) Human 55S mitochondrial ribosome (PDB ID: 6ZM6). Light Yellow: SSU ribosomal protein; Dark Yellow: SSU rRNA; Light Blue: LSU ribosomal protein; Dark Blue: LSU rRNA. UCSF-ChimeraX (Goddard et al., 2018) and PyMOL (Schrödinger, LLC) are used for preparation of illustrations.

additional r-proteins and r-protein extensions (Wilson & Doudna Cate, 2012). Eukaryotic cells also have organellar ribosomes, in the mitochondria and the chloroplast, which were predicted to be similar to the bacterial ribosomes from which they are said to have originated (endosymbiotic theory) (Martin, Garg, & Zimorski, 2015). However, there are marked differences between the organelle ribosomes from their bacterial and eukaryotic counterparts, including differences in size, rRNA/protein ratio and several structural differences (Table 1).

The nascent chain starts partial folding within the peptide exit tunnel (Bhushan et al., 2010), and it is followed by the recruitment of myriad of proteins, chaperones and other factors for post-translational processing, folding and localization (Akbar, Bhakta, & Sengupta, 2021; Lakshmipathy, Gupta, Pinkert, Etchells, & Hartl, 2010). Different classes of proteins are processed and targeted differently. Secretory and membrane proteins are one such specialized class of molecules, which have an elaborate system for their processing and localization. An in-depth understanding of the

Table 1 Comparison of Ribosomes from different species.

Properties		**Ribosome origin**				
		Bacterial (*E.coli*)	**Eukaryotic cytosolic (human)**	**Mammalian mitochondria (human)**	**Leishmania mitochondria (*Leishmania taremtolae*)**	**Plant chloroplast (*Spinacia oleracea*)**
Molecular mass		2.3 MDa	3.3–4.3 MDa	2.7 MDa	2.2 MDa	2.6 MDa
Sedimentation coefficient	Whole	70S	80S	55S	50S	70S
	Large	50S	60S	39S	40S	50S
	Small	30S	40S	28S	28–30S	30S
Diameter		260 Å	300 Å	320 Å	245 Å	265 Å
Large subunit composition		23S & 5S rRNA + 33 proteins	28S, 5.8S & 5S rRNA + 47 proteins	16S rRNA + 50 proteins	12S rRNA + 77 proteins	23S, 5S & 4.8S rRNA + 33 proteins
Small subunit composition		16S rRNA + 21 proteins	18S rRNA + 33 proteins	12S rRNA + 29 proteins	9S rRNA + 56 proteins	16S rRNA + 25 proteins
RNA: Protein ratio		2: 1	1: 1	1: 2	1: 3	3: 2
SRP or membrane interacting partners		uL29 23S rRNA L23b L32	eL35 28S rRNA uL23e L19	L47 L28 L23 L24 L45		uL4

transport pathways of these proteins, that translocate through or insert into the membrane, is very important.

Biological membranes play a crucial role in maintaining the integrity of a cell. All cells are enclosed by a phospholipid bilayer called the cell membrane or the plasma membrane, which excludes the constituents of the cell from the external environment. Eukaryotic cells have several membrane-enclosed organelles, each of which is involved in specialized functions. While a lot of the proteins function in the cytosol, some proteins are designed to cross membrane and execute functions outside the cytoplasm. These non-cytosolic proteins are found in eukaryotic organelles such as Endoplasmic Reticulum, Golgi Apparatus, lysosomes etc.; as well in the outer membrane and in the periplasmic space (the space between the plasma membrane and the outer membrane) in bacteria. Some proteins function completely outside the cell, and such proteins are often secreted by cells to regulate their neighboring cells/environment. On the other hand, a significant number of proteins are integral part of the membrane (membrane proteins). Roughly 30% of the entire proteome of humans is composed of membrane proteins (Fagerberg, Jonasson, Von Heijne, Uhlen, & Berglund, 2010).

Both secretory and integral membrane proteins have to traverse the plasma membrane, and utilize similar tools for the transport. Secretory proteins have to be transported across the membrane, bypassing the lipid phase, whereas, integral membrane proteins have to be embedded into the lipid bilayer of the plasma membrane. The interactions of many integral membrane proteins with other organelles and cytoskeleton elements are responsible for maintaining the shape and stability of such components. Several of these membrane proteins execute vital functions for the cell's survival (Kulandaisamy et al., 2018). The quality control (proper translation, precise localization and folding) of membrane proteins and their accurate insertion into the membrane is essential for their optimal function. Mutations or aberrant activity in membrane proteins have been implicated in many different developmental disorders and diseases, including several types of cancer, neurodegenerative diseases, etc. (Sanders & Myers, 2004). Even the misassembly of these proteins is known to cause several diseases, which further highlights the importance of these proteins and their regulation.

Apart from taking active part in protein synthesis, the ribosome plays a crucial role in targeting secretory and membrane proteins to membranes. The rRNA and proteins surrounding the tunnel exit offer platforms for the recruitment of factors involved in protein translocation (Fig. 2). Ever since the discovery of signal sequence-mediated localization of proteins to

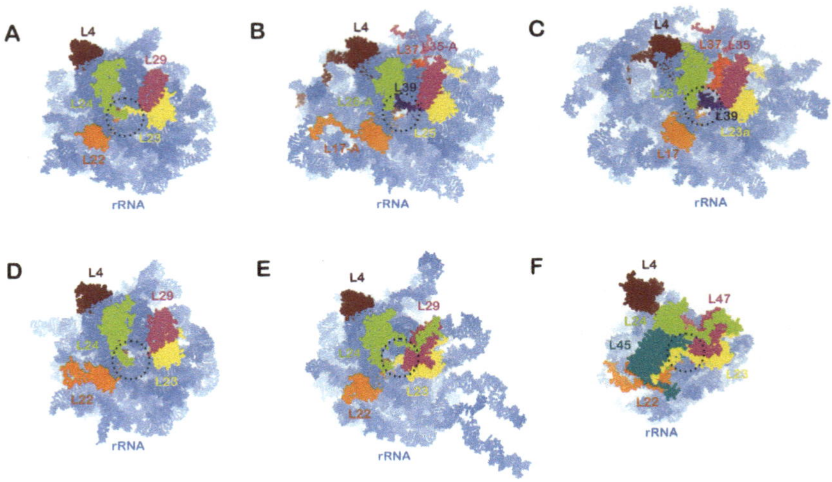

Fig. 2 Cartoon representations of the ribosomal exit tunnel and its periphery. (A) *E. coli* LSU rRNA with the L4, L22, L23, L24 and L29 proteins near the exit tunnel. (B) Yeast LSU rRNA with the L4, L17A, L25, L26A, L35A, L37 and L39 proteins. (C) Human cytosolic LSU rRNA with the L4, L17, L23a, L26, L35, L37 and L39 proteins. (D) Spinach LSU rRNA with the L4, L22, L23, L24 and L29 proteins. (E) Arabidopsis LSU rRNA with the L4, L22, L23, L24 and L29 proteins. (F) Human mitochondrial LSU rRNA with the L4, L22, L23, L24, L45 and L47 proteins. In (A) to (F), the exit tunnel is marked with a dotted circle. Homologous proteins are shown in identical colors.

the endoplasmic reticulum membrane by the research groups of G. Blobel and C. Milstein (Milstein, Brownlee, Harrison, & Mathews, 1972; Sabatini, Blobel, Nonomura, & Adelman, 1971), researchers have become curious to understand the mechanisms of protein transport to/across membranes. Structural, genetic and biochemical studies in the past several years have led to a significant progress towards a mechanistic understanding of protein translocation.

Protein translocation may take place during protein synthesis in a concerted manner (co-translational), or following completion of synthesis of the protein (post-translational); primarily using a conserved protein-conducting channel (eukaryotic Sec61 or prokaryotic SecYEG channels) (Mitra, Frank, & Driessen, 2006). Co-translational protein targeting in prokaryotes and eukaryotes, where actively translating ribosomes are directly involved in coordinating the process, by and large relies on the signal recognition system, which is a conserved ribonucleoprotein complex in all the kingdoms of life (Steinberg, Knupffer, Origi, Asti, & Koch, 2018). Secretory and membrane proteins are synthesized with additional cleavable N-terminal 'signal' peptide sequences serving as tags, and are sorted through recognition of these N-terminal tags by specialized molecules to classify their cellular destination (Blobel et al., 1979). Interestingly, sequences of signal peptides are not conserved, but they contain specific structural elements organized in a similar pattern for both prokaryotic and eukaryotic proteins (von Heijne, 1990).

Although the process of co-translational protein translocation is reasonably similar, it is far from identical in different life forms. The differences in the ribosome, the membrane, the cargo-protein being transported and the factors involved in translocation all add up to similar, yet unique mechanisms for localizing the proteins at the targets. In this book chapter, we review the recent findings that help us to shape up the mechanistic models of co-translational protein transports in bacteria and eukaryotes, and highlight still existing gaps in our understanding of these mechanisms.

2. Membrane translocation via a conserved channel mechanism

A prerequisite for proper functioning of proteins is translocation to their site of action. Various protein factors and complexes play a synergistic role to ensure the integrity of this crucial process. Membranes, being barrier between organelles or extra cellular matrix, is fundamental for accurate integration and/or transportation of proteins. Proteins that associate with the membrane are termed membrane proteins.

Membrane proteins are broadly of two kinds: integral membrane proteins that have segments embedded into the phospholipid bilayer; and peripheral membrane proteins which are anchored to the membrane either via contacts with the lipid or certain membrane bound proteins. Likewise, distinct pathways exist for incorporation of integral membrane proteins into the plasma membrane and the transport of proteins into the periplasmic space. Integration of membrane proteins is predominantly co-translational, while transport of periplasmic proteins is majorly post-translational; although there is evidence of both co-translational and post-translational transport in both (Lu et al., 1998).

A proteinaceous channel called the translocon, is deployed by cells to transport proteins into or across the plasma membrane (Fig. 3). The primary translocon in all life-forms is the Sec translocon, which is a heterotrimeric membrane protein complex (Brundage, Hendrick, Schiebel, Driessen, & Wickner, 1990) composed of α, β, and γ-subunits (Mitra et al., 2005).

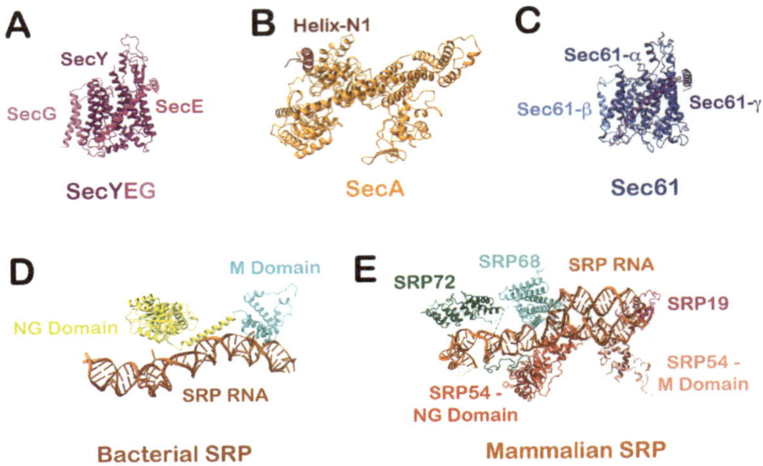

Fig. 3 Atomic structures of the translocon and accessory components involved in co-translational transport. (A) Bacterial Sec YEG (PDB ID: 5NCO). (B) Bacterial Sec YEG associated ATPase; Sec A (PDB ID: 6S0K). (C) Eukaryotic Sec61 (PDB ID: 4CG7). (D) Bacterial SRP consisting of RNA and protein (PDB ID: 5NCO). (E) Mammalian SRP consisting of RNA and proteins (SRP 19, SRP 54, SRP 68 and SRP 72) (PDB ID: 6FRK).

The α-subunit is hourglass shaped and comprised of 10 transmembrane domains, while the β and γ subunit are single-pass transmembrane proteins that remain associated with the α-subunit (Itskanov & Park, 2019). The α-subunit creates a gated channel in the lipid bilayer, while the β and γ subunits interact with the α-subunit within the lipid bilayer to allow for the protein transport.

The functioning of the translocon is critical, but alone it is passive in nature, and it needs to be assisted by several other binding partners to drive the translocation process. When a protein is recruited to the channel by appropriate factors, the channel undergoes conformational changes to allow the protein to pass through. Transmembrane proteins, which are to be embedded into the lipid bilayer, are rich in hydrophobic sequences, and move laterally into the lipid layer. Proteins targeted for release into the periplasm have a higher proportion of hydrophilic sequences, and pass through the channel completely out into the periplasmic space (Facey & Kuhn, 2004).

For co-translational integration of proteins into the membrane, the translating ribosomes that are synthesizing membrane proteins are recruited to the plasma membrane via a complex pathway that involving several components (Fig. 3). Unsurprisingly, an elaborate system is in place to facilitate proper and accurate insertion of proteins into the membrane, which involve both the translating ribosome and the plasma membrane. Post-translational protein transport on the other hand, involves the recognition of target protein by specific chaperone proteins, which then recruit it to the translocon channel for export into the periplasm (Castanie-Cornet, Bruel, & Genevaux, 2014; Ngosuwan, Wang, Fung, & Chirico, 2003).

3. Molecular machinery of bacterial protein transport

The prokaryotic bacteria are often considered model organisms for the study of universally conserved pathways, such as protein translocation across membranes. Bacteria lack membrane-bound organelles, and all essential functions are executed within the cytoplasm. The transport of proteins into and across the plasma membrane is mediated by the SecYEG translocon (Fig. 3A) (Lycklama & Driessen, 2012).

The core of the translocon channel (SecYEG) consists of the integral membrane proteins SecY (α subunit) and SecE (γ-subunit), which interact with the membrane protein SecG (β subunit) and a peripheral ATPase called SecA (Fig. 3B). Initially, it was not clear as to whether SecYEG heterotrimer acts as a single unit or as multimeric assemblies of itself. A very recent total internal reflection fluorescence microscopy study by Koch et al. showed that SecYEG exists predominantly as a single copy, and remains so even in the presence of ribosome (co-translational protein transport) (Koch et al., 2021). Using mutant *E. coli* cell lines, Ryabichko et al. showed that the presence of the dianionic lipid cardiolipin in the membrane is necessary for the Sec translocon stability as well as proper protein translocation and folding (Ryabichko et al., 2020). An elaborate mechanism is in place to recruit the translating ribosome to the Sec translocon and facilitate protein insertion, the most predominant method being the Signal Sequence (*SS*)-Signal Recognition Particle (SRP)-FtsY based pathway.

3.1 SRP-mediated protein targeting

Secretory and integral membrane proteins bear an N-terminal signal sequence (SS) (Lee & Bernstein, 2001), which is co-translationally identified and bound by the Signal Recognition Particle (SRP) as the nascent chain exits the peptide exit tunnel on the bacterial 50S ribosomal subunit (Saraogi & Shan, 2014; Shao & Hegde, 2011). Subsequently, SRP delivers ribosome-nascent chain complexes (RNCs) from the cytosol to the translocon (Poritz et al., 1990).

SRP consists of a hairpin-shaped 4.5S RNA (SRP RNA) and Ffh protein comprising an N-terminal Ras-like GTPase (NG domain) and a methionine-rich M domain. SRP recognizes SS through a "fingerloop motif," a hydrophobic groove sealed by a flanking loop (Fig. 3D) (Keenan, Freymann, Walter, & Stroud, 1998; Zopf, Bernstein, Johnson, & Walter, 1990). Cryo-EM analysis of the RNC-SRP-SR-SecYEG quaternary complex reveals several interesting observations (Fig. 4). The SRP-ribosome interaction is stabilized via interactions at three distinct sites in the vicinity of the polypeptide exit tunnel: the NG domain binds the ribosomal protein uL29, M domain interacts with 23S rRNA of the large subunit, while the SRP RNA contacts the bL32 protein (Fig. 4A) (Jomaa, Boehringer, Leibundgut, & Ban, 2016).

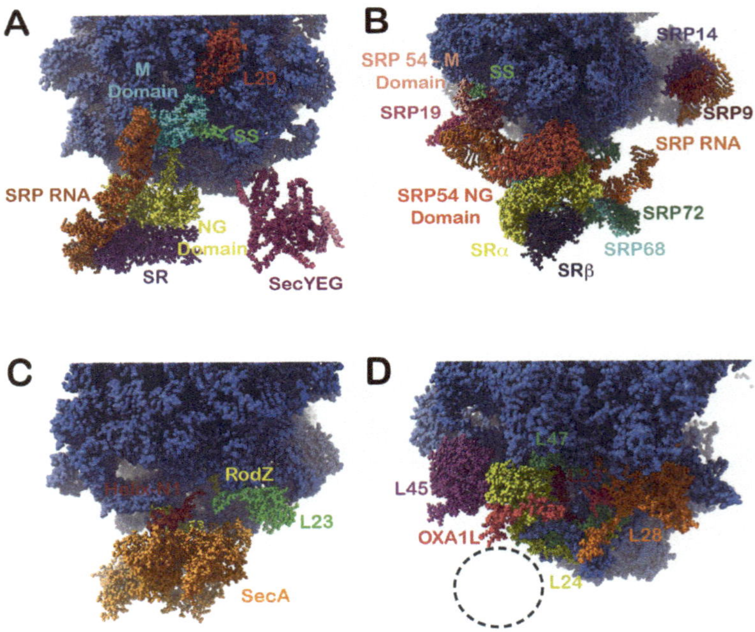

Fig. 4 Mapping the interactions of ribosome with protein translocation associated complexes. (A) Interaction of Sec YEG and SRP with *E.coli* LSU. SR binds to the NG domain of SRP to form a heterodimer. The M domain of SRP interacts with L29 and Sec YEG interacts with 23S rRNA (PDB ID: 5NCO). (B) The interaction of SRP-SR complex with eukaryotic cytosolic LSU (PDB ID: 5NCO). (C) Interaction of Sec A with E.coli LSU. Helix N1 of Sec A binds to L23 of the LSU (PDB ID: 6S0K). (D) Human mitochondrial LSU in complex with Oxa1L insertase. L23, L24, L28, L45 and L47 are the interacting partners of Oxa1L. The atomic coordinates of Oxa1L are incomplete (the missing coordinates indicated by dotted circle) (PDB ID: 6ZM5).

The SRP-bound translating ribosomes in turn bind to the SRP receptor (SR) (FtsY in bacteria), which in turn interacts with the SecYEG translocon. FtsY comprises of a NG domain (analogous to Ffh) and an unstructured "A" domain. Recruitment of the RNC-SRP to the membrane is initiated by the GTP dependent heterodimerization of the NG domains of SRP and FtsY; and this stage is termed the "early state" of SRP-FtsY complex (Siu, Spanggord, & Doudna, 2007). FtsY then binds to the fifth cytosolic loop of the SecY protein of the SecYEG translocon, thereby bringing the RNC in contact with the translocon.

FtsY-SecY interaction is GTP-independent, but requires a phospholipid environment, as proved by photo-crosslinking and FRET experiments (Kuhn et al., 2015). Interaction of the SecYEG with the RNC-SRP complex reduces the affinity of SRP towards the RNC as well as towards FtsY; as confirmed by FRET based studies (Draycheva, Lee, & Wintermeyer, 2018). Conformational changes then lead to the dissociation of SRP-FtsY NG dimer from the RNC, forming the "closed state" (Zhang, Schaffitzel, Ban, & Shan, 2009). The A-domain of FtsY establishes contacts with both the membrane and the translocon, and helps in positioning of the RNC (Jomaa et al., 2017). This leads to the displacement the SRP-FtsY NG dimer, which in turn causes the SRP RNA to adopt a kinked and untwisted conformation. These conformational changes subsequently trigger a GTP hydrolysis event which leads to the dissociation of SRP and SR, thereby transferring the RNC-SS to the Sec translocon, and is called the "activated state" (Akopian, Dalal, Shen, Duong, & Shan, 2013).

The first cryo-EM study of the RNC-SecYEG complex showed that the binding of SecYEG to the large subunit (50S) creates a large frontal opening between the polypeptide exit tunnel and the protein-conducting channel (PPC) complex to accommodate the nascent chain. Interactions between ribosomal RNA (rRNA) hairpins of helix 24 and helix 59 and the cytosolic factor-associating domains (CFAD) of SecY seem to facilitate ribosome-translocon binding, and the binding is further stabilized by the interaction of ribosomal proteins L29 and L23 with the cytoplasmic region of the SecG protein (Mitra et al., 2005; Mitra & Frank, 2006).

The interaction with ribosomal components allows the creation of a continuous channel from the ribosomal peptide exit tunnel to the SecYEG translocation pore, allowing direct release of nascent proteins into the membrane (Saraogi & Shan, 2014), thus preventing the exposure of these highly hydrophobic protein from the aqueous environment of the cytosol, an event which could lead to their misfolding and aggregation.

While it is known that the inner membrane of bacteria is negatively charged, it is still not clear as to whether the lipid composition of the membrane played any role in protein translocation.

3.2 Non-SRP mediated pathways

While the SRP-FtsY mediated protein transport is the most predominant and well-studied pathway for protein insertion into the membrane, it is not the only one. SecYEG often utilizes the assistance of various other protein/protein complexes for its function. Apart from the SRP receptor FtsY, SecYEG interacts with the membrane-associated ATPase SecA, membrane integral proteins such as YidC, the SecDF-YajC complex, PpiD or YfgM (Petriman et al., 2018). The periplasmic proteins PpiD and YfgM are associated with release of secretory protein from SecYEG complex into the periplasmic space (Antonoaea, Furst, Nishiyama, & Muller, 2008; Gotzke et al., 2014).

YidC helps in release of proteins into the lipid membrane by interacting with SecY (Sachelaru et al., 2015). YidC was shown to interact with the translating ribosome, and the RNC-YidC interaction is strongly enhanced when the protein is embedded in a lipid bilayer; indicating that YidC could well be associated in anchoring the ribosome to the membrane during protein translocation (Kedrov et al., 2013).

One key interaction partner is the peripheral ATPase protein SecA (Fig. 3B). SecA was almost exclusively associated with post-translational protein transport of secretory proteins. However, SecA was also found to interact with the 70S ribosome (Huber et al., 2011), as well as with the ribosome-bound nascent proteins (Wang, Yang, & Shan, 2017). There are reports of SecA being involved in the membrane insertion of certain membrane proteins with large periplasmic domains (Gebert, Overhoff, Manson, & Boos, 1988; Traxler & Murphy, 1996). Furthermore, certain inner membrane protein and single span-membrane proteins have been show to be dependent on SecA rather than SRP-FtsY for membrane insertion (Kihara & Ito, 1998; Rawat, Zhu, Lindner, Dalbey, & White, 2015).

Our understanding was further refined by Wang et al., who demonstrated that SecA, and not the SRP machinery, is responsible for the co-translational targeting and translocation of an inner membrane protein (Wang et al., 2017). There seems to be a competitive binding of SRP and SecA in the vicinity of the polypeptide exit tunnel of the RNC, and the nascent chain seems to determine the binding partner; i.e., SecYEG and SecA have different cargoes that they are expected to shuttle to the SecYEG complex co-translationally (integral membrane proteins and inner membrane proteins/single pass membrane proteins, respectively).

SecA seems to bind to the signal sequence of target non-membrane proteins co-translationally and recruit it to the translocon for export (Karamyshev & Johnson, 2005). Structural elucidation of SecA-RNC complex reveals that the N-terminal domain of SecA interacts with the ribosome, and in tandem with ribosomal protein uL23, creates a pocket for nascent chain binding of the transmembrane domain of target inner membrane proteins; and this binding precedes the transfer of the RNC to the SecYEG translocon (Fig. 4C) (Wang et al., 2019). SecA establishes contacts with the cytosolic domain of integral membrane protein SecY, and then inserts itself into the membrane by establishing further contacts with the transmembrane helices and even periplasmic regions of SecY. (Banerjee, Zheng, Abolafia, Harper, & Oliver, 2017).

4. Co-translational protein transport in eukaryotes:

Eukaryotes, being more complex in design and function, possess several membrane-bound organelles which serve to compartmentalize different functions. Eukaryotes have two types of ribosomes: the cytosolic 80S ribosomes involved in translation of all nucleus-encoded genes; and ribosomes of the organelles (mitochondria and chloroplast) which translate a few genes that are encoded in the organelle DNA.

Like bacteria, eukaryotic membrane proteins often undergo co-translational targeting. While the Sec translocon is similar in structure and function to the bacterial counterpart, the integration of proteins into the membrane have several key differences (Denks et al., 2014). There are different pathways for plasma membrane and organelle protein insertion. The most important is perhaps the fact that there is no direct interaction with the plasma membrane. Instead, membrane proteins are injected into the membrane of the Endoplasmic Reticulum (ER) (Ito, 1996).

The endoplasmic reticulum (ER) is the largest organelle in the cell and it is involved in protein synthesis and transport, lipid metabolism, carbohydrate metabolism and Ca^{2+} storage (Clapham, 2007; Reid & Nicchitta, 2015). To execute such diverse functions, the ER requires many proteins and unique physical structures (Schwarz & Blower, 2016). Based on the difference in appearance under an electron microscope, the ER can be classified as Rough ER or Smooth ER; the "roughness" being the presence of ribosomes on the outer membrane of the ER. The secretory pathway proteins are typically synthesized by the rough ER-associated ribosomes and co-translationally translocated across the membrane via the translocon channel (Blobel & Dobberstein, 1975a, 1975b).

The organelles, namely mitochondria and chloroplast, also possess ribosomes of their own. Like bacteria, these organelles are double-membrane bound, with several membrane-bound proteins serving various functions. Mitochondrial ribosomes interact with the mitochondrial inner membrane, and serves to insert proteins into the membrane. Interaction of the ribosomes with the mitochondrial inner membrane is necessary for the effective functioning of the organelles, and for the organism as a whole (De Silva, Tu, Amunts, Fontanesi, & Barrientos, 2015). Lesser information is available on chloroplast ribosomes. Nevertheless, interaction of the chlororibosomes with chloroplast inner membrane, and its role in membrane protein translocation have been reported (Zoschke & Barkan, 2015).

4.1 Cytosolic ribosome

The eukaryotic cytoplasmic ribosome bound to the membrane ER sheets are considered as the primary domain for functions such as protein maturation, synthesis, membrane translocation and insertion, post-translational modification, folding and assembly, quality control and degradation (Araki & Nagata, 2011). Rough ERs are involved in co-translational protein transport, in a process similar to the bacterial pathways discussed earlier and involves the interaction between the ribosomes and the ER membrane.

Similar to the bacterial counterpart, both secretory and integral membrane-proteins follow the Sec pathway, and are secreted into the ER lumen or integrated into the ER lipid membrane, respectively (Shao & Hegde, 2011). Following proper insertion and post-translational modifications, the membrane pinches off into vesicles (secretory proteins are entrapped within the vesicle, while integral-membrane proteins are embedded in the inner wall of the membrane) to fuse with the Golgi apparatus. After further modifications and processing, they again pinch off as vesicles from the Golgi apparatus and targeted to their destination, where they fuse with the membrane to either release the secretory protein or embed the integral-membrane protein.

The Signal Sequence emerging out of a translating ribosome is recognized by the Signal Recognition Particle (SRP) (Fig. 3E). The Ribosome Nascent Chain (RNC)-SRP complex binds to the ER membrane by interaction between the SRP and the SRP-Receptor (SR), and stabilized by interactions between the ribosome and the Sec61 complex (Fig. 4B). Sec61, homologous to the SecYEG complex in bacteria, is a heterotrimeric protein which forms a protein conducting channel found in the ER membrane (Fig. 3C) (Egea, Stroud, & Walter, 2005, Walter & Johnson, 1994). The translating ribosome remains bound to the translocation channel (translocon) during the entire transport process (Mothes et al., 1997).

4.1.1 SRP-SRP receptor (SR) mediated targeting

The mammalian protein translocation is similar to the bacterial system. Nevertheless, mammalian SRP is more complex, and composed of one 7S RNA molecule (SRP RNA) and six polypeptides having different molecular masses (9, 14, 19, 54, 68, 72 kDa) (Walter & Johnson, 1994). The overall dimensions and shape of the SRP RNA's secondary structure is conserved from eukaryotes to archaea, despite the remarkable evolutionary drift in the SRP RNA sequences (Schmitz et al., 1999).

The mammalian SRP resembles the shape of an elongated rod of 240 Å long X 60 Å wide, while the 300 nucleotide long SRP RNA extends throughout the length of the rod (Zwieb, Van Nues, Rosenblad, Brown, & Samuelsson, 2005). The central S domain, located on the SRP RNA, serves to recognize the signal sequence in the nascent chain of the translating protein. It recruits the cargo to the Sec translocon by interacting with the SR present on the ER membrane (Luirink & Sinning, 2004).

The SRP54 protein, homologous to the Ffh protein in bacteria, contains a central GTPase domain flanked by the N domain (together called NG domain) and the M domain, both linked by a flexible hinge region (Fig. 4). This M domain serves as bridge, interacting with both the SRP RNA and the signal sequence (Wild et al., 2019). SRP54 undergoes significant conformational change during the process of co-translational translocation of protein and has a two-step binding to the ribosome. The first high affinity binding corresponds to the tight contact of the M domain at the ribosomal exit tunnel, while the second weaker binding corresponds to the N-domain when it binds across the exit in the scanning and engaged pre-handover states.

Co-translational protein transport begins with the SRP binding to the ribosome and scanning the nascent chain carrying a signal sequence. The SRP binds to the signal sequence resulting in the formation of SRP-RNC pre-handover complex (Voorhees & Hegde, 2015; Wild, Halic, Sinning, & Beckmann, 2004). SRP54 is the key player in the signal sequence recognition and it undergoes significant structural changes. The M domain of SRP54 is positioned into the funnel of the polypeptide exit tunnel of the large ribosomal subunit engaging the signal sequence, while the NG domain is locked in place across the tunnel exit (Halic et al., 2004). Following this, the SRP-RNC complex binds to the SR on the ER membrane and the RNC is finally handed over to the ER translocon (Kobayashi et al., 2018; Wild et al., 2019).

Mammalian SR is a heterotrimeric integral membrane protein present in the ER membrane, composed of SRα and SRβ having a molecular mass of 72 and 30 kD, respectively, both subunits bearing GTPase domains (Fig. 4). It is located in the ER membrane in which the SRα is associated with the membrane through an N-terminal domain via interaction with the SRβ (Tajima, Lauffer, Rath, & Walter, 1986). MicroScale Thermophoresis analysis (MST analysis) reveals that a highly positively charged Ribosome Binding Region (RBR) is identified within SRα and is believed to be a major contributor to the ribosome interaction (Wild et al., 2019).

4.1.2 Structural characteristics of the ribosome-bound Sec61 complex

The Sec61 channel is a heterotrimeric protein consisting of the Sec61α, Sec61β, Sec61γ subunits (Fig. 3). The Sec61α subunit lies in the centre and contains ten transmembrane helices arranged in two pseudo-symmetrical N and C-terminal halves around a central pore that is sealed from passive ion flux by a ring comprising of bulky hydrophobic amino acid residues and a plug helix. The Sec61β and γ subunits are comprised of one transmembrane helix each and are located on the periphery of the Sec61 complex. Two adjacent transmembrane helices 2 and 7 of Sec61α forms a lateral gate which is responsible for the insertion of nascent transmembrane helices or signal sequences emerging from the translating ribosomes into the lipid layer (Pfeffer, Dudek, Zimmermann, & Forster, 2016).

The ribosome during co-translational protein translocation and membrane insertion binds to the cytosolic loops of the Sec61α. Comparison of the cryo-EM structure of the mammalian 80S ribosome-Sec61 complex in both idle and translating states provided detailed structural insights into their interactions (Cheng, Jiang, Mandon, & Gilmore, 2005; Voorhees, Fernandez, Scheres, & Hegde, 2014). Based on these structural studies, it was observed that Sec61 interacts with the ribosome through the evolutionary conserved loop 6/7 and loop 8/9 in the α subunit of Sec61, as well as the N terminal helix of Sec61γ. The interacting surface is extensively composed of loop 8/9 and Sec61γ which together contact the backbone of the 28S rRNA and the ribosomal proteins uL23 and eL29 (located in the proximity of the polypeptide exit tunnel). Several conserved basic residues in loop 8/9 including His 404 interacts with Thr 82 of uL23 and a universally conserved Arg405 forms a stacking interaction with rRNA residue C2526. The hydroxyl group of Thr 407 in helix 10 is within hydrogen bonding distance of the side chain of Asn 36 of eL19, while the Arg 20 of the γ-subunit forms a salt bridge with Asp 148 of uL23. The hydrogen bonding interactions stabilizes the conformation of loop 8/9 and anchors the ER translocon at the ribosomal tunnel exit.

A similar heterotrimeric complex, Sec61p complex is found in yeast which is composed of the Sec61p, Sbh1p and Sss1p (Panzner, Dreier, Hartmann, Kostka, & Rapoport, 1995). The Sec61p, Sbh1p and Sss1p are homologous to Sec61α, Sec61β and Sec61γ respectively of the mammalian Sec61p complex. In yeast the Sec61p complex is involved in both co-translational and post-translational protein translocation, by associating with the tetrameric Sec62-Sec63 complex and forming a functional heptameric Sec complex (Hartmann et al., 1994).

4.1.3 Interaction between the ribosome and the accessory translocon components

The translocon complex contains several accessory components, which assists the Sec61 channel or are involved in maturation of the nascent chains by covalent modifications and chaperone like functions. One such component is the Translocon Associated Protein Complex (TRAP) (Wang & Dobberstein, 1999). The translocon component TRAP complex in mammals is a hetero-tetrameric membrane protein complex (Pfeffer et al., 2016) which is thought to assist the Sec61 complex in protein insertion and in particular topogenesis of polytopic membrane proteins (Sommer, Junne, Kalies, Spiess, & Hartmann, 2013). Structural studies using CET and subtomogram analysis suggested that TRAP consists of a bundle of transmembrane helices flanking Sec61γ and C-terminal half of Sec61α. TRAP is connected to the ribosome via a short cytosolic domain linked to the ribosomal protein L38 and the large subunit ribosomal RNA expansion segment (ES) ES20L/ES26L, as well as ER luminal segment.

The Oligosaccharyl Transferase Complex (OST) is another accessory component that catalyzes the N-linked glycosylation of proteins (Kelleher & Gilmore, 2006; Mohorko, Glockshuber, & Aebi, 2011), which has been shown to interact with the ribosome. It is a multi-subunit complex containing atleast seven membrane proteins with a large luminal domain (Kelleher & Gilmore, 2006). It has been reported that the cytosolic

C-terminal domain of RPN1 (a subunit of the OST complex) interacts with the ribosome. It forms a four-helix bundle aligning in a cavity made from rRNA helices H19/H20, rRNA expansion segment ES7a (H25) and ribosomal protein eL28 (Braunger et al., 2018). However, the absence of detailed structural information for most of the OST subunits with large ER luminal domains remains a roadblock that limits our understanding of these interactions and their implications. Detailed structural and functional analysis could help us gain more clarity.

4.1.4 ER membrane-ribosome association

Following the "handover," the RNC is ready to insert the nascent chain into the Sec translocon. While the ribosome-Sec61 tarnslocon interactions position and stabilize the RNC onto the Sec translocon, the ribosome establishes further contacts to stabilize the translating ribosome on the membrane. Pfeffer et al. carried out cryo-electron tomography (Cryo-ET) based 3D reconstruction of the 80S ribosome bound to the complete ER translocation machinery in a native membrane environment using functional pancreatic ER microsomes, and obtained several insights into the ribosome-membrane interaction (Pfeffer et al., 2012).

A marked difference between the ribosomes of prokaryotes and eukaryotes is the fact that Eukaryotic rRNA contains various insertions referred to as Expansion Segments (ES) (Heinonen, Schnare, & Gray, 1990). One of the largest insertions is the ES27 of the large subunit rRNA referred to as ES27L. It has been shown to have a direct contact with the ER membrane. ES27L has been shown to stabilize the association of the ribosomes to the ER membrane by direct interactions between negatively charged phosphate groups of the rRNA backbone and the positively charged head groups of the membrane lipids. Additionally, interactions with integral membrane proteins, for example Sec62/Sec63 (Muller et al., 2010) complex and ERj1 (Dudek et al., 2005) also serve to stabilize the binding. ES27L binds to the cytosolic domain of ERj1, as shown by single particle study of purified 80S ribosomes (Blau et al., 2005). Another possible candidate for binding is a highly abundant protein residing in the ER membrane, the P180/Ribosome Receptor Protein (RRP). The C-terminal cytosolic domain of P180 has 54 repeats of a 10 amino acid consensus motif which is believed to be responsible for the rRNA binding (Wanker, Sun, Savitz, & Meyer, 1995).

After the transport of proteins into the ER, conformational changes of ES27L help the dissociation of ribosomes from the ER membrane and hence play a pivotal role in the whole process (Potter & Nicchitta, 2002). The role of the extension segment in both the binding and the dissociation is a unique phenomenon, and unlike the prokaryotic counterpart.

4.1.5 Alternative targeting routes

While protein translocation is predominantly mediated by the SRP-SR-Sec61 complex mediated pathways, other pathways, both SRP-dependent and independent, also exist. A Sec61p complex homolog, the Ssh1p complex was discovered in the yeast ER membrane (Finke et al., 1996) that is exclusively involved in the co-translational translocation. It is composed of Ssh1p which is a distant homolog of Sec61p; and Sbh2p, a homolog of Sbh1p and Sss1p. It is reported that the Ssh1p recognizes a subset of signal sequences that are also recognized by Sec61p but not by Sec62p (Wittke, Dunnwald, Albertsen, & Johnsson, 2002). Ssh1p complex showed preferential binding to highly hydrophobic signal sequences while avoiding lesser hydrophobic sequences, indicating a subset of the proteins that are translocated into the membrane maybe transported using these transporters; in a pathway that possibly involves the SRP and SR (Ng, Brown, & Walter, 1996).

Although the SRP-dependent protein translocation pathway remains the predominant pathway, an SRP independent targeting system was also discovered in yeast, *Saccharomyces cerevisiae*, termed as the SND (SRP-independent targeting) system (Ast, Cohen, & Schuldiner, 2013). It is suggested that this system functions in parallel with the SRP and GET (the guided entry of tail anchored proteins (GET) pathway which is involved in the post-translational entry of tail anchored proteins (TA) to the ER) (Zhang et al., 2021) pathway to target a broad range of substrates to the ER and hence can also act as a backup targeting system (Aviram et al., 2016).

The SND system comprises of three proteins namely, SND1, SND2, SND3. The SND1 is proposed to be a cytosolic peripheral ribosomal protein (Huh et al., 2003), while SND2 and SND3 are localized in the ER membrane and appear to contain four transmembrane domains and one transmembrane domain, respectively (Aviram et al., 2016; Zhao et al., 2013). It is shown that the three pathways SRP, SND and GET work in parallel in targeting of proteins to the ER with the SND pathway targeting a diverse range of substrates; the complex inter-connection of these pathways being a primary reason as to why the SND pathway evaded detection by researchers until recently. Till date, mechanistic details of the SND pathway remain unknown. According to the predicted mode of function, the SND1 is believed to interact with the ribosomes and is involved in capturing the nascent substrates. The ER transmembrane proteins SND2 and SND3 form a complex with the Sec translocon, working in tandem to act as a receptor for the substrate and promote the handoff to the translocon (Aviram et al., 2016). The exact mechanism how the ribosome interacts with the SND proteins and the ER translocon is still obscure and requires further investigation.

4.1.6 Cytoplasmic ribosome-mitochondria outer membrane interaction

Conventionally, it was considered that proteins shuttle through the ER and Golgi before reaching their target organelles. The presence of the ribosomes in the vicinity of the mitochondria provided circumstantial evidence for the existence of co-translational protein import (Crowley & Payne, 1998; Kellems, Allison, & Butow, 1975) and that nuclear encoded mitochondrial proteins can be synthesized locally on its outer membrane. However, it was a theory still viewed with skepticism,

Recently, Gold et al. showed that the ribosome does indeed interact with the surface of the mitochondria as confirmed by cryo-electron tomography (Gold, Chroscicki, Bragoszewski, & Chacinska, 2017). The study indicated that specific interactions occur between the ribosome and the TOM complex and the binding of the nascent chain is an important factor for ribosome recruitment and stabilization on the outer membrane of the mitochondria.

Apart from the well-documented ribosome-ER interactions, the report of ribosome-mitochondrial interactions opens new avenues. Further research in these lines could help us understand the details of the interaction, the cargo proteins transported by this pathway etc. It is also tempting to think that the ribosome could possibly interact with other organelles and membranes in a similar fashion, and the whole ribosome-membrane interaction cascade could be much more complicated and intricate than the current perception.

4.2 Mitochondrial ribosome

Mitochondria are double membrane bound organelles present in the cytoplasm of most eukaryotic cells. Their main function is to generate ATP, the energy currency of cell. Apart from that, mitochondria are also involved in calcium homeostasis, lipid metabolism and apoptosis (Vafai & Mootha, 2012). Mitochondrial inner membrane is bent into folds called the cristae which is the primary site of energy production system, the electron transport chain (ETC). The folding ensures the enhancement of surface area which facilitates the incorporation of the oxidative phosphorylation complexes that are essential for energy production (Schenkel & Bakovic, 2014).

ETC uses a series of oxidation-reduction reaction to move electrons from one protein complex to another and finally generate free energy that is utilized for the phosphorylation of ADP to ATP. More than 99% of the proteins and other molecules that make up the mitochondria originate at the nucleus. However, a few genes (approx 37) are translated at the mitochondrial matrix by the mitochondrial ribosomes (mitoribosomes). 13 of these mitochondrial genes encode proteins that are integral part of the ETC (Anderson et al., 1981). Hence, efficient translation of these proteins and integration within the inner membrane is critical for proper functioning of mitochondria and thus survival of the cell.

It is widely believed that mitochondria originated from α-proteobacterial ancestor by the process of endosymbiosis. In due course of time, the mitoribosome diverged from its bacterial counterpart in terms of structure, function and composition (Martin et al., 2015). In general, mitoribosomes were found to contain a higher number of proteins that cover the surface of rRNA core; with varying length of rRNA between species. Interestingly, mitochondrial r-proteins are also significantly elongated in some species compared to their bacterial counterparts, including those that surround the exit tunnel (Fig. 2F). It is extended in fungi and plants but reduced in metazoans. The structural alteration ensured functional supremacy allowing mitoribosomes to mostly synthesize membrane proteins and hydrophobic components of ETC. One unique feature of mammalian mitochondrial mRNA is the absence of 5′ untranslated region (Temperley, Wydro, Lightowlers, & Chrzanowska-Lightowlers, 2010). Bacterial 5′ UTR contains the Shine-Dalgarno sequence which interacts with the small ribosomal subunit to select the start codon. The absence of 5′ UTR in mitochondria means additional ribosome alteration and/or initiation factors is required for start codon selection.

4.2.1 Structural diversity in mitoribosomes

There is a lot of diversity in the shape and size of mitoribosomes among species. The yeast 74S mitoribosome consists of 54S large subunit and 37S small subunit (Desai, Brown, Amunts, & Ramakrishnan, 2017). The presence of additional ribosomal proteins and rRNA expansion resulted in 30% increase in molecular weight when compared to 70S bacterial ribosome. In contrast, mammals possess 55S mitoribosome consisting of 39S large and 28S small subunit (Greber et al., 2015). They lack the rRNA expansions resulting in rRNA size of almost half of that of bacterial rRNA and the rRNA core remains completely covered by protein (Amunts, Brown, Toots, Scheres, &

Ramakrishnan, 2015). The protozoan Leishmania mitoribosomes possess one of the lowest amounts of rRNA. The 28-30S SSU and the 40S LSU consists of 610 nucleotide long 9S and 1173 nucleotide long 12S rRNA, respectively, which are considerably smaller than their bacterial or mammalian counterparts (de la Cruz, Lake, Simpson, & Simpson, 1985; Maslov et al., 2006). Cryo-EM maps highlighted that the overall structure is highly porous with a diameter of 245 Å (Sharma, Booth, Simpson, Maslov, & Agrawal, 2009).

Mitoribosomes synthesize proteins that are components of oxidative phosphorylation complex. To perform their function, these proteins need to be integrated into the inner membrane. In order to facilitate the inner membrane transport, mitoribosomes attach to the inner mitochondrial membrane and perform co-translational protein insertion. Meiqin Liu and Linda Spremulli highlighted that almost 50% of the mitoribosomes co-purify with the mitochondrial inner membrane which is consistent with the notion that mitoribosomes bind to the inner membrane (Liu & Spremulli, 2000). The attachment of mitoribosome to the inner membrane is facilitated by various proteins.

These membrane proteins (either integral or peripheral) interact with the ribosomal subunit and guide the macromolecular complex to the exact location from where the insertion of the nascent polypeptide can be carried out in an efficient manner. B.V. Möller-Hergt et al. demonstrated the synergistic role of three proteins (Mrx15, Mba1 and Oxa1) that play a crucial role in co-translational insertion of cytochrome *c* oxidase subunit 2(Cox2) in yeast mitochondria (Moller-Hergt, Carlstrom, Stephan, Imhof, & Ott, 2018). Mrx15, an integral membrane protein and Mba1, a peripheral membrane protein bind to the LSU of the mitoribosome and tether it to the inner membrane. These two proteins ensure the proper alignment of the LSU at the site of insertion so that Oxa1, the insertase, can fulfill efficient membrane insertion. It may be noted here that bacterial YidC discussed earlier is the homologous counterpart of Oxa1 of mitochondria.

Upon completion of Cox2 synthesis, Mba1 dissociates from the ribosome and assists the newly synthesized protein in downstream processes. Absence of Mrx15 and Mba1 results in a comparable decrease in association of mitoribosome with inner membrane and thus the complete process of co-translational insertion is hampered (Moller-Hergt et al., 2018). Cryo-ET studies revealed two distinct contact points between the mitoribosome and the inner membrane. One of them is the rRNA expansion segment 96-ES1. Whether this rRNA binds to the positively charged phospholipids of the membrane or specific proteins is still unclear. The second contact is mediated by Mba1, which binds to the mitoribosome such that it connects the peptide exit tunnel to the insertion site. Biochemical studies indicated the involvement of other proteins like Mdm33 and Cox11 as ribosomal tethers (Frazier et al., 2006; Khalimonchuk, Ostermann, & Rodel, 2005). Deletion of Mba1 and Oxa1 didn't completely dissociate mitoribosomes from the inner membrane (Bauerschmitt et al., 2010) but it resulted in improper insertion of newly synthesized proteins into the inner membrane (Ott et al., 2006).

While analyzing the mitoribosome assembly, Zeng et al. depicted that the LSU remained attached to the inner membrane during the assembly and this attachment was not nullified by the simultaneous deletion of Mba1, Oxa1 (Zeng, Smith, & Barrientos, 2018). This finding cemented the role of rRNA in membrane association.

In mammals, protein mL45, homolog of yeast Mba1, is an integral part of the LSU. It is one of the crucial proteins that connect the mitoribosome to the inner membrane. The rRNA mediated contact that is prominent in yeast is absent in mammals (Englmeier, Pfeffer, & Forster, 2017). N-terminal of mL45 resides within the exit tunnel and prevents the helix formation of the nascent polypeptide (Itoh et al., 2021). It is released only when the mitoribosome attains an elongated confirmation. This movement guarantees the proper targeting of the mitoribosome to the membrane insertion region on the inner membrane (Kummer et al., 2018). The positioning of mL45 at the exit tunnel allows it to interact with the nascent polypeptide and adjust the orientation to facilitate co-translational membrane insertion (Kummer et al., 2018; Ott & Herrmann, 2010).

High resolution cryo-EM map revealed 3 distinct sites through which Oxa1 binds to the LSU of mitoribosome. This includes three mitoribosomal proteins and rRNA. The first site comprises of bL28m, uL29m and the rRNA bypass element H4-H8 that collectively interacts with C-terminal of Oxa1 (Fig. 4D). The second site includes the sole mitoribosomal protein uL24m and the third site is formed by the core domain of mL45 (Itoh et al., 2021). The inner membranes possess a high amount of cardiolipin, an anionic phospholipid. It helps to maintain membrane integrity and support the functioning of large protein complexes (Ban, Kohno, Ishihara, & Ishihara, 2018). Reduction in cardiolipin content results in defective mitoribosome assembly and impaired association with Oxa1. Mutant cells with defective cardiolipin synthesis gene lack proper association of mitoribosome with Oxa1 which results in impaired insertion of nascent protein (Lee et al., 2020).

4.2.2 Plant mitoribosome

Plants possess one of the largest mitoribosomes (76S–78S) in the living kingdom. The most striking feature of plant mitoribosome is the increased size of SSU compared to the LSU, which is credited to the presence of plant specific proteins and rRNA expansions (Waltz et al., 2019). These proteins are mostly pentatricopeptide repeat proteins (PPR) that extensively interact with plant specific rRNA expansions to alter the shape of the mitoribosome (Waltz, Soufari, Bochler, Giege, & Hashem, 2020). mtLSU has two types of RNA; 26S and 5S (Unseld, Marienfeld, Brandt, & Brennicke, 1997). 5S rRNA, an integral part of bacterial, cytosolic and chloroplast ribosome, is absent in the mitoribosome of yeast and mammals.

When compared to the bacterial ribosomes, the peptide exit tunnel of yeast and mammalian mitoribosome is quite different; on account of them being remodeled by Mba1 or mL45. However, the plant mitoribosomes have near-identical exit tunnel to the bacterial ribosomes, barring slight changes in the surrounding rRNA and a slight extension of uL29m (Waltz et al., 2019). The tunnel makes contact with Oxa1 with the help of a plant-specific linker protein as Mba1 or mL45 are absent (Waltz, Corre, Hashem, & Giege, 2020). Some of the proteins encoded by plant mitochondria (7 mitoribosomal proteins) are soluble in nature which raises the assumption that the mitoribosomes may not be integrally associated with the inner membrane as their animal cell counterparts. This also raises the possibility of the presence of specific-factors or pathways to recruit the ribosome to the inner membrane in a nascent chain-dependent pathway. These could involve proteins homologous to the bacterial system, or completely novel factors or pathways. Nonetheless, these avenues remain unexplored till date, and significant portions of the plant mitoribosomes and their pathways still remain studded with mystery.

4.3 Chloroplast ribosome

Chloroplasts are double membrane bound organelles present within the cells of plant and green algae. Presence of chlorophyll *a* and chlorophyll *b* accounts for the green coloration which distinguishes chloroplasts from other plastids. Like mitochondria, chloroplasts are also believed to originate by endosymbiosis of photosynthetic cyanobacteria (Reyes-Prieto, Weber, & Bhattacharya, 2007). Due to this, chloroplasts possess their own genome as well as the machineries of transcription and translation to decode the genetic information into proteins (Sun & Zerges, 2015; Tiller & Bock, 2014).

Inside the chloroplasts there are closed disk like structures called the thylakoids which are enclosed by thylakoid membrane. The thylakoid membrane is the host to various proteins like cytochrome *b*/f, ATP synthase and photosystem I and II which facilitates electron transport and finally ATP synthesis during photosynthesis (Schmitz-Linneweber et al., 2001). The chloroplast genome encodes majority of the proteins required for building these complexes which are responsible for carrying out photosynthesis. Among them, many subunits are integral membrane components of the thylakoid that are synthesized on thylakoid bound ribosomes and favors the co-translational insertion into the thylakoid membrane (Kroliczewski, Piskozub, Bartoszewski, & Kroliczewska, 2016; Zoschke & Barkan, 2015).

The chloroplast ribosomes or chlororibosomes are specialized in synthesizing and targeting these proteins to the thylakoid membrane, along with the synthesis of RNA polymerase subunits, NADH dehydrogenase and some ribosomal proteins. Sequence analysis confirms that chlororibosomes bear a distinct resemblance to eubacteria such as cyanobacteria and even *E. coli*. The 16S rRNA of chlororibosome SSU is slightly shorter compared to *E. coli* 16S rRNA and the LSU is made up of 3 rRNAs (5S, 4.8S and 23S), unlike *E. coli* LSU which is composed of 2 rRNA subunits (5S and 23S) (Yamaguchi & Subramanian, 2000; Yamaguchi, Von Knoblauch, & Subramanian, 2000).

There is only scanty information available on the molecular details of co-translational protein targeting and transport in chloroplasts. In seeded plants, chloroplast possesses an idiosyncratic SRP system consisting of chloroplast homologs of prokaryotic counterparts like cpSecY1 (SecY), cpSRP54 (Ffh subunit of SRP), cpFtsY (SRP receptor) and Alb3 (YidC). There are some differences between chloroplast and cytosolic SRP (Ziehe, Dunschede, & Schunemann, 2017). Interestingly, in chloroplasts, the RNA component of SRP, a crucial component of the bacterial and cytosolic counterparts is lost during evolution (Trager et al., 2012). cpSRP54 maintains a pool by forming a high affinity complex with cpSRP43 (Dunschede et al., 2015).

Similar to cytosolic counterpart, chloroplasts contain a ribosome associated pool of cpSRP54 which plays a crucial role in transport of some thylakoid membrane proteins (Schuenemann et al., 1998). The cytosolic SRP contacts the ribosome close to the peptide exit tunnel. In contrast to that, a photo crosslinking assay in cpSRP54/ribosome complex recently revealed that the interaction occurs with uL4 which is not localized near the tunnel exit but is a component of the internal wall of peptide tunnel

(Hristou et al., 2019). The C-terminal extension of cpSRP54 likely plays the pivotal role in uL4 binding and thus co-translational insertion into the thylakoid membrane (Kroliczewski, Bartoszewski, & Kroliczewska, 2017). Once cpSRP54/ribosome complex is transferred to the thylakoid membrane, cpSRP54 interacts with the SRP receptor (FtsY) to position the translating ribosome on top of the translocon. Post-translational insertion of membrane proteins also takes place with coordinated action of cpFtsY, cpSRP54 and the insertase Alb3 (Kroliczewski et al., 2017). Due to the lack of high resolution structural information, it is unclear how the translating ribosome is positioned on the translocon and the mechanism of co-translational insertion.

5. Concluding remarks

Emerging structural as well as functional studies have provided an overview of the mechanistic models of ribosome-assisted co-translational protein translocation across membranes in prokaryotes and eukaryotes (Fig.5). We now understand minute structural details of pivotal interactions, and observe that the mechanisms employed by different types of organisms are indeed quite different. Nonetheless, as discussed earlier, several unresolved issues remain to be addressed. More studies combining different approaches of structural and molecular techniques need to be done in order to fully understand the complex mechanisms involved to a greater accuracy.

Furthermore, structural studies on co-translational translocation have been mostly restricted to model prokaryotic organisms such as *Escherichia coli*, or the eukaryotic cytoplasmic ribosomes. Not much is known about the same in other prokaryotes such as gram-positive bacteria or archaebacteria. Barring some structural studies on the mitoribosomes in recently years, the mechanisms involved in the organellar ribosomes are largely unexplored. Structural studies have indeed shown that there are significant differences between processes that were considered to be almost similar previously, such as those between mitochondrial ribosomes and the bacterial ribosomes. Further studies on different species and the organelle ribosomes engaged in co-translational translocation could open up further details of the process and open up new avenues in research and medicine.

While we have discussed the mechanisms of normal co-translational translocation, there could be several other spatial or temporal factors which may be involved in these processes. There could be specific pathways associated in particular conditions, such as stress response or in response to a

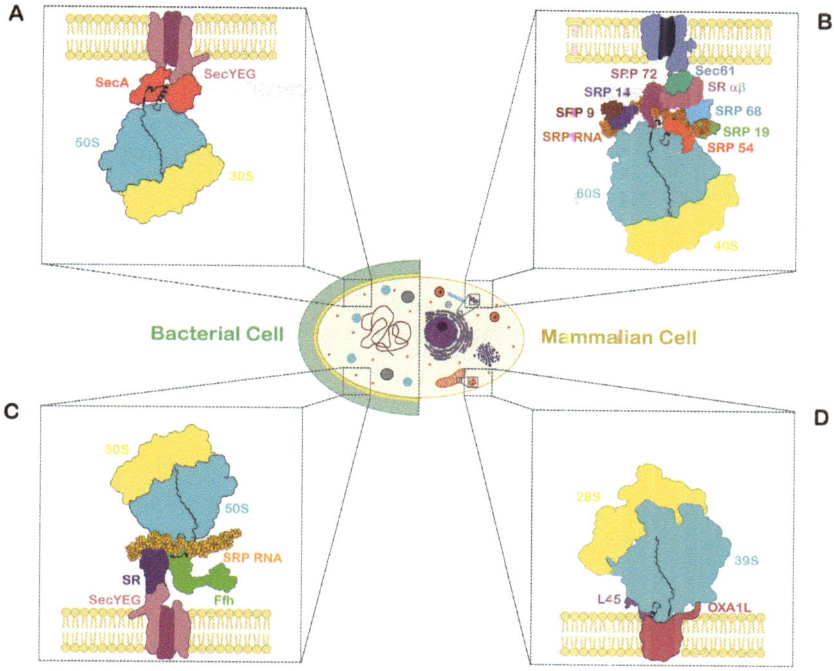

Fig. 5 Schematic representation of co-translational protein transport pathways in prokaryotes and eukaryotes. (A) and (C) represent bacterial translocation pathways mediated by SecA and SRP-SR complex, respectively, through SecYEG translocon. The SRP-SR mediated protein transport in mammalian cytoplasm through Sec61 translocon is shown in (B), whereas, the OXA1L mediated co-translational protein insertion by mitoribosome in mammals is shown in (D).

stimulus. One interesting study in this regard suggested that an *E. coli* protein (YqjD) localizes a part of the ribosome to the membrane during the stationary phase (Yoshida et al., 2012). While the exact role of this sequestration is not known, it is predicted to sequester the ribosomes and inactivate them as the cell moves towards dormancy.

Co-translational translocation is probably the most prominent example of the interplay between ribosome and membrane. Other instances may exist where ribosome-membrane communication occurs. There have been several evidences of factors associated with the ribosomes and the membrane, whose functions have not been explored yet. Even though the work of ribosomes in regulating the membrane and its function by regulating membrane-protein insertion has been well studied, the idea that membrane (and membrane-bound factors) could regulate the ribosome seems even more intriguing.

Another such instance is the eukaryotic protein RACK1. RACK1 is a ribosomal protein that binds to the 40S ribosomal subunit in the vicinity of the mRNA exit tunnel. RACK1 has also been shown to interact with several membrane-associated binding partners such as β-integrins or Focal Adhesion Kinases (FAK). Focal adhesions are sites where cells attach to the extracellular matrix, and RACK1 has been shown to play an active role in their regulation (Nielsen, Flygaard, & Jenner, 2017). Although it is not entirely clear if RACK1 is enable to carry out these interactions when bound to the ribosome, circumstantial evidences do indicate a possibility that RACK1 may recruit the ribosomes to these specific sites of the membrane, possibly for local translation or regulation (Nilsson, Sengupta, Frank, & Nissen, 2004).

A similar argument may be made for the process of axon outgrowth in nerve cells, where local translation near the axon tip is vital for the ability of the axons to respond to stimuli (Campbell & Holt, 2001). These could possibly be scenarios of the membrane regulating and guiding translation for the benefit of the cell. However, the evidences till date; although intriguing, are far from convincing. Rigorous experimentation, both structural and biochemical, could provide new insights into such pathways.

Acknowledgments

We acknowledge financial supports from SERB, DST (India) sponsored projects (SB/SO/BB-0025/2014 and CRG/2019/001788), and CSIR-Indian Institute of Chemical Biology, Kolkata, India. KS, PB, AB and AD thank CSIR, DBT and UGC, India for research fellowships.

Conflict of interest statement

The authors declare no conflict of interest.

Author contributions

All authors searched the literature, contributed to data interpretation, preparation of the manuscript, and approved it for publication.

References

Akbar, S., Bhakta, S., & Sengupta, J. (2021). Structural insights into the interplay of protein biogenesis factors with the 70S ribosome. *Structure*, *29*(7), 755–767.e4.

Akopian, D., Dalal, K., Shen, K., Duong, F., & Shan, S. O. (2013). SecYEG activates GTPases to drive the completion of cotranslational protein targeting. *The Journal of Cell Biology*, *200*, 397–405.

Amunts, A., Brown, A., Toots, J., Scheres, S. H. W., & Ramakrishnan, V. (2015). Ribosome. The structure of the human mitochondrial ribosome. *Science*, *348*, 95–98.

Anderson, S., Bankier, A. T., Barrell, B. G., De Bruijn, M. H., Coulson, A. R., Drouin, J., et al. (1981). Sequence and organization of the human mitochondrial genome. *Nature, 290*, 457–465.
Antonoaea, R., Furst, M., Nishiyama, K., & Muller, M. (2008). The periplasmic chaperone PpiD interacts with secretory proteins exiting from the SecYEG translocon. *Biochemistry, 47*, 5649–5656.
Araki, K., & Nagata, K. (2011). Protein folding and quality control in the ER. *Cold Spring Harbor Perspectives in Biology, 3*, a007526.
Ast, T., Cohen, G., & Schuldiner, M. (2013). A network of cytosolic factors targets SRP-independent proteins to the endoplasmic reticulum. *Cell. 152*, 1134–1145.
Aviram, N., Ast, T., Costa, E. A., Arakel, E. C., Chuartzman, S. G., Jan, C. H., et al. (2016). The SND proteins constitute an alternative targeting route to the endoplasmic reticulum. *Nature, 540*, 134–138.
Ban, T., Kohno, H., Ishihara, T., & Ishihara, N. (2018). Relationship between OPA1 and cardiolipin in mitochondrial inner-membrane fusion. *Biochimica et Biophysica Acta - Bioenergetics, 1859*, 951–957.
Banerjee, T., Zheng, Z., Abolafia, J., Harper, S., & Oliver, D. (2017). The SecA protein deeply penetrates into the SecYEG channel during insertion, contacting most channel transmembrane helices and periplasmic regions. *The Journal of Biological Chemistry, 292*, 19693–19707.
Bauerschmitt, H., Mick, D. U., Deckers, M., Vollmer, C., Funes, S., Kehrein, K., et al. (2010). Ribosome-binding proteins Mdm38 and Mba1 display overlapping functions for regulation of mitochondrial translation. *Molecular Biology of the Cell, 21*, 1937–1944.
Bhushan, S., Gartmann, M., Halic, M., Armache, J. P., Jarasch, A., Mielke, T., et al. (2010). Alpha-helical nascent polypeptide chains visualized within distinct regions of the ribosomal exit tunnel. *Nature Structural & Molecular Biology, 17*, 313–317.
Blau, M., Mullapudi, S., Becker, T., Dudek, J., Zimmermann, R., Penczek, P. A., et al. (2005). ERj1p uses a universal ribosomal adaptor site to coordinate the 80S ribosome at the membrane. *Nature Structural & Molecular Biology, 12*, 1015–1016.
Blobel, G., & Dobberstein, B. (1975a). Transfer of proteins across membranes. I. Presence of proteolytically processed and unprocessed nascent immunoglobulin light chains on membrane-bound ribosomes of murine myeloma. *J Cell Biol, 67*, 835–851.
Blobel, G., & Dobberstein, B. (1975b). Transfer of proteins across membranes. II. Reconstitution of functional rough microsomes from heterologous components. *J Cell Biol, 67*, 852–862.
Blobel, G., Walter, P., Chang, C. N., Goldman, B. M., Erickson, A. H., & Lingappa, V. R. (1979). Translocation of proteins across membranes: The signal hypothesis and beyond. *Symposia of the Society for Experimental Biology, 33*, 9–36.
Braunger, K., Pfeffer, S., Shrimal, S., Gilmore, R., Berninghausen, O., Mandon, E. C., et al. (2018). Structural basis for coupling protein transport and N-glycosylation at the mammalian endoplasmic reticulum. *Science, 360*, 215–219.
Brundage, L., Hendrick, J. P., Schiebel, E., Driessen, A. J., & Wickner, W. (1990). The purified E. coli integral membrane protein SecY/E is sufficient for reconstitution of SecA-dependent precursor protein translocation. *Cell, 62*, 649–657.
Campbell, D. S., & Holt, C. E. (2001). Chemotropic responses of retinal growth cones mediated by rapid local protein synthesis and degradation. *Neuron, 32*, 1013–1026.
Castanie-Cornet, M. P., Bruel, N., & Genevaux, P. (2014). Chaperone networking facilitates protein targeting to the bacterial cytoplasmic membrane. *Biochimica et Biophysica Acta, 1843*, 1442–1456.
Cech, T. R. (2000). Structural biology. The ribosome is a ribozyme. *Science, 289*, 878–879.
Cheng, Z., Jiang, Y., Mandon, E. C., & Gilmore, R. (2005). Identification of cytoplasmic residues of Sec61p involved in ribosome binding and cotranslational translocation. *The Journal of Cell Biology, 168*, 67–77.

Clapham, D. E. (2007). Calcium signaling. *Cell, 131*, 1047–1058.
Crowley, K. S., & Payne, R. M. (1998). Ribosome binding to mitochondria is regulated by GTP and the transit peptide. *The Journal of Biological Chemistry, 273*, 17278–17285.
de la Cruz, V. F., Lake, J. A., Simpson, A. M., & Simpson, L. (1985). A minimal ribosomal RNA: Sequence and secondary structure of the 9S kinetoplast ribosomal RNA from Leishmania tarentolae. *Proceedings of the National Academy of Sciences of the United States of America, 82*, 1401–1405.
De Silva, D., Tu, Y. T., Amunts, A., Fontanesi, F., & Barrientos, A. (2015). Mitochondrial ribosome assembly in health and disease. *Cell Cycle, 14*, 2226–2250.
Denks, K., Vogt, A., Sachelaru, I., Petriman, N. A., Kudva, R., & Koch, H. G. (2014). The sec translocon mediated protein transport in prokaryotes and eukaryotes. *Molecular Membrane Biology, 31*, 58–84.
Desai, N., Brown, A., Amunts, A., & Ramakrishnan, V. (2017). The structure of the yeast mitochondrial ribosome. *Science, 355*, 528–531.
Draycheva, A., Lee, S., & Wintermeyer, W. (2018). Cotranslational protein targeting to the membrane: Nascent-chain transfer in a quaternary complex formed at the translocon. *Scientific Reports, 8*, 9922.
Dudek, J., Greiner, M., Muller, A., Hendershot, L. M., Kopsch, K., Nastainczyk, W., et al. (2005). ERj1p has a basic role in protein biogenesis at the endoplasmic reticulum. *Nature Structural & Molecular Biology, 12*, 1008–1014.
Dunschede, B., Trager, C., Schroder, C. V., Ziehe, D., Walter, B., Funke, S., et al. (2015). Chloroplast SRP54 was recruited for posttranslational protein transport via complex formation with chloroplast SRP43 during land plant evolution. *The Journal of Biological Chemistry, 290*, 13104–13114.
Egea, P. F., Stroud, R. M., & Walter, P. (2005). Targeting proteins to membranes: Structure of the signal recognition particle. *Current Opinion in Structural Biology, 15*, 213–220.
Englmeier, R., Pfeffer, S., & Forster, F. (2017). Structure of the human mitochondrial ribosome studied in situ by Cryoelectron tomography. *Structure, 25*(1574–1581), e2.
Facey, S. J., & Kuhn, A. (2004). Membrane integration of E. coli model membrane proteins. *Biochimica et Biophysica Acta, 1694*, 55–66.
Fagerberg, L., Jonasson, K., Von Heijne, G., Uhlen, M., & Berglund, L. (2010). Prediction of the human membrane proteome. *Proteomics, 10*, 1141–1149.
Finke, K., Plath, K., Panzner, S., Prehn, S., Rapoport, T. A., Hartmann, E., et al. (1996). A second trimeric complex containing homologs of the Sec61p complex functions in protein transport across the ER membrane of S. cerevisiae. *The EMBO Journal, 15*, 1482–1494.
Frazier, A. E., Taylor, R. D., Mick, D. U., Warscheid, B., Stoepel, N., Meyer, H. E., et al. (2006). Mdm38 interacts with ribosomes and is a component of the mitochondrial protein export machinery. *The Journal of Cell Biology, 172*, 553–564.
Gebert, J. F., Overhoff, B., Manson, M. D., & Boos, W. (1988). The Tsr chemosensory transducer of Escherichia coli assembles into the cytoplasmic membrane via a SecA-dependent process. *The Journal of Biological Chemistry, 263*, 16652–16660.
Genuth, N. R., & Barna, M. (2018). Heterogeneity and specialized functions of translation machinery: From genes to organisms. *Nature Reviews. Genetics, 19*, 431–452.
Goddard, T. D., Huang, C. C., Meng, E. C., Pettersen, E. F., Couch, G. S., Morris, J. H., et al. (2018). UCSF ChimeraX: Meeting modern challenges in visualization and analysis. *Protein Science, 27*, 14–25.
Gold, V. A., Chroscicki, P., Bragoszewski, P., & Chacinska, A. (2017). Visualization of cytosolic ribosomes on the surface of mitochondria by electron cryo-tomography. *EMBO Reports, 18*, 1786–1800.
Gotzke, H., Palombo, I., Muheim, C., Perrody, E., Genevaux, P., Kudva, R., et al. (2014). YfgM is an ancillary subunit of the SecYEG translocon in Escherichia coli. *The Journal of Biological Chemistry, 289*, 19089–19097.

Greber, B. J., Bieri, P., Leibundgut, M., Leitner, A., Aebersold, R., Boehringer, D., et al. (2015). Ribosome. The complete structure of the 55S mammalian mitochondrial ribosome. *Science*, *348*, 303–308.
Halic, M., Becker, T., Pool, M. R., Spahn, C. M., Grassucci, R. A., Frank, J., et al. (2004). Structure of the signal recognition particle interacting with the elongation-arrested ribosome. *Nature*, *427*, 808–814.
Hartmann, E., Sommer, T., Prehn, S., Gorlich, D., Jentsch, S., & Rapoport, T. A. (1994). Evolutionary conservation of components of the protein translocation complex. *Nature*, *367*, 654–657.
Heinonen, T. Y., Schnare, M. N., & Gray, M. W. (1990). Sequence heterogeneity in the duplicate large subunit ribosomal RNA genes of Tetrahymena pyriformis mitochondrial DNA. *The Journal of Biological Chemistry*, *265*, 22336–22341.
Hristou, A., Gerlach, I., Stolle, D. S., Neumann, J., Bischoff, A., Dunschede, B., et al. (2019). Ribosome-associated chloroplast SRP54 enables efficient Cotranslational membrane insertion of key photosynthetic proteins. *Plant Cell*, *31*, 2734–2750.
Huber, D., Rajagopalan, N., Preissler, S., Rocco, M. A., Merz, F., Kramer, G., et al. (2011). SecA interacts with ribosomes in order to facilitate posttranslational translocation in bacteria. *Molecular Cell*, *41*, 343–353.
Huh, W. K., Falvo, J. V., Gerke, L. C., Carroll, A. S., Howson, R. W., Weissman, J. S., et al. (2003). Global analysis of protein localization in budding yeast. *Nature*, *425*, 686–691.
Ito, K. (1996). The major pathways of protein translocation across membranes. *Genes to Cells*, *1*, 337–346.
Itoh, Y., Andrell, J., Choi, A., Richter, U., Maiti, P., Best, R. B., et al. (2021). Mechanism of membrane-tethered mitochondrial protein synthesis. *Science*, *371*, 846–849.
Itskanov, S., & Park, E. (2019). Structure of the posttranslational sec protein-translocation channel complex from yeast. *Science*, *363*, 84–87.
Jomaa, A., Boehringer, D., Leibundgut, M., & Ban, N. (2016). Structures of the *E. coli* translating ribosome with SRP and its receptor and with the translocon. *Nat Commun*, *7*, 10471.
Jomaa, A., Fu, Y. H., Boehringer, D., Leibundgut, M., Shan, S. O., & Ban, N. (2017). Structure of the quaternary complex between SRP, SR, and translocon bound to the translating ribosome. *Nature Communications*, *8*, 15470.
Karamyshev, A. L., & Johnson, A. E. (2005). Selective SecA association with signal sequences in ribosome-bound nascent chains: A potential role for SecA in ribosome targeting to the bacterial membrane. *The Journal of Biological Chemistry*, *280*, 37930–37940.
Kedrov, A., Sustarsic, M., De Keyzer, J., Caumanns, J. J., Wu, Z. C., & Driessen, A. J. (2013). Elucidating the native architecture of the YidC: Ribosome complex. *Journal of Molecular Biology*, *425*, 4112–4124.
Keenan, R. J., Freymann, D. M., Walter, P., & Stroud, R. M. (1998). Crystal structure of the signal sequence binding subunit of the signal recognition particle. *Cell*, *94*, 181–191.
Kelleher, D. J., & Gilmore, R. (2006). An evolving view of the eukaryotic oligosaccharyltransferase. *Glycobiology*, *16*, 47R–62R.
Kellems, R. E., Allison, V. F., & Butow, R. A. (1975). Cytoplasmic type 80S ribosomes associated with yeast mitochondria. IV. Attachment of ribosomes to the outer membrane of isolated mitochondria. *J Cell Biol*, *65*, 1–14.
Khalimonchuk, O., Ostermann, K., & Rodel, G. (2005). Evidence for the association of yeast mitochondrial ribosomes with Cox11p, a protein required for the Cu(B) site formation of cytochrome c oxidase. *Current Genetics*, *47*, 223–233.
Kihara, A., & Ito, K. (1998). Translocation, folding, and stability of the HflKC complex with signal anchor topogenic sequences. *The Journal of Biological Chemistry*, *273*, 29770–29775.
Kobayashi, K., Jomaa, A., Lee, J. H., Chandrasekar, S., Boehringer, D., Shan, S. O., et al. (2018). Structure of a prehandover mammalian ribosomal SRP.SRP receptor targeting complex. *Science*, *360*, 323–327.

Koch, S., Seinen, A. B., Kamel, M., Kuckla, D., Monzel, C., Kedrov, A., et al. (2021). Single-molecule analysis of dynamics and interactions of the SecYEG translocon. *The FEBS Journal*, *288*, 2203–2221.

Kroliczewski, J., Bartoszewski, R., & Kroliczewska, B. (2017). Chloroplast PetD protein: Evidence for SRP/Alb3-dependent insertion into the thylakoid membrane. *BMC Plant Biology*, *17*, 213.

Kroliczewski, J., Piskozub, M., Bartoszewski, R., & Kroliczewska, B. (2016). ALB3 Insertase mediates cytochrome b6 co-translational import into the thylakoid membrane. *Scientific Reports*, *6*, 34557.

Kuhn, P., Draycheva, A., Vogt, A., Petriman, N. A., Sturm, L., Drepper, F., et al. (2015). Ribosome binding induces repositioning of the signal recognition particle receptor on the translocon. *The Journal of Cell Biology*, *211*, 91–104.

Kulandaisamy, A., Binny Priya, S., Sakthivel, R., Tarnovskaya, S., Bizin, I., Honigschmid, P., et al. (2018). MutHTP: mutations in human transmembrane proteins. *Bioinformatics*, *34*, 2325–2326.

Kummer, E., Leibundgut, M., Rackham, O., Lee, R. G., Boehringer, D., Filipovska, A., et al. (2018). Unique features of mammalian mitochondrial translation initiation revealed by cryo-EM. *Nature*, *560*, 263–267.

Lake, J. A. (1985). Evolving ribosome structure: Domains in archaebacteria, eubacteria, eocytes and eukaryotes. *Annual Review of Biochemistry*, *54*, 507–530.

Lakshmipathy, S. K., Gupta, R., Pinkert, S., Etchells, S. A., & Hartl, F. U. (2010). Versatility of trigger factor interactions with ribosome-nascent chain complexes. *The Journal of Biological Chemistry*, *285*, 27911–27923.

Lee, H. C., & Bernstein, H. D. (2001). The targeting pathway of Escherichia coli presecretory and integral membrane proteins is specified by the hydrophobicity of the targeting signal. *Proceedings of the National Academy of Sciences of the United States of America*, *98*, 3471–3476.

Lee, R. G., Gao, J., Siira, S. J., Shearwood, A. M., Ermer, J. A., Hofferek, V., et al. (2020). Cardiolipin is required for membrane docking of mitochondrial ribosomes and protein synthesis. *J Cell Sci*, 133.

Liu, M., & Spremulli, L. (2000). Interaction of mammalian mitochondrial ribosomes with the inner membrane. *The Journal of Biological Chemistry*, *275*, 29400–29406.

Lu, Y., Xiong, X., Helm, A., Kimani, K., Bragin, A., & Skach, W. R. (1998). Co- and post-translational translocation mechanisms direct cystic fibrosis transmembrane conductance regulator N terminus transmembrane assembly. *The Journal of Biological Chemistry*, *273*, 568–576.

Luirink, J., & Sinning, I. (2004). SRP-mediated protein targeting: Structure and function revisited. *Biochimica et Biophysica Acta*, *1694*, 17–35.

Lycklama, A. N. J. A., & Driessen, A. J. (2012). The bacterial sec-translocase: Structure and mechanism. *Philosophical Transactions of the Royal Society of London. Series B, Biological Sciences*, *367*, 1016–1028.

Martin, W. F., Garg, S., & Zimorski, V. (2015). Endosymbiotic theories for eukaryote origin. *Philosophical Transactions of the Royal Society of London. Series B, Biological Sciences*, *370*, 20140330.

Maslov, D. A., Sharma, M. R., Butler, E., Falick, A. M., Gingery, M., Agrawal, R. K., et al. (2006). Isolation and characterization of mitochondrial ribosomes and ribosomal subunits from Leishmania tarentolae. *Molecular and Biochemical Parasitology*, *148*, 69–78.

Milstein, C., Brownlee, G. G., Harrison, T. M., & Mathews, M. B. (1972). A possible precursor of immunoglobulin light chains. *Nature: New Biology*, *239*, 117–120.

Mitra, K., & Frank, J. (2006). Ribosome dynamics: Insights from atomic structure modeling into cryo-electron microscopy maps. *Annual Review of Biophysics and Biomolecular Structure*, *35*, 299–317.

Mitra, K., Frank, J., & Driessen, A. (2006). Co- and post-translational translocation through the protein-conducting channel: Analogous mechanisms at work? *Nature Structural & Molecular Biology, 13*, 957–964.
Mitra, K., schaffitzel, C., shaikh, T., Tama, F., Jenni, S., Brooks, C. L., et al. (2005). Structure of the E. coli protein-conducting channel bound to a translating ribosome. *Nature, 438*, 318–324.
Mohorko, E., Glockshuber, R., & Aebi, M. (2011). Oligosaccharyltransferase: The central enzyme of N-linked protein glycosylation. *Journal of Inherited Metabolic Disease, 34*, 869–878.
Moller-Hergt, B. V., Carlstrom, A., Stephan, K., Imhof, A., & Ott, M. (2018). The ribosome receptors Mrx15 and Mba1 jointly organize cotranslational insertion and protein biogenesis in mitochondria. *Molecular Biology of the Cell, 29*, 2386–2396.
Moore, P. B., Traut, R. R., Noller, H., Pearson, P., & Delius, H. (1968). Ribosomal proteins of Escherichia coli. II. Proteins from the 30 s subunit. *Journal of Molecular Biology, 31*, 441–461.
Mothes, W., Heinrich, S. U., Graf, R., Nilsson, I., Von Heijne, G., Brunner, J., et al. (1997). Molecular mechanism of membrane protein integration into the endoplasmic reticulum. *Cell, 89*, 523–533.
Muller, L., De Escauriaza, M. D., Lajoie, P., Theis, M., Jung, M., Muller, A., et al. (2010). Evolutionary gain of function for the ER membrane protein Sec62 from yeast to humans. *Molecular Biology of the Cell, 21*, 691–703.
Ng, D. T., Brown, J. D., & Walter, P. (1996). Signal sequences specify the targeting route to the endoplasmic reticulum membrane. *The Journal of Cell Biology, 134*, 269–278.
Ngosuwan, J., Wang, N. M., Fung, K. L., & Chirico, W. J. (2003). Roles of cytosolic Hsp70 and Hsp40 molecular chaperones in post-translational translocation of presecretory proteins into the endoplasmic reticulum. *The Journal of Biological Chemistry, 278*, 7034–7042.
Nielsen, M. H., Flygaard, R. K., & Jenner, L. B. (2017). Structural analysis of ribosomal RACK1 and its role in translational control. *Cellular Signalling, 35*, 272–281.
Nilsson, J., Sengupta, J., Frank, J., & Nissen, P. (2004). Regulation of eukaryotic translation by the RACK1 protein: A platform for signalling molecules on the ribosome. *EMBO Reports, 5*, 1137–1141.
Ott, M., & Herrmann, J. M. (2010). Co-translational membrane insertion of mitochondrially encoded proteins. *Biochimica et Biophysica Acta, 1803*, 767–775.
Ott, M., Prestele, M., Bauerschmitt, H., Funes, S., Bonnefoy, N., & Herrmann, J. M. (2006). Mba1, a membrane-associated ribosome receptor in mitochondria. *The EMBO Journal, 25*, 1603–1610.
Panzner, S., Dreier, L., Hartmann, E., Kostka, S., & Rapoport, T. A. (1995). Posttranslational protein transport in yeast reconstituted with a purified complex of sec proteins and Kar2p. *Cell, 81*, 561–570.
Petriman, N. A., Jauss, B., Hufnagel, A., Franz, L., Sachelaru, I., Drepper, F., et al. (2018). The interaction network of the YidC insertase with the SecYEG translocon, SRP and the SRP receptor FtsY. *Scientific Reports, 8*, 578.
Pfeffer, S., Brandt, F., Hrabe, T., Lang, S., Eibauer, M., Zimmermann, R., et al. (2012). Structure and 3D arrangement of endoplasmic reticulum membrane-associated ribosomes. *Structure, 20*, 1508–1518.
Pfeffer, S., Dudek, J., Zimmermann, R., & Forster, F. (2016). Organization of the native ribosome-translocon complex at the mammalian endoplasmic reticulum membrane. *Biochimica et Biophysica Acta, 1860*, 2122–2129.
Poritz, M. A., Bernstein, H. D., Strub, K., Zopf, D., Wilhelm, H., & Walter, P. (1990). An E. coli ribonucleoprotein containing 4.5S RNA resembles mammalian signal recognition particle. *Science, 250*, 1111–1117.

Potter, M. D., & Nicchitta, C. V. (2002). Endoplasmic reticulum-bound ribosomes reside in stable association with the translocon following termination of protein synthesis. *The Journal of Biological Chemistry*, *277*, 23314–23320.

Rawat, S., Zhu, L., Lindner, E., Dalbey, R. E., & White, S. H. (2015). SecA drives transmembrane insertion of RodZ, an unusual single-span membrane protein. *Journal of Molecular Biology*, *427*, 1023–1037.

Reid, D. W., & Nicchitta, C. V. (2015). Diversity and selectivity in mRNA translation on the endoplasmic reticulum. *Nature Reviews. Molecular Cell Biology*, *16*, 221–231.

Reyes-Prieto, A., Weber, A. P., & Bhattacharya, D. (2007). The origin and establishment of the plastid in algae and plants. *Annual Review of Genetics*, *41*, 147–168.

Ryabichko, S., Ferreira, V. M., Vitrac, H., Kiyamova, R., Dowhan, W., & Bogdanov, M. (2020). Cardiolipin is required in vivo for the stability of bacterial translocon and optimal membrane protein translocation and insertion. *Scientific Reports*, *10*, 6296.

Sabatini, D. D., Blobel, G., Nonomura, Y., & Adelman, M. R. (1971). Ribosome-membrane interaction: Structural aspects and functional implications. *Advances in Cytopharmacology*, *1*, 119–129.

Sachelaru, I., Petriman, N. A., Kudva, R., Kuhn, P., Welte, T., Knapp, B., et al. (2015). YidC occupies the lateral gate of the SecYEG translocon and is sequentially displaced by a nascent membrane protein. *The Journal of Biological Chemistry*, *290*, 14492.

Sanders, C. R., & Myers, J. K. (2004). Disease-related misassembly of membrane proteins. *Annual Review of Biophysics and Biomolecular Structure*, *33*, 25–51.

Saraogi, I., & Shan, S. O. (2014). Co-translational protein targeting to the bacterial membrane. *Biochimica et Biophysica Acta*, *1843*, 1433–1441.

Schenkel, L. C., & Bakovic, M. (2014). Formation and regulation of mitochondrial membranes. *Int J Cell Biol*, *2014*, 709828.

Schmitz, U., Behrens, S., Freymann, D. M., Keenan, R. J., Lukavsky, P., Walter, P., et al. (1999). Structure of the phylogenetically most conserved domain of SRP RNA. *RNA*, *5*, 1419–1429.

Schmitz-Linneweber, C., Maier, R. M., Alcaraz, J. P., Cottet, A., Herrmann, R. G., & Mache, R. (2001). The plastid chromosome of spinach (Spinacia oleracea): Complete nucleotide sequence and gene organization. *Plant Molecular Biology*, *45*, 307–315.

Schuenemann, D., Gupta, S., Persello-Cartieaux, F., Klimyuk, V. I., Jones, J. D., Nussaume, L., et al. (1998). A novel signal recognition particle targets light-harvesting proteins to the thylakoid membranes. *Proceedings of the National Academy of Sciences of the United States of America*, *95*, 10312–10316.

Schwarz, D. S., & Blower, M. D. (2016). The endoplasmic reticulum: Structure, function and response to cellular signaling. *Cellular and Molecular Life Sciences*, *73*, 79–94.

Shao, S., & Hegde, R. S. (2011). Membrane protein insertion at the endoplasmic reticulum. *Annual Review of Cell and Developmental Biology*, *27*, 25–56.

Sharma, M. R., Booth, T. M., Simpson, L., Maslov, D. A., & Agrawal, R. K. (2009). Structure of a mitochondrial ribosome with minimal RNA. *Proceedings of the National Academy of Sciences of the United States of America*, *106*, 9637–9642.

Siu, F. Y., Spanggord, R. J., & Doudna, J. A. (2007). SRP RNA provides the physiologically essential GTPase activation function in cotranslational protein targeting. *RNA*, *13*, 240–250.

Sommer, N., Junne, T., Kalies, K. U., Spiess, M., & Hartmann, E. (2013). TRAP assists membrane protein topogenesis at the mammalian ER membrane. *Biochimica et Biophysica Acta*, *1833*, 3104–3111.

Steinberg, R., Knupffer, L., Origi, A., Asti, R., & Koch, H. G. (2018). Co-translational protein targeting in bacteria. *FEMS Microbiology Letters*, *365*.

Sun, Y., & Zerges, W. (2015). Translational regulation in chloroplasts for development and homeostasis. *Biochimica et Biophysica Acta, 1847*, 809–820

Tajima, S., Lauffer, L., Rath, V. L., & Walter, P. (1986). The signal recognition particle receptor is a complex that contains two distinct polypeptide chains. *The Journal of Cell Biology, 103*, 1167–1178.

Temperley, R. J., Wydro, M., Lightowlers, R. N., & Chrzanowska-Lightowlers, Z. M. (2010). Human mitochondrial mRNAs--like members of all families, similar but different. *Biochimica et Biophysica Acta, 1797*, 1081–1085.

Tiller, N., & Bock, R. (2014). The translational apparatus of plastids and its role in plant development. *Molecular Plant, 7*, 1105–1120.

Trager, C., Rosenblad, M. A., Ziehe, D., Garcia-Petit, C., Schrader, L., Kock, K., et al. (2012). Evolution from the prokaryotic to the higher plant chloroplast signal recognition particle: The signal recognition particle RNA is conserved in plastids of a wide range of photosynthetic organisms. *Plant Cell, 24*, 4819–4836.

Traxler, B., & Murphy, C. (1996). Insertion of the polytopic membrane protein MalF is dependent on the bacterial secretion machinery. *The Journal of Biological Chemistry, 271*, 12394–12400.

Unseld, M., Marienfeld, J. R., Brandt, P., & Brennicke, A. (1997). The mitochondrial genome of Arabidopsis thaliana contains 57 genes in 366,924 nucleotides. *Nature Genetics, 15*, 57–61.

Vafai, S. B., & Mootha, V. K. (2012). Mitochondrial disorders as windows into an ancient organelle. *Nature, 491*, 374–383.

Von Heijne, G. (1990). The signal peptide. *The Journal of Membrane Biology, 115*, 195–201.

Voorhees, R. M., Fernandez, I. S., Scheres, S. H., & Hegde, R. S. (2014). Structure of the mammalian ribosome-Sec61 complex to 3.4 A resolution. *Cell, 157*, 1632–1643.

Voorhees, R. M., & Hegde, R. S. (2015). Structures of the scanning and engaged states of the mammalian SRP-ribosome complex. *Elife*, 4.

Walter, P., & Johnson, A. E. (1994). Signal sequence recognition and protein targeting to the endoplasmic reticulum membrane. *Annual Review of Cell Biology, 10*, 87–119.

Waltz, F., Corre, N., Hashem, Y., & Giege, P. (2020). Specificities of the plant mitochondrial translation apparatus. *Mitochondrion, 53*, 30–37.

Waltz, F., Nguyen, T. T., Arrive, M., Bochler, A., Chicher, J., Hammann, P., et al. (2019). Small is big in Arabidopsis mitochondrial ribosome. *Nat Plants, 5*, 106–117.

Waltz, F., Soufari, H., Bochler, A., Giege, P., & Hashem, Y. (2020). Cryo-EM structure of the RNA-rich plant mitochondrial ribosome. *Nat Plants, 6*, 377–383.

Wang, L., & Dobberstein, B. (1999). Oligomeric complexes involved in translocation of proteins across the membrane of the endoplasmic reticulum. *FEBS Letters, 457*, 316–322.

Wang, S., Jomaa, A., Jaskolowski, M., Yang, C. I., Ban, N., & Shan, S. O. (2019). The molecular mechanism of cotranslational membrane protein recognition and targeting by SecA. *Nature Structural & Molecular Biology, 26*, 919–929.

Wang, S., Yang, C. I., & Shan, S. O. (2017). SecA mediates cotranslational targeting and translocation of an inner membrane protein. *The Journal of Cell Biology, 216*, 3639–3653.

Wanker, E. E., Sun, Y., Savitz, A. J., & Meyer, D. I. (1995). Functional characterization of the 180-kD ribosome receptor in vivo. *The Journal of Cell Biology, 130*, 29–39.

Wild, K., Halic, M., Sinning, I., & Beckmann, R. (2004). SRP meets the ribosome. *Nature Structural & Molecular Biology, 11*, 1049–1053.

Wild, K., Juaire, K. D., Soni, K., Shanmuganathan, V., Hendricks, A., Segnitz, B., et al. (2019). Reconstitution of the human SRP system and quantitative and systematic analysis of its ribosome interactions. *Nucleic Acids Research, 47*, 3184–3196.

Wilson, D. N., & Doudna Cate, J. H. (2012). The structure and function of the eukaryotic ribosome. *Cold Spring Harbor Perspectives in Biology, 4*.

Wittke, S., Dunnwald, M., Albertsen, M., & Johnsson, N. (2002). Recognition of a subset of signal sequences by Ssh1p, a Sec61p-related protein in the membrane of endoplasmic reticulum of yeast Saccharomyces cerevisiae. *Molecular Biology of the Cell, 13*, 2223–2232.

Yamaguchi, K., & Subramanian, A. R. (2000). The plastid ribosomal proteins. Identification of all the proteins in the 50 S subunit of an organelle ribosome (chloroplast). *The Journal of Biological Chemistry, 275*, 28466–28482.

Yamaguchi, K., Von Knoblauch, K., & Subramanian, A. R. (2000). The plastid ribosomal proteins. Identification of all the proteins in the 30 S subunit of an organelle ribosome (chloroplast). *The Journal of Biological Chemistry, 275*, 28455–28465.

Yoshida, H., Maki, Y., Furuike, S., Sakai, A., Ueta, M., & Wada, A. (2012). YqjD is an inner membrane protein associated with stationary-phase ribosomes in Escherichia coli. *Journal of Bacteriology, 194*, 4178–4183.

Zeng, R., Smith, E., & Barrientos, A. (2018). Yeast Mitoribosome large subunit assembly proceeds by hierarchical incorporation of protein clusters and modules on the inner membrane. *Cell Metabolism, 27*(645–656), e7.

Zhang, Y., De Laurentiis, E., Bohnsack, K. E., Wahlig, M., Ranjan, N., Gruseck, S., et al. (2021). Ribosome-bound Get4/5 facilitates the capture of tail-anchored proteins by Sgt2 in yeast. *Nature Communications, 12*, 782.

Zhang, X., Schaffitzel, C., Ban, N., & Shan, S. O. (2009). Multiple conformational switches in a GTPase complex control co-translational protein targeting. *Proceedings of the National Academy of Sciences of the United States of America, 106*, 1754–1759.

Zhao, Y., Hu, J., Miao, G., Qu, L., Wang, Z., Li, G., et al. (2013). Transmembrane protein 208: A novel ER-localized protein that regulates autophagy and ER stress. *PLoS One, 8*, e64228.

Ziehe, D., Dunschede, B., & Schunemann, D. (2017). From bacteria to chloroplasts: Evolution of the chloroplast SRP system. *Biological Chemistry, 398*, 653–661.

Zopf, D., Bernstein, H. D., Johnson, A. E., & Walter, P. (1990). The methionine-rich domain of the 54 kd protein subunit of the signal recognition particle contains an RNA binding site and can be crosslinked to a signal sequence. *The EMBO Journal, 9*, 4511–4517.

Zoschke, R., & Barkan, A. (2015). Genome-wide analysis of thylakoid-bound ribosomes in maize reveals principles of cotranslational targeting to the thylakoid membrane. *Proceedings of the National Academy of Sciences of the United States of America, 112*, E1678–E1687.

Zwieb, C., Van Nues, R. W., Rosenblad, M. A., Brown, J. D., & Samuelsson, T. (2005). A nomenclature for all signal recognition particle RNAs. *RNA, 11*, 7–13.

CHAPTER SIX

Mechanisms of long-distance allosteric couplings in proton-binding membrane transporters

Ana-Nicoleta Bondar[a,b,*]
[a]University of Bucharest, Faculty of Physics, Măgurele, Romania
[b]Forschungszentrum Jülich, Institute of Computational Biomedicine, Jülich, Germany
*Corresponding author: e-mail address: nbondar@fizica.unibuc.ro

Contents

1. Introduction 200
2. Proton sensing and proton transfer in human physiology and disease 200
3. Titratable protein sidechains in protein environments 201
4. Protonation-coupled changes in membrane protein structure and dynamics 203
   4.1 Conformational couplings in protonation-coupled membrane transporters 203
   4.2 From pump to channel: Inter-helical interactions and conformational couplings in bacteriorhodopsin, halorhodopsin, and channelrhodopsin 205
   4.3 Protonation-coupled conformational dynamics of the *Escherichia coli* lactose permease, LacY, and in the xylose permease XylE 219
   4.4 A carboxylate-hydroxyl H-bond motif of membrane transporters 225
5. Conclusions 228
Acknowledgments 229
References 229

Abstract

Membrane transporters that use proton binding and proton transfer for function couple local protonation change with changes in protein conformation and water dynamics. Changes of protein conformation might be required to allow transient formation of hydrogen-bond networks that bridge proton donor and acceptor pairs separated by long distances. Inter-helical hydrogen-bond networks adjust rapidly to protonation change, and ensure rapid response of the protein structure and dynamics. Membrane transporters with known three-dimensional structures and proton-binding groups inform on general principles of protonation-coupled protein conformational dynamics. Inter-helical hydrogen bond motifs between proton-binding carboxylate groups and a polar sidechain are observed in unrelated membrane transporters, suggesting common principles of coupling protonation change with protein conformational dynamics.

Advances in Protein Chemistry and Structural Biology, Volume 128
ISSN 1876-1623
https://doi.org/10.1016/bs.apcsb.2021.09.002

1. Introduction

Reversible post-translational protonation regulates protein function by altering conformational dynamics (Schönichen, Webb, Jacobson, & Barber, 2013). At pH range relevant to physiology, protonation, i.e., the covalent binding of an H^+ ion, typically occurs at a minority of protein sites (Damaghi, Wojktowiak, & Gillies, 2013; Schönichen et al., 2013). When an amino acid residue sidechain changes its protonation, its electrostatic interactions with other protein groups and nearby water molecules will change, and as a response protein groups and water molecules could change coordinates and/or local dynamics. Depending on interactions at the proton-binding site and elsewhere in the protein, on longer timescale local changes in protein structure and/or dynamics may propagate at remote sites in the protein; in the case of membrane proteins, changes in protein structure and dynamics may also propagate to the surrounding lipid. The reaction coordinate of a protonation-coupled membrane transporter or receptor is thus highly complex, as it involves rapid electronic structure change at the proton-binding site, and long timescale changes in the coordinates of atoms of the protein, internal waters, and lipid molecules near the protein. Mechanisms by which the local electronic structure change at the proton-binding site couples to changes in protein structure and dynamics are central to our understanding of bioenergetics processes.

This review is focused on mechanisms by which protonation change couples to long-distance protein conformational change and water dynamics in membrane transporters and receptors of interest as model systems, and/or because of their implications in human physiology and disease. Inter-helical hydrogen(H)-bonds, particularly a carboxylate-hydroxyl H-bond motif commonly found at functionally important sites of membrane transporters and receptors, are key elements of dynamic hydrogen(H)-bond networks that may shape protein dynamics.

2. Proton sensing and proton transfer in human physiology and disease

pH values inside various compartments of biological cells are tightly regulated, as it is essential for the proper functioning of different organelles. Proton transfer across cell membranes, such as by symporters, antiporters, and proton-pumping ATPases, is a key part of the cellular apparatus that ensures distinct pH values in cell compartments, from ~4.7 in the lysosome to ~7.2 in the cytosol and ~8 in the mitochondria (Casey, Grinstein, & Orlowski, 2010). Local variations in pH are, however, possible, and indeed part of normal physiological processes. For example, the plasma membrane Na^+/H^+ exchanger, which is a major regulator of intracellular pH, is located at the leading edge of a migrating cell, where it contributes to establishing a pH gradient that is important for cell motility and cell adhesion (Casey et al., 2010; Stock & Schwab, 2006). In folded mitochondrial membranes, the local pH at the F_0F_1 ATP synthase, which uses protons to generate ATP, is ~0.30 units less acidic than at cytochrome c oxidase of the respiratory chain, which pumps the protons (Rieger, Jung, & Busch, 2014). Dis-regulation of pH homeostasis has been implicated in disease—cytoplasmic acidification in ischemia, and mutations in a channel that controls pH of early endosomes, in kidney disease (Casey et al., 2010).

Protonation dynamics is of direct interest for drug design and drug binding, such as opioid drugs that would bind to opioid receptors only when protonated at the acidic pH of an inflamed tissue, and thus have reduced toxicity (Spahn et al., 2018), or prescription drugs that bind to a coordination cage host in a pH-dependent manner, with potential applications for drug release (Cullen, Turega, Hunter, & Ward, 2015). Protonation-coupled transporters ensure transport of glutamate neurotransmitter in the brain (Edwards, 2007), and a microbial protonation-coupled ion channel is used to control neuronal cells with light (Zhang, Wang, Boyden, & Deisseroth, 2006).

3. Titratable protein sidechains in protein environments

The protonation of a titratable protein group is determined by its pK_a, which indicates the pH value at which the group is 50% protonated (Srivastava, Barber, & Jacbson, 2007). Carboxylate and histidine sidechains are of primary interest as potential proton-binding sites: Aspartic (Asp) and glutamic (Glu) acids in solution have nominal pK_a values of 3.67 ± 0.4 and, respectively, 4.25 ± 0.5 (Thurlkill, Grimsley, Scholtz, & Pace, 2006), and histidine (His) has a nominal pK_a of 6.54 ± 0.4 (Schönichen et al., 2013). Asp and Glu would thus be negatively charged at pH above their nominal pK_a values, whereas His sidechains would be positively charged at pH below their nominal pK_a (Pace, Grimsley, & Scholtz, 2009).

Cysteine (Cys) sidechains, with nominal pK_a of 8.5, are also of potential interest as sites where protonation might change (Srivastava et al., 2007);

tyrosine (Tyr) and lysine (Lys) sidechains, and arginine (Arg) sidechains, have somewhat higher nominal pK_a values of 10 and 12, respectively (Srivastava et al., 2007).

Nominal pK_a values may, however, have rather limited use in predicting the protonation state of a titratable protein sidechain in a particular protein environment: The local pH in the cell may vary as part of normal physiological processes, or during disease—for example, local changes in intracellular pH contribute to cell migration, and pH be acidic at the surface of a cancer cell (White, Grillo-Hill, & Barber, 2017), or in inflammation. At enzyme active sites, ionizable catalytic groups often have largely perturbed pK_a values (Harris & Turner, 2002).

The pK_a value of a titratable protein sidechain depends on its local dielectric environment, Coulomb interactions with other charged groups in the protein, and H bonding (Pace et al., 2009). For example, experimental measurements indicated that Glu sidechains located at interior sites of staphylococcal nuclease can have pK_a values between 5.2 and 9.4 (Isom, Castañeda, Cannon, Velu, & García-Moreno, 2010), i.e., significantly higher than the nominal value; in the sugar-proton transporter LacY, the pK_a of a Glu sidechain required for sugar binding is as high as 10.5 ± 01 (Grytsyk, Sugihara, Kaback, & Hellwig, 2017). In cytochrome *c* oxidase, a carboxylate group that functions as internal proton donor has a pK_a of $\sim$9.4 (Namslauer, Aagard, Kotsonouri, & Przezinski, 2003) that increases to >11 in an uncoupled mutant (Brändén, Pawate, Gennis, & Brzezinski, 2005). The pK_a of a buried Lys sidechain can be as low as 5.3 (Isom, Castañeda, Cannon, & García-Moreno, 2011), which appears compatible with the suggestion that an internal Lys sidechain might act as proton donor in *Exiguobacterium sibiricum* rhodopsin (Balashov et al., 2013; Gushchin et al., 2013). On the surface of PsbO, an extrinsic subunit of the photosystem II complex, the pK_a values of Asp sidechains range from <2 to >7 (Gerland et al., 2020). Likewise, large variations in pK_a values for all titratable sidechains were observed for a dataset of 78 folded proteins (Pace et al., 2009).

Since proteins contain numerous titratable sidechains (Bondar & Lemieux, 2019; Pace et al., 2009; Samatey, Xu, & Popot, 1995) whose pK_a values could depend on local protein interactions, and since charged groups of a protein might serve different adaptations in the cell (García-Moreno, 2009), identifying those few titratable sidechains that change protonation during the functioning of a particular protein is often a difficult task. Studies with model proteins, some of which are discussed here, provide clues about interaction motifs at sites where proteins are more likely to bind protons, and about interactions that could couple protonation change with changes in protein structure and dynamics. Lessons from simple model systems could guide studies of more complex proteins of interest to bio-medical applications.

4. Protonation-coupled changes in membrane protein structure and dynamics

Changing the protonation state at an internal protein site—such as deprotonation of an internal Asp or Glu sidechain, alters drastically the local electrostatic environment. Though, as a response, changes in protein structure and dynamics may occur, the functioning of the protein would require it to tolerate such ionization. The notion that the protein indeed can tolerate ionization at an internal sidechain was demonstrated with experiments on 25 variants of staphylococcal nuclease with Glu sidechains introduced at various internal sites: most variants tolerated the Glu sidechain without major conformational rearrangements (Isom et al., 2010). This robustness of the protein structure is needed for allosteric conformational change upon protonation change—e.g., for a membrane transporter to undergo a reaction cycle whereby it passes through intermediate conformations with distinct protonation states, and then it returns to its starting, resting state.

Proteins from the family of microbial rhodopsins are valuable as model systems for the architecture of protonation-coupled membrane transporters and receptors: Microbial rhodopsins function as pumps, channels, or receptors, and three-dimensional structures of each of these types of proteins have been solved. In what follows, microbial rhodopsins will be used to introduce general features of the architecture of protonation-coupled membrane transporters and receptors, and to illustrate common sequence and structure motifs likely important for protonation-coupled protein function.

4.1 Conformational couplings in protonation-coupled membrane transporters

Proton pumps use an external energy source to transfer protons against the electrochemical ion gradient; by contrast, proton channels open in response to an external input, such as ligand binding, to allow protons to pass down their electrochemical gradient. Which elements of the protein architecture are determinant for the pump vs. channel function is essential to understand, as it could guide the rational design of proteins and mimetic systems for bio-medical and bio-technological applications.

A key difference between ion pumps and channels consists in how they gate access to ions: Whereas pumps are thought to use two gates, never opened simultaneously, to ensure that ions enter the pump from one side of the membrane, and exit at the opposite side, ion channels are thought to use only one gate which, when open, allows ions to pass (Gadsby, 2009). Identifying structural elements that function as gate(s) is thus of central interest, but often challenging (Gadsby, 2009), and affected by imprecise definitions of what constitutes a membrane transporter's gate.

According to one definition centered on the structural basis of gates of enzymes, a gate consists of a structural element—amino acid residue or a larger fragment or even domain of the protein- that samples two conformations, open vs closed (Gora, Brezovsky, & Damborsky, 2013). A survey of enzyme structures indicated aromatic sidechains (Trp, Phe, and Tyr) occur most frequently in relatively simple gates that involve one or two amino acid residues, and opening via a motion described as wing or, respectively, as swinging door, with structural rearrangements restricted to about 2–3 Å (Gora et al., 2013); synchronized motions of protein backbone groups may control opening of an aperture-type gate, and motions of loops and larger structural elements might be needed to allow access of large substrates (Gora et al., 2013).

Given the importance of gates for mechanistic descriptions of ion transporters, the important question is how are gates identified, in other words, how is a gating function assigned to a specific structural element of the transporter. As illustrated with the examples below, evidence that one or more specific amino acid residue participate in gating may be derived from mutations that stall the progress of the pump along the reaction coordinate, or from structural biology/molecular dynamics observations that motions of the residue(s) would be are required to make space for water or solute to pass.

For the plasma membrane proton pump AHA2, a gating role was assigned to N106, whose mutation blocks the pump at different intermediate conformations along the reaction coordinate (Ekberg, Wielandt, Buch-Pedersen, & Palmgren, 2013). The strong impact of the N106 mutations on the protein conformational dynamics could be explained by perturbed local interactions: in the wild-type pump, N106 can H-bond directly to the primary proton donor D684, and nearby there are other charged and polar sidechains (Ekberg et al., 2013; Guerra & Bondar, 2015; Pedersen, Buch-Pedersen, Morth, Palmgren, & Nissen, 2007). That mutating a single sidechain, E148, can stall the transporter in a particular intermediate conformation, was also observed for a ClC chloride ion channel, suggesting that a swing motion of E148 could gate the channel (Dutzler, Campbell, & MacKinnon, 2003). Computer simulations on the proton pump cytochrome *c* oxidase suggested motions of N106 allow formation of an internal water chain for proton conduction (Henry, Yu, Rodinger, & Pomès, 2009).

The three examples above suggest that charged and polar sidechains within the ion-transfer pore are particularly important for the control of structural dynamics of the transporter, as local interactions of these sidechains might govern conformational dynamics along the reaction coordinate of the pump. It is thus instructive to consider briefly the case of bacteriorhodopsin, halorhodopsin, and channelrhodopsin, as channelrhodopsins might have evolved 'by loss or uncoupling of a gate', from bacteriorhodopsin and halorhodopsin (Gadsby, 2009). An increase in the volume of internal protein cavities, associated with higher mobility of internal waters, might have contributed to the evolution of a pump into a channel (Volkov et al., 2017).

4.2 From pump to channel: Inter-helical interactions and conformational couplings in bacteriorhodopsin, halorhodopsin, and channelrhodopsin

Microbial rhodopsins, of which more than 7000 exist in prokaryotes and eukarya (Govorunova, Sineschchekov, Li, & Spudich, 2017), have diverse sequences, however functional interconversions between a few microbial rhodopsins have been achieved, suggesting the protein scaffold is compatible with different functionalities.

The outward proton pump bacteriorhodopsin was the first microbial rhodopsin to be identified as the only protein of the purple membrane, which owes its purple color to the retinal chromophore that is covalently bound to the opsin protein (Lanyi, 1978; Oesterhelt & Stoeckenius, 1971). Halorhodopsin, an inward chloride pump (Schobert & Lanyi, 1982), has about 25–35% sequence identity with bacteriorhodopsin, and an overall similar structure (Song & Gunner, 2014). A key difference between halorhodopsin and bacteriorhodopsin is that a Thr replaces the D85 primary proton acceptor of bacteriorhodopsin—yet the D85T mutation converts bacteriorhodopsin into an inward chloride pump (Sasaki et al., 1995), which suggests that the sidechain at position 85 provides ion specificity (Sasaki et al., 1995), though the relatively efficiency of chloride pumping by the bacteriorhodopsin mutant indicates the chloride binding site is suboptimal (Váró, 2000). In the presence of the anion azide (N_3^-), halorhodopsin pumps protons, being thought that two azide molecules, one in the extracellular half, the other in the cytoplasmic half, could perform the function of the bacteriorhodopsin proton-transfer groups D85 and D96 (Váró, Brown, Needleman, & Lanyi, 1996). A proton-pumping rhodopsin from the alga *Coccomyxa subellipsoidea* that shares 37% sequence homology with bacteriorhodopsin could be converted into a channel by mutating Y57 to Lys and R83 to Gln (corresponding to Y57 and R82 in bacteriorhodopsin) (Vogt et al., 2015).

Taken together, the observations summarized above on functional interconversions among ion-transport rhodopsins underline common principles of their functional mechanisms, and their usefulness as model systems to decipher how ion transporters work. Importantly, functional interconversions among microbial rhodopsins are asymmetric (Inoue, Nomura, & Kandori, 2016), for example, a halorhodopsin mutant with Asp at position of bacteriorhodopsin D85 failed to allow retinal deprotonation (Váró, Brown, et al., 1996), such that only mutations that reverse an evolutionary sequence change are thought to enable interconversions (Inoue et al., 2016).

4.2.1 Protonation-coupled conformational change of bacteriorhodopsin

Bacteriorhodopsin is one of the best-studied membrane transporters (Balashov & Ebrey, 2001; Edmonds & Luecke, 2004; Gerwert, Freier, & Wolf, 2014; Herzfeld & Lansing, 2002; Kandori, 2004; Lanyi, 1999; Lozier, Bogomolni, & Stoeckenius, 1975; Yamamoto, Hayakawa, Murakami, & Kouyama, 2009). The first three-dimensional structure of bacteriorhodopsin in the purple membrane of *H. salinarium* was solved in 1975 with cryo-electron microscopy (cryo-EM) at a resolution of 7 Å, when it was also suggested that lipid molecules would fill spaces between molecules ofbacteriorhodopsin (Henderson & Unwin, 1975). Some 24 years later, in 1999, two X-ray crystallography structures—one solved at 1.9 Å resolution (Belrhali et al., 1999), and the other at 1.5 Å resolution (Luecke, Schobert, Richter, Cartailler, & Lanyi, 1999), provided a detailed view of how bacteriorhodopsin interacts with lipids of the purple membrane, and with internal water molecules. Structures solved subsequently for intermediate conformations of bacteriorhodopsin provided valuable glimpses about molecular interactions along the reaction path of the pump, and have also led to controversies about details of the proton-transfer mechanism (Wickstrand, Dods, Royant, & Neutze, 2015). In recent years, the structure of the resting state of the pump solved at 1.3 Å resolution (Hasegawa, Jonotsuka, Miki, & Takeda, 2018), and time-resolved serial femtosecond crystallography of bacteriorhodopsin intermediate states (Kovacs et al., 2019; Nango et al., 2016; Weinert et al., 2019), provided a detailed view of the protonation-coupled motions of the pump along its reaction coordinate.

Briefly, in the resting state of bacteriorhodopsin the primary proton-transfer group, the Schiff base of the all-trans retinal, interacts with the primary proton acceptor, D85, via a water molecule that further bridges to D212 (Fig. 1A). Carboxylate groups at three distinct sites of the resting-state protein are protonated (Brown et al., 1995; Dioumaev et al., 1998; Metz, Siebert, & Engelhardt, 1992): at the cytoplasmic side, D96, at the extracellular side, the E194/E204 dyad, and at the core of the protein, D115. The sidechains of D96 and D115 are within inter-helical H-bond distance of T90 and, respectively, T46 (del Val, Bondar, & Bondar, 2014). The E194/E204 dyad (Fig. 1A) is largely exposed to extracellular bulk water, and details of how protein sidechains and nearby water molecules contribute to proton storage have been controversial (Bondar & Smith, 2009; Garczareck, Brown, Lanyi, & Gerwert, 2005; Garczareck & Gerwert, 2006; Kalaidzidis, Belevich, & Kaulen, 1998; Phatak, Ghosh, Yu, Cui, & Elstner, 2008). The recent crystal structure of bacteriorhodopsin solved at 1.25 Å resolution indicates the carboxylate groups of E194 and E204 within short distance of 2.1–2.4 Å (Hasegawa et al., 2018), which would be compatible with the two carboxylates sharing a proton.

Upon photo-isomerization of the retinal to 13-cis and formation of the intermediate state denoted as L, the retinal Schiff base assumes a cytoplasmic interaction whereby it can H-bond to the hydroxyl group of T89 (Fig. 1A), or via a water molecule that also H-bonds to T89, and the Schiff base proton may be transferred to D85 via the T89 sidechain (Bondar, Baudry, Suhai, Fischer, & Smith, 2008; Bondar, Elstner, Suhai, Smith, & Fischer, 2004; Bondar, Fischer, Smith, Elstner, & Suhai, 2004; Bondar, Smith, & Fischer, 2006; Nango et al., 2016; Ni et al., 2018). Protonation at the D85 and E194/E204 sites impacts the orientational dynamics of R85: Whereas in the resting ground state of the pump R85 is oriented towards D85/D212, in the static structure of the M intermediate, in which the Schiff base is deprotonated and D85 is neutral, R85 orients towards the extracellular proton release site E194/E204. When a protonated E204

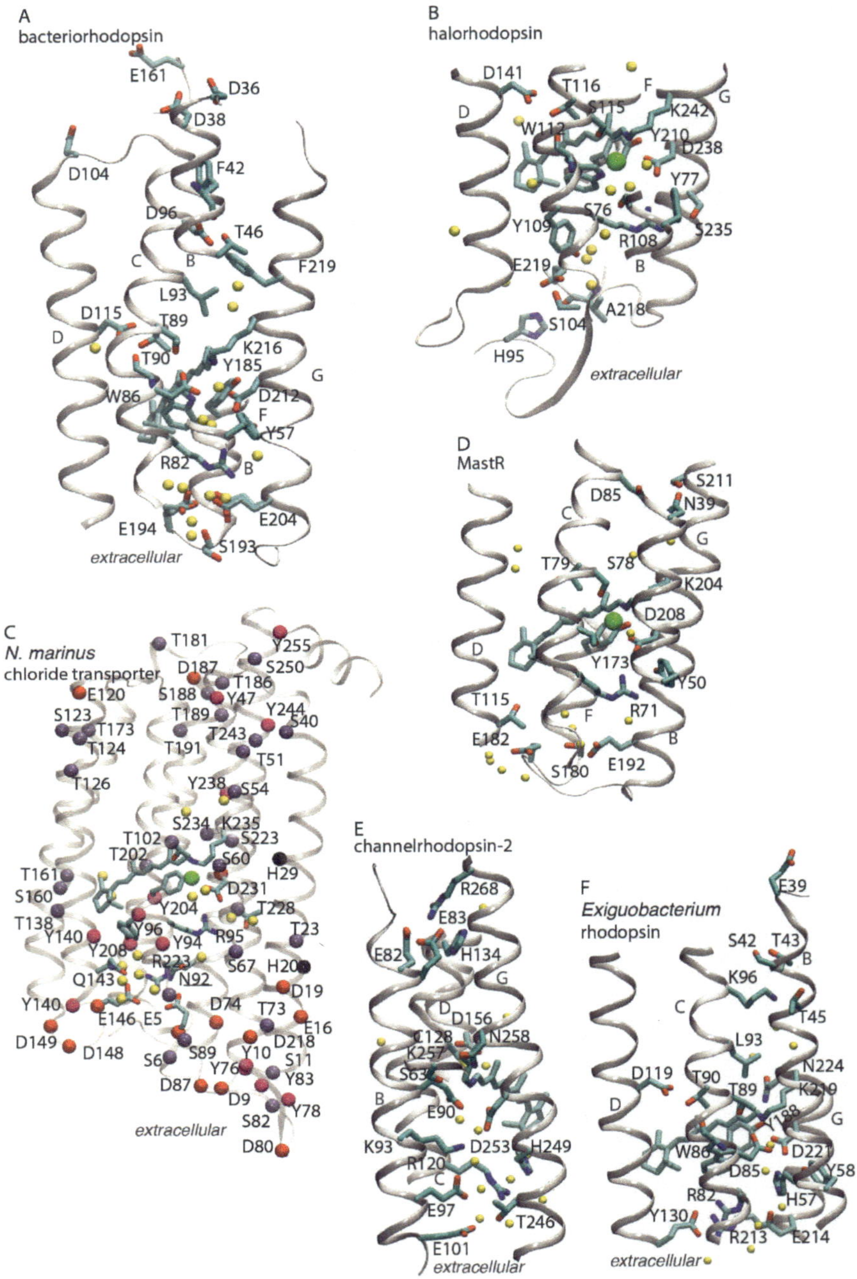

Fig. 1 Proton-binding and conformational coupling in retinal proteins. For clarity, only selected transmembrane helical segments and connecting loops are shown; the seven transmembrane helical segments of retinal proteins are commonly labeled with letters *(Continued)*

H-bonds with E194, R82 oscillates between orientations towards D85/D212 and towards E194/E204 (Ge & Gunner, 2016) when both E194 and E204 are negatively charged following proton release to the bulk, R82 orients toward the extracellular side (Ge & Gunner, 2016).

During the decay of M into the N intermediate, the third proton-transfer step occurs from D96 (Fig. 1A) (Gerwert, Hess, Soppa, & Oesterhelt, 1989; Otto et al., 1989) to the cytoplasmic-oriented retinal Schiff base. Early electron crystallography data on the N intermediate of the F219L mutant indicated protein conformational changes occur at the cytoplasmic side—helices E and F tilt away from the protein, and helix G undergoes smaller rearrangement, opening a channel proposed to be accessible to water (Vonck, 2000). The more recent time-resolved serial crystallography data captured the movie of structural changes of bacteriorhodopsin over 200 ms, including at a time snapshot where the protein sampled an N-like conformation (Weinert et al., 2019). The data were interpreted to suggest that conformational changes of the protein, including reorientations of the sidechains of L93 and F219, generate the space needed for a chain of three water molecules that bridges D96 to the retinal Schiff base (Weinert et al., 2019). Absence of the waters chain in the structure of the triple mutant D96G/F171C/F219L (Wang et al., 2018), which is constitutively open (Subramaniam et al., 1999), was suggested to indicate the importance of conformational change to enable water chain formation (Weinert et al., 2019).

Fig. 1—Cont'd A–G. Water molecules within 12 Å of the retinal molecule are shown as small yellow spheres. (A) Proton-transfer groups of the bacteriorhodopsin proton pump. (B–D) Selected interactions within the transmembrane region of the *H. salinarium* halorhodopsin chloride pump, and the chloride pump from *N. marinus*, and the *M. repens* chloride pump MastR. The chloride ion near the retinal Schiff base is shown as a green sphere; for simplicity, the chloride ion bound at the cytoplasmic side of the *N. marinus* chloride pump is not shown. In panel C, Cα atoms of Asp/Glu are shown as red spheres, Ser/Thr, iceblue, Tyr, magenta, and His, black. (E) Selected interactions of the wild-type *C. reinhardtii* channelrhodopsin-2. (F) Selected interactions of the rhodopsin proton pump from the permafrost bacterium *Exiguobacterium sibiricum*, EXR; note that K96, thought to be internal proton donor, is close to a stretch of three Ser/Thr groups of the nearby helix B—S42, T43 and T45. The molecular graphics were prepared with VMD based on structures with PDB IDs 5ZIM, 1.25 Å resolution (Hasegawa et al., 2018), panel A; 1E12, 1.8 Å resolution (Kolbe, Besir, Essen, & Oesterhelt, 2000), panel B; 5G28, 1.57 Å resolution (Kim et al., 2016), panel C; chain A of 6XL3, 2.33 Å resolution (Besaw et al., 2020), panel D; chain A of 6EID, 2.39 Å resolution, panel E; chain A of PDB ID: 4HYJ, 2.3 Å resolution (Gushchin et al., 2013).

The nature of the coupling between protein conformational change and formation of a water chain is, however, somewhat unclear, as coordinate snapshots lack direct information about the sequence of structural changes along the continuous path between two protein intermediate states. Internal water molecules that are mobile or are positionally disordered may be difficult to solve in a protein crystal structure (Levitt & Park, 1993; Quillin, Wingfield, & Matthews, 2006), limited resolution X-ray crystallography data on water coordinates inside an apolar protein cavity may have artifacts due to the electron densities of atoms that line that cavity (Quillin et al., 2006), and whether a not a protein cavity is occupied by water depends on the temperature (Bondar et al., 2008; Gottschalk, Dencher, & Halle, 2001).

In the particular case of bacteriorhodopsin, a chain of four water molecules connecting D96 to the retinal is thermodynamically allowed already in the resting state (Roux, Nina, Pomes, & Smith, 1996); this column of four waters is stabilized by favorable interactions between these waters, between waters and backbone groups, and between the first and fourth waters with D96 and the retinal C15H bond (Roux et al., 1996). The observation that a water chain is allowed between D96 and the retinal in the resting state of the pump is compatible with simulations on the D85N mutant modeled from the structure of the resting state: On the relatively short, ~62.3 ns timescale of the simulations, a chain of three waters could bridge D96 to the $C_{15}H$ retinal bond (del Val, Bondar, & Bondar, 2014). Likewise, for the L intermediate, FTIR spectroscopy data were interpreted to suggest that a cluster of water molecules is present on the cytoplasmic side of the protonated retinal Schiff base; the water cluster would then relocate from the vicinity of the retinal Schiff base to a site closer to the region of T46 and F219 in the M intermediate, and then back to the vicinity of the retinal Schiff base in the N intermediate (Morgan, Gennis, & Maeda, 2008).

The findings that the protein matrix of the resting state of the pump can allow a chain of waters between D96 and the C15H retinal bond, and that a water cluster present in the L intermediate relocates as the pump changes conformation to M suggests that, in the N intermediate, rearrangements of amino acid residues and the cytoplasmic orientation of the retinal Schiff base, might help order or otherwise stabilize interactions within an already existing chain of internal water molecules, instead of being required for cytoplasmic water to enter and bridge D96 to the retinal Schiff base.

In the fourth proton-transfer step, D96 re-protonates from the cytoplasmic bulk during decay of N into the O intermediate. The rate constant for the transition is pH-independent below pH 7, and it decreases linearly with pH for values above 7, suggesting proton uptake from the bulk (Ames & Mathies, 1989). Proton transfer is thought to involve the bulk-exposed carboxylate groups D36, D104, E161, and E234 (Fig. 1A), which function as a proton-collecting antenna (Checover, Marantz, Nachliel, & Gutman, 2001; Riesle, Oesterhelt, Dencher, & Heberle, 1996), and D38—which might pick up the proton from D36 and serve as an intermediate carrier for proton transfer to D96 (Schätzler et al., 2003). The transfer of a proton from D38 to D96 involves transient expansion and solvation of a hydrophobic shaft centered at F42 (Friedman, Nachliel, & Gutman, 2003), and protonation of D96 impacts both protein structure and internal water dynamics (Wang et al., 2018).

Once re-protonated, during the transition from N to the O intermediate the retinal molecule undergoes a thermal back isomerization from 13-cis, 15-anti to all-trans. The intrinsic barrier for thermal isomerization around the C13=C14 bond of protonated retinal is ~28 kcal/mol (Bondar, Fischer, Suhai, & Smith, 2005); in the protein environment, the isomerization barrier must be reduced significantly, because the entire reaction cycle takes milliseconds to complete. Contribution to lowering the barrier for thermal isomerization of the retinal in the protein environment are brought by retinal twisting and changes of the retinal H-bond environment (Bondar et al., 2005; Elghobashi-Meinhardt, Phatak, Bondar, Elstner, & Smith, 2018), and hydrophobic contacts between the C13 retinal methyl group and the sidechain of L93 (Delaney, Schweiger, & Subramaniam, 1995).

In the fifth and last proton-transfer step, the D85 proton is transferred to the extracellular E194/E204 site (Fig. 1A). This is a long-distance proton transfer: In the resting state of bacteriorhodopsin (Hasegawa et al., 2018), the distance between the carboxyl oxygen atoms of D85 and E194/E204 is ~13 Å, i.e., too long for a direct proton jump. The proton is transferred from D85 to E194/E204 via internal water molecules, and likely via D212 as an intermediate carrier (Dioumaev, Brown, Needleman, & Lanyi, 1999; Phatak et al., 2009). Transient protonation of D212 could weaken its inter-helical H-bond with Y57. Since simulations of Y57F and D212N mutants indicate that perturbations of the Y57-D212 inter-helical H-bond cause rapid changes in dynamics—localized in Y57F, and more extended in D212N (del Val, Bondar, & Bondar, 2014), passage of the wild-type protein through an intermediate with D212 protonated likely associates with protein conformational change.

In summary, the reaction cycle of bacteriorhodopsin includes several discrete proton-transfer steps within the inter-helical domain. The first proton-transfer step, from the retinal Schiff base to D85, is on a relatively short distance of ~4 Å that may be overcome via a single intermediate proton carrier; the remaining proton transfer steps involve distances of ~10–12 Å. To pick up a proton from the cytoplasmic bulk, the protein might use a cluster of closely spaced carboxylate groups. Although details of the coupling mechanism remain unclear, local protonation change associates with changes in protein conformational dynamics, and in internal hydration.

4.2.2 Conformational couplings of the chloride pump halorhodopsin

Halorhodopsin, initially thought to function as a sodium pump (Lanyi & Weber, 1980; Lindley & MacDonald, 1979), was recognized as an inward light-driven chloride pump based on biochemistry measurements using cell envelope vesicles (Schobert & Lanyi, 1982). Then, the halo-opsin gene was isolated and cloned from *H. halobium* (*salinarium*) (Hegemann, Blanck, Vogelsang-Wenke, Lottspeich, & Oesterhelt, 1987). A similar arrangement of the transmembrane helices in bacteriorhodopsin and halorhodopsin was proposed based on sequence similarity, and on functional similarities in that both proteins use retinal as a chromophore, and pump ions (Blanck & Oesterhelt, 1987).

Perhaps the best studied halorhodopsins are from *H. salinarium* and from *Natronobacterium pharaonis*; these two halorhodopsins have a high sequence homology of about 66% (Lanyi, Duschl, Hatfield, May, & Oesterhelt, 1990) and similar, though not identical, reaction cycles (Essen, 2002; Kulcsár, Groma, Lanyi, & Váró, 2000) and three-dimensional structures (Kolbe et al., 2000; Kouyama et al., 2010; Kouyama, Kawaguchi, Nakanishi, Kubo, & Murakami, 2015). For the sake of simplicity, in what follows halorhodopsin is preferentially referred to without indicating the organism.

Halorhodopsin differs from bacteriorhodopsin at key sites where bacteriorhodopsin binds protons—bacteriorhodopsin D85 is replaced by Thr in halorhodopsin, and D96, by Ala (Fig. 1B) (Tittor, Soell, Oesterhelt, Butt, & Bamberg, 1989). The resting ground state of halorhodopsin binds a chloride ion within ~4 Å of the retinal Schiff base, at a site that would correspond to bacteriorhodopsin D85 (Kolbe et al., 2000). In the extracellular half of the protein, a network of four charged or polar sidechains and seven water molecules could serve as chloride uptake pathway (R123, E234, R176, and

H100 in *pharaonis* halorhodopsin numbering) (Kouyama et al., 2010); E234 (which corresponds to bacteriorhodopsin E194) is protonated in the resting state of the pump and deprotonated a later intermediate (Shibata, Saito, Demura, & Kandori, 2006), such that E234 might be a gate of the extracellular chloride uptake path (Kouyama et al., 2010). The carboxylate group corresponding to bacteriorhodopsin D115 (D141 in *salinarium* halorhodopsin, Fig. 1B) might be protonated (Chon et al., 1999).

Under physiological conditions, the retinal Schiff base of wild-type halorhodopsin remains protonated throughout the reaction cycle (Oesterhelt, Tittor, & Bamberg, 1992). Halorhodopsin can, however, undergo deprotonation of the retinal Schiff base, and pump protons to the cytoplasmic side, when azide is present and retinal absorbs two photons of different wavelength: first, upon illumination with green light, the retinal Schiff base releases its proton to the cytoplasmic side, then, upon subsequent illumination with blue light, the Schiff base reprotonates from the extracellular side (Bamberg, Tittor, & Oesterhelt, 1993; Oesterhelt et al., 1992). Both the deprotonation and reprotonation of the retinal Schiff base require presence of azide, being thought that one azide plays the role of bacteriorhodopsin D85, and the second, of D96 (Váró, Brown, et al., 1996).

The azide proton-pumping reaction cycle of halorhodopsin has similar kinetics with that of bacteriorhodopsin, including passage through an M intermediate with deprotonated retinal Schiff base—though an N intermediate could not be detected (Kulcsár et al., 2000). The observation that usage of azide and light to control the orientation of the retinal Schiff base can suffice to make halorhodopsin pump protons (although in the direction opposite to bacteriorhodopsin), suggests similarities in the mechanisms of action of the two proteins.

The notion of similar ion-pumping mechanisms being used by bacteriorhodopsin and halorhodopsin is supported by the crystallographic analysis of the intermediate states of halorhodopsin in the presence of bromide: Translocation of the anion from the extracellular to the cytoplasmic side of the retinal Schiff base segment associates with movements of helices C and F, and with the generation of a transient water channel at the cytoplasmic side (Kouyama et al., 2015). Flexible deformation of the protein, particularly of helices C and G, is required to lower the activation energy for chloride translocation past the retinal Schiff base segment (Gruia, Bondar, Smith, & Fischer, 2005); protein flexibility, and particularly of helix C, is also required to lower the activation energy and the reaction energy of the primary proton-transfer step in bacteriorhodopsin (Bondar et al., 2006).

In spite of these similarities between the coupling of protein conformational dynamics and ion pumping in bacteriorhodopsin and halorhodopsin, conformational dynamics might play somewhat different roles in the reaction cycles of the two pumps: Thermodynamic analyses of the entropy vs. enthalpy changes led to the conclusion that the reaction cycle of bacteriorhodopsin is driven by conformational change, and of halorhodopsin, by changes of electrostatic interactions (Kulcsár et al., 2000). In the resting state of halorhodopsin, the chloride ion near the retinal Schiff experiences a favorable, attractive potential, for which the sidechain of R123 is essential; further away from the initial binding site, the electrostatic potential is anion repulsive (Jardón-Valadez, Bondar, & Tobias, 2014).

A light-activated inward chloride-pumping rhodopsin denoted as NM-R3 was identified in the marine flavobacterium *Nonlabens marinus* S1-08^{T}, being thought that marine and halobacterial chloride-pumping microbial rhodopsins have evolved independently (Yoshizawa et al., 2010). Key differences between the amino acid residue sequences of the *H. salinarium* halorhodopsin and NM-R3 are observed at the sites corresponding to bacteriorhodopsin D85 and D96—unlike halorhodopsin, which has here Thr and, respectively Ala, NM-R3 has an Asn and, respectively, a Gln (Yoshizawa et al., 2010). Intriguingly, the sequence of NM-3 shares ~36% sequence identity with the sodium-pump microbial rhodopsin from *Krokinobacter eikastus*, as compared to ~20–21% sequence identity with the *H. salinarium* and *N. pharaonis* halorhodopsins (Hosaka et al., 2016).

The crystal structure of the *N. marinus* chloride pump from ref. (Kim et al., 2016) has a chloride ion interacting directly with the protonated Schiff base, and a second chloride ion bound at the cytoplasmic side of the protein; structural rearrangement of N92 (Fig. 1C) was hypothesized as important for the formation of a water-containing internal cavity through which the chloride ion might pass (Kim et al., 2016). Overall, however, the mechanisms for chloride pumping used by the halobacterial vs the flavobacterium transporters might be similar (Kim et al., 2016).

A peculiar feature of the structure of the *N. marinus* chloride pump is that the extracellular half has a cluster of Asp/Glu sidechains, partially overlapping with a cluster of Ser/Thr and Tyr sidechains, whereas the cytoplasmic half hosts multiple Ser/Thr groups (Fig. 1C). N92 is within 2.8 Å distance of E5 of the extracellular cluster (Fig. 1C). Whether the extracellular carboxylate cluster of the *N. marinus* chloride pump functions as a proton-antenna cluster as suggested for the cytoplasmic cluster of bacteriorhodopsin, or whether it might help guide chloride ions through the protein, is unclear. Likewise, the functional role of Ser/Thr groups located close to each other, S123, T124 and T126; S160 and T161; T186, S188, T189, and T191; S223 and S224 (Fig. 1C), is unclear. Presence of multiple intra-helical Ser/Thr motifs in a membrane protein is relatively infrequent, and it can shape protein dynamics by, e.g., causing local flexibility of the transmembrane helical segment that hosts multiple Ser/Thr groups, and/or to enhance local hydration (del Val, White, & Bondar, 2012).

The recent structure of the chloride-pumping rhodopsin MastR from the cyanobacterium *Mastigocladopsis repens* (Besaw et al., 2020) underscored similarities in the architecture and functional mechanisms of halorhodopsin and bacteriorhodopsin. At the extracellular side of MastR, E182 (corresponding to bacteriorhodopsin E194, Fig. 1A) was suggested to contribute to a gate that controls chloride entrance, and to potentially undergo protonation change during the reaction cycle (Besaw et al., 2020). In the crystal structure, E182 is part of a local cluster of charged and polar protein groups, and water molecules (Fig. 1D). At the cytoplasmic side of MastR, D85 makes an inter-helical H-bond to N39 (Fig. 1D), and protonation change of D85 associates with chloride translocation and might be required to ensure directional ion pumping (Besaw et al., 2020; Harris et al., 2018). T74 (corresponding to bacteriorhodopsin D85) coordinates, together with S78 and a water molecule, the chloride ion that is bound at the starting location near the protonated retinal Schiff base; mutating T74 to Asp suffices to transform MastR into an outward proton pump (Besaw et al., 2020).

4.2.3 Long-distance H-bond paths of channelrhodopsin: From static crystal structures to graph analyses of dynamic protein-water H-bond networks

The light-driven cation ion channel channelrhodopsin-2 from *Chlamidomonas reinhardtii*, which is perhaps one of the best characterized channelrhodopsins, has only ~16% sequence identity with bacteriorhodopsin (del Val, Royuela-Flor, Milenkovic, & Bondar, 2014). In various channelrhodopsins whose sequences were inspected, helices B, C, D, and G, contain more polar groups than bacteriorhodopsin, whereas helix F tends to be less polar (del Val, Royuela-Flor, et al., 2014). The higher polarity of channelrhodopsin as compared to bacteriorhodopsin explains the observation that, during molecular dynamics simulations of a channelrhodopsin model, numerous water molecules visited the inter-helical region of at least transiently (Siemers et al., 2019). Indeed, the sequence of channelrhodopsin-2 contains so many charged and polar groups that, a hydropathy analysis based on the translocon free energy of membrane insertion (Snider, Jayasinghe, Hristova, & White, 2009) predicted only five transmembrane helices, instead of seven (del Val, Royuela-Flor, et al., 2014).

The first three-dimensional structure of a channelrhodopsin was solved at 6 Å- resolution for a mutant of *C. renhardtii* channelrhodopsin-2 (Müller, Bamann, Bamberg, & Kühlbrandt, 2011), followed by the crystal structure of a *C. reinhardtii* channelrhodopsin-1/channelrhodopsin-2 chimera solved at a resolution of 2.3 Å (Kato et al., 2012). In recent years, the significant interest in using channelrhodopsins for optogenetics applications has likely contributed to research efforts leading to detailed characterization of channelrhodopsins by structural biology: Three-dimensional structures were solved for *C. reinhardtii* channelrhodopsin-2 (Oda et al., 2021; Volkov et al., 2017), for the anion channelrhodopsin-1 from *Guillardia theta* (Kim et al., 2018; Li et al., 2019), and for mutants of the *C. reinhardtii* channelrhodopsin-1/channelrhodopsin-2 chimera (Kato et al., 2015, 2018).

Based on crystal structures of the *C. reinhardtii* channelrhodopsin-2, an ion conduction path was proposed to consist of two extracellular and two cytoplasmic cavities; these cavities are separated by an extracellular gate centered at R120 (Fig. 1E), a central gate that includes the Schiff base, D253, E90, S63, and N258, and a third, cytoplasmic gate, which is centered at R268 (Volkov et al., 2017). An additional gate, denoted as the DC gate consists of the interaction between C128 and D156 (Nack et al., 2010); D156 (Fig. 1E), which corresponds to bacteriorhodopsin D115 (Fig. 1A), was proposed to function as proton donor for the retinal Schiff base of channelrhodopsin-2 (Lórenz-Fonfría et al., 2013). Presence on helix-B of a Lys group impacts the energetics of primary proton transfer from the retinal Schiff base (Adam & Bondar, 2018) (K123 in the channelrhodopsin chimera, K93 in channelrhodopsin-2, see Fig. 1E). Three carboxylate groups—E90, E97 and E101, which are part of an extracellular H-bond network (Fig. 1A), are among groups suggested to be involved in gating (Volkov et al., 2017). The extracellular, central, and cytoplasmic gates might need to be open at the same time for ions to move through the four internal cavities (Volkov et al., 2017).

The two structures of *G. theta* channelrhodopsin-1 are overall rather similar (Li et al., 2019), however they were interpreted to suggest different constrictions—a single central constriction (Kim et al., 2018), vs three constrictions (Li et al., 2019).

As discussed above, static structures of channelrhodopsins have identified pores that may be occupied by water, and solved coordinates for discrete water molecules. Internal water molecules of ion channels can, however, be rather dynamic—waters are intrinsically mobile, and rearrangements of structural elements thought to function as gates could associate with changes in the number and mobility of water molecules in inner cavities of the channel. Static coordinate structures provide limited information about these dynamics of the internal waters, which, unlike amino acid residues covalently bound to each other, are also indistinguishable. Atomistic molecular dynamics simulations allow the motions of the channel and its internal water to be studied at room temperature, in a fluid, hydrated lipid membrane environment. When many water molecules visit the interior of the protein at least transiently, studying protein-water interactions brings about the challenge of the large number of interactions that might need to be accounted for.

A solution that was utilized for the protein-water H-bond network sampled during simulations of the channelrhodopsin-1/channelrhodopsin-2 chimera was to rely on graph-based analyses of time-dependent coordinates extracted from simulations (Siemers et al., 2019). Briefly, the graph representation of a dynamic protein-water H-bond network consists of nodes, which typically are the charged and polar protein sidechains that can H-bond or salt-bridge, and edges that connect pairs of these nodes—the edges are H-bonds either directly between the protein groups, or water-mediated bridges. Once the graph of protein-water H-bonds has been computed, it can be queried to derive parameters used to characterize the protein of interest (Karathanou & Bondar, 2019; Lazaratos, Karathanou, & Bondar, 2020; Siemers et al., 2019). For example, the graph of H-bonds computed for a retinal protein may be queried to identify all connections that pass via the retinal Schiff base, or all shortest-distance paths that connect a proton donor and acceptor pair. In the case of proton donor-acceptor pairs separated by distances too long for direct H-bonds, graphs may be queried to extract the percentage of time during which all individual H-bonds of a long-distance path are sampled simultaneously (Siemers et al., 2019).

Graph analyses of molecular dynamics simulations of the channelrhodopsin-1/channelrhodopsin-2 chimera indicated that the transmembrane helical region of the protein hosts an extensive network of protein-water H-bonds, such that the retinal Schiff base connects frequently to E274 at the extracellular side (Siemers et al., 2019). The retinal Schiff bases of the dimeric ion channel are part of a common network of protein-water H-bonds that contains no fewer than 46 charged and polar groups; within this extensive H-bond network, the two retinal Schiff bases transiently connect via path passing via 12 protein groups and some 34 water molecules, with an estimated length of ~27 Å (Siemers et al., 2019).

A transient connection between two primary proton donors of a dimeric ion channel would suggest that, when one proton donor deprotonates, and thus its local H-bond network rearranges in response to the electronic structure change, the perturbation might be relayed to the second proton donor, potentially leading to cooperativity between the functioning of the two protomers of channelrhodopsin (Siemers et al., 2019). Cooperativity was discussed, e.g., for pairs of protomers of the bacteriorhodopsin trimer (Tokaji, 1993), for which it was suggested that cooperative effects in the purple membrane might influence the pK_a of the cytoplasmic proton donor D96 (Váró, Needleman, & Lanyi, 1996); cooperativity between $K_V7.4$ potassium ion channels clustered in cochlear outer hair cells is thought to increase mechanosensitivity (Perez-Flores et al., 2020); cooperativity between protomers of the dimeric voltage-gated sodium ion channel $Na_V1.5$ was suggested to lead to coupled gating (Clatot et al., 2017); the protomers of the dimeric voltage-gated Hv1 proton channel have individual permeation paths that open in a cooperative manner (Qiu, Rebolledo, Gonzalez, & Larsson, 2013).

The internal network of protein-water H-bonds that inter-connects the retinal Schiff bases of the dimeric channelrhodopsin chimera responds rapidly to mutations that alter H-bonding (Siemers et al., 2019), suggesting the H-bond network can propagate perturbations to long distances. A similar observation was made for the unrelated SecY protein translocon of the *Sec* protein secretion path: In the closed state of the protein translocon, an extensive network of >70H-bonds inter-connects different segments of the translocon, and likely helps the translocon respond rapidly to perturbations caused by mutation (Bondar, del Val, Freites, Tobias, & White, 2010; Bondar & White, 2012). Likewise, in GPCRs, an inter-helical interaction known as the ionic lock must be disrupted—such as by agonist binding, for the receptor to sample activated conformations (Kobilka, 2007).

The complexity of dynamic protein-water H-bond networks that can be hosted by an ion channel would be difficult to describe based on static coordinate snapshots that capture the protein in a crystallographic environment. Given the essential role of water molecules in mediating proton-transfer paths, information on the fluctuating protein-water H-bond networks hosted by the ion channel is essential to decipher mechanisms of long-distance proton transfers and protein conformational dynamics.

4.3 Protonation-coupled conformational dynamics of the *Escherichia coli* lactose permease, LacY, and in the xylose permease XylE

LacY is a lactose/proton symporter from the large Major Facilitator Superfamily (MFS), which includes proteins that mediate transport of molecules such as sugar molecules, drugs, or metabolites of the Krebs cycle (Pao, Paulsen, & Saier Jr., 1998). LacY has 12 transmembrane helical segments, arranged as two six-helix bundles that delineate a large hydrophilic cavity where the sugar molecule binds (Abramson et al., 2003). During sugar transport LacY alternates between conformations open to the periplasmic vs the cytoplasmic side, with passage via a transient occluded intermediate (Stelzl, Fowler, Sansom, & Beckstein, 2014).

Site-directed mutagenesis data and the static structure of LacY were interpreted to suggest that R302 and E325 (Fig. 2A) are directly involved in proton transfer across the membrane, whereas E269 and H322 couple proton transfer to the binding of sugar substrate (Abramson et al., 2003). LacY mutants with a non-ionazable group replacing E269 could not accumulate sugar (Franco & Brooker, 1994). The E325A mutant allows sugar exchange and counter-flow, but proton transfer is abolished (Carrasco, Antes, Poonian, & Kaback, 1986); E325D can catalyze sugar-proton symport, but at a rate much reduced relative to the wild-type protein, suggesting a carboxylate group at position 325 is important for the transporter functionality (Carrasco et al., 1989).

E325 has a large apparent pK_a, of ~10.5, being thought that sugar binding requires E325 to be protonated, and that deprotonation of E325 following dissociation of the sugar molecule would be facilitated by interactions with R302 (Grytsyk et al., 2017).

The H322R, H322K, H322N, and H322Q mutants allow lactose influx without proton symport (Püttner, Sarkar, Padan, Lolkema, & Kaback, 1989), whereas H322A cannot bind the substrate, and H322F has significantly reduced binding (He & Kaback, 1997). Neither E262D nor E269Q can bind a substrate molecule (He & Kaback, 1997).

D240 (Fig. 2A), though not directly involved in proton transfer across the membrane, could impact protein conformational dynamics via inter-helical interactions with K302 (Abramson et al., 2003)—the D240E

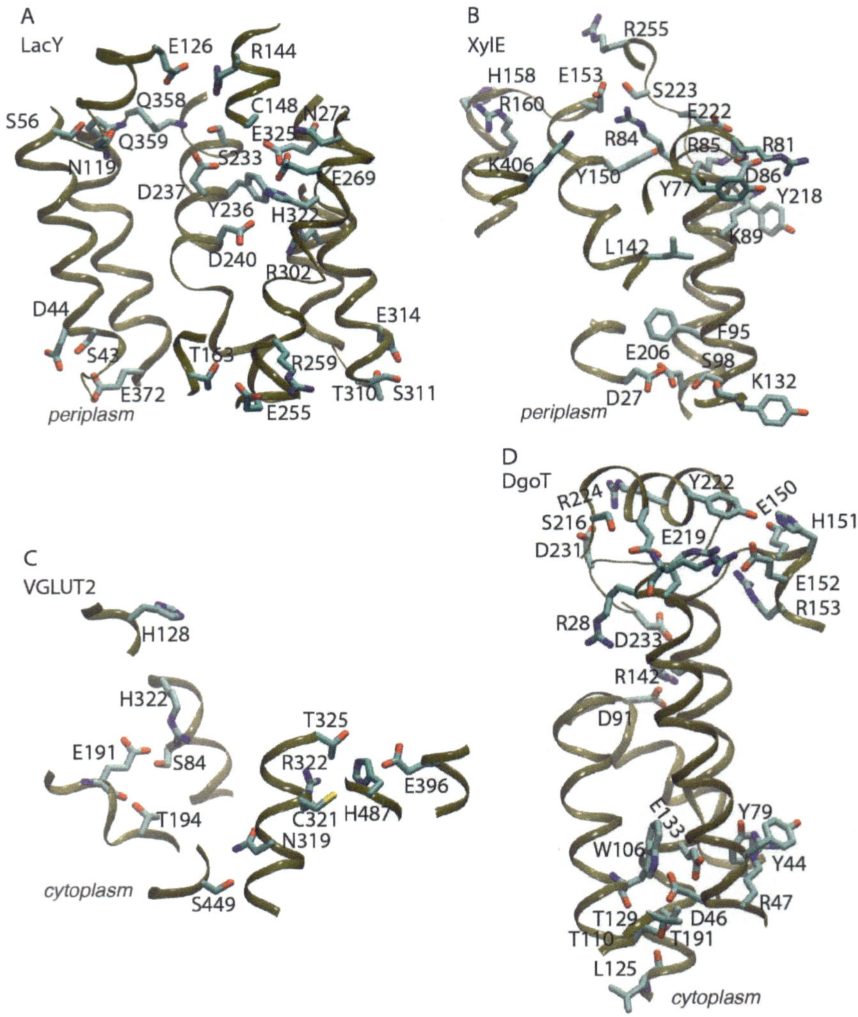

Fig. 2 Inter-helical H-bonds and proton-binding sites of MFS transporters. For clarity of the interactions shown, only the protein segments that contain the groups of interest are shown. (A) The *E. coli* lactose permease LacY, PDB ID:6V8N, chain A, solved at 3.6 Å resolution (Guan, Mirza, Verner, Iwata, & Kaback, 2007). H322 is within 3.0 Å distance of Y236, which further H-bonds to E325; S56 H-bonds to N119 and Q359 (Bondar, 2021). (B) The D-xylose-proton symporter XylE from *E. coli*, PDB ID:4QIQ (Wisedchaisri, Park, Iadanza, Zheng, & Gonen, 2014). (C) Selected interactions of the rat protonation-coupled glutamate transporter VGLUT2, PDB ID:6V4D, 3.5 Å resolution (Li et al., 2020). (D) Selected interactions of the inward-facing *E. coli* D-galactonate-proton symporter DgoT, PDB ID:6E9N, solved at 2.9 Å resolution, chain A (Leano et al., 2019).

mutant is inactive, suggesting the length of the carboxylic sidechain at position 240 is important (Sahin-Tóth & Kaback, 1993). The decreased sugar-binding affinity and fast sugar efflux of D240E were explained by a lower affinity for protons (Guan & Kaback, 2009).

Within one helical turn of D240, D237 makes with K358 a charge pair required for the transport of lactose, as neutral substitution of either D237 or K358 inhibits transport, whereas interchanging these two groups allows transport (Dunten, Sahin-Tóth, & Kaback, 1993). At the cytoplasmic side of D237, E126 is involved in substrate binding and can salt bridge to R144 (Guan et al., 2007). Proton pulse experiments on LacY bound to fluorescein at position C148 (Fig. 2B) indicated proton exchange between E126 and fluorescein is fast at pH >7.4, and slow at pH < 7.1, suggesting changes in internal hydration accompanying LacY protein dynamics (Nachliel & Gutman, 2001).

Atomistic simulations of LacY in a hydrated lipid membrane environment, with protonated vs. unprotonated E325, revealed that changes in the protonation state of E325 couple with remarkable protein conformational changes: whereas protonated E325 preserved local interactions observed in the starting structure of inward-facing LacY, when E325 was negatively charged protein dynamics was enhanced, and a transmembrane helical segment tilted towards closing the protein at the cytoplasmic cleft (Andersson et al., 2012).

Tight coupling between protonation state and protein conformation was also proposed for xylose permease XylE, for which protonation change of D27 is thought to couple with conformational change, whereas the nearby E206 might stay protonated (Wisedchaisri et al., 2014). The structure of inward-facing XylE from ref. (Wisedchaisri et al., 2014) indicates D27 within H-bond distance from E206, and within 3.6 Å from the hydroxyl group of S98 (Fig. 2B, Table 1). Since local hydrophobic contacts reduce dynamics of H-bonding groups (del Val, Bondar, & Bondar, 2014), hydrophobic contacts between the D27-E206 pair and F95 (Fig. 1B) could help stabilize the H-bond between the two carboxylates. A structure of outward-facing, partially occluded XylE bound to D-xylose (PDB ID:4GBI, (Sun et al., 2012)) has D47 and E206 within 2.8 Å distance, but S98 is 4.5 Å away from D47 and H-bonds instead to the backbone of Y132, suggesting dynamics at the D47 inter-helical H-bond cluster during conformational transitions of XylE.

Table 1 Illustration of carboxylate-hydroxyl inter-helical H-bond motifs. References reporting protonation changes of carboxylate groups are given in the main text and as notes of the table, and three-dimensional structures used to measure distances are indicated. H-bond distances between the carboxylate oxygen atoms and Ser/Thr hydroxyl oxygen atoms are reported in Å, as measured with Visual Molecular Dynamics (VMD) (Humphrey, Dalke, & Schulten, 1996) for the Protein Data Bank (PDB) coordinate file indicated in the table. In the case where the Asp/Glu carboxylate and Ser/Thr hydroxyl groups have bifurcated H-bonding, both H-bond distances are reported. Some of the Ser/Thr hydroxyl groups engaged in inter-helical H-bonding with an Asp/Glu carboxylate group also have intra-helical H-bonding with the backbone carbonyl group of the *i-4* residue of the preceding helical turn; when present, the intra-helical H-bond of the Ser/Thr hydroxyl group is also reported.

Protein	Inter-helical pair		Protonation change/local H-bond interaction	Distance (Å)	PDB/figure
Microbial rhodopsins					
Bacteriorhodopsin	D96	T46	D96 is proton donor	2.7, 3.2	5ZIM, Fig. 1A
			T46 H-bonds to F42	2.7	
	D115	T90	D115 stays protonated	2.5, 3.1	
			T90 H-bonds to W86	2.7	
Halorhodopsin	D141	T116	D141 likely protonated	2.6, 3.2	1E12, Fig. 1B
			T116 H-bonds to W112	2.7	
	E219	S104	Local H-bond cluster[a]	3.1	
Chloride pump MastR	D85	N39[b]	D85 changes protonation	2.7	6XL3, Fig. 1D
Channelrhodopsin-2	E101	T246	E101 is part of a gate	2.7	6EID, Fig. 1E
			T246 H-bonds to A242	2.8	
C1C2 chimera[c]	E129	S102	E129 deprotonates[d]	3.3, 3.6	3UG9
			S102 H-bonds to T98	2.5	
Anabaena sensory rhodopsin[e]	D75	T47	D75 corresponds to bacteriorhodopsin D85	2.7	1XIO
Sodium pump KR2	D116	S70	D116 is primary proton acceptor	2.8[f]	3X3C
Gloeobacter rhodopsin[g]	E132	S77	E132 corresponds to bacteriorhodopsin D96	3.5	6NWD
Exiguobacterium rhodopsin[h]	D119	T90	D119 likely protonated	2.5, 3.0	4HYJ, Fig. 1F
			T90 H-bonds to W86	2.8	

MFS transporters					
LacY permease	E255	T163	–	2.4	2V8N, Fig. 2A
	E374	S43	–	3.1	
XylE symporter	D27	S98	D27 changes protonation	3.6	4QIQ, Fig. 2B
	E153	S223	Important for transport[i]	3.1	
VGLUT-2	E191	S84	E191 might protonate	2.7	6V4D, Fig. 2C
			E191-T194 H-bond[j]	2.1	
DGOT	E219	S216[k]	E219 is conserved[l]	3.0	6E9N, Fig. 2D
			E219 H-bonds to R27	2.5	
Other transporters					
P-type AHA2	E114	S294	–	2.9	5KSD, Fig. 3A
	D684	N106	D684 is proton donor	2.8	
			N106 H-bonds to I287	3.5	
AcrB	D407	T978	D407-D408 implicated in proton transfer[m]	3.1	2DHH, Fig. 3B
	D408	S481		3.3	
			S481 H-bonds to M478	3.1	
	E417	N361	–	3.1	

[a]The E219 carboxylate is within 2.8 Å distance of Y109, and the backbone carbonyl group of A218 is within 3.0 Å distance of H95 (Fig. 1B).
[b]N39 is also within 3.1 Å distance from S211.
[c]*C. reinhardii* channelrhodopsin-1/channelrhodopsin-2 chimera structure (Kato et al., 2012).
[d]E129 is protonated in dark-state (Takemoto et al., 2015), and it corresponds to channelrhodopsin-2 E90 (Fig. 1E), which is also protonated in the dark state (Eisenhauer et al., 2012). The H-bond motif of E129 was discussed in ref. (Bondar & Smith, 2017).
[e]In the crystal structure of *Anabaena* rhodopsin (Vogeley et al., 2004), S47 is also within 2.6 Å distance of the hydroxyl group of Y11 (Bondar & Smith, 2017).
[f]The crystal structure PDB ID 3X3C (Kato et al., 2015) indicates two conformations for N116, here only the shortest H-bond distance is reported. D116 was suggested to function as proton acceptor for the retinal Schiff base proton (Kato, Inoue, et al., 2015).
[g]E132 was suggested as the putative cytoplasmic proton donor for the cyanobacterial proton pump *Gloeobacter* rhodopsin (Morizumi et al., 2019).
[h]D119 of *Exiguobacterium* rhodopsin is likely protonated (Dioumaev et al., 2013).
[i]The E153A mutant has significantly reduced transport activity (Sun et al., 2012).
[j]The backbone of E191 has an intra-helical H-bond with T194 (Bondar, 2021).
[k]E210 and S215 are located on the same protein segment, see Fig. 2D.
[l]It was noted that E219 and the nearby Y222 are conserved (Leano et al., 2019). E219 is within H-bond distance of the S216 sidechain, and of the R27 backbone amide group.
[m]D407, D408 and K940 are part of the proton-transfer path (Murakami et al., 2006).

4.3.1 Protonation-coupled substrate transport by the SLC17 transporters VGLUT and DgoT

The vesicular glutamate transporters VGLUT1–3 that belong to the SLC17 family of the MFS are essential for neurotransmission (Drew, North, Nagarathinan, & Tanabe, 2021; Wojcik et al., 2004). VGLUT uses the transmembrane potential generated by the vesicular-type ATPase to translocate the neurotransmitter glutamate into synaptic vesicles against the pH gradient, and has an additional channel-like activity for chloride ions (Drew et al., 2021; Leano et al., 2019). The recent cryo-EM structure of the rat VGLUT2 indicated inter-helical H-bond distances of 2.9–3.0 Å between E396 and H487 (Fig. 2B), and between H487 and C321, the site where the glutamate substrate might bind (Li et al., 2020). At the lumen side of VGLUT2 R88 (Fig. 2B), whose mutation to Ala affects synaptic transmission, is close to E191 (Li et al., 2020); E191 was hypothesized to function as a pH sensor, with the nearby H128 potentially the initial proton-binding group (Li et al., 2020).

Earlier computer simulations performed on a homology model of the human VGLUT1 based on the *E. coli* glycerol-3-phosphate transporter GlpT structure indicated interactions between R80 and E183, and between E388 and H479 (Almqvist, Huang, Laaksonen, Wang, & Hovmöller, 2007); these inter-helical interactions would be reminiscent of those at sites E191 and, respectively, E396, of VGLUT2.

It had been noted that mammalian SLC17 transporters have relatively high sequence identity with the bacterial homologs, of up to 27% (Leano et al., 2019). The crystal structure of the *E. coli* galactonate transporter DgoT, captured in an inward-facing conformation, was interpreted to suggest D46 could be protonated (Leano et al., 2019). Near D46, E133 is thought to be protonated in the outward-facing conformation of DgoT, and protonated E133 facilitates interactions between R47 and galactonate; without substrate bound, E133 would deprotonate and interact with R47 (Leano et al., 2019).

The D46-E133 pair is close to T110 and T129, both of which make intra-helical H-bonds with the carbonyl group one helical turn down the amino acid residue sequence—W106 and, respectively, L125 (Fig. 2D); as D46 H-bonds to the W106 sidechain, and the adjacent R47 H-bonds to two sidechains of Y44 and Y79 (Fig. 2D), it follows that a putative proton-binding site of DgoT is part of an H-bond cluster that inter-connects polar and charged sidechains, and two protein backbone groups.

At the other side of the membrane in DgoT, and extended H-bond cluster includes bifurcated H-bonding of Y222 with H151 and E152, and

H-bonding of the E219 carboxylate group to S216 and to the backbone of R27 (Fig. 2D, Table 1); it was noted that the nearby R29, which salt bridges to E152, corresponds to an Arg sidechain whose mutation in the homologs transporter sialin causes disease (Leano et al., 2019).

4.4 A carboxylate-hydroxyl H-bond motif of membrane transporters

Knowledge of the proton-bonding group and its H-bonding environment are of paramount importance for describing mechanisms of action of membrane transporters. The discussion above of protonation-coupled conformational dynamics in microbial transporters and MSF transporters highlights the central role of inter-helical H-bonding at the proton-binding site, as altered local dynamics upon changes in the protonation state couple to changes in protein conformational dynamics. Given the wealth of knowledge that exists for membrane transporters that have been studied extensively, the question arises as to whether proton binding and protonation coupling might involve common motifs that could then be identified in new transporters and, e.g., help short-list amino acid residues for site-directed mutagenesis experiments.

An inter-helical carboxylate-hydroxyl H-bond motif was identified at sites that are functionally important for several membrane transporters and receptors (Bondar, 2021; Bondar & Lemieux, 2019; Bondar & Smith, 2017; del Val, Bondar, & Bondar, 2014; Siemers et al., 2019).

Such a carboxylate-hydroxyl H-bond motif could couple protonation change to changes in local helix dynamics, which can then propagate to remote regions of the protein: The protonation state of the carboxylate group impacts the strength of its inter-helical H-bond to the Ser/Thr hydroxyl group, which in turn influences the availability of the hydroxyl group to engage in intra-helical H-bonding to a backbone carbonyl group (Bondar & Smith, 2017). Intra-helical H-bonding of Ser/Thr to the *i-4* backbone carbonyl group of the preceding helical turn is common in α-helical proteins, as it lowers the energetic cost associated with inserting the hydrophilic sidechain into the membrane (Gray & Matthews, 1984). Depending on the sequence-structure context, intra-helical H-bonding between a Ser/Thr hydroxyl and a protein backbone group can associate with enhanced local dynamics of the helix, or with enhanced local hydration (del Val et al., 2012).

The survey of protein structures illustrated in Figs. 1 and 2, and summarized in Table 1, suggests that carboxylate-hydroxyl H-bonding is rather common to protonation-coupled membrane transporters from various organisms. Among microbial rhodopsins, for example, bacteriorhodopsin has inter-helical carboxyl-hydroxyl H-bond pairs at the proton-binding sites D96 and D115 (Fig. 1A, Table 1); at the extracellular proton-release site, E194 and E204 may H-bond to S193, though here the high-resolution structure indicates structural dynamics (Hasegawa et al., 2018). Inter-helical hydroxyl H-bonding of a likely protonated carboxylate is also observed in halorhodopsin (D141, Fig. 1B) and in *Exiguobacterium* rhodopsin (D119, Fig. 1F) (Chon et al., 1999; Dioumaev et al., 2013).

XylE has inter-helical carboxylate-hydroxyl interactions at the putative proton-binding site D27, and at E153 (Fig. 2B, Table 1). In the cryo-EM structure of the glutamate transporter VGLUT-2 E191, which is thought to function as proton-binding group (Li et al., 2020), is within H-bond distance of S84 (Fig. 2C) (Bondar, 2021); in the AHA2 plasma membrane proton pump E114, which is within two helical turns from the proton-binding group D684, H-bonds to S94 (Fig. 3A, Table 1). The multidrug transporter

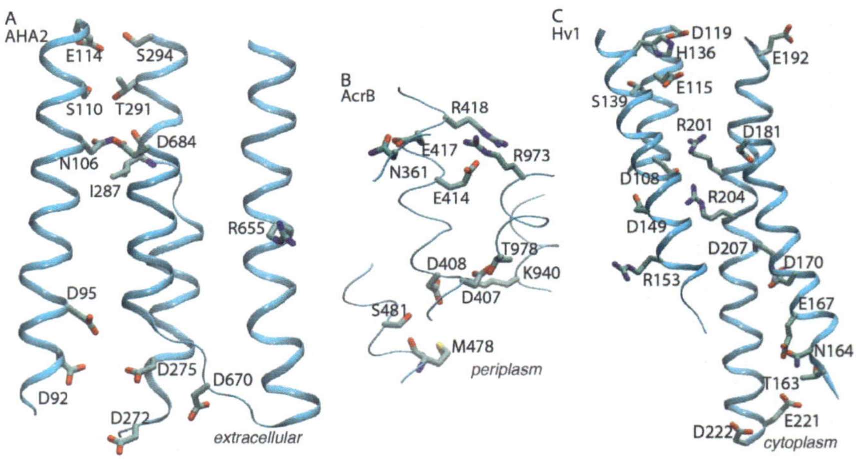

Fig. 3 Illustration of H-bonding at proton-binding sites of other membrane transporters. For clarity, only some of the protein segments are shown. (A) The AHA2 plasma membrane proton pump, PDB ID:5KSD, chain A, 3.5 Å resolution (Focht, Croll, Pedersen, & Nissen, 2017). D92 and D95 might be involved in proton release at the extracellular side (Pedersen et al., 2007). (B) The multidrug transporter AcrB; the molecular graphics is based on chain A from the crystal structure PDB ID:2DHH, solved at 2.8 Å resolution (Murakami, Nakashima, Yamashita, & Yamaguchi, 2006). The carboxylate-hydroxyl inter-helical H-bonding of D407 was discussed in ref. (del Val, Bondar, & Bondar, 2014). (C) The mouse voltage-gated proton channel Hv1, solved at 3.45 Å resolution (Takeshita et al., 2014). D108 is important for proton selectivity (Takeshita et al., 2014).

AcrB has at the D407-D408 proton-transfer site (Murakami et al., 2006) inter-helical H-bonds between D407 and S978, and between D408 and S481 (Fig. 3B, Table 1). H-bonding between a Glu sidechain involved in proton transfer, and a Ser hydroxyl group, was also suggested for the green fluorescent protein (Agmon, 2005).

Though common to receptors discussed here, the carboxyl-hydroxyl H-bond motif is likely only one of the motifs utilized by proton-binding membrane transporters. In bacteriorhodopsin D212, which might serve as intermediate proton carrier, H-bonds to Y57 and Y185 (Fig. 1A). MastR and AHA2 have proton-binding Asp H-bonded to an Asn of the nearby helix (see D85-N39 in Fig. 1D, and D684-N106 in Fig. 2A). LacY has E325 and H322 bridged to each other via Y236, and part of a larger inter-helical H-bond cluster (Fig. 2A); a somewhat similar arrangement is observed for VGLUT-2 (Bondar, 2021), though here E396 and H487 H-bond directly to each other and H487 is within 2.9–3.3 Å distance of S321 and T325 (Fig. 2C).

The examples presented here thus suggest that a proton-binding carboxylate is rather likely to have inter-helical H-bonding with hydroxyl groups of Ser, Thr, or Tyr, with an Asn carboxyamide group, or may be part of a local H-bond cluster with a His imidazole group; a His sidechain involved in proton transfer may be part of an inter-helical H-bond cluster with a carboxylate group, and with polar sidechains. Other H-bond environments of protonated carboxylate groups and of imidazole groups that might be involved in proton binding include backbone carbonyl H-bonding of a protonated carboxylate (Andersson, Hauser, Karjalainen, & Barth, 2008), and H-bonding between His sidechains (Fu, Miao, Qin, & Cross, 2020; Ludwig et al., 2003).

Taken together, observations discussed above suggest that carboxylate and imidazole groups where protonation state may change tend to be involved in inter-helical H-bonding with a polar sidechain, and can also be part of larger inter-helical H-bond clusters. Such an H-bond arrangement would enable allosteric coupling between protonation state, water dynamics, and protein conformational dynamics.

4.4.1 Proton transfer to an internal carboxylate group

Proton transfers to/from internal carboxylate groups typically involves protein-water H-bond networks sampled transiently as the transporter passes through its reaction cycle. Inspection of static structures suggests similarities between structural arrangements of some of these transporters.

In bacteriorhodopsin, a proton antenna consisting of a few carboxylate sidechains could collect the proton from the cytoplasmic bulk and deliver it to D38 (Checover et al., 2001; Riesle et al., 1996). The distance between the internal carboxylate, D38, and the bulk-exposed carboxylates, is about 10 Å (Fig. 1A); this distance is similar to that between D96 and the retinal Schiff base (~11 Å), and to that between D85 and E194-E204 (~12–13 Å, Fig. 1A).

Similar distances between proton donor and acceptor pairs are observed in other proteins. In the mouse Hv1 structure, the internal D108 is within ~11 Å of E115 at the periplasmic side (Fig. 3C). In VGLUT2, H128 (Fig. 2C) is thought to pick up a proton from the lumen and deliver it E191 (Li et al., 2020); in the cryo-EM structure, the distance between H128-Nε2 and E191-Oε1 is ~13 Å. Similarly, in AHA2 the distance between the internal D684 and E114 at the cytoplasmic side is ~13 Å (Fig. 3A). The ~17 Å distance between D684 and D92/D95, which might serve as a proton-release site (Pedersen et al., 2007), is somewhat long, suggesting proton transfer might involve an intermediate proton carrier whose identity is unclear.

Distances of ~10-13 Å between internal and bulk-exposed proton-transfer groups, as observed for some of the donor-acceptor pairs discussed here, could be reached by 3–4 intervening water molecules; rotation of one or both carboxylate sidechains could also help shorten the distance between the proton-transfer pair. Sidechains that line the cavity along which proton transfer will occur might also need to reorient to allow formation of the proton-transfer wire.

In the example of the well-studied bacteriorhodopsin proton pump, the cavity between the proton donor D96 and the retinal Schiff base proton acceptor already hosts two discrete water molecules in the resting state of the pump (Fig. 1A), and presence of a chain of H-bonded waters would be thermodynamically allowed (Roux et al., 1996). These features of the internal cavity of bacteriorhodopsin could be compatible with its being utilized for proton transfer—the enthalpy change associated with hydration of an internal protein cavity is favorable when there are already waters in the cavity, and polar groups line the cavity, entropy changes depending on the size of the cavity (Henry et al., 2009).

5. Conclusions

Membrane transporters move ions, water, and larger substrates across membranes, and couple to an energetically favorable reaction for uphill transport. Transport of a substrate thus involves communication between regions of the protein exposed to the two sides of the cell membrane that hosts the protein. For a protein embedded in the cytoplasmic membrane of *E. coli*, for example, that would imply a distance of about 38 Å (Mitra, Ubarratxena-Belandia, Taguchi, Warren, & Engelman, 2004). Transporters that translocate protons across membranes may only do so by coupling protonation change with changes in water accessibility and protein structural rearrangements. Titratable protein sidechains directly involved in proton binding and proton transfer are located strategically at sites that ensure control of their pK_a values, and formation of proton-transfer paths at the adequate moment along the reaction path of the pump.

The functioning of membrane transporters is often discussed in terms of gates that control accessibility of the transporter to the inner vs. the outer side of the membrane. Although different definitions of gates might be used, the examples of membrane transporters discussed here suggest inter-helical H-bonds and inter-helical H-bond clusters at proton binding sites can provide efficient means of coupling protonation change with changes in protein and water dynamics. Carboxylate groups that bind protons appear to be typically involved in inter-helical H-bonding with a polar sidechain—such as with the hydroxyl group of Ser/Thr, or Asn; carboxylate sidechains also appear to be part of local H-bond clusters with His sidechains that might be part of proton-binding sites. Knowledge of such common proton-binding motifs could inform studies of the general principles of action of membrane transporters in biology.

Acknowledgments

Part of the work reviewed here was supported in part by funding from the German Research Foundation, the DFG, Collaborative Research Center Protonation Dynamics in Protein Function, Project C4, and from the European Union's Horizon 2020 Research and Innovation Program, Marie Sklodowska-Curie grant agreement No 860592, Innovative Training Network "Proton and proton-coupled transport".

References

Abramson, J., Smirnova, I., Kasho, V., Verner, G., Kaback, H. R., & Iwata, S. (2003). Structure and mechanism of the lactose permease of *Escherichia coli*. *Science*, *301*, 601–615. https://doi.org/10.1126/science.1088196.

Adam, S., & Bondar, A.-N. (2018). Mechanism by which water and protein electrostatic interactions control proton transfer at the active site of channel rhodopsin. *PLoS One*, *13*, e0201298. https://doi.org/10.1371/journal.pone.0201298.

Agmon, N. (2005). Proton pathways in green fluorescent protein. *Biophysical Journal, 88,* 2452–2461. https://doi.org/10.1529/biophysj.104.055541.

Almqvist, J., Huang, Y., Laaksonen, A., Wang, D.-N., & Hovmöller, S. (2007). Docking and homology model explain inhibition of the human vesicular glutamate transporters. *Protein Science, 16,* 1819–1829. https://doi.org/10.1110/ps.07294470.

Ames, J. B., & Mathies, R. A. (1989). Bacteriorhodopsin's photocycle: A kinetic resonance Raman study. *Biochemistry, 29,* 7181–7190.

Andersson, M., Bondar, A.-N., Freites, J. A., Tobias, D. J., Kaback, H. R., & White, S. H. (2012). Proton-coupled dynamics in lactose permease. *Structure, 20,* 1893–1904.

Andersson, J., Hauser, K., Karjalainen, E.-L., & Barth, A. (2008). Protonation and hydrogen bonding of Ca^{2+} site residues in the E2P phosphoenzyme intermediate of sarcoplasmic reticulum Ca^{2+}-ATPase studied by a combination of infrared spectroscopy and electrostatic calculations. *Biophysical Journal, 94,* 600–611. https://doi.org/10.1529/biophysj.107.114033.

Balashov, S. P., & Ebrey, T. G. (2001). Trapping and spectroscopic indentification of the photointermediates of bacteriorhodopsin at low temperatures. *Photochemistry and Photobiology, 73,* 453–462. https://doi.org/10.1562/0031-8655(2001)0730453TASIOT2.0.CO2.

Balashov, S. P., Petrovskaya, L., Imasheva, E., Lukashev, E. P., Dioumaev, A. K., Wang, J. M., et al. (2013). Breaking the carboxyl rule. Lysine 96 facilitates reprotonation of the Schiff base in the photocycle of a retinal protein from *Exiguobacterium sibiricum. The Journal of Biological Chemistry, 288,* 21254–21265. https://doi.org/10.1074/jbc.M113.465138.

Bamberg, E., Tittor, J., & Oesterhelt, D. (1993). Light-driven proton or chloride pumping by halorhodopsin. *Proceedings of the National Academy of Sciences of the United States of America, 90,* 639–643.

Belrhali, H., Nollert, P., Royant, A., Menzel, C., Rosenbusch, J. P., Landau, E. M., et al. (1999). Protein, lipid and water organization in bacteriorhodopsin crystals: A molecular view of the purple membrane at 1.9 Å resolution. *Structure,* 7, 909–917. https://doi.org/10.1016/S0969-2126(99)80118-X.

Besaw, J. E., Ou, W.-L., Morizumi, T., Eger, B. T., Sanchez Vasquez, J. D., Chi, J. H. Y., et al. (2020). The crystal structures of a chloride-pumping microbial rhodopsin and its proton-pumping mutant illuminate proton transfer determinants. *The Journal of Biological Chemistry, 295,* 14793–14804.

Blanck, A., & Oesterhelt, D. (1987). The halo-opsin gene. II. Sequence, primary structure of halorhodopsin and comparison with bacteriorhodopsin. *The EMBO Journal, 6,* 265–273.

Bondar, A.-N. (2021). Proton-binding motifs of membrame-bound proteins: From bacteriorhodopsin to spike protein S. *Frontiers in Chemistry, 9,* 685761. https://doi.org/10.3389/fchem.2021.685761.

Bondar, A.-N., Baudry, J., Suhai, S., Fischer, S., & Smith, J. C. (2008). Key role of active-site water molecules in bacteriorhodopsin proton-transfer reactions. *The Journal of Physical Chemistry B, 112,* 14729–14741.

Bondar, A.-N., del Val, C., Freites, J. A., Tobias, D. J., & White, S. H. (2010). Dynamics of SecY translocons with translocation-defective mutations. *Structure, 18,* 847–857.

Bondar, A.-N., Elstner, M., Suhai, S., Smith, J. C., & Fischer, S. (2004). Mechanism of primary proton transfer in bacteriorhodopsin. *Structure, 12,* 1281–1288.

Bondar, A.-N., Fischer, S., Smith, J. C., Elstner, M., & Suhai, S. (2004). Key role of electrostatic interactions in bacteriorhodopsin proton transfer. *Journal of the American Chemical Society, 126,* 14668–14677.

Bondar, A. N., Fischer, S., Suhai, S., & Smith, J. C. (2005). Tuning of retinal twisting in bacteriorhodopsin controls the directionality of the early photocycle steps. *The Journal of Physical Chemistry B, 109,* 14786–14788.

Bondar, A.-N., & Lemieux, H. J. (2019). Reactions at membrane interfaces. *Chemical Reviews, 119*, 6162–6183. https://doi.org/10.1021/acs.chemrev.8b00596.
Bondar, A.-N., & Smith, J. C. (2009). Water molecules in short- and long-distance proton transfer steps of bacteriorhodopsin proton pumping. *Israel Journal of Chemistry, 49*(2), 155–161.
Bondar, A.-N., & Smith, J. C. (2017). Protonation-state coupled conformational dynamics in reaction mechanisms of channel and pump rhodopsins. *Photochemistry and Photobiology, 93*, 1336–1344. https://doi.org/10.1111/php.12790.
Bondar, A.-N., Smith, J. C., & Fischer, S. (2006). Structural and energetic determinants of primary proton transfer in bacteriorhodopsin. *Photochemical & Photobiological Sciences, 5*, 547–552.
Bondar, A.-N., & White, S. H. (2012). Hydrogen bond dynamics in membrane protein function. *Biochimica et Biophysica Acta, 1818*, 942–950.
Brändén, G., Pawate, A. S., Gennis, R. B., & Brzezinski, P. (2005). Controlled uncoupling and recoupling of proton pumping in cytochrome $_c$ oxidase. Proceedings of the National Academy of Sciences of the United States of America, 103, 317–322. 10.1073%01pnas.0507734103.
Brown, L. S., Sasaki, J., Kandori, H., Maeda, A., Needleman, R., & Lanyi, J. K. (1995). Glutamic acid 204 is the terminal release group at the extracellular surface of bacteriorhodopsin. *The Journal of Biological Chemistry, 270*, 27122–27126.
Carrasco, N., Antes, L. M., Poonian, M. S., & Kaback, H. R. (1986). *Lac* permease of *Escherichia coli*: Histidine-322 and glutamic acid-322 may be components of a charge-relay system. *Biochemistry, 25*, 4484–4486. https://doi.org/10.1021/bi00364a004.
Carrasco, N., Püütner, I. B., Antes, L. M., Lee, J. A., Larigan, J. D., Lolkema, J. S., et al. (1989). Characterization of site-directed mutants in the *lac* permease of *Escherichia coli*. 2. Glutamate-325 replacements. *Biochemistry, 28*, 2533–2539. https://doi.org/10.1021/bi00432a028.
Casey, J. R., Grinstein, S., & Orlowski, J. (2010). Sensors and regulators of intracellular pH. *Nature Reviews Molecular Cell Biology, 11*, 50–61. https://doi.org/10.1038/nrm2820.
Checover, S., Marantz, Y., Nachliel, E., & Gutman, M. (2001). Dynamics of the proton transfer reaction on the cytoplasmic surface of bacteriorhodopsin. *Biochemistry, 40*, 4281–4292.
Chon, Y.-S., Kandori, H., Sasaki, J., Lanyi, J. K., Needleman, R., & Maeda, A. (1999). Existence of two L photointermediates of halorhodopsin from *Halobacterium salinarium*, differing in their protein and water FTIR bands. *Biochemistry, 38*, 9445–9449. https://doi.org/10.1021/bi9903042.
Clatot, J., Hoshi, M., Wan, X., Liu, H., Jain, A., Shinlapawittayatorn, K., et al. (2017). Voltage-gated sodium channels assemble and gate as dimers. *Nature Communications, 8*, 2077. https://doi.org/10.1038/s41467-017-02262-0.
Cullen, W., Turega, S., Hunter, C. A., & Ward, M. D. (2015). pH-dependent binding of guests in the cavity of a polyhedral coordination cage: Reversible uptake and release of drug molecules. *Chemical Science, 6*, 625. https://doi.org/10.1039/c4sc02090a.
Damaghi, M., Wojktowiak, J. W., & Gillies, R. J. (2013). pH sensing and regulation in cancer. *Frontiers in Physiology, 4*, 1–10. https://doi.org/10.3389/fphys.2013.00370.
del Val, C., Bondar, L., & Bondar, A.-N. (2014). Coupling between inter-helical hydrogen bonding and water dynamics in a proton transporter. *Journal of Structural Biology, 186*, 95–111. https://doi.org/10.1016/j.jsb.2014.02.010.
del Val, C., Royuela-Flor, J., Milenkovic, S., & Bondar, A.-N. (2014). Channel rhodopsins—A bioinformatics perspective. *Biochimica et Biophysica Acta Bioenergetics, 1837*, 643–655.
del Val, C., White, S. H., & Bondar, A.-N. (2012). Ser/Thr motifs in transmembrane proteins: Conservation patterns and effects on local protein structure and dynamics. *The Journal of Membrane Biology, 245*, 717–730. https://doi.org/10.1007/s00232-012-9452-4.

Delaney, J. K., Schweiger, U., & Subramaniam, S. (1995). Molecular mechanism of protein-retinal coupling in bacteriorhodopsin. *Proceedings of the National Academy of Sciences of the United States of America*, *92*, 11120–11124.

Dioumaev, A. K., Brown, L. S., Needleman, R., & Lanyi, J. K. (1999). Fourier transform infrared spectra of a late intermediate of the bacteriorhodopsin photocycle suggest transient protonation of Asp-212. *Biochemistry*, *38*, 10070–10078. https://doi.org/10.1021/bi990873+.

Dioumaev, A. K., Petrovskaya, L., Wang, J. M., Balashov, S. P., Dolgikh, D. A., Kirpichnikov, M. P., et al. (2013). Photocycle of Exiguobacterium sibiricum rhodopsin characterized by low-temperature trapping in the IR and time-resolved studies in the visible. *The Journal of Physical Chemistry. B*, *117*, 7235–7253. https://doi.org/10.1021/jp402430w.

Dioumaev, A. K., Richter, H.-T., Brown, L. S., Tanio, M., Tuzi, S., Saito, H., et al. (1998). Existence of a proton transfer chain in bacteriorhodopsin: Participation of Glu-194 in the release of protons to the extracellular side. *Biochemistry*, *37*, 2496–2506.

Drew, D., North, R. A., Nagarathinan, K., & Tanabe, M. (2021). Structures and general transport mechanisms by the major facilitator Superfamily (MFS). *Chemical Reviews*, *121*, 5289–5335. https://doi.org/10.1021/acs.chemrev.0c00983.

Dunten, R. L., Sahin-Tóth, M., & Kaback, H. R. (1993). Role of the charge pair aspartic acid-237-Lysine-358 in the lactose permease of *Escherichia coli*. *Biochemistry*, *32*, 3139–3145.

Dutzler, R., Campbell, E. B., & MacKinnon, R. (2003). Gating the selectivity filter in ClC chloride channels. *Science*, *300*, 108–112. https://doi.org/10.1126/science.1082708.

Edmonds, B. W., & Luecke, H. (2004). Atomic resolution structures and the mechanism of ion pumping in bacteriorhodopsin. *Frontiers in Bioscience*, *9*, 1556–1566.

Edwards, R. E. (2007). The neurotransmitter cycle and quantal size. *Neuron*, *55*, 835–858. https://doi.org/10.1016/j-neuron.2007.09.001.

Eisenhauer, K., Kuhne, J., Ritter, E., Berndt, A., Wolf, S., Freier, E., et al. (2012). In channelrhodopsin-2 Glu-90 is crucial for ion selectivity and is deprotonated during the photocycle. *The Journal of Biological Chemistry*, *287*, 6904–6911.

Ekberg, K., Wielandt, A. G., Buch-Pedersen, M. J., & Palmgren, M. G. (2013). A conserved asparagine in a P-type proton pump is required for efficient gating of protons. *The Journal of Biological Chemistry*, *288*, 9610–9618.

Elghobashi-Meinhardt, N., Phatak, P., Bondar, A.-N., Elstner, M., & Smith, J. C. (2018). Catalysis of ground state cis-trans isomerization of bacteriorhodopsin's retinal chromophore by a hydrogen-bond network. *Journal of Membrane Biology*. https://doi.org/10.1007/s00232-018-0027-x.

Essen, L.-O. (2002). Halorhodopsin: Light-driven ion pumping made simple? *Current Opinion in Structural Biology*, *12*, 516–522.

Focht, D., Croll, T. I., Pedersen, B. P., & Nissen, P. (2017). Improved model of proton pump crystal structure obtained by interactive molecular dynamics flexible fitting expands the mechanistic model for proton translocation in P-type ATPases. *Frontiers in Physiology*, *8*, 202. https://doi.org/10.3389/fphys.2017/00202.

Franco, P. J., & Brooker, R. J. (1994). Functional roles of Glu-269 and Glu-325 within the lactose permease of *Escherichia coli*. *The Journal of Biological Chemistry*, *269*, 7379–7386.

Friedman, R., Nachliel, E., & Gutman, M. (2003). The role of intraprotein cavities in the catalytic cycle of bacteriorhodopsin. *Biophysical Journal*, *85*, 886–896.

Fu, R., Miao, Y., Qin, H., & Cross, T. A. (2020). Observation of the imidazole-imidazolium hydrogen bonds responsible for selective proton conductance in the influenza A M2 channel. *Journal of the American Chemical Society*, *142*, 2115–2119. https://doi.org/10.1021/jacs.9b09985.

Gadsby, D. C. (2009). Ion channels versus ion pumps: The principal difference, in principle. *Nature Reviews Molecular Cell Biology*, *10*, 344–352.
García-Moreno, B. (2009). Adaptations of proteins to cellular and subcellular pH. *Journal of Biology*, *8*, 98. https://doi.org/10.1186/jbiol199.
Garczareck, F., Brown, L. S., Lanyi, J. K., & Gerwert, K. (2005). Proton binding within a membrane protein by a protonated water cluster. *Proceedings of the National Academy of Sciences of the United States of America*, *102*, 3633–3638.
Garczareck, F., & Gerwert, K. (2006). Functional waters in intraprotein proton transfer monitored by FTIR difference spectroscopy. *Nature*, *439*, 109–112.
Ge, X., & Gunner, M. R. (2016). Unraveling the mechanism of proton translocation in the extracellular half-channel of bacteriorhodopsin. *Proteins*, *84*, 639–654. https://doi.org/10.1002/prot.25013.
Gerland, L., Friedrich, D., Hopf, L., Donovan, E. J., Wallmann, A., Erdmann, N., et al. (2020). pH-dependent protonation of surface carboxylates in PsbO enables local buffering and triggers structural changes. *ChemBioChem*, *21*, 1597–1604. https://doi.org/10.1002/cbic.201900739.
Gerwert, K., Freier, E., & Wolf, S. (2014). The role of protein-bound water molecules in microbial rhodopsins. *Biochimica et Biophysica Acta*, *1837*, 606–613.
Gerwert, K., Hess, B., Soppa, J., & Oesterhelt, D. (1989). Role of aspartate-96 in proton translocation by bacteriorhodopsin. *Proceedings of the National Academy of Sciences of the United States of America*, *86*, 4943–4947.
Gora, A., Brezovsky, J., & Damborsky, J. (2013). Gates of enzymes. *Chemical Reviews*, *113*, 5871–5923.
Gottschalk, M., Dencher, N. A., & Halle, B. (2001). Microsecond exchange of internal water molecules in bacteriorhodopsin. *Journal of Molecular Biology*, *311*, 605–621.
Govorunova, E. G., Sineschchekov, O. A., Li, H., & Spudich, E. N. (2017). Microbial rhodopsins: Diversity, mechanisms, and optogenetic applications. *Annual Review of Biochemistry*, *86*, 845–872. https://doi.org/10.1146/annurev-biochem-101910-144233.
Gray, T. M., & Matthews, B. W. (1984). Intrahelical hydrogen bonding of serine, threonine and cysteine residues within α-helices and its relevance to membrane-bound proteins. *Journal of Molecular Biology*, *175*, 75–81. https://doi.org/10.1016/0022-2836(84)90446-7.
Gruia, A. D., Bondar, A.-N., Smith, J. C., & Fischer, S. (2005). Mechanism of a molecular valve in the halorhodopsin chloride pump. *Structure*, *13* 617–627. https://doi.org/10.1016/j.str.2005.01.021.
Grytsyk, N., Sugihara, J., Kaback, H. R., & Hellwig, P. (2017). pK_a of Glu325 in LacY. *Proceedings of the National Academy of Sciences of the United States of America*, *114*, 1530–1535. https://doi.org/10.1073/pnas.1621431114.
Guan, L., & Kaback, H. R. (2009). Properties of LacY efflux mutant. *Biochemistry*, *48*, 9250–9255. https://doi.org/10.1021/bi9014067.
Guan, L., Mirza, O., Verner, G., Iwata, S., & Kaback, H. R. (2007). Structural determination of wild-type lactose permease. *Proceedings of the National Academy of Sciences of the United States of America*, *104*, 15294–15298. https://doi.org/10.1073/pnas.0707688104.
Guerra, F., & Bondar, A.-N. (2015). Dynamics of the plasma membrane proton pump. *The Journal of Membrane Biology*, *248*, 443–453. https://doi.org/10.1007/s00232-014-9732-2.
Gushchin, I., Chervakov, P., Kuzmichev, P., Popov, A. N., Round, E., Borshchevskiy, V., et al. (2013). Structural insights into the proton pumping by the unusual proteorhodopsin from nonmarine bacteria. *PNAS*, *110*, 12631–12636. https://doi.org/10.1073/pnas.1221629110.

Harris, A., Saita, M., Resler, T., Hughes-Visentin, A., Maia, R., Pranga-Sellnau, F., et al. (2018). Molecular details of the unique mechanism of chloride transport by a cyanobacterial rhodopsin. *Physical Chemistry Chemical Physics*, *20*, 3184–3199. https://doi.org/10.1039/c7cp06068h.

Harris, T. K., & Turner, G. J. (2002). Structural basis of perturbed pK_a values of catalytic groups in enzyme active sites. *IUBMB Life*, *53*, 85–98. https://doi.org/10.1080/10399710290038972.

Hasegawa, N., Jonotsuka, H., Miki, K., & Takeda, K. (2018). X-ray structure analysis of bacteriorhodopsin at 1.3 Å resolution. *Scientific Reports*, *8*, 13123. https://doi.org/10.1063/s41598-018-31370-0.

He, M. M., & Kaback, H. R. (1997). Interaction between residues Glu269 (helix VIII) and His322 (helix X) of the lactose permease of *Escherichia coli* is essential for substrate binding. *Biochemistry*, *36*, 13688–13692. https://doi.org/10.1021/bi9715324.

Hegemann, P., Blanck, A., Vogelsang-Wenke, H., Lottspeich, F., & Oesterhelt, D. (1987). The halo-opsin gene. I. Identification and isolation. *EMBO Journal.*, *6*, 259–264.

Henderson, R., & Unwin, P. N. T. (1975). Three-dimensional model of purple membrane obtained by electron microscopy. *Nature*, *257*, 28–32.

Henry, R. M., Yu, C.-H., Rodinger, T., & Pomès, R. (2009). Functional hydration and conformational gating of proton uptake in cytochrome *c* oxidase. *Journal of Molecular Biology*, *387*, 1165–1185. https://doi.org/10.1016/j.jmb.2009.02.042.

Herzfeld, J., & Lansing, J. C. (2002). Magnetic resonance studies of the bacteriorhodopsin pump cycle. *Annual Review of Biophysics and Biomolecular Structure*, *31*, 73–95.

Hosaka, T., Yoshizawa, Y., Ohsawa, N., Hato, M., DeLong, E. F., Kogure, K., et al. (2016). Structural mechanism for light-driven transport by a new type of chloride pump, *Nonlabens marinus* rhodopsin-3. *The Journal of Biological Chemistry*, *291*, 17488–17495. https://doi.org/10.1074/jbc.M116.728220.

Humphrey, W., Dalke, W., & Schulten, K. (1996). VMD: Visual molecular dynamics. *Journal of Molecular Graphics*, *14*, 33–38. https://doi.org/10.1016/0263-7855(96)00018-5.

Inoue, K., Nomura, Y., & Kandori, H. (2016). Asymmetric functional conversion of eubacterial light-driven ion pumps. *The Journal of Biological Chemistry*, *291*, 9883–9893. https://doi.org/10.1074/jbc.M116.716498.

Isom, D. G., Castañeda, C. A., Cannon, B. R., & García-Moreno, B. (2011). Large shifts in pK_a values of lysine residues buried inside a protein. *Proceedings of the National Academy of Sciences of the United States of America*, *108*, 5260–5265. https://doi.org/10.1073/pnas.1010750108.

Isom, D. G., Castañeda, C. A., Cannon, B. R., Velu, P. D., & García-Moreno, B. (2010). Charges in the hydrophobic interior of proteins. *Proceedings of the National Academy of Sciences of the United States of America*, *107*, 16096–16100. https://doi.org/10.1073/pnas.1004213107.

Jardón-Valadez, E., Bondar, A.-N., & Tobias, D. A. (2014). Electrostatic and hydrogen bond dynamics in chloride pumping by halorhodopsin. *Biochimica et Biophysica Acta*, *1837*, 1964–1972.

Kalaidzidis, I. V., Belevich, I. N., & Kaulen, A. D. (1998). Photovoltage evidence that Glu-204 is the intermediate proton donor rather than the terminal proton release in bacteriorhodopsin. *FEBS Letters*, *434*, 197–200. https://doi.org/10.1016/s0014-5793(98)00980-6.

Kandori, H. (2004). Role of internal water molecules in bacteriorhodopsin. *Biochimica et Biophysica Acta*, *1460*, 177–191. https://doi.org/10.1016/S0005-2728(00)00138-9.

Karathanou, K., & Bondar, A.-N. (2019). Using graphs of dynamic hydrogen-bond networks to dissect conformational coupling in a protein motor. *Journal of Chemical Information and Modeling*, *59*, 1882–1896. https://doi.org/10.1021/acs.jcim.8b00979.

Kato, H. E., Inoue, K., Abe-Yoshizumi, R., Kato, Y., Ono, H., Konno, M., et al. (2015). Structural basis for Na^+ transport mechanism by a light-driven Na^+ pump. *Nature*, *521*, 48–53.

Kato, H. E., Kamiya, M., Sugo, S., Ito, J., Taniguchi, R., Orito, A., et al. (2015). Atomistic design of microbial opsin-based blue-shifted optogenetics tools. *Nature Communications*, *6*, 7177. https://doi.org/10.1038/ncomms8177.

Kato, H. E., Kim, Y. S., Paggi, J. M., Evans, K. E., Allen, W. E., Richardson, C., et al. (2018). Structural mechanisms of selectivity and gating in anion channel rhodopsins. *Nature*, *561*, 349–354. https://doi.org/10.1038/s41586-018-0504-5.

Kato, H. E., Zhang, F., Yizhar, O., Ramakhrishnan, C., Nishizawa, T., Hirata, K., et al. (2012). Crystal structure of the channelrhodopsin light-gated cation channel. *Nature*, *482*, 369–374.

Kim, Y. S., Kato, H. E., Yamashita, K., Ito, S., Inoue, K., Ramakrishnan, C., et al. (2018). Crystal structure of the natural anion-conducting channel rhodopin GtACR1. *Nature*, *561*, 343–348. https://doi.org/10.1038/s41586-018-0511-6.

Kim, K., Kwon, S.-K., Jun, S.-H., Cha, J. S., Kim, H., Lee, W., et al. (2016). Crystal structure and functional characterization of a light-driven chloride pump having an NTQ motif. *Nature Communications*, 7, 12677. https://doi.org/10.1038/ncomms12677.

Kobilka, B. K. (2007). G protein coupled receptor structure and activation. *Biochimica et Biophysica Acta*, *1768*, 794–807. https://doi.org/10.1016/j.bbamem.2006.10.021.

Kolbe, M., Besir, H., Essen, L.-O., & Oesterhelt, D. (2000) Structure of the light-driven chloride pump halorhodopsin at 1.8 Å resolution. *Science*, *288*, 1390–1396.

Kouyama, T., Kanada, S., Takeguchi, Y., Naurusawa, A., Murakami, M., & Ihara, K. (2010). Crystal structure of the light.driven chloride pump halorhodopsin from *Natromonas pharaonis*. *Journal of Molecular Biology*, *396*, 564–579. https://doi.org/10.1016/j.jmb.2009.11.061.

Kouyama, T., Kawaguchi, H., Nakanishi, T., Kubo, H., & Murakami, M. (2015). Crystal structures of the L_1, L_2, N, and O states of *pharaonis* halorhodopsin. *Biophysical Journal*, *108*, 2680–2690. https://doi.org/10.1016/j.bpj.2015.04.027.

Kovacs, G. N., Colletier, J.-P., Grünbein, M. L., Yang, Y., Stensitzki, T., Batyuk, A., et al. (2019). Three-dimensional view of ultrafast dynamics in photoexcited bacteriorhodopsin. *Nature Communications*, *10*, 3177. https://doi.org/10.1038/s41467-019-10758-0.

Kulcsár, Á., Groma, G. I., Lanyi, J. K., & Váró, G. (2000). Characterization of the proton-transporting photocycle of *pharaonis* halorhodopsin. *Biophysical Journal*, *79*, 2705–2713.

Lanyi, J. K. (1978). Light energy conversion in *Halobacterium halobium*. *Microbiological Reviews*, *42*, 682–706.

Lanyi, J. K. (1999). Bacteriorhodopsin. *Internatioanl Review of Cytology*, *187*, 161–202.

Lanyi, J. K., Duschl, A., Hatfield, G. W., May, K., & Oesterhelt, D. (1990). The primary structure of a halorhodopsin from *Natronibacterium pharaonis*. Structural, functional and evolutionary implications for bacterial rhodopsins and halorhodopsins. *The Journal of Biological Chemistry*, *265*, 1253–1260.

Lanyi, J. K., & Weber, H. J. (1980). Spectrophotometric identification of the pigment associated with light-driven primary sodium translocation in *Halobacterium halobium*. *The Journal of Biological Chemistry*, *255*, 243–250.

Lazaratos, M., Karathanou, K., & Bondar, A.-N. (2020). Graphs of dynamic H-bond networks: From model proteins to protein complexes in cell signaling. *Current Opinion in Structural Biology*, *64*, 79–87. https://doi.org/10.1016/j.sbi.2020.06.006.

Leano, J. B., Batarni, S., Eriksen, J., Juge, N., Pak, J. E., Kimura-Someya, T., et al. (2019). Structures suggest a mechanism for energy coupling by a family of organic anion transporters. *PLoS Biology*, *17*, e30000260. https://doi.org/10.1371/journal.pbio.3000260.

Levitt, M., & Park, B. H. (1993). Water: Now you see it, now you don't. *Structure, 1*, 223–226.
Li, F., Eriksen, J., Finer-Moore, J., Chang, R., Nguyen, P., Bowen, A., et al. (2020). Ion transport and regulation in a synaptic vesicle glutamate transporter. *Science, 368*, 893–897. https://doi.org/10.1126/science.aba9202.
Li, H., Huang, C.-Y., Govorunova, E. G., Schafer, C. T., Sineschchekov, O. A., Wang, M., et al. (2019). Crystal structure of a natural light-gated anion channelrhodopsin. *eLife, 8*, e41741. https://doi.org/10.7554/eLife.41741.
Lindley, E. V., & MacDonald, R. E. (1979). A second mechanism for sodium extrusion in *Halobacterium halobium*: A light-driven sodium pump. *Biochimical and Biophysical Research Communications, 88*, 491–499.
Lórenz-Fonfría, V. A., Resler, T., Krause, N., Nack, M., Gossing, M., von Mollard, G. F., et al. (2013). Transient protonation changes in channelrhodpsin-2 and their relevance to channel gating. *Proceedings of the National Academy of Sciences of the United States of America, 110*, 1273–1281.
Lozier, R. H., Bogomolni, R. A., & Stoeckenius, W. (1975). Bacteriorhodopsin: A light-driven proton pump in *Halobacterium halobium*. *Biophysical Journal, 15*, 955–962. https://doi.org/10.1016/S0006-3495(75)85875-9.
Ludwig, M.-G., Vanek, M., Guerini, D., Gasser, J. A., Jones, C. E., Junker, U., et al. (2003). Proton-sensing G-protein-coupled receptors. *Nature, 425*, 93–98.
Luecke, H., Schobert, B., Richter, H.-T., Cartailler, J.-P., & Lanyi, J. K. (1999). Structure of bacteriorhdopsin at 1.55 Å resolution. *Journal of Molecular Biology, 291*, 899–911.
Metz, G., Siebert, F., & Engelhardt, M. (1992). Asp85 is the only internal aspartic acid that gets protonated in the M intermediate and the purple-to-blue transition of bacteriorhodopsin. A solid-state 13C CP-MAS NMR investigation. *FEBS Letters, 303*, 237–241.
Mitra, K., Ubarratxena-Belandia, I., Taguchi, T., Warren, G., & Engelman, D. M. (2004). Modulation of the bilayer thickness of exocytic pathway membranes by membrane proteins rather than cholesterol. *Proceedings of the National Academy of Sciences of the United States of America, 101*, 4083–4088. https://doi.org/10.1073/pnas.0307332101.
Morgan, J. E., Gennis, R. B., & Maeda, A. (2008). A role for internal water molecules in proton affinity changes in the Schiff base and Asp85 for the one-way proton transfer in bacteriorhodopsin. *Photochemistry and Photobiology, 84*, 1038–1045.
Morizumi, T., Ou, W.-L., van Eps, N., Inoue, K., Kandori, H., Brown, L. S., et al. (2019). X-ray crystallographic structure and oligomerization of *Gloeobacter* rhodopsin. *Scientific Reports, 9*, 11283. https://doi.org/10.1038/s41598-019-47445-5.
Müller, M., Bamann, C., Bamberg, E., & Kühlbrandt, W. (2011). Projection structure of channelrhodopsin-2 at 6Å resolution by electron crystallography. *Journal of Molecular Biology, 414*, 86–95.
Murakami, S., Nakashima, R., Yamashita, E., & Yamaguchi, A. (2006). Crystal structure of bacterial multidrug efflux transporter AcrB. *Nature, 419*, 587–593.
Nachliel, E., & Gutman, M. (2001). Probing the substrate binding domain of lactose permease by a proton pulse. *Biochimica et Biophysica Acta, 1514*, 33–50.
Nack, M., Radu, I., Gossing, M., Bamann, C., Bamberg, E., von Mollard, G. F., et al. (2010). The DC gate in channelrhodopsin-2: Crucial hydrogen bonding interaction between C128 and D156. *Photochemical & Photobiological Sciences, 9*, 194–198.
Namslauer, A., Aagard, A., Kotsonouri, A., & Przezinski, P. (2003). Intramolecular proton-transfer reactions in a membrane-bound proton pump: The effect of pH on the peroxy to ferryl transition in cytochrome *c* oxidase. *Biochemistry, 42*, 1488–1498. https://doi.org/10.1021/bi026524o.
Nango, E., Royant, A., Kubo, M., Nakane, T., Wickstrand, C., Kimura, T., et al. (2016). A three-dimensional movie of structural changes in bacteriorhodopsin. *Science, 354*, 1552–1557.

Ni, Q. Z., Can, T. V., Daviso, E., Belenky, M., Griffin, R. G., & Herzfeld, J. (2018). Primary transfer step in the light-driven ion pump bacteriorhodopsin: An irreversible U-turn revealed by dynamic nuclear polarization-enhanced magic angle spinning NMR. *Journal of the American Chemical Society*, *140*, 4085–4091. https://doi.org/10.1021/jacs.8b00022.

Oda, K., Nomura, T., Nakane, T., Yamashita, K., Inoue, K., Ito, S., et al. (2021). Time-resolved serial femtosecond crystallography reveals early structural changes in channel rhodopsin. *eLife*, *10*, e62389. https://doi.org/10.7554/eLife.62389.

Oesterhelt, D., & Stoeckenius, W. (1971). Rhodopsin-like protein from the purple membrane of *Halobacterium halobium*. *Nature*, *233*, 149–152.

Oesterhelt, D., Tittor, J., & Bamberg, E. (1992). A unifying concept for ion translocation by retinal proteins. *Journal of Bioenergetics and Biomembranes*, *24*, 181–191.

Otto, H., Marti, T., Holz, M., Mogi, T., Lindau, M., Khorana, H. G., et al. (1989). Aspartic acid-96 is the internal proton donor in the reprotonation of the Schiff base of bacteriorhodopsin. *Proceedings of the National Academy of Sciences of the United States of America*, *86*, 9228–9232.

Pace, C. N., Grimsley, G. R., & Scholtz, J. M. (2009). Protein ionizable groups: pK values and their contribution to protein stability and solubility. *Journal of Biological Chemistry*, *284*, 13285–13289. https://doi.org/10.1016/j.jmb.2006.07.056.

Pao, S. S., Paulsen, I. T., & Saier, M. H., Jr. (1998). Major facilitator superfamily. *Microbiology and Molecular Biology Reviews*, *62*, 1–34.

Pedersen, B. P., Buch-Pedersen, M. J., Morth, J. B., Palmgren, M. G., & Nissen, P. (2007). Crystal structure of the plasma membrane proton pump. *Nature*, *450*, 1111–1114.

Perez-Flores, M., Lee, J. H., Park, S., Zhang, X.-D., Sihn, C.-R., Ledford, H. A., et al. (2020). Cooperativity of Kv7.4 channels confers ultrafast electromechanical sensitivity and emergent properties in cochlear outer hair cells. *Science Advances*, *6*, eaba1104. https://doi.org/10.1126/sciadv.aba1104.

Phatak, P., Frähmcke, J. S., Wanko, M., Hoffmann, M., Strodel, P., Smith, J. C., et al. (2009). Long-distance proton transfer with a break in the bacteriorhodopsin active site. *Journal of the American Chemical Society*, *131*, 7064–7078.

Phatak, P., Ghosh, N., Yu, H., Cui, Q., & Elstner, M. (2008). Amino acids with an intermolecular proton bond as proton storage site in bacteriorhodopsin. *Proceedings of the National Academy of Sciences of the United States of America*, *105*, 19672–19677.

Püttner, I. B., Sarkar, H. K., Padan, E., Lolkema, J. S., & Kaback, H. R. (1989). Characterization of site-directed mutants in the lac permease of *Escherichia coli*. 1. Replacement of histidine residues. *Biochemistry*, *28*, 2525–2533.

Qiu, F., Rebolledo, S., Gonzalez, C., & Larsson, H. P. (2013). Subunit interactions during cooperative opening of voltage-gated proton channels. *Neuron*, *77*, 288–298. https://doi.org/10.1016/j.neuron.2012.12.021.

Quillin, M. L., Wingfield, P. T., & Matthews, B. W. (2006). Determination of solvent cavities in IL-1β using experimentally phased electron density. *Proceedings of the National Academy of Sciences of the United States of America*, *103*, 19749–19753. https://doi.org/10.1073%02pnas.0609442104.

Rieger, B., Jung, W., & Busch, K. B. (2014). Lateral pH gradient between OXPHOS complex and F_0F_1 ATP-sytnhase in folded mitochondrial membranes. *Nature Communications*, *5*, 3103. https://doi.org/10.1038/ncomms4103.

Riesle, J., Oesterhelt, D., Dencher, N. A., & Heberle, J. (1996). D38 is an essential part of the proton translocation pathway in bacteriorhodopsin. *Biochemistry*, *35*, 6635–6643.

Roux, B., Nina, M., Pomes, R., & Smith, J. C. (1996). Thermodynamic stability of water molecules in the bacteriorhodopsin proton channel: A molecular dynamics free energy perturbation study. *Biophysical Journal*, *71*, 670–681.

Sahin-Tóth, M., & Kaback, H. R. (1993). Properties of interacting aspartic acid and lysine residues in the lactose permease of *Escherichia coli*. *Biochemistry*, *32*, 10027–10035. https://doi.org/10.1021/bi00089a019.

Samatey, F. A., Xu, C., & Popot, J.-L. (1995). On the distribution of amino acid residues in transmembrane α-helix bundles. *Proceedings of the National Academy of Sciences of the United States of America, 92*, 4577–4581.

Sasaki, J., Brown, L. S., Chon, Y.-S., Kandori, H., Maeda, A., Needleman, R., et al. (1995). Conversion of bacteriorhodopsin into a chloride ion pump. *Science, 269*, 73–75.

Schätzler, B., Dencher, N. A., Tittor, J., Oesterhelt, D., Yaniv-Checover, S., Nachliel, E., et al. (2003). Subsecond proton-hole propagation in bacteriorhodopsin. *Biophysical Journal, 84*, 671–686.

Schobert, B., & Lanyi, J. K. (1982). Halorhodopsin is a light-driven chloride pump. *The Journal of Biological Chemistry, 257*, 10306–10313.

Schönichen, A., Webb, B. A., Jacobson, M. P., & Barber, D. L. (2013). Considering protonation as a posttranslational modification regulating protein structure and function. *Annual Review of Biophysics, 42*, 289–314. https://doi.org/10.1146/annurev-biophys-050511-102349.

Shibata, M., Saito, Y., Demura, M., & Kandori, H. (2006). Deprotonation of E234 during the photocycle of *Natronomonas pharaonis* halorhodopsin. *Chemical Physics Letters, 432*, 545–547. https://doi.org/10.1016/j.cplett.2006.10.111.

Siemers, M., Lazaratos, M., Karathanou, K., Guerra, F., Brown, L. S., & Bondar, A.-N. (2019). Bridge: A graph-based algorithm to analyze dynamic H-bond networks in membrane proteins. *Journal of Chemical Theory and Computation, 15*, 6781–6798. https://doi.org/10.1021/acs/jctc.9b00697.

Snider, C., Jayasinghe, S., Hristova, K., & White, S. H. (2009). MPeX: A tool for exploring membrane proteins. *Protein Science, 18*, 2624–2628.

Song, Y., & Gunner, M. R. (2014). Halorhodopsin pumps Cl^- and bacteriorhodopsin pumps protons by a common mechanism that uses conserved electrostatic interactions. *Proceedings of the National Academy of Sciences of the United States of America, 111*, 16377–16382.

Spahn, V., del Vecchio, G., Rodriguez-Gaztelumendi, A., Temp, J., Labuz, D., Kloner, M., et al. (2018). Opioid receptor signaling, analgesic and side effects induced by a computationally designed pH-dependent agonist. *Scientific Reports, 8*, 8965. https://doi.org/10.1038/s41598-018-27313-4.

Srivastava, J., Barber, D. L., & Jacbson, M. P. (2007). Intracellular pH sensors: Design principles and functional significance. *Physiology, 22*, 30–39. https://doi.org/10.1152/physiol.00035.2006.

Stelzl, L. S., Fowler, P. W., Sansom, M. S. P., & Beckstein, O. (2014). Flexible gates generate occluded intermediates in the transport cycle of LacY. *Journal of Molecular Biology, 426*, 735–751. https://doi.org/10.1016/j.jmb.2013.10.024.

Stock, C., & Schwab, A. (2006). Role of the Na+/H+ exchanger NHEI in cell migration. *Acta Physiologica, 187*, 149–157. https://doi.org/10.1111/j.1748-1716.2006.01543.x.

Subramaniam, S., Lindahl, M., Bullough, P. A., Faruqi, A. R., Tittor, J., Oesterhelt, D., et al. (1999). Protein conformational changes in the bacteriorhodopsin photocycle. *Journal of Molecular Biology, 287*, 145–161. https://doi.org/10.1006/jmbi.1999.2589.

Sun, L., Zeng, X., Yan, C., Sun, X., Gong, X., Rao, Y., et al. (2012). Crystal structure of a bacterial homologie of glucose transporters GLUT1-4. *Nature, 490*, 361–366. https://doi.org/10.1038/nature11524.

Takemoto, M., Kato, H. E., Koyama, M., Ito, J., Kamiya, M., Hayashi, S., et al. (2015). Molecular dynamics of channelrhodopsin at the early stages of channel opening. *PLoS One*, e031094. https://doi.org/10.1371/journal.pnone.0131094.

Takeshita, K., Sakata, S., Yamashita, E., Fujiwara, Y., Kawanabe, A., Kurokawa, T., et al. (2014). X-ray crystal structure of voltage-gated proton channel. *Nature Structural & Molecular Biology, 21*, 352–357.

Thurlkill, R. A., Grimsley, G. R., Scholtz, J. M., & Pace, C. N. (2006). pK values of the ionizable groups of proteins. *Protein Science, 15*, 1214–1218. https://doi.org/10.1110/ps.051840806.

Tittor, J., Soell, C., Oesterhelt, D., Butt, H. J., & Bamberg, E. (1989). A defective proton pump, point-mutated bacteriohdopsin Asp96Asn is fully reactivated by azide. *The EMBO Journal*, *8*, 3477–3482.

Tokaji, Z. (1993). Dimeric-like kinetic cooperativity of the bacteriorhodopsin molecules in purple membranes. *Biophysical Journal*, *65*, 1130–1134. https://doi.org/10.1016/S0006-3495(93)81165-2.

Váró, G. (2000). Analogies between halorhodopsin and bacteriorhodopsin. *Biochimica et Biophysica Acta*, *1460*, 220–229. https://doi.org/10.1016/S0005-2728(00)00141-9.

Váró, G., Brown, L. S., Needleman, R., & Lanyi, J. K. (1996). Proton transport by halorhodopsin. *Biochemistry*, *35*, 6604–6611.

Váró, G., Needleman, R., & Lanyi, J. K. (1996). Protein structural change at the cytoplasmic surface as the cause of cooperativity in the bacteriorhodopsin photocycle. *Biophysical Journal*, *70*, 461–467. https://doi.org/10.1016/S0006-3495(96)79589-9.

Vogeley, L., Sineshchekov, O. A., Trivedi, V. D., Sasaki, J., Spudich, J. L., & Luecke, H. (2004). *Anabaena* sensory rhodopsin: A photochromic color sensor at 2.0 Å. *Science*, *306*, 1390–1393.

Vogt, A., Guo, Y., Tsunoda, S. P., Kateriya, S., Elstner, M., & Hegemann, P. (2015). Conversion of a light-driven proton pump into a light-gated ion channel. *Scientific Reports*, *5*, 16450. https://doi.org/10.1038/srep16450.

Volkov, O., Kovalev, K., Polovinkin, V., Borschchevskiy, V., Bamann, C., Astashkin, R., et al. (2017). Structural insights into ion conduction by channelrhodopsin 2. *Science*, *358*, eaan8862.

Vonck, J. (2000). Structure of the bacteriorhodopsin mutant F219L N intermediate revelaed by electron crystallography. *The EMBO Journal*, *19*, 2152–2160.

Wang, T., Sessions, R. B., Lunde, C. S., Rouhani, S., Glaesser, R. M., Duan, Y., et al. (2018). Deptoronation of D96 in bacteriorhodopsin opens the proton uptake pathway. *Structure*, *21*, 290–297. https://doi.org/10.1016/j.str.2012.12.018.

Weinert, T., Skopintsev, P., James, D., Dworkowski, F., Panepucci, E., Kekilli, D., et al. (2019). Proton uptake mechanism in bacteriorhodopsin captured by serial synchrothron crystallography. *Science*, 365. https://doi.org/10.1126/science.aaw8634.

White, K. A., Grillo-Hill, B. K., & Barber, D. L. (2017). Cancer cell behaviors mediated by disregulated pH dynamics at glance. *Journal of Cell Science*, *130*, 663–669. https://doi.org/10.1242/jcs.195297.

Wickstrand, C., Dods, R., Royant, A., & Neutze, R. (2015). Bacteriorhodopsin: Would the real structural intermediates please stand up? *Biochimica et Biophysica Acta*, *1850*, 536–553.

Wisedchaisri, G., Park, M.-S., Iadanza, M. G., Zheng, H., & Gonen, T. (2014). Proton-coupled sugar transport in the prototypical major facilitator superfamiliy protein XylE. *Nature Communications*, *5*, 4521. https://doi.org/10.1038/ncomms5521.

Wojcik, S. M., TRhee, J. S., Herzog, E., Sigler, A., Jahn, R., Takamori, S., et al. (2004). An essential role for vesicular glutamate transporter 1 (VGLUT1) in postnatal development and control of quantal size. *Proceedings of the National Academy of Sciences of the United States of America*, *101*, 7158–7163.

Yamamoto, M., Hayakawa, N., Murakami, M., & Kouyama, T. (2009). Crystal structure of different substates of bacteriorhodopsin's M intermediate at various pH levels. *Journal of Molecular Biology*, *393*, 559–573.

Yoshizawa, S., Kumagai, Y., Kim, H., Ogura, Y., Hayashi, T., Iwasaki, W., et al. (2010). Functional characterization of flavobacteria rhodopsins reveals a unique class of light-driven chloride pump in bacteria. *Proceedings of the National Academy of Sciences of the United States of America*, *111*, 6732–6737. https://doi.org/10.1073/pnas.1403051111.

Zhang, F., Wang, L.-P., Boyden, E. S., & Deisseroth, K. (2006). Channel rhodopsin-2 and optical control of excitable cells. *Nature Methods*, *3*, 785–792. https://doi.org/10.1038/NMETH936.

CHAPTER SEVEN

Structures and functions of the membrane-damaging pore-forming proteins

Anish Kumar Mondal and Kausik Chattopadhyay*

Department of Biological Sciences, Indian Institute of Science Education and Research Mohali, Mohali, Punjab, India

*Corresponding author: e-mail address: kausik@iisermohali.ac.in

Contents

1. Introduction 242
2. α-PFPs 246
   2.1 Colicin family 246
   2.2 Cytolysin A family 248
   2.3 Actinoporin family 250
3. β-PFPs 253
   3.1 Hemolysin family 253
   3.2 Aerolysin family 258
   3.3 Cholesterol-dependent cytolysins 260
4. Membrane-attack complex and perforin family 264
5. Gasdermins 268
6. PFPs from multicellular eukaryotes 270
7. Conclusion 273

Acknowledgments 275

References 275

Abstract

Pore-forming proteins (PFPs) of the diverse life forms have emerged as the potent cell-killing entities owing to their specialized membrane-damaging properties. PFPs have the unique ability to perforate the plasma membranes of their target cells, and they exert this functionality by creating oligomeric pores in the membrane lipid bilayer. Pathogenic bacteria employ PFPs as toxins to execute their virulence mechanisms, whereas in the higher vertebrates PFPs are deployed as the part of the immune system and to generate inflammatory responses. PFPs are the unique dimorphic proteins that are generally synthesized as water-soluble molecules, and transform into membrane-inserted oligomeric pore assemblies upon interacting with the target membranes. In spite of sharing very little sequence similarity, PFPs from diverse organisms display incredible structural similarity. Yet, at the same time, structure-function mechanisms of the PFPs document remarkable versatility. Such notions establish PFPs as the

Advances in Protein Chemistry and Structural Biology, Volume 128
ISSN 1876-1623
https://doi.org/10.1016/bs.apcsb.2021.07.001

fascinating model system to explore variety of unsolved issues pertaining to the structure-function paradigm of the proteins that interact and act in the membrane environment. In this article, we discuss our current understanding regarding the structural basis of the pore-forming functions of the diverse class of PFPs. We attempt to highlight the similarities and differences in their structures, membrane pore-formation mechanisms, and their implications for the various biological processes, ranging from the bacterial virulence mechanisms to the inflammatory immune response generation in the higher animals.

1. Introduction

Plasma membrane plays crucial roles in the survival and functioning of all the living cells. It acts as a physical boundary to compartmentalize the intra-cellular processes. At the same time, it serves as a selective permeability barrier that enables the exchange of critical biomolecules between the cell and the extracellular milieu. Therefore, any damage to the integrity of the plasma membranes is detrimental for the cellular physiology, and it can even cause cell-death. Remarkably, Nature's diverse life forms have acquired this strategy of membrane-permeabilization to execute various functions that are crucial for their survival, growth, and defense mechanisms. Wide variety of organisms from almost all the kingdoms of life achieve this purpose by employing an ancient, yet conserved mechanism that involves utilization of a unique class of membrane-damaging pore-forming proteins (PFPs) (Gonzalez, Bischofberger, Pernot, van der Goot, & Freche, 2008; Iacovache, Bischofberger, & van der Goot, 2010). In their generalized mechanism of action, PFPs act by forming oligomeric pores in the plasma membranes of the target cells. In spite of sharing very little sequence similarity, they display remarkable similarity in their overall structure and mechanism of membrane pore formation (Dal Peraro & van der Goot, 2016; Mondal & Chattopadhyay, 2020; Mondal et al., 2018).

Different organisms utilize PFPs for different purposes, but mostly as the part of their virulence or defense strategies. For example, pathogenic bacteria utilize PFPs as the virulence factors (Los, Randis, Aroian, & Ratner, 2013). Eukaryotic organisms such as sea anemones also use PFPs to paralyze and/or kill their prey (Kristan, Viero, Dalla Serra, Macek, & Anderluh, 2009). PFPs from the higher vertebrates play crucial roles in their defense mechanisms against the pathogens, or as the key executioners for the generation of the inflammatory responses (Kovacs & Miao, 2017; Rosado et al., 2008).

Members of the PFP family can be classified into various distinct groups based on the following criteria: (i) source organisms producing the toxin protein, (ii) stoichiometry and/or pore size, and (iii) structural organization. On basis of the structural characteristics of the pore-forming scaffolds, PFPs are classified into two broad categories: α-PFPs and β-PFPs (Iacovache et al., 2010; Mondal & Chattopadhyay, 2020). PFPs that utilize α-helical bundles to generate the transmembrane pores are designated as the α-PFPs, whereas those which form β-barrel pores are categorized as the β-PFPs (Gouaux, 1997; Heuck, Tweten, & Johnson, 2001; Mondal et al., 2020; Ros & Garcia-Saez, 2015). β-barrel pore-forming protein toxins can be further classified based on the size of their pore architecture: (i) small β-PFPs that generate small pores of ~1–2 nm diameter, utilizing ~7–8 protomers in their assembly (Song et al., 1996), and (ii) large β-PFPs that create large pores of ~35–50 nm diameter, utilizing 30–50 protomers (Lukoyanova, Hoogenboom, & Saibil, 2016).

PFPs exhibit remarkable diversity in their structures. However, the mechanism of membrane pore-formation and the specific folds/motifs implicated for the process, appear to remain conserved even in the evolutionarily distant family members. PFPs are generally secreted as the water-soluble molecules that transform into the transmembrane oligomeric pore assemblies upon interacting with the target cell membrane (Dal Peraro & van der Goot, 2016; Gonzalez et al., 2008; Iacovache et al., 2010; Mondal & Chattopadhyay, 2020; Mondal et al., 2018). PFPs first recognize the target cells through binding to the specific receptors or receptor-like molecules, which could be lipid, sugar, or protein (Kathuria & Chattopadhyay, 2018). Binding of the PFP molecules onto the cell surface concentrates the toxin molecules on the two-dimensional space of the membrane lipid bilayer, restricting their free diffusion. Increase in the local concentration of the toxin molecules in the membranes is believed to promote their self-assembly, which is the most critical step of the pore-formation (Ros & Garcia-Saez, 2015). PFPs generally follow one of the two pathways of membrane pore-formation. In one pathway, oligomerization precedes membrane-insertion of the pore-forming scaffolds. Alternatively, oligomerization and membrane-insertion steps could occur concomitantly to create the membrane pores (Dal Peraro & van der Goot, 2016). Studies on large number of PFPs from diverse categories have revealed mechanistic insights regarding the general mode of action of these cell-killing proteins. However, intricate details of the molecular mechanism of the pore-formation remains obscure in many cases, as the detailed mechanism can vary from one family member to the other, even among the close relatives. Transmembrane scaffolds of the pores generated by the PFPs can be formed entirely by the "protein-only" motifs. However, in the case of certain PFPs lipids from the cell membranes can also get involved with the protein motifs to generate the so-called "toroidal pore" (Barlic et al., 2004; Mesa-Galloso et al., 2019). Therefore, PFPs can exhibit diverse mechanism of membrane pore-formation, even though the overall general scheme of the process remains similar.

The binding process of the PFPs with the target cell membranes is one of the most crucial steps. Interactions with the target membranes trigger all the necessary structural rearrangements required for the pore-formation process. It is evident from the numerous studies that PFPs employ versatile mechanisms to interact with the target cell membrane. Membrane lipid components are definitely one of the major candidates that can act as the specific receptor-like moieties (Ros & Garcia-Saez, 2015). Most of the PFPs are capable of showing pore-forming activity in the artificial lipid bilayer of the lipid vesicles or liposomes, thus proving the notion that the PFPs can exhibit pore-forming ability even in the absence of any carbohydrate or protein receptor. Whatsoever, it is evident that the PFPs require obligatory presence of certain specific lipids to perform their membrane-damaging activity (Kulma et al., 2010). Cholesterol-dependent cytolysins (CDCs) are a family of large pore-forming β-barrel PFPs. Pore-forming activity of the CDCs critically depends on the presence of cholesterol in their target membranes (Christie, Johnstone, Tweten, Parker, & Morton, 2018; Giddings, Johnson, & Tweten, 2003). Moreover, actinoporin family of α-PFPs shows dependence on the presence of sphingomyelin in the membranes (Soto et al., 2017). However, exact role of cholesterol and other lipids in regulating the pore-formation mechanism varies from one PFP to another, and remain obscure in many cases. Apart from membrane lipids, interactions of the PFPs with the proteinaceous receptors are also well documented. Intermedilysin, a β-barrel PFP in the CDC family, has been shown to display interaction with human CD59 on the target cell membranes (Giddings, Zhao, Sims, & Tweten, 2004). Some of the bacterial PFPs have been shown to display specific interaction with the glycan moieties present on the cell surface (Shewell et al., 2020). Therefore, PFPs appear to employ diverse strategies for binding to the target cells. Also, the ability of the PFPs to recognize specific cell-surface receptors or receptor-like molecules confers their cell and tissue specificity in many cases.

Once the membrane-binding step is achieved, membrane-bound PFP molecules undergo massive structural and conformational rearrangement to form the membrane-inserted pores (Gouaux, 1997). Efficient membrane interaction of the PFPs is prerequisite to initiate these structural rearrangements. However, it still remains unclear how exactly these rearrangements are triggered in the PFP structures after the membrane interaction process. Two prominent and irreversible structural changes that occur after the membrane-binding of the PFPs are: (i) oligomerization of the membrane-bound PFP monomers, and (ii) insertion of the pore-forming motifs into the lipid bilayer of the target membrane to generate the transmembrane scaffold of the pore. Several studies have revealed that the membrane-binding step is followed by the self-assembly of the PFPs on the membrane surface that generates oligomers of defined stoichiometry. Stoichiometry of the oligomers can vary depending on the specific PFP: (i) it can be as small as a dimeric assembly, as suggested for colicin A of *Escherichia coli* (Dunkel, Pulagam, Steinhoff, & Klare, 2015); (ii) tetrameric assembly is formed by the actinoporin sticholysin, produced by *Stichodactyla helianthus* (Mancheno, Martin-Benito, Martinez-Ripoll, Gavilanes, & Hermoso, 2003); (iii) *E. coli* cytolysin A forms dodecameric assembly (Mueller, Grauschopf, Maier, Glockshuber, & Ban, 2009); (iv) CDC family of PFPs form oligomers composed of 30–50 protomers (Morton, Sani, Parker, & Separovic, 2019). Depending on the stoichiometry of the oligomeric assembly, the PFP pore size can vary from 1–2 nm to 30–50 nm, in diameter.

Insertion of the pore-forming motifs of the PFPs into the membrane lipid bilayer is another crucial structural rearrangement in the pore-formation mechanism. In case of α-PFTs and small pore-forming β-PFTs, there are designated pre-formed pore-forming motifs composed of α-helices and β-strands, respectively. Prior to membrane interaction, these motifs remain packed against the central protein domain. After membrane-binding and oligomerization, these pore-forming motifs reorganize to finally breach the lipid bilayer. Interestingly, in the case of large pore-forming β-PFPs such as CDCs, designated α-helical regions participate and transform to generate membrane-inserted β-barrels during the pore-formation process (Christie et al., 2018; Mondal & Chattopadhyay, 2020; Mondal et al., 2020). A striking feature of the PFPs' pore-forming motifs is their ability to be compatible with the physicochemical environment of the membrane lipid bilayer upon pore-formation. In the water-soluble state, most of the hydrophobic residues of the pore-forming motifs generally remain packed against the protein core. In the transmembrane form, these residues face toward the lipid bilayer, while most of the hydrophilic residues of the pore-forming motifs face toward the water-filled lumen of the pore.

Oligomerization of the membrane-bound PFP protomers and membrane-insertion of the pore-forming motifs are mechanistically the most crucial steps. These two processes can either be two discrete events where membrane-insertion occurs only upon completion of the oligomerization event, or these two processes can take place in a concerted manner. However, the precise sequence of these events may vary depending on the particular PFP under consideration.

Several studies have revealed that oligomerization of the PFPs in the membrane often generates transient metastable intermediates, in which the pore-forming motifs partially collapse but do not fully insert into the bilayer. These intermediates are termed as the "pre-pores" (Bayley, Jayasinghe, & Wallace, 2005). Once the pre-pore assemblies are created, pore-forming motifs fully insert into the membranes and generate the functional transmembrane pores. Distinct arc-like pre-pore assemblies are observed in the case of large pore-forming β-PFPs such as CDCs, where sequential addition of the protomers with concomitant membrane-insertion of the pore-forming motifs is documented (Jiao, Ruan, & Scheuring, 2021).

One common consequence of the pore-forming functionality of the PFPs is the damage to the cell membranes on which they act upon. However, apart from the direct membrane-damaging cell-killing activity, PFPs can also evoke wide range of cellular responses (Verma, Gandhi, Lata, & Chattopadhyay, 2021). Exact nature and extent of such responses depend on the specific PFP and the target cell types. Also, depending on the particular context of the functioning of a specific PFP, such cellular responses can either have pathophysiological consequences, or they may have beneficial effects (Khilwani & Chattopadhyay, 2015; Khilwani, Mukhopadhaya, & Chattopadhyay, 2015).

In this review, we discuss the structural and biochemical mechanisms associated with the membrane-damaging pore-forming functionality of some of the prominent class of PFPs. We attempt to highlight the versatility in the structure-function mechanisms of wide variety of PFPs that are implicated in diverse biological processes, ranging from bacterial virulence mechanisms to the generation of inflammatory/immune responses in higher vertebrates.

2. α-PFPs

2.1 Colicin family

The colicins are a group of PFPs secreted by some of the species of *E. coli*. Colicins generate pores in the inner bacterial membranes, thereby killing other bacterial species (Cascales et al., 2007). These PFPs regulate the microbial colonies by killing other related competing bacterial species. However, colicins are also found in other bacterial species such as *Shigella boydii* (Dolejsova et al., 2019).

Pore-formation mechanism of colicins remains only partly understood. The first enigmatic aspect is the translocation of colicin across the bacterial outer membranes and periplasmic space to reach the site of its action, that is the inner cell membranes. The proposed model suggests binding of colicin to the receptor on the bacterial outer membranes, followed by translocation of its unfolded region into the periplasm, where it can connect to the protein complex system such as Tol-Pal, which appear to drive entry of the colicin, presumably by exploiting the proton-motive force (Cascales et al., 2007; Jansen et al., 2020).

One of the first PFP structures was solved for the pore-forming domain of colicin (Parker, Pattus, Tucker, & Tsernoglou, 1989; Parker, Postma, Pattus, Tucker, & Tsernoglou, 1992) (Fig. 1). The pore-forming domains of colicins are made up of amphipathic α-helices, with a hydrophobic helical-hairpin at the non-polar core. Exact mechanism of membrane pore-formation by colicins is unclear. Also, the high-resolution structural detail of the colicin pores has not been elucidated till date. In the absence

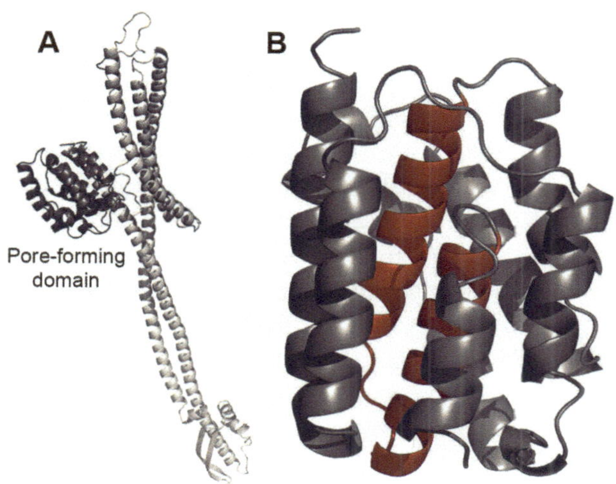

Fig. 1 Colicins are the archetypical α-PFPs. (A) Crystal structure of full length colicin Ia (PDB Id: 1CII). (B) Crystal structure of the pore-forming domain of colicin A (PDB Id: 1COL). The hydrophobic helical-hairpin at the non-polar core of the pore-forming domain is colored in red.

of the high-resolution structural model, it remains unknown how exactly the pore-forming domains of the colicins create the pores. Studies utilizing artificial planar lipid bilayer systems have revealed that one of the colicin family members, colicin E1, inserts into the membrane when the transmembrane potential is negative, and it generates a voltage-gated channel that has reversible open and close conformations (Su, Ho, Merrill, & Lipkowski, 2019). Based on biochemical, biophysical, and computational studies, it is proposed that the amphipathic helices of the colicin pore-forming domain extend and move away from the hydrophobic helical hairpins, and eventually insert into the membranes (Kienker, Qiu, Slatin, Finkelstein, & Jakes, 1997; Lakey et al., 1991; Ridley, Johnson, & Lakey, 2010). Nevertheless, the exact orientation of these helices in the membrane-inserted configuration still remains unclear.

Interestingly, colicin-fold is found in other toxins outside the colicin family that include translocation domain of diphtheria toxin and cry toxin from *Bacillus thuringiensis* (Choe et al., 1992; Xu, Wang, Yu, & Sun, 2014). Colicin-fold is also documented in the mammalian apoptotic proteins, such as BAX and BCL-2 that act to permeabilize the mitochondrial membranes upon activation (Westphal, Dewson, Czabotar, & Kluck, 2011). Therefore, it is possible that these proteins might have common evolutionary relationship with the bacterial colicins.

2.2 Cytolysin A family

Cytolysin A (ClyA) family is a well-characterized family of bacterial α-PFPs, secreted by *E. coli*, as well as *Salmonella* and *Shigella* species (Hunt, Green, & Artymiuk, 2010; Wallace et al., 2000). ClyA is one of the few α-PFPs for which high-resolution structures have been determined for both the water-soluble form and transmembrane pore state (Mueller et al., 2009; Wallace et al., 2000) (Fig. 2). Water-soluble form of ClyA is an elongated molecule, composed of five helical regions, along with a short β-hairpin (termed as β-tongue) near one end (Wallace et al., 2000) (Fig. 2A). The pore-formation process involves massive structural rearrangements in the ClyA structure (Fig. 2B). The critical structural changes occur in the N-terminal helical region (so-called αA1 motif), which rearranges to form the pore lumen. Also, the β-tongue hairpin converts into a helix-turn-helix to embed into the lipid bilayer during pore-formation (Giri Rao, Desikan, Ayappa, & Gosavi, 2016; Sathyanarayana, Visweswariah, & Ayappa, 2021).

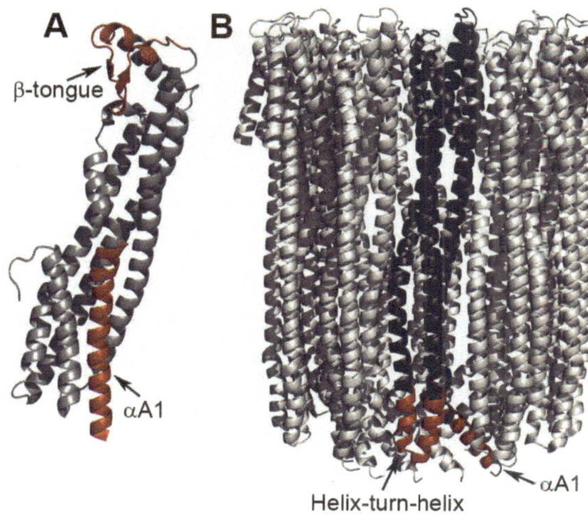

Fig. 2 ClyA is another prototype α-PFP. (A) Crystal structure of the monomeric form of ClyA (PDB Id: 1QOY). Regions corresponding to the pore-forming motif is shown in red. (B) Crystal structure of the oligomeric pore assembly of ClyA (PDB Id: 2WCD). One of the protomers is highlighted as dark gray, along with its pore-forming motifs shown in red.

Upon interaction with the membranes, the β-tongue swings out of the protein core and makes the initial contact with the membranes. Subsequent structural rearrangements triggers oligomerization and insertion of the N-terminal helices to generate a dodecameric pore assembly. The pore size of ClyA is found to be ~2.8 nm (Mueller et al., 2009). Oligomerization mechanism is proposed to involve sequential addition of the protomers, which allows the amphipathic pore-forming helices to accumulate to form the transmembrane helical barrel after insertion through a wedge-like mechanism (Vaidyanathan, Sathyanarayana, Maiti, Visweswariah, & Ayappa, 2014). An alternate "growing-pore" model has also been proposed for the pore-formation mechanism of ClyA. In this model, the conversion of the transmembrane state precedes the oligomerization, and the intermediates can also form transmembrane channels (Agrawal, Apoorva, & Ayappa, 2017; Sathyanarayana et al., 2021; Vaidyanathan et al., 2014). Another non-sequential mode of pore-formation is proposed for ClyA, in which protomers form assembly competent intermediates that undergo transition into oligomers of various stoichiometries, and then subsequently assemble to form the dodecameric pore (Benke et al., 2015). It has been shown in one study that ClyA can form oligomers in solution that could generate transmembrane pores upon membrane binding (Fahie et al., 2013). However, these soluble oligomers of ClyA could possibly represent off-pathway abortive assemblies (Roderer, Benke, Schuler, & Glockshuber, 2016).

Transition of the ClyA monomer in the course of pore-formation has also been analyzed using Molecular Dynamics (MD) simulation-based study. MD simulations reveal that the transition of ClyA into pore requires an obligatory presence of the membranes. Furthermore, two pathways of monomer transition have been proposed. The first model suggests insertion of the β-tongue into the membranes followed by the insertion of the helix (designated as αA1) to complete the transition. The second model suggests insertion of the αA1 helix followed by the insertion of the β-tongue motif (Giri Rao et al., 2016).

Cholesterol plays a critical role in the pore-formation process of ClyA. It is proposed that cholesterol possibly stabilizes (i) the formation of the transmembrane protomer states by interacting with the cholesterol recognition and consensus (CRAC) motifs present in ClyA, as well as (ii) the pore complex by interacting with the binding pockets of the β-tongues (Sathyanarayana et al., 2018).

2.3 Actinoporin family

Actinoporins are a group of eukaryotic PFPs produced by the sea anemones. Sea anemones use actinoporins to either kill their prey or as the part of their defense mechanism against the predators. The most prominent examples of the actinoporins include Fragaceatoxin, Equinatoxin II and Sticholysin II (Athanasiadis, Anderluh, Macek, & Turk, 2001; Kristan et al., 2009; Mancheno et al., 2003; Mechaly, Bellomio, Morante, Gonzalez-Manas, & Guerin, 2009).

High-resolution structures have been determined for a number of actinoporins, such as Fragaceatoxin (FraC; from *Actinia fragacea*), Equinatoxin II (EqtII; from *Actinia equina*) and Sticholysin II (StnII; from *Stichodactyla helianthus*), in their water-soluble monomeric forms (Hinds, Zhang, Anderluh, Hansen, & Norton, 2002; Mancheno et al., 2003; Tanaka, Caaveiro, Morante, Gonzalez-Manas, & Tsumoto, 2015). Actinoporin structures are composed of a core β-sandwich domain, flanked by a ~30 residue-long α-helical region at its N-terminal side (Fig. 3A). This N-terminal helix plays crucial roles in the membrane-damaging pore-formation mechanism. In the water-soluble state, this N-terminal helical region remains packed against the core β-sandwich domain. Upon membrane interaction, and in the

Fig. 3 Actinoporins are the distinct class of eukaryotic α-PFPs that share structural similarity with bacterial toxin TDH. (A) Crystal structure of the monomeric form of FraC (PDB Id: 4TSL). N-terminal region corresponding to the pore-forming motif is colored in red. (B) Crystal structure of the octameric pore form of FraC (PDB Id: 4TSY). Pore-forming motif of one of the protomers in its transmembrane configuration is shown in red. Lipid molecules associated with the structure is shown as yellow spheres. (C) Crystal structure of the TDH tetramer (PDB Id: 3A57). C-terminal region tethered by the disulfide bond against the β-sandwich domain of one of the protomers is highlighted.

course of membrane pore-formation, this helical region inserts into the membranes. Indeed, the structure of the transmembrane oligomeric pore state of FraC confirms that this N-terminal α-helices of the contributing protomers insert into the membranes to create the transmembrane α-helical bundle as the pore scaffold (Fig. 3B) (Kristan et al., 2009; Rojko et al., 2013; Tanaka et al., 2015).

Stoichiometry and the architecture of the oligomeric pore assembly varies among the actinoporin family members. For StnII and EqtII, 3–4 protomers have been proposed to form the pore. In such cases, pore-forming motifs of the PFP along with the membrane lipids are thought to constitute the pore lining, and represent the typical toroidal model of membrane pore formation (Baker, Rojko, Cronin, Anderluh, & Wallace, 2014; Mancheno et al., 2003). Detail insights regarding the structural mechanism of pore-formation by the actinoporin family of PFPs have been obtained from the pore structure of FraC. FraC has been shown to form octameric pore with transmembrane scaffold of α-helical bundle constituted from the N-terminal region of the protein. Interestingly, pore structure shows presence of associated lipid molecules at the pore lining (Fig. 3B). Such lipids have been proposed to serve as the assembly cofactors, stabilizing the membrane pore state of FraC (Mechaly et al., 2011, 2009; Tanaka et al., 2015).

Formation of any intermediate assembly states by the actinoporins during pore-formation process remains unclear. One recent study has suggested that sticholysins can form oligomers in solution and can generate higher order stoichiometric assembly, such as heptamer (Palacios-Ortega et al., 2021). Another study has suggested that StnI can form pores of octameric or non-americ stoichiometry and can have a pre-pore like intermediate states of 8-mer and/or 9-mer assembly (Antonini et al., 2014). Even in the case of FraC, intermediate pre-pore-like structures have been captured (Mechaly et al., 2011). Whatsoever, any of the structural changes in the pore-forming motifs is not very clear in such intermediates. However, some insight was obtained from the structures determined for the lipid-bound monomeric form of FraC. These structures indicate possible unfolding of some of the residues within the N-terminal helical region upon interaction with the membrane lipids. This, in turn, may act to trigger detachment of this pore-forming motif from the core β-sandwich domain for subsequent membrane insertion and pore formation (Tanaka et al., 2015).

Pore-formation efficacy of the actinoporins depends critically on the presence of specific lipids. Actinoporins have been shown to target sphingomyelin (SM)-containing membranes (Kristan et al., 2009; Schon et al., 2008; Soto et al., 2017). Actinoporin structures also highlight the presence of distinct and conserved pockets that can bind to the phosphocholine headgroup (Mancheno et al., 2003). Such ability of the actinoporins to recognize and bind to the lipid moieties can potentially act as the possible mechanism of membrane targeting by the actinoporins.

In the recent years, a group of actinoporin-like toxins have been discovered in Hydra (Ameirika, Sha, & Hwang, 2017). These Hydra actinoporin-like toxins (HALTs) share only about 30% amino acid sequence similarity with the actinoporins, and lack some of the sequence signatures of documented in the actinoporins (Glasser, Rachamim, Aharonovich, & Sher, 2014). Nevertheless, HALTs display membrane-damaging pore-forming property with cell-killing activity. Interestingly, HALTs do not display any binding propensity toward sphingomyelin; rather, they bind to sulfatides (Yap, Tan, & Hwang, 2019). More detail structure-function studies would be required to establish the evolutionary relationship of these HALTs with the actinoporins.

It is important to note that one of the bacterial PFPs, thermostable direct hemolysin (TDH) of *Vibrio parahaemolyticus*, shows remarkable structural similarity with the actinoporins (Yanagihara et al., 2010). TDH harbors core β-sandwich domain similar to that present in the actinoporins (Fig. 3C). However, TDH exists as tetramer in solution, even before encountering the target membranes. Tetrameric assembly state has been shown to be crucial for the pore-forming activity of TDH (Kundu, Tichkule, Pandit, & Chattopadhyay, 2017). However, the structure of the membrane pore state of TDH is not known at present, and it remains unclear whether TDH indeed forms tetrameric pores in the membranes. TDH structure also differs from those of the structurally characterized actinoporins with respect to the following aspects: (i) TDH harbors much shorter N-terminal region with high degree of conformational fluctuations, and (ii) considerably larger C-terminal region that remains tethered against the core β-sandwich domain (Fig. 3C). Extended C-terminal region has been shown to stabilize the tetrameric assembly state of TDH, presumably via formation of inter-protomer contacts. This in turn explains unique ability of TDH to form tetramers in solution, that is otherwise not generally observed with the actinoporins (Kundu et al., 2017). Although shorter in length, 11-residue long N-terminal region of TDH plays crucial roles in the membrane-damaging function of TDH, as observed with the actinoporins. Therefore, consistent with the actinoporin structural mechanism, this N-terminal sequence could possibly serve as the pore-forming motif of TDH. However, it remains unknown how such short stretch of the N-terminal sequence of TDH could act as the pore-forming motif (Kundu et al., 2020). It is possible that TDH employs toroidal model of membrane pore-formation, in which its N-terminal motif collaborates with the membrane lipids to generate the pore lining. Once again, it remains intriguing how the structure-function mechanisms have evolved in the eukaryotic actinoporins, and its structural homolog TDH in the family of bacterial PFPs.

3. β-PFPs

3.1 Hemolysin family

Hemolysin family consists of small pore-forming β-PFPs, generally produced by the pathogenic bacteria. Some of the prominent examples include α-hemolysin produced by *Staphylococcus aureus*, *Vibrio cholerae* cytolysin (VCC) from *Vibrio cholerae*, and necrotic enteritis toxin B (NetB) from *Clostridium perfringens* (Olson & Gouaux, 2005; Savva et al., 2013; Song et al., 1996). Conventionally, these PFPs have been designated as hemolysins due to their potent ability to cause lysis of erythrocytes from diverse species. Pore-formation leads to free diffusion of water and ions across the membrane pores, that in turn results into colloid-osmotic lysis of the erythrocytes, and release of hemoglobin from these cells. Although the exact physiological relevance of erythrocyte-lysis depends on the particular context of the PFP mode of action, hemolytic activity assay is considered as one of the most conventional assays for most of the PFPs. Whatsoever, bacteria use these hemolysins for diverse purposes during pathogenesis that include cell-killing, epithelial barrier disruption for bacterial dissemination, as well as immune evasion (Los et al., 2013).

Hemolysins are one of the best characterized groups of PFPs. Although hemolysins are produced by a variety of Gram-negative bacteria, *Staphylococcus aureus* produces several of these PFPs and use those as their virulence factors (Schreiber, Chan, & Shorr, 2011). Hemolysins of *S. aureus* can either be composed of single component that generates heptameric pore (such as α-hemolysin) (Song et al., 1996), or they can be composed of two components (the bi-component toxins), which generally generate octameric pores (such as γ-hemolysin, leukocidin AB (LukAB), and LukED) (Vasquez et al., 2020; Yamashita et al., 2014). High-resolution structures have been determined for both the soluble form and pore state of some of these toxins (Fig. 4). These hemolysins generally adopt an overall similar β-strand-rich fold in their water-soluble monomeric state. For example, *S. aureus* α-hemolysin is an archetypical small pore-forming β-barrel PFP. In its water-soluble form, it adopts a compact structural domain rich in β-strands. This domain is commonly referred as the cytolysin domain in the case of the small pore-forming β-barrel PFPs. Cytolysin domain of *S. aureus* α-hemolysin and related PFPs harbor a specific designated motif of ~40 amino acid residues termed as the pre-stem (Fig. 4A). In the water-soluble monomeric form, pre-stem remains attached against the

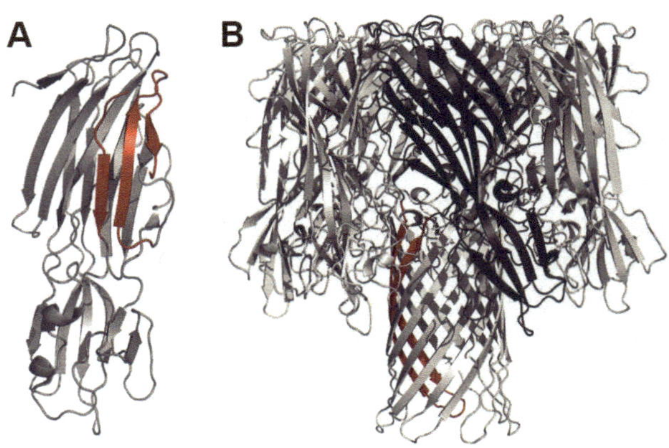

Fig. 4 *S. aureus* α-hemolysin is a prototype β-PFP. (A) Crystal structure of the monomeric form of α-hemolysin (based on the PDB Id: 4IDJ). Pre-stem is colored in red. (B) Crystal structure of the heptameric pore form of α-hemolysin (PDB Id: 7AHL). One β-strand pair of the transmembrane β-barrel contributed by one of the protomers is shown in red.

cytolysin domain in the form of β-strands. In the course of membrane pore-formation, this pre-stem serves as the pore-forming motif to create the transmembrane scaffold. Interaction with the target membranes leads to massive structural and organizational changes leading to the formation of ring-like heptameric structures. Subsequently, the pre-stem motifs from the protomers detach themselves form the cytolysin core in a concerted manner, transform into anti-parallel β-hairpins, and insert into the membrane lipid bilayer to generate the 14-stranded β-barrel pore lining (Fig. 4B). Membrane pore-formation involves formation of the pre-pore oligomeric intermediates, in which the pore-forming motifs possibly undergo partial collapse from their original position, presumably to resolve the steric class between the interacting protomers (Song et al., 1996).

Bi-component β-barrel PFPs such as leukocidins follow a slightly different mechanism of membrane pore-formation. Bacteria secrete the two components, the "S" and "F" subunits, as the water-soluble molecules. The S-component first recognizes a proteinaceous receptor on the target cell surface, and then recruits the F-component. This is followed by the oligomerization process generating an octameric pre-pore structure, containing alternating S and F-protomers. Interestingly, one of the leukocidins, LukAB, is secreted as preformed dimers, which upon binding to the receptor on the cell membranes assemble into the octameric pores. Coordinated inter-protomer interactions are crucial for the oligomerization of these PFPs (Alonzo & Torres, 2014; DuMont et al., 2014).

Hemolysin family of PFPs commonly exhibit requirement of specific lipids on the target membranes. Common membrane lipids such as phosphatidylcholine (PC) have been implicated as the target for *S. aureus* α-hemolysin and γ-hemolysin (Galdiero & Gouaux, 2004; Potrich et al., 2009). Proteinaceous receptors have also been identified for some of the hemolysin family members. *S. aureus* α-hemolysin binds to ADAM10 (a disintegrin and metalloproteinase 10) on the human epithelial cells (Wilke & Bubeck Wardenburg, 2010). It is also suggested that α-hemolysin binding to ADAM10 elevates its metalloproteinase activity, thereby leading to the vascular endothelial-cadherin cleavage that has implications for the endothelial barrier disruption (Wilke & Bubeck Wardenburg, 2010). Staphylococcal leukocidins have been shown to employ a diverse group of chemokine receptors as the receptors to interact with the target cell surface. For example, LukAB targets CD11b subunit of integrin as it's receptor while killing human neutrophils (DuMont et al., 2014).

Many pathogenic strains of the gram-negative bacterium *V. cholerae* produce a potent hemolysin toxin that is termed as the *V. cholerae* cytolysin (VCC) (Olson & Gouaux, 2005). *V. cholerae* is the causative agent of the severe diarrheal disease cholera (Clemens, Nair, Ahmed, Qadri, & Holmgren, 2017). Cholera Toxin (CT) is the major virulence factor *V. cholerae* (Vanden Broeck, Horvath, & De Wolf, 2007). Apart from CT, *V. cholerae* produces a number of accessory toxins that seem to contribute to the pathogenesis process. Due to its potent cytolytic and cytotoxic activity, VCC is considered as a potent virulence factor of *V. cholerae*, particular in the strains that lack the ability produce CT (Khilwani & Chattopadhyay, 2015).

Unlike other conventional bacterial hemolysins, VCC is secreted as an inactive precursor termed as Pro-VCC that harbors an N-terminal pro-domain (Fig. 5A). Presence of the pro-domain maintains an optimal structural plasticity that is required for efficient secretion and proper folding of the toxin (Paul & Chattopadhyay, 2011). It also keeps toxin in an inactive state, thereby blocking its pre-mature oligomeric pore-formation during secretion through the bacterial membranes. Only upon secretion, and by the action of the secreted bacterial proteases, pro-domain is removed and mature VCC is generated. Regulation of the pore-forming activity due to the presence of the pro-domain is quite unique for VCC, and is not there

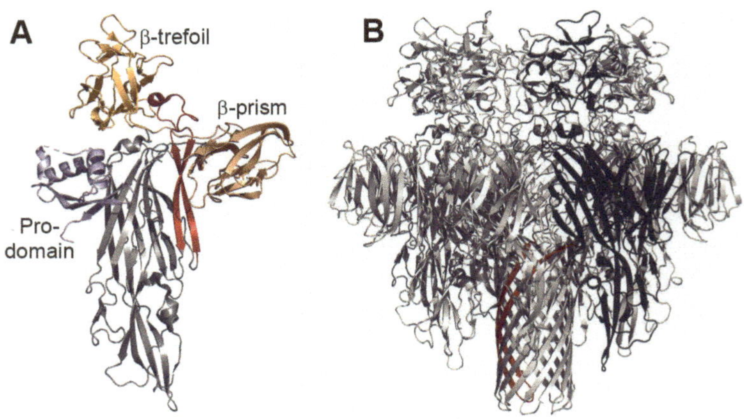

Fig. 5 VCC is an atypical β-PFP. (A) Crystal structure of the Pro-VCC monomer shows the presence of additional domains that are not commonly present in the prototype β-PFPs. Pre-stem motif is shown in red (PDB Id: 1XEZ). (B) Crystal structure of the heptameric pore formed by VCC (PDB Id: 3O44). One β-strand pair of the transmembrane β-barrel contributed by one of the protomers is shown in red.

in the structurally-related hemolysins. Upon interacting with the target cell membranes, mature VCC generates transmembrane heptameric β-barrel pores (Fig. 5B). VCC is one of the few PFPs, for which high-resolution structures have been determined for both the water-soluble monomeric state of Pro-VCC, as well as for the transmembrane heptameric pore assembly (Fig. 5) (De & Olson, 2011; Olson & Gouaux, 2005). Analysis of the structures reveal overall similarity of the cytolysin domain of VCC with those of the related β-PFPs such as *S. aureus* α-hemolysin and leukocidins. Pre-stem of VCC remains packed against the cytolysin domain as anti-parallel β-strand pair, while those of leukocidins are generally composed of three β-strands. Whatsoever, in the process of oligomeric pore-formation, pre-stem motifs from seven VCC protomers undergo structural rearrangement and insert into the membranes to create the 14-stranded transmembrane β-barrel (Fig. 5). One very recent study on VCC has provided important new insight regarding how this structural rearrangement of the pore-forming pre-stem motif could be initiated in the membrane environment (Mondal et al., 2021). It is proposed that the interaction of VCC with the membranes is sensed by the PFP structure to initiate the domain reorganizations. Structural constraint at the hinge region of the pore-forming motif plays crucial role to connect such event, presumably via mediating important communications across the neighboring structural modules/domains. Presence of a conserved tyrosine residue within a hydrophobic pocket at this site appears to maintain this structural constraint, and such a structural feature is found to be conserved in the related β-PFP such as *S. aureus* α-hemolysin and LukF. Therefore, this could act as a conserved mechanism associated with the pore-formation processes of this related group of β-PFPs.

One remarkable feature of VCC is that it harbors two additional lectin-like domains that are not present in the archetypical β-PFPs. At the C-terminal boundary of the cytolysin domain, there is a β-trefoil lectin-like domain followed by a β-prism lectin-like domain (Fig. 5A) (Olson & Gouaux, 2005). Functional implication of the β-trefoil domain of VCC is not clearly known at present, as no specific lectin activity could be assigned to this domain yet. In contrast, β-prism domain has been shown to confer specific binding ability toward the cell surface glycans. Specific residues responsible for the β-prism domain's lectin activity have also been mapped. β-prism domain's carbohydrate-binding activity not only enhances binding of VCC toward the target cell surface, but it also plays a regulatory role in the pore-formation mechanism (Rai, Paul, & Chattopadhyay, 2013). It has been proposed that the lectin-like activity of the β-prism domain acts as a regulatory switch to trigger reorganization of this domain that is otherwise required for the pore-formation (Rai et al., 2013). Similar lectin-like activity has also been documented in another closely related hemolysin of Vibrio species, *Vibrio vulnificus* hemolysin (VVH) (Kaus, Lary, Cole, & Olson, 2014). Unlike VCC, however, VVH harbors only one β-trefoil lectin-like domain at the C-terminus of the cytolysin domain. This β-trefoil domain of VVH has been shown to display specific affinity toward the glycan moieties. VVH also associates with the membrane gangliosides, and such interaction could regulate the pore-forming activity of the toxin (Kashimoto et al., 2020). It is indeed intriguing how such accessory interactions with cell surface entities such as glycans/glycoconjugates have been acquired in the structure-function mechanisms of these Vibrio hemolysins.

VCC employs multiple interaction mechanisms toward the membrane lipids (Kathuria & Chattopadhyay, 2018). VCC can recognize membrane phospholipid headgroups using the loop regions located at the bottom tip of the cytolysin domain (Rai & Chattopadhyay, 2015). VCC also recognizes membrane cholesterol, and specific interaction of the toxin with the membrane cholesterol is extremely crucial for the pore-formation mechanism. Consistent with such cholesterol-binding propensity, VCC is targeted toward the cholesterol-rich membrane microdomains in erythrocytes (Kathuria, Mondal, Sharma, Bhattacharyya, & Chattopadhyay, 2018).

Altogether, these group of small β-barrel pore-forming hemolysins highlight remarkable similarities in their overall structure-function mechanisms; yet, at the same time, they differ from each other in terms of displaying distinct regulatory mechanisms. It is fascinating to note how fine-tuning within the overall similar structural scaffold results into evolution of distinctly regulated pore-formation mechanisms in these closely related protein toxins.

3.2 Aerolysin family

Another well-characterized family of small pore-forming β-PFPs is the aerolysin family. The aerolysin protein was first described in the *Aeromonas* spp. Since then, several aerolysin-like proteins have been discovered in other bacterial family, such as monalysin from *Pseudomonas entomophila*, α-toxin from *Clostridium septicum*, ε-toxin from *Clostridium perfringens* and parasporins from *Bacillus anthracis* (Ballard, Sokolov, Yuan, Kagan, & Tweten, 1993; Opota et al., 2011; Popoff, 2011; Xu et al., 2014). Aerolysin-like proteins are also found in the eukaryotic organisms, such as hemolytic lectin from the mushroom, *Laetiporus sulphureus*, lysenin from earthworm *Eisenia fetida* and hydralysin from *Cnidaria* spp. (Bokori-Brown et al., 2016; Mancheno, Tateno, Goldstein, Martinez-Ripoll, & Hermoso, 2005; Mancheno, Tateno, Sher, & Goldstein, 2010; Sher et al., 2005). Interestingly, higher-order animals also produce aerolysin-like proteins that include biomphalysin from the snail, *Biomphalaria glabrata*, and βγ-CAT from frog *Bombina maxima*. Another novel aerolysin-like protein Dln1 has been discovered in *Danio rerio* (Galinier et al., 2013; Jia et al., 2016; Zhao et al., 2014).

Aerolysin is secreted as soluble monomeric precursor, pro-aerolysin. Pro-aerolysin structure is composed of multiple distinct domains, along with an extended C-terminal pro-peptide. Removal of the pro-peptide generates the active mature toxin (Howard & Buckley, 1985; Iacovache et al., 2011; Parker et al., 1994). Aerolysin-like proteins share an overall similar domain architecture. They harbor variable receptor-binding domain(s) and a conserved core pore-forming module. Pore-forming module of aerolysin is composed of 5-stranded β-sheets, and an insertion-competent pre-stem loop between two of the five β-strands (Fig. 6A) (Iacovache et al., 2016).

Similar to the small pore-forming β-PFPs such as α-hemolysin and VCC, aerolysin also forms heptameric pore assembly in the target membranes (Fig. 6B and C). However, the heptameric pore architecture and the pore-formation mechanism of aerolysin differ considerably from those of the hemolysin family. The pre-stem loop along with the two β-strands from the contributing protomers participate to create a considerably elongated transmembrane β-barrel (Fig. 6C) (Degiacomi et al., 2013; Iacovache et al., 2016).

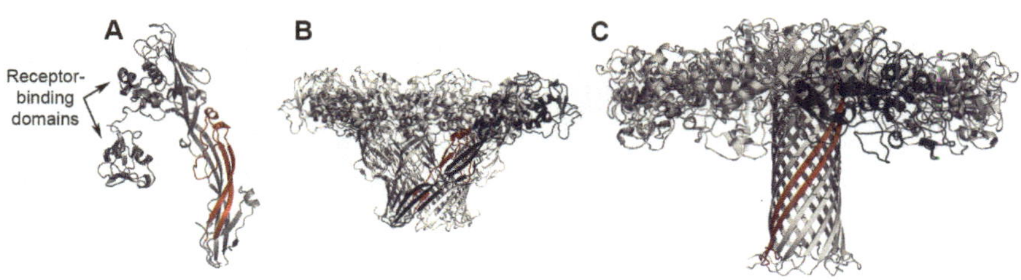

Fig. 6 Aerolysin forms an elongated β-barrel pore. (A) Crystal structure of the pro-aerolysin monomer (PDB Id: 1PRE). (B) Cryo-EM structure of aerolysin pre-pore assembly (PDB Id: 5JZH; EMD-8185). (C) Cryo-EM structure of aerolysin pore (PDB Id: 5JZT; EMD-8187). Region corresponding to the pore-forming motif (in A–C) is shown in red. Only one of the protomers is highlighted in the structures shown in the panels (B) and (C).

High-resolution structures have been elucidated for the pre-pore and pore states of aerolysin proteins (Fig. 6B and C). Based on these structural models, membrane interaction and oligomerization first generate the pre-pore intermediates, followed by the final structural transitions that involve refolding of the pre-stem loop into the β-strand pair. The pore-formation mechanism of aerolysin reveals a unique swirling movement, accompanied by the vertical collapse of the complex, causing the insertion of the pre-pore complex to generate the heptameric pore (Degiacomi et al., 2013). Aerolysin pre-pore intermediates highlight a novel structural fold of two concentric β-barrels, generated through contribution of the rearranged β-strands from each of the contributing protomers, and glued together through hydrophobic interactions. Subsequently, the aerolysin oligomeric assembly undergoes vertical collapse, along with movement of the concentric β-barrels toward the target membranes. This is also accompanied by the unfolding of the pre-stem(s), and extension of the inner β-barrel that eventually inserts into the membrane to create an elongated β-barrel stem. Aerolysin pore exhibits a flat umbrella-like extracellular region which is elevated from the membrane and the β-barrel stem region is much longer compared to the hemolysin family. The tips of the aerolysin β-strands are amphipathic in nature and they anchor the β-barrel stem into the membrane bilayer with a rivet-like conformation (Iacovache et al., 2006).

Aerolysin has been shown to bind to the N-glycosylated glycophosphatidylinositol (GPI)-anchored proteins. The receptor-binding domain(s) of aerolysin show dual binding capacity for the glycan part of the GPI-anchor as well as for the N-linked sugar moieties of the protein receptor (Hong et al., 2002).

3.3 Cholesterol-dependent cytolysins

Cholesterol-dependent cytolysins (CDCs) are the distinct subclass of β-PFPs, which generate large-size pores on their target membrane. The name CDC is derived from their obligatory requirement for the presence of membrane cholesterol for the pore-forming activity (Christie et al., 2018). CDCs are mostly secreted by gram-positive bacteria of *Listeria*, *Clostridium* and *Streptococcus* species (Bryant et al., 2005; Dramsi & Cossart, 2002; Koster et al., 2014; Rossjohn, Feil, McKinstry, Tweten, & Parker, 1997). In the past decade, CDCs from the gram-negative bacteria have also been identified from *Desulfobulbus* and *Enterobacter* species (Hotze et al., 2013).

Owing to their large size, determination of the crystal structure of the pore configuration of the CDCs remains challenging. However, crystal structures have been determined for several CDCs in their monomeric form representing the water-soluble states prior to the membrane interaction. These include listeriolysin O (LLO) from *Listeria monocytogenes*, streptolysin O (SLO) from *Streptococcus pyogenes*, and perfringolysin O (PFO) from *Clostridium perfringens* (Feil, Ascher, Kuiper, Tweten, & Parker, 2014; Koster et al., 2014; Rossjohn et al., 1997). However, transmission electron microscope (TEM), AFM and cryo-EM-based studies have provided important insights regarding the pre-pore and pore states of some of the CDCs. It is worth mentioning that the CDCs exhibit overall structural fold similar to those found in the mammalian membrane attack complex/perforin (MACPF) family of PFPs (Anderluh, Kisovec, Krasevec, & Gilbert, 2014).

CDCs play diverse roles in the bacterial pathogenesis. For example, LLO assists *L. monocytogenes* to escape from the phagosome into the cytosol of the target host cells (Dramsi & Cossart, 2002). Pneumolysin (PLY) is required for bacterial invasion and dissemination, whereas SLO helps the bacteria to spread infection and develop gangrene (Bryant et al., 2005; Nishimoto, Rosch, & Tuomanen, 2020). However, it remains unclear whether the pore-forming activity of the CDCs is responsible for all these pathophysiological activities.

Pore-formation mechanism of the CDCs involve binding of the toxin to the cholesterol-containing membranes, followed by structural changes to form stable CDC oligomers (Fig. 7). The molecular details of the oligomerization vary among the CDC family members. In case of PFO, there are key residues which act as electrostatic switch regulating the pre-pore assembly (Wade et al., 2015). However, such regulation is absent in PLY. In the PLY pore structure, adjacent loops of the domain D4 interact with each other to maintain the stability of the pre-pore (van Pee et al., 2017). Pre-pore intermediate structures of the CDCs are predominantly in the form of arcs, and incomplete rings as revealed by the TEM and AFM-based studies (Leung et al., 2014). Pre-pore structure undergoes further structural rearrangement to bring the domain D3 and transmembrane regions in close proximity to generate the final pore (Ramachandran, Tweten, & Johnson, 2005). Final pore formation requires complete unfolding of the pore-forming motifs, and their structural rearrangement and insertion into the membrane lipid bilayer. CDC monomer contains three α-helical regions that unfold and refold to form two transmembrane hairpins (TMHs) that generate the β-barrel pore (Hotze & Tweten, 2012;

Fig. 7 Structural disposition of a prototype CDC. (A) Crystal structure of the monomeric form of pneumolysin (PDB Id: 4ZGH). Two of the helical motifs that would transform into the transmembrane hairpins are shown in red. Location of the tryptophan-rich undecapeptide motif in the domain D4 is marked with asterisk. (B) Cryo-EM structure of the pneumolysin pore (PDB Id: 5LY6; EMD-4118). One of the protomers in the pore assembly is highlighted. (C) Pneumolysin protomer in the transmembrane oligomeric pore state. Transmembrane β-hairpins in the protomer (in the panels B and C) are shown in red.

Mondal & Chattopadhyay, 2020). Studies on PFO suggest that each pair of β-strands is strictly aligned, and they are ~20° tilted with respect to the perpendicular to the membranes (Morton et al., 2019; van Pee et al., 2017). It is proposed that these structural transitions during pore-formation are orchestrated by the D1 domain (Kacprzyk-Stokowiec et al., 2014). Another study has shown that the activation energy required for the pre-pore to pore transition arises from a salt-bridge between the adjacent monomers in PFO (Wade et al., 2015). Time-lapse AFM based study has suggested that PFO undergoes a vertical collapse that brings the TMHs closer to the membrane after which they extend to penetrate the lipid bilayer (Czajkowsky, Hotze, Shao, & Tweten, 2004). Cryo-EM based structure combining with AFM and molecular modeling have elucidated the general mechanism for pore-formation by the CDCs. The cryo-EM structure of PLY reveals that the lipids are simply pushed aside when TMHs insert into the membrane (Fig. 7B and C) (van Pee et al., 2017). However, solid-state NMR and freeze-fracture EM of PLY pores suggest that lipids are ejected while the TMHs insert into the membrane (Bonev, Gilbert, Andrew, Byron, & Watts, 2001). The latter model also suggests possible toroidal pore formation by PLY.

The D4 domain of the CDCs is identified as the region that interacts with the target membranes (Farrand, LaChapelle, Hotze, Johnson, &

Tweten, 2010). Domain D4 is composed of β-sandwich structure linked by three loops and a tryptophan-rich undecapeptide (UDP) motif (Fig. 7A). The threonine-leucine pair within the loop region is considered to be the cholesterol-recognition motif. After the recognition of the membrane cholesterol, several residues from the loop regions and UDP interact with the membranes, and anchor the protomers, but do not deeply insert into the bilayer core (Soltani, Hotze, Johnson, & Tweten, 2007). Interestingly, cholesterol recognition is also affected by the phospholipid packing in the membrane. Tightly packed phospholipids in the raft-like membrane micro-domains have been shown to weaken the interaction of PFO with the membranes, probably by decreasing the bio-availability of the membrane cholesterol (Nelson, Johnson, & London, 2008). In contrast, less tighter be the packing of the phospholipids, less would be the critical concentration of cholesterol required for the membrane pore-formation (Bakht, Pathak, & London, 2007). Also, the size of the lipid head-groups can affect the CDC-membrane interactions. The presence of phosphatidylethanolamine (PE) and ceramide reduces the requirement of cholesterol for PFO to bind to the lipid vesicles (Morton et al., 2019). Similarly, it is observed in case of LLO that the smaller head-groups like that in palmitoyl-oleyl-phosphatidylethanolamine (POPE) alleviate the cholesterol-LLO interaction (Kozorog, Sani, Separovic, & Anderluh, 2018). Apart from cholesterol recognition, proteins and glycan moieties have also been identified as the cognate receptors for the CDCs. Human CD59 has been characterized as the receptor for the CDCs such as intermedilysin (ILY) from *Streptococcus intermedius* and vaginolysin (VLY) from *Gardnerella vaginalis* (Giddings et al., 2004; Zilnyte, Venclovas, Zvirbliene, & Pleckaityte, 2015). Whatsoever, cholesterol remains critically important for their functionality. For example, ILY exhibits compromised pore-formation in the CD59-containing target membranes that lack cholesterol (Giddings et al., 2004).

Apart from cholesterol, additional interactions are also evident in some of the CDCs. For example, lectinolysin (LLY) from *Streptococcus mitis* has an additional N-terminal domain with lectin-like activity, that provides an extra level of regulation during its pore-forming activity. This region in lectinolysin has been shown to interact with the fucose-rich regions on the cell-surface (Farrand et al., 2008; Feil et al., 2012). In another report, SLO and PLY have been shown to display lectin-like activity through their D4 domain, and such activity appears to be critical for the pore-forming functionality (Shewell et al., 2020). However, molecular mechanism of such interactions needs to be elucidated in more detail.

4. Membrane-attack complex and perforin family

Membrane-attack complex and perforin family (MACPF) is a group of mammalian PFPs that performs wide array of functions, mostly in the vertebrate immunity and defense mechanisms (Liu & Lieberman, 2020). However, MACPF share significant structural similarity with the bacterial CDCs, and form giant β-barrel pores of ~300–350 Å diameter, with distinct stoichiometry of ~35–50 protomer (Figs. 8 and 9) (Anderluh et al., 2014; Reboul, Whisstock, & Dunstone, 2016; Rosado et al., 2008). Interestingly, structural homologs of MACPF family are found in all kingdoms of life which performs diverse functions. The examples include pleurotolysin from fungi *Pleurotus ostreatus* that acts as cytolysin, and a perforin-like protein reported to be involved in *Plasmodium* egress/invasion (Tomita et al., 2004; Wade & Tweten, 2015).

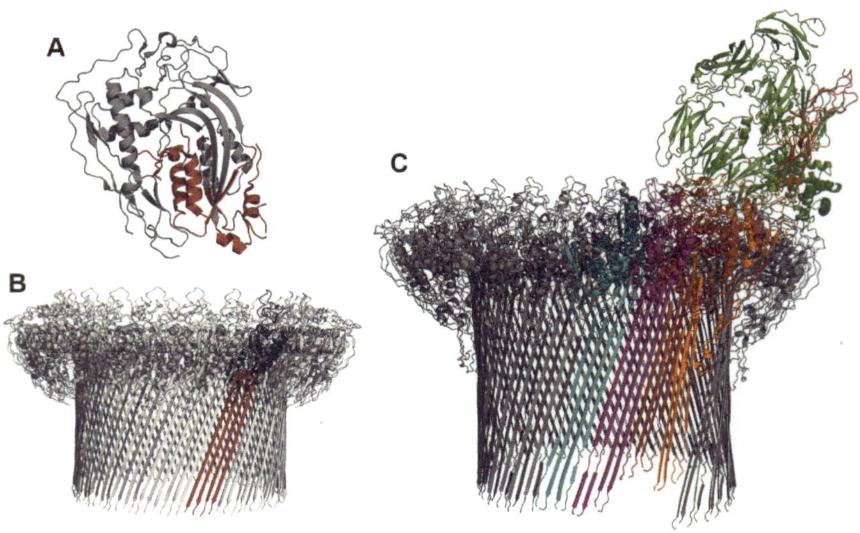

Fig. 8 Structural disposition of the MAC. (A) Crystal structure of the complement component C9 monomer (PDB Id: 6CXO). Helical motifs that would transform into the transmembrane hairpins are shown in red. (B) Cryo-EM structure of the pore assembly formed by the complement component C9 (PDB Id: 6DLW; EMD-7773). One of the protomers is highlighted, with its transmembrane β-strands shown in red. (C) Cryo-EM structure of the MAC pore assembly (PDB Id: 6H04; EMD-0107) constituted with the complement components C9 (gray), C5 (green), C8α (cyan), C8β (purple), C7 (red), C6 (orange), and C8γ (blue).

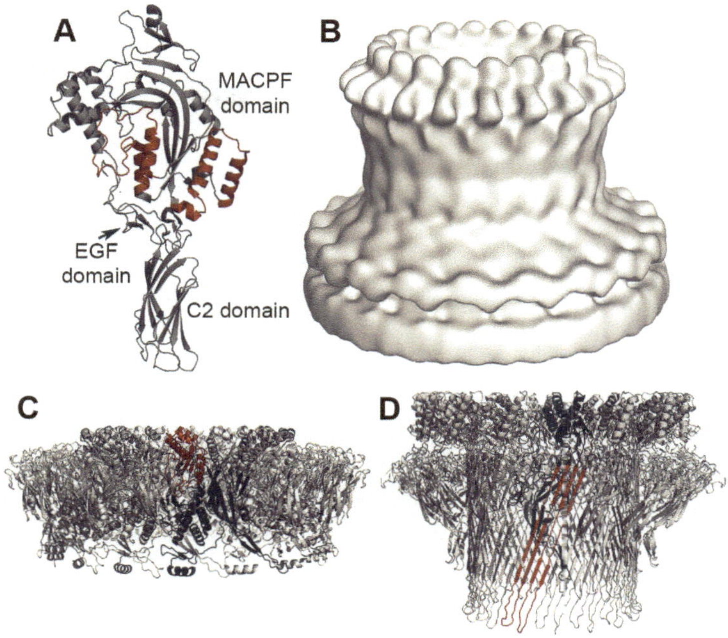

Fig. 9 Perforin structure. (A) Crystal structure of perforin in its monomer state (PDB Id: 3NSJ). Helical regions that would transform into the transmembrane pore-forming motifs are shown in red. (B) Cryo-EM structure of the perforin pore (EMD-1769). (C) Cryo-EM structure of the pre-pore assembly of MPEG1/perforin 2 (PDB Id: 6SB3; EMD-10134). Helical regions of one of the protomers that would transform into the transmembrane pore-forming motifs are shown in red. (D) Cryo-EM structure of the pore assembly of MPEG1/perforin 2 (PDB Id: 6SB5; EMD-10135). Transmembrane β-strands of one of the protomers are shown in red.

MAC (Membrane attack complex) of the complement system can directly kill Gram-negative bacteria, by generating pores on the cell membranes (Rosado et al., 2008; Serna, Giles, Morgan, & Bubeck, 2016). Defects in MAC components lead to the recurring infections by Gram-negative bacteria (Ram, Lewis, & Rice, 2010). However, MAC components are also shown to assemble on the parasites and Gram-positive bacteria (Berends et al., 2013; Hoover, Berger, Nacy, Hockmeyer, & Meltzer, 1984). Perforins on the other hand, are secreted by the cytotoxic T lymphocytes and Natural Killer cells (NK cells) (Podack & Konigsberg, 1984). Perforins generate pores on the virus-infected cells and cancerous host cells (Bolitho, Voskoboinik, Trapani, & Smyth, 2007). Owing to its large pore size, perforins assist in delivering the pro-apoptotic proteases, granzymes, into the target host cells leading to its programmed cell-death (Bird et al., 2005; Metkar et al., 2015; Voskoboinik, Smyth, & Trapani, 2006).

Molecular assembly of MAC requires involvement of six proteins. Out of which C6, C7, C8α, C8β and C9 possess MACPF/CDC domain (Fig. 8) (Menny et al., 2018; Reboul et al., 2016; Rosado et al., 2008). The protein C5 undergoes proteolytic cleavage to generate C5b. This initiates a complex but orchestrated assembly of multiple proteins of MAC pore architecture. C5b binds to C6 to generate C5b6 which recruits C7 to generate C5b7 (Marshall, Hasegawa, Davidson, Nussenzweig, & Whitlow, 1996; Preissner, Podack, & Muller-Eberhard, 1985). The C5b7 is suggested to drive the membrane-interaction mechanism of MAC (Marshall et al., 1996). Then, further recruitment of C8 leads to formation of C5b8. TMH regions of C8α then penetrate the lipid bilayer for final membrane anchoring (Bubeck et al., 2011). C5b8 then recruits 12-to-18 C9 protomers into the complex to generate the final pore (Reboul et al., 2016; Serna et al., 2016; Spicer et al., 2018). It is interesting to note that C8α exhibits the membrane penetration before the complete ring formation occurs (Bubeck et al., 2011; Hadders et al., 2012). It has been shown that C8α and perforin share a significant sequence similarity in their pore-forming region, but the molecular details of perforin pore assembly still remains obscure (Mondal et al., 2020). The assembly of perforin requires calcium in the membrane anchoring step (Baran et al., 2009; Voskoboinik et al., 2005). Similarity with C8α and presence of the MACPF/CDC domain suggest formation of the giant-sized pore, which has been confirmed by TEM and cryo-EM based analyses (Fig. 9) (Law et al., 2010). It is not clear, however, whether the pre-pore intermediates are generated in such case (Praper et al., 2011). The perforin pore has been found to display selectivity for positively charged proteins, as evidenced by the well-established delivery of positively charged granzymes by the perforin pore (Bird et al., 2005; Stewart et al., 2014).

Negative staining electron micrographs of C9 polymerization have revealed overall shape and structure of the MAC several decades ago (Tschopp, 1984). However, because of the low resolution of these structural analyses, molecular details of the MAC assembly could not be elucidated. Whatsoever, in the recent years, with the advancement of cryo-EM, the molecular details of the MAC assembly and architecture have been determined. Single-particle cryo-EM structure of the poly C9 pore reveal the 22-mer C9 complex, with 88 β-strands penetrating the membrane (Fig. 8A and B) (Dudkina et al., 2016). It has been shown that C5b7 and

C5b8 display membrane binding and oligomerization that finally generate ~120 Å pore with poly-C9 molecules (Dudkina et al., 2016; Parsons et al., 2019). Another recent report utilized cryo-EM along with flicker spectroscopy to determine the pore-formation mechanism by MAC assembly (Menny et al., 2018). C6 and C7 perform the membrane-interaction role whereas C8 and C9 traverse the membrane bilayer (Fig. 8C) (Reboul et al., 2016). Interestingly, it is observed that the C5b6 acts as a platform for the β-barrel pore, and the pore architecture is supported by a glycan scaffold (Menny et al., 2018).

Perforin mediated cytotoxicity in the virus-infected cells and tumors cells are critical in the context of the host defense mechanisms. However, pore-formation mechanism of perforin is still not fully understood in molecular details. AFM on planar bilayers and deep-etch EM on mammalian cells have shown that perforin forms arc-like structures on the target membranes (Metkar et al., 2015). X-ray crystal structure of murine perforin along with the cryo-EM reconstruction of the perforin pore have revealed some of the details of the perforin pore-forming mechanism (Fig. 9A and B) (Law et al., 2010). The monomer structure of perforin appears like a "key" with an N-terminal MACPF/CDC domain, followed by epidermal growth factor (EGF) domain. The C2 domain at the C-terminal mediates calcium binding, and is responsible for the membrane-interaction (Fig. 9A). Single-particle cryo-EM structure of human perforin pore (Fig. 9B) suggest that the MACPF domains fit in the opposite orientation as compared to the CDC pore structures. Similar to the MAC complex, heights of the monomer and the pore structures of perforin are similar, suggesting that there is no vertical collapse of the complex during pore-formation. This is also supported by the length of the TMHs of perforin that are similar to those of C8α, and twice as long as those in the CDCs (Law et al., 2010). It is also noticed that the perforin pore is lined by oligosaccharide moieties; however, exact requirement of such modifications for the perforin function is not clear (Law et al., 2010).

Macrophage-expressed gene 1 (MPEG1) or perforin 2 is another crucial and prominent member in the family of MACPF (Fig. 9C) (Bayly-Jones, Pang, Spicer, Whisstock, & Dunstone, 2020). It protects the cells against the intracellular pathogens and helps in the destruction of the engulfed pathogens (McCormack et al., 2015). MPEG1/perforin 2 has been shown to be associated with the intracellular vesicles, and it exerts its function in the phagosomes encapsulating bacterial pathogens. Recent structural studies have provided critical insights regarding the pore-formation mechanism of MPEG1 (Fig. 9C and D) (Ni et al., 2020; Pang et al., 2019). Based on the structural models, it is proposed that MPEG1 forms hexadecameric pre-pore oligomeric assembly, under the condition of endosomal pH (Fig. 9C). In this pre-pore state, transmembrane hairpins of the MACPF domain are yet to open; rather, they remain positioned opposite to the membrane-binding domain. Upon further acidification in the phagosomal pH condition, pre-pore is transformed into pore on the bacterial membranes that involves nearly 180° transition of the MACPF domain and the transmembrane hairpins which eventually insert into the target membranes (Fig. 9D) (Ni et al., 2020; Pang et al., 2019).

5. Gasdermins

Another distinct family of mammalian PFPs is the gasdermins (GSDMs), one of which (gasdermin D; GSDMD) acts as the key executioner molecule for pyroptosis, an inflammatory mode of cell-death associated with the loss of membrane integrity and release of inflammatory mediators (Broz, Pelegrin, & Shao, 2020). There are six genes coding for the GSDMs in humans: GSDMA-E and DFNB59 (Liu & Lieberman, 2020). Initially, GSDMs have been observed to be expressed in skin (GSDMA), lungs (GSDMB) and gut (GSDMC) (Tamura et al., 2007). However, GSDMD is found to be highly expressed in antigen-presenting cells such as dendritic cells and macrophages, where it regulates inflammatory immune responses (Evavold et al., 2018).

GSDMs share significantly conserved architecture of N-terminal and C-terminal domains that are connected to each other via a flexible linker (Fig. 10). The linker contains the proteolytic cleavage sites, where cleavage leads to the activation of the N-terminal domain (NTD) that binds to the membranes and generate pores (Broz et al., 2020; Kovacs & Miao, 2017; Liu & Lieberman, 2020). C-terminal domain (CTD) acts as the autoinhibitory module for the action of the GSDM NTD (Ding et al., 2016; Kuang et al., 2017; Liu et al., 2018, 2019). Proteolytic cleavage in the linker region induces a conformational transition within NTD of GSDMs that exposes the critical residues for membrane binding, leading to membrane interaction and oligomerization. Structural studies have revealed that the relative position of the CTD with respect to NTD can be significantly different in different GSDMs. Therefore, it is possible that GSDMs may have different inter-domain interactions for the mechanism of autoinhibition imposed by the presence of their corresponding CTDs (Liu et al., 2018).

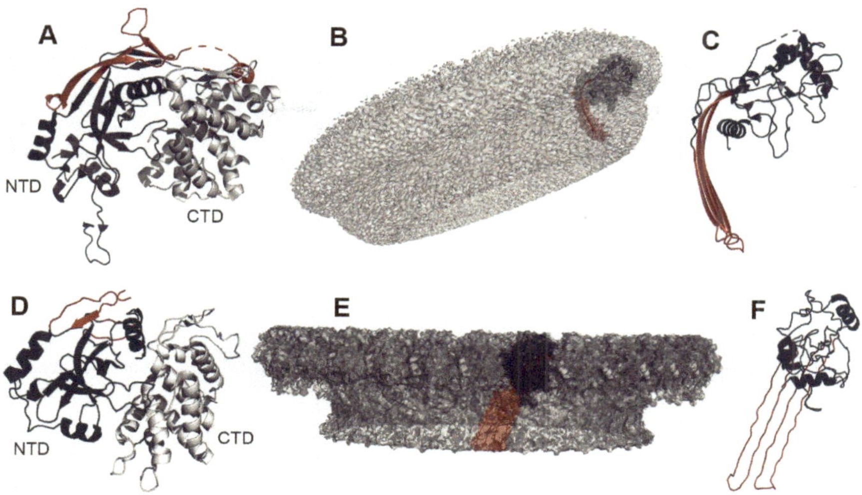

Fig. 10 GSDM structure. (A) Crystal structure of full length GSDMA3 monomer (PDB Id: 5B5R). (B) Cryo-EM structure of the pore formed by GSDMA3 NTD (PDB Id: 6CB8; EMD-7449). One of the protomers in the pore assembly is highlighted. (C) GSDMA3 NTD protomer in the transmembrane oligomeric pore state. (D) Crystal structure of full length GSDMD monomer (PDB Id: 6N9O). (E) Cryo-EM structure of the pore formed by GSDMD NTD (PDB Id: 6VFE; EMD-21160). One of the protomers in the pore assembly is highlighted. (F) GSDMD NTD protomer in the transmembrane oligomeric pore state. Transmembrane β-strands in the protomers, and corresponding regions in the monomeric states of the GSDMs are shown in red.

GSDMD is one of the most important family members, and it is critically implicated in the process of inflammatory cell-death or pyroptosis (Evavold et al., 2018; He et al., 2015). Cytosolic sensing of pathogenic determinants leads to the activation of the multiprotein complex, inflammasomes, which in turn activate inflammatory caspases like caspase-1 and caspase-11 (Kayagaki et al., 2015; Lamkanfi & Dixit, 2014). These caspases cleave the linker region of GSDMD, liberating NTD to generate pores on the cell membranes form the cytosolic side (Lieberman, Wu, & Kagan, 2019; Liu & Lieberman, 2020; Liu et al., 2016). This will allow the inflammatory cytokines to release and activate the immune cells.

GSDM family members do not share any sequence similarity with the CDCs or MACPF family. Nevertheless, GSDMs form giant β-barrel pores that resemble those formed by the CDCs and MACPFs (Fig. 10). Cryo-EM structure has been determined for the pore formed by the NTD of GDSMA3 that reveals stoichiometric assembly of 27 protomers with 108 antiparallel β-strands (Fig. 10B) (Ruan, Xia, Liu, Lieberman, & Wu, 2018). The inner and outer pore diameters are ~180 Å and ~280 Å, respectively, with a height of 70 Å. The study also identifies a positively charged helix that interacts with the acidic membrane lipid, cardiolipin, which is present in the membranes of most of the bacteria. This cryo-EM structure also highlights the structural changes in GSDMA3 NTD upon membrane interaction that form the extended β-barrel pores.

Very recently, the cryo-EM structures of the GSDMD pore assemblies have been elucidated (Fig. 10E) (Xia et al., 2021). In contrast to the GSDMA3 pore, the GSDMD pore assembly is found to be larger in size (Ruan et al., 2018; Xia et al., 2021). GSDMD appears to form 33-mer pore with inner and outer diameter of 215 Å and 310 Å. This suggests variability in the pore architectures among the GSDM family, even though they share overall similar monomer structures. GSDMD pre-pore appears to be 40 Å shorter in height compared to the pore assembly. The protomers in the pre-pore state appear to mimic the structure of the GSDMD NTD in its autoinhibited form. The pore structure of GSDMD also reveals the membrane-interacting motifs. However, the most fascinating feature of the GSDMD pore structure is the predominantly acidic nature of the pore conduit. This presumably acts as an electrostatic filter to prevent the release of the precursor cytokines, while allowing the secretion of mature IL-1β and IL-18 (Xia et al., 2021).

6. PFPs from multicellular eukaryotes

We have discussed so far that all the kingdoms of life secrete proteins/toxins that can generate pores in their target membranes. In this section, we are going to highlight some of the PFPs that are secreted by a wide range of unrelated higher-order eukaryotes. These proteins are either newly discovered or less frequently discussed in the previous literatures.

Lysenin is a PFP secreted by the earthworm *Eisenia fetida*. It belongs to the aerolysin family of PFPs, and has been shown to display hemolytic and cytotoxic activities (Yamaji et al., 1998). Lysenin specifically interacts with sphingomyelin (SM) (Kulma et al., 2010; Yamaji et al., 1998; Yamaji-Hasegawa et al., 2003; Yilmaz et al., 2013; Yilmaz, Yamaji-Hasegawa, Hullin-Matsuda, & Kobayashi, 2018). TEM and AFM-based imaging have shown that the lysenin oligomers form honeycomb-like structures on the SM-cholesterol liposomes (Yamaji-Hasegawa et al., 2003; Yilmaz & Kobayashi, 2015). X-ray crystallography and single-particle cryo-EM structures of lysenin have revealed critical insights about its pore assembly

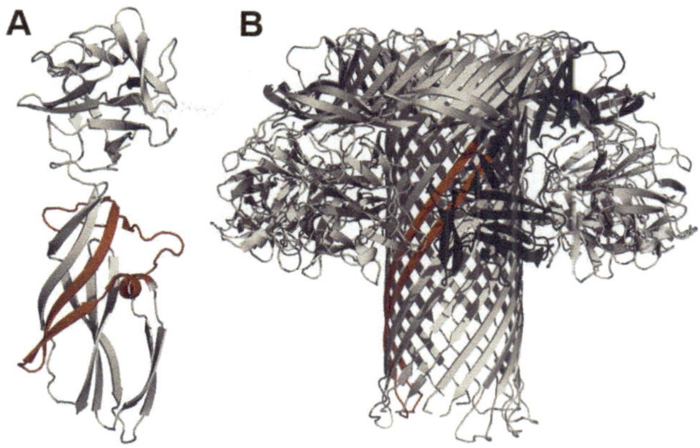

Fig. 11 Lysenin shows structural similarity to aerolysins. (A) Crystal structure of the monomeric form of lysenin (PDB Id: 3ZXG). (B) Crystal structure of 9-mer pore assembly of lysenin (PDB id: 5EC5). Transmembrane β-strands in one of the protomers of the pore, (in A) and corresponding regions in the monomeric state (in B) are shown in red.

(Bokori-Brown et al., 2016; De Colibus et al., 2012). The lysenin pore consists of nine protomers and it generates a mushroom-shaped structure (Bokori-Brown et al., 2016). The C-terminal domain interacts with PC headgroups, and one β-hairpin from each protomer participates to generates the nonameric β-barrel pore (Bokori-Brown et al., 2016; Podobnik et al., 2016) (Fig. 11). Interestingly, a few arc-shaped oligomers have also been observed for lysenin that possibly represent the pre-pore state (Podobnik et al., 2016; Yilmaz & Kobayashi, 2015; Yilmaz et al., 2013). It has been observed that these pre-pore arc-shaped oligomers are proteolipidic in nature, and can exhibit membrane-damaging activity suggesting that pre-pore state can convert into pores once they get a cue for the vertical collapse in the membranes (Yilmaz et al., 2018).

One new β-PFP has been reported from the freshwater snail *Biomphalaria glabrata* that is biomphalysin, which do not share much sequence similarity with aerolysins, but displays structural similarity with the aerolysin family members (Galinier et al., 2013). Natterin is another aerolysin-like PFP present in the venom of the fish *Thalassophryne nattereri* (Magalhaes et al., 2005). Natterin-like proteins have been identified in others fishes, such as *Gadus morhua*, *Lampetra japonica* and *Lampetra morii*, *Salvelinus alpinus*, *Plotosus lineatus* and *Danio rerio*, and other aquatic organisms such as *Crassostrea gigas* and *Acropora digitifera* (Gacesa et al., 2015; Gudbrandsson et al., 2015; Jia et al., 2016; Lima et al., 2021; Rajan, Patel, Kitani, Viswanath, &

Brinchmann, 2017; Tamura, Yamakawa, & Shiomi, 2011; Unno et al., 2016; Wu et al., 2017; Xue et al., 2012). Although natterins are not well-characterized structurally, it is known that they possess C-terminal domain, similar to the conserved core domain present in the aerolysins. This aerolysin-like domain has been implicated in the toxicity and membrane penetration. EM micrographs of nattering-like Dln1 protein from zebrafish *Danio rerio* shows octameric ring-like pore structure (Jia et al., 2016).

Another recently characterized toxin is the stonefish toxin, stonustoxin (SNTX), which is a heterodimeric cytolytic protein. SNTX is comprised of two structurally related α and β subunits (Chen, Kini, Yuen, & Khoo, 1997; Ghadessy et al., 1996). Each subunit has an N-terminal CDC/MACPF pore-forming domain, a central focal-adhesion-targeting domain, a thioredoxin domain, and a C-terminal tripartite motif family-like PRYSPRY domain (PRY SPla and the RYanodine Receptor immune recognition domain) (Fig. 12). It is believed that the PRYSPRY domain mediates the interaction

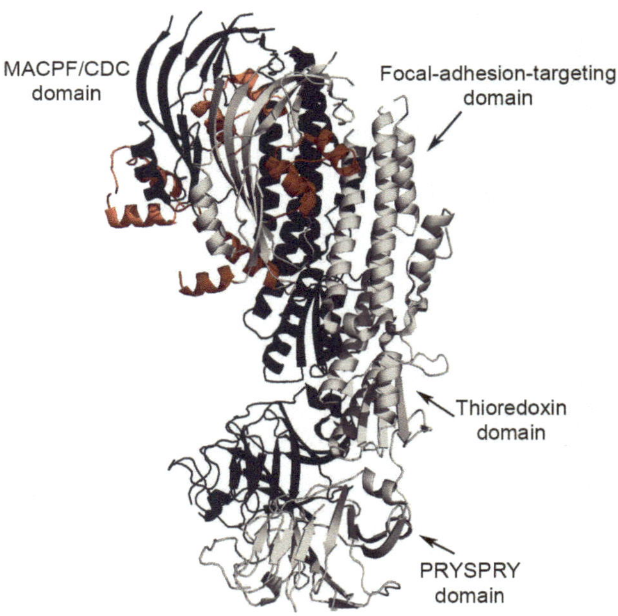

Fig. 12 Crystal structure of the α/β heterodimeric assembly of stonustoxin (SNTX) (PDB Id: 4WVM). Two different subunits are shown in light and dark gray. Their α-helical regions that would transform into the corresponding transmembrane hairpins in the oligomeric pore state are shown in light and dark red color.

of SNTX with cell-surface molecules. Phylogenetic analysis suggests that SNTX represents an ancient branch of CDC/MACPF family (Ellisdon et al., 2015). Crystal structure of the SNTX dimer suggests a model of pre-pore assembly consisting of 20 SNTX subunits (10 α/β heterodimers), with a pore lumen of ~112 Å (Ellisdon et al., 2015). SNTX-like CDC/MACPF family proteins are distributed in eukaryotic life-forms, although they are known to play crucial roles in immune-related functions rather than as venom constituent.

Enterolobin is a plant-based cytolytic protein. Although it was discovered a few decades ago, very few studies have tried to uncover the pore-forming mechanism of this unique plant-based PFP. Enterolobin is secreted by the tropical tree, *Enterolobium contortisiliquum*, and is toxic to the larvae of bruchid *Callosobruchus maculatus* (de Sousa & Morhy, 1989). Interestingly, amino-acid sequence of enterolobin is found to be similar to aerolyins of *Aeromonas* species (Sousa, Richardson, Fontes, & Morhy, 1994). Biochemical analysis has revealed that it exists as dimers in solution (Bittencourt et al., 2003). At pH below 7.0, entrolobin oligomerizes as hexamer, as suggested by the light-scattering studies. High ionic strength also causes dimerization of the toxin (Bittencourt et al., 2003). A few more PFPs have been reported in the plants. However, their structural details and mode of action mostly remain unclear. Another protein RCD-1 from fungi *Neurospora crassa* has been characterized recently. RCD-1 has been implicated in the programmed cell death of the asexual spores of the fungi (Daskalov, Gladieux, Heller, & Glass, 2019; Daskalov, Mitchell, Sandstrom, Vance, & Glass, 2020). Structural analysis has revealed RCD-1 to be a structural homolog of the N-terminal pore-forming domain of mammalian gasdermins. Even overexpression of RCD-1 in the human 293T cells leads to pyroptosis-like cell death (Daskalov et al., 2020). Although this protein is found to exist in the cell periphery, and interact with the negatively charged membrane phospholipids, the mode of action and pore-forming mechanism remains unexplored at present.

7. Conclusion

PFPs are the distinct class of membrane-damaging proteins that are equipped with the unique ability to form multimeric pores in the membrane lipid bilayer. These proteins are found in all kingdoms of life. PFPs act as the potent toxins, and serve as the virulence factors in many pathogenic bacteria. Bacterial PFPs generally kill host cells, trigger pathophysiological responses, disrupt epithelial barriers and thus contribute to the overall bacterial pathogenesis process. Several PFPs have also been implicated in eukaryotic organisms as a part of their venom to kill or catch their prey. PFPs play crucial roles in the vertebrate immune system and defense mechanism. MACPF family of mammalian PFPs generate pores on the pathogens, infected and malignant host cells, thereby killing them. There exist remarkable similarities in the structure-function mechanisms of the PFPs from all the forms of life. One such similarity is the overall resemblance in the mechanism of membrane pore-formation. All PFPs are generally secreted as the water-soluble molecules, and they undergo structural metamorphosis into the transmembrane oligomeric pores. However, prominent differences can be observed in each of the PFPs, in terms of their molecular architecture and mechanism of structural rearrangements. PFPs are being extensively studied for the past few decades, but recent advances in structural biology have enabled determination of the molecular architecture of large pore complexes of several PFPs. Combining the tools of structural biology with biochemical, biophysical, and computational approaches have led to more critical understanding of the pore-formation mechanism of the PFPs. However, intricate details of each of the steps of membrane pore-formation in many PFPs still remain obscure.

One of the most enigmatic aspects of the PFP mode of action is the driving force that triggers oligomerization of the PFP units in the membrane platform. It has been suggested that clustering of the monomeric proteins on the membrane surface increases local concentration that possibly triggers oligomerization. However, it remains an open question whether membrane interaction could provide any structural cues that induce oligomerization. Also, it is not fully understood how the membrane-interaction leads to the overall structural reorganization in the PFP architecture that generate the transmembrane pore. Many PFPs such as actinoporins have a very small membrane-penetrating region, and they are suggested to generate toroidal pores, where along with the protein motifs, lipids also participate to form the pore-lining. Nevertheless, it is not clear how the shape and size of the phospholipids regulate the toroidal pore-formation. Currently, vast efforts are being made to understand the structure-function mechanisms associated with the membrane-damaging action of the PFPs. Also, another aspect of the PFP mode of action is being explored in the recent years that attempt to connect the membrane-damaging action of these proteins with their ability to evoke more programmed cellular responses. Such responses can have much broader impact to shape up the outcome of diverse processes such as bacterial virulence mechanisms, host-pathogen interactions, and inflammatory immune response generation. Therefore, comprehensive knowledge of the structures and functions of the PFPs is not only important to have better understanding of the structural paradigm in this unique class of dimorphic proteins, but it is also essential to apprehend their roles in the diverse physiological and pathophysiological processes.

Acknowledgments

Protein structural coordinates and EM maps were retrieved from the protein data bank (PDB; available at rcsb.org) and EMDataResource (EMD; available at emdataresource.org). Protein structural models were generated with the PyMOL Molecular Graphics System, version 2.5.0, Schrodinger, LLC.

We acknowledge Indian Institute of Science Education and Research Mohali for support.

References

Agrawal, A., Apoorva, K., & Ayappa, K. G. (2017). Transmembrane oligomeric intermediates of pore forming toxin Cytolysin A determine leakage kinetics. *RSC Advances*, 7(82), 51750–51762. https://doi.org/10.1039/C7RA07304F.

Alonzo, F., 3rd, & Torres, V. J. (2014). The bicomponent pore-forming leucocidins of Staphylococcus aureus. *Microbiology and Molecular Biology Reviews*, *78*(2), 199–230. https://doi.org/10.1128/MMBR.00055-13.

Ameirika, Sha, H. X., & Hwang, J. S. (2017). Identification of a target protein of Hydra actinoporin-like toxin-1 (HALT-1) using GST affinity purification and SILAC-based quantitative proteomics. *Toxicon*, *133*, 153–161. https://doi.org/10.1016/j.toxicon.2017.05.007.

Anderluh, G., Kisovec, M., Krasevec, N., & Gilbert, R. J. (2014). Distribution of MACPF/CDC proteins. *Sub-Cellular Biochemistry*, *80*, 7–30. https://doi.org/10.1007/978-94-017-8881-6_2.

Antonini, V., Perez-Barzaga, V., Bampi, S., Penton, D., Martinez, D., Dalla Serra, M., et al. (2014). Functional characterization of sticholysin I and W111C mutant reveals the sequence of the actinoporin's pore assembly. *PLoS One*, *9*(10), e110824. https://doi.org/10.1371/journal.pone.0110824.

Athanasiadis, A., Anderluh, G., Macek, P., & Turk, D. (2001). Crystal structure of the soluble form of equinatoxin II, a pore-forming toxin from the sea anemone Actinia equina. *Structure*, *9*(4), 341–346. https://doi.org/10.1016/s0969-2126(01)00592-5.

Baker, M. A., Rojko, N., Cronin, B., Anderluh, G., & Wallace, M. I. (2014). Photobleaching reveals heterogeneous stoichiometry for equinatoxin II oligomers. *Chembiochem*, *15*(14), 2139–2145. https://doi.org/10.1002/cbic.201300799.

Bakht, O., Pathak, P., & London, E. (2007). Effect of the structure of lipids favoring disordered domain formation on the stability of cholesterol-containing ordered domains (lipid rafts): Identification of multiple raft-stabilization mechanisms. *Biophysical Journal*, *93*(12), 4307–4318. https://doi.org/10.1529/biophysj.107.114967.

Ballard, J., Sokolov, Y., Yuan, W. L., Kagan, B. L., & Tweten, R. K. (1993). Activation and mechanism of Clostridium septicum alpha toxin. *Molecular Microbiology*, *10*(3), 627–634. https://doi.org/10.1111/j.1365-2958.1993.tb00934.x.

Baran, K., Dunstone, M., Chia, J., Ciccone, A., Browne, K. A., Clarke, C. J., et al. (2009). The molecular basis for perforin oligomerization and transmembrane pore assembly. *Immunity*, *30*(5), 684–695. https://doi.org/10.1016/j.immuni.2009.03.016.

Barlic, A., Gutierrez-Aguirre, I., Caaveiro, J. M., Cruz, A., Ruiz-Arguello, M. B., Perez-Gil, J., et al. (2004). Lipid phase coexistence favors membrane insertion of equinatoxin-II, a pore-forming toxin from Actinia equina. *The Journal of Biological Chemistry*, *279*(33), 34209–34216. https://doi.org/10.1074/jbc.M313817200.

Bayley, H., Jayasinghe, L., & Wallace, M. (2005). Prepore for a breakthrough. *Nature Structural & Molecular Biology*, *12*(5), 385–386. https://doi.org/10.1038/nsmb0505-385.

Bayly-Jones, C., Pang, S. S., Spicer, B. A., Whisstock, J. C., & Dunstone, M. A. (2020). Ancient but not forgotten: New insights into MPEG1, a macrophage perforin-like immune effector. *Frontiers in Immunology*, *11*, 581906. https://doi.org/10.3389/fimmu.2020.581906.

Benke, S., Roderer, D., Wunderlich, B., Nettels, D., Glockshuber, R., & Schuler, B. (2015). The assembly dynamics of the cytolytic pore toxin ClyA. *Nature Communications*, *6*, 6198. https://doi.org/10.1038/ncomms7198.

Berends, E. T., Dekkers, J. F., Nijland, R., Kuipers, A., Soppe, J. A., van Strijp, J. A., et al. (2013). Distinct localization of the complement C5b-9 complex on Gram-positive bacteria. *Cellular Microbiology*, *15*(12), 1955–1968. https://doi.org/10.1111/cmi.12170.

Bird, C. H., Sun, J., Ung, K., Karambalis, D., Whisstock, J. C., Trapani, J. A., et al. (2005). Cationic sites on granzyme B contribute to cytotoxicity by promoting its uptake into target cells. *Molecular and Cellular Biology*, *25*(17), 7854–7867. https://doi.org/10.1128/MCB.25.17.7854-7867.2005.

Bittencourt, S. E., Silva, L. P., Azevedo, R. B., Cunha, R. B., Lima, C. M., Ricart, C. A., et al. (2003). The plant cytolytic protein enterolobin assumes a dimeric structure in solution. *FEBS Letters*, *549*(1–3), 47–51. https://doi.org/10.1016/s0014-5793(03)00763-4.

Bokori-Brown, M., Martin, T. G., Naylor, C. E., Basak, A. K., Titball, R. W., & Savva, C. G. (2016). Cryo-EM structure of lysenin pore elucidates membrane insertion by an aerolysin family protein. *Nature Communications*, 7, 11293. https://doi.org/10.1038/ncomms11293.

Bolitho, P., Voskoboinik, I., Trapani, J. A., & Smyth, M. J. (2007). Apoptosis induced by the lymphocyte effector molecule perforin. *Current Opinion in Immunology*, *19*(3), 339–347. https://doi.org/10.1016/j.coi.2007.04.007.

Bonev, B. B., Gilbert, R. J., Andrew, P. W., Byron, O., & Watts, A. (2001). Structural analysis of the protein/lipid complexes associated with pore formation by the bacterial toxin pneumolysin. *The Journal of Biological Chemistry*, *276*(8), 5714–5719. https://doi.org/10.1074/jbc.M005126200.

Broz, P., Pelegrin, P., & Shao, F. (2020). The gasdermins, a protein family executing cell death and inflammation. *Nature Reviews. Immunology*, *20*(3), 143–157. https://doi.org/10.1038/s41577-019-0228-2.

Bryant, A. E., Bayer, C. R., Chen, R. Y., Guth, P. H., Wallace, R. J., & Stevens, D. L. (2005). Vascular dysfunction and ischemic destruction of tissue in Streptococcus pyogenes infection: The role of streptolysin O-induced platelet/neutrophil complexes. *The Journal of Infectious Diseases*, *192*(6), 1014–1022. https://doi.org/10.1086/432729.

Bubeck, D., Roversi, P., Donev, R., Morgan, B. P., Llorca, O., & Lea, S. M. (2011). Structure of human complement C8, a precursor to membrane attack. *Journal of Molecular Biology*, *405*(2), 325–330. https://doi.org/10.1016/j.jmb.2010.10.031.

Cascales, E., Buchanan, S. K., Duche, D., Kleanthous, C., Lloubes, R., Postle, K., et al. (2007). Colicin biology. *Microbiology and Molecular Biology Reviews*, *71*(1), 158–229. https://doi.org/10.1128/MMBR.00036-06.

Chen, D., Kini, R. M., Yuen, R., & Khoo, H. E. (1997). Haemolytic activity of stonustoxin from stonefish (Synanceja horrida) venom: Pore formation and the role of cationic amino acid residues. *The Biochemical Journal*, *325*(Pt. 3), 685–691. https://doi.org/10.1042/bj3250685.

Choe, S., Bennett, M. J., Fujii, G., Curmi, P. M., Kantardjieff, K. A., Collier, R. J., et al. (1992). The crystal structure of diphtheria toxin. *Nature*, *357*(6375), 216–222. https://doi.org/10.1038/357216a0.
Christie, M. P., Johnstone, B. A., Tweten, R. K., Parker, M. W., & Morton, C. J. (2018). Cholesterol-dependent cytolysins: From water-soluble state to membrane pore. *Biophysical Reviews*, *10*(5), 1337–1348. https://doi.org/10.1007/s12551-018-0448-x.
Clemens, J. D., Nair, G. B., Ahmed, T., Qadri, F., & Holmgren, J. (2017). Cholera. *Lancet*, *390*(10101), 1539–1549. https://doi.org/10.1016/S0140-6736(17)30559-7.
Czajkowsky, D. M., Hotze, E. M., Shao, Z., & Tweten, R. K. (2004). Vertical collapse of a cytolysin prepore moves its transmembrane beta-hairpins to the membrane. *The EMBO Journal*, *23*(16), 3206–3215. https://doi.org/10.1038/sj.emboj.7600350.
Dal Peraro, M., & van der Goot, F. G. (2016). Pore-forming toxins: Ancient, but never really out of fashion. *Nature Reviews. Microbiology*, *14*(2), 77–92. https://doi.org/10.1038/nrmicro.2015.3.
Daskalov, A., Gladieux, P., Heller, J., & Glass, N. L. (2019). Programmed cell death in neurospora crassa is controlled by the allorecognition determinant rcd-1. *Genetics*, *213*(4), 1387–1400. https://doi.org/10.1534/genetics.119.302617.
Daskalov, A., Mitchell, P. S., Sandstrom, A., Vance, R. E., & Glass, N. L. (2020). Molecular characterization of a fungal gasdermin-like protein. *Proceedings of the National Academy of Sciences of the United States of America*, *117*(31), 18600–18607. https://doi.org/10.1073/pnas.2004876117.
De Colibus, L., Sonnen, A. F., Morris, K. J., Siebert, C. A., Abrusci, P., Plitzko, J., et al. (2012). Structures of lysenin reveal a shared evolutionary origin for pore-forming proteins and its mode of sphingomyelin recognition. *Structure*, *20*(9), 1498–1507. https://doi.org/10.1016/j.str.2012.06.011.
De, S., & Olson, R. (2011). Crystal structure of the Vibrio cholerae cytolysin heptamer reveals common features among disparate pore-forming toxins. *Proceedings of the National Academy of Sciences of the United States of America*, *108*(18), 7385–7390. https://doi.org/10.1073/pnas.1017442108.
de Sousa, M. V., & Morhy, L. (1989). Enterolobin, a hemolytic protein from Enterolobium contortisiliquum seeds (Leguminosae- -Mimosoideae). Purification and characterization. *Anais da Academia Brasileira de Ciências*, *61*(4), 405–412. http://www.ncbi.nlm.nih.gov/pubmed/2486896.
Degiacomi, M. T., Iacovache, I., Pernot, L., Chami, M., Kudryashev, M., Stahlberg, H., et al. (2013). Molecular assembly of the aerolysin pore reveals a swirling membrane-insertion mechanism. *Nature Chemical Biology*, *9*(10), 623–629. https://doi.org/10.1038/nchembio.1312.
Ding, J., Wang, K., Liu, W., She, Y., Sun, Q., Shi, J., et al. (2016). Pore-forming activity and structural autoinhibition of the gasdermin family. *Nature*, *535*(7610), 111–116. https://doi.org/10.1038/nature18590.
Dolejsova, T., Sokol, A., Bosak, J., Smajs, D., Konopasek, I., Mikusova, G., et al. (2019). Colicin U from Shigella boydii forms voltage-dependent pores. *Journal of Bacteriology*, *201*(24), e00493-19. https://doi.org/10.1128/JB.00493-19.
Dramsi, S., & Cossart, P. (2002). Listeriolysin O: A genuine cytolysin optimized for an intracellular parasite. *Journal of Cell Biology*, *156*(6), 943–946. https://doi.org/10.1083/jcb.200202121.
Dudkina, N. V., Spicer, B. A., Reboul, C. F., Conroy, P. J., Lukoyanova, N., Elmlund, H., et al. (2016). Structure of the poly-C9 component of the complement membrane attack complex. *Nature Communications*, 7, 10588. https://doi.org/10.1038/ncomms10588.
DuMont, A. L., Yoong, P., Liu, X., Day, C. J., Chumbler, N. M., James, D. B., et al. (2014). Identification of a crucial residue required for Staphylococcus aureus LukAB cytotoxicity and receptor recognition. *Infection and Immunity*, *82*(3), 1268–1276. https://doi.org/10.1128/IAI.01444-13.

Dunkel, S., Pulagam, L. P., Steinhoff, H. J., & Klare, J. P. (2015). In vivo EPR on spin labeled colicin A reveals an oligomeric assembly of the pore-forming domain in E. coli membranes. *Physical Chemistry Chemical Physics*, *17*(7), 4875–4878. https://doi.org/10.1039/c4cp05638h.

Ellisdon, A. M., Reboul, C. F., Panjikar, S., Huynh, K., Oellig, C. A., Winter, K. L., et al. (2015). Stonefish toxin defines an ancient branch of the perforin-like superfamily. *Proceedings of the National Academy of Sciences of the United States of America*, *112*(50), 15360–15365. https://doi.org/10.1073/pnas.1507622112.

Evavold, C. L., Ruan, J., Tan, Y., Xia, S., Wu, H., & Kagan, J. C. (2018). The pore-forming protein gasdermin D regulates interleukin-1 secretion from living macrophages. *Immunity*, *48*(1), 35–44 e36. https://doi.org/10.1016/j.immuni.2017.11.013.

Fahie, M., Romano, F. B., Chisholm, C., Heuck, A. P., Zbinden, M., & Chen, M. (2013). A non-classical assembly pathway of Escherichia coli pore-forming toxin cytolysin A. *The Journal of Biological Chemistry*, *288*(43), 31042–31051. https://doi.org/10.1074/jbc.M113.475350.

Farrand, S., Hotze, E., Friese, P., Hollingshead, S. K., Smith, D. F., Cummings, R. D., et al. (2008). Characterization of a streptococcal cholesterol-dependent cytolysin with a lewis y and b specific lectin domain. *Biochemistry*, *47*(27), 7097–7107. https://doi.org/10.1021/bi8005835.

Farrand, A. J., LaChapelle, S., Hotze, E. M., Johnson, A. E., & Tweten, R. K. (2010). Only two amino acids are essential for cytolytic toxin recognition of cholesterol at the membrane surface. *Proceedings of the National Academy of Sciences of the United States of America*, *107*(9), 4341–4346. https://doi.org/10.1073/pnas.0911581107.

Feil, S. C., Ascher, D. B., Kuiper, M. J., Tweten, R. K., & Parker, M. W. (2014). Structural studies of Streptococcus pyogenes streptolysin O provide insights into the early steps of membrane penetration. *Journal of Molecular Biology*, *426*(4), 785–792. https://doi.org/10.1016/j.jmb.2013.11.020.

Feil, S. C., Lawrence, S., Mulhern, T. D., Holien, J. K., Hotze, E. M., Farrand, S., et al. (2012). Structure of the lectin regulatory domain of the cholesterol-dependent cytolysin lectinolysin reveals the basis for its lewis antigen specificity. *Structure*, *20*(2), 248–258. https://doi.org/10.1016/j.str.2011.11.017.

Gacesa, R., Chung, R., Dunn, S. R., Weston, A. J., Jaimes-Becerra, A., Marques, A. C., et al. (2015). Gene duplications are extensive and contribute significantly to the toxic proteome of nematocysts isolated from Acropora digitifera (Cnidaria: Anthozoa: Scleractinia). *BMC Genomics*, *16*, 774. https://doi.org/10.1186/s12864-015-1976-4.

Galdiero, S., & Gouaux, E. (2004). High resolution crystallographic studies of alpha-hemolysin-phospholipid complexes define heptamer-lipid head group interactions: Implication for understanding protein-lipid interactions. *Protein Science*, *13*(6), 1503–1511. https://doi.org/10.1110/ps.03561104.

Galinier, R., Portela, J., Mone, Y., Allienne, J. F., Henri, H., Delbecq, S., et al. (2013). Biomphalysin, a new beta pore-forming toxin involved in Biomphalaria glabrata immune defense against Schistosoma mansoni. *PLoS Pathogens*, *9*(3), e1003216. https://doi.org/10.1371/journal.ppat.1003216.

Ghadessy, F. J., Chen, D., Kini, R. M., Chung, M. C., Jeyaseelan, K., Khoo, H. E., et al. (1996). Stonustoxin is a novel lethal factor from stonefish (Synanceja horrida) venom. cDNA cloning and characterization. *Journal of Cell Biology*, *271*(41), 25575–25581. https://doi.org/10.1074/jbc.271.41.25575.

Giddings, K. S., Johnson, A. E., & Tweten, R. K. (2003). Redefining cholesterol's role in the mechanism of the cholesterol-dependent cytolysins. *Proceedings of the National Academy of Sciences of the United States of America*, *100*(20), 11315–11320. https://doi.org/10.1073/pnas.2033520100.

Giddings, K. S., Zhao, J., Sims, P. J., & Tweten, R. K. (2004). Human CD59 is a receptor for the cholesterol-dependent cytolysin intermedilysin. *Nature Structural & Molecular Biology*, *11*(12), 1173–1178. https://doi.org/10.1038/nsmb862.

Giri Rao, V. V., Desikan, R., Ayappa, K. G., & Gosavi, S. (2016). Capturing the membrane-triggered conformational transition of an alpha-helical pore-forming toxin. *The Journal of Physical Chemistry B*, *120*(47), 12064–12078. https://doi.org/10.1021/acs.jpcb.6b09400.

Glasser, E., Rachamim, T., Aharonovich, D., & Sher, D. (2014). Hydra actinoporin-like toxin-1, an unusual hemolysin from the nematocyst venom of Hydra magnipapillata which belongs to an extended gene family. *Toxicon*, *91*, 103–113. https://doi.org/10.1016/j.toxicon.2014.04.004.

Gonzalez, M. R., Bischofberger, M., Pernot, L., van der Goot, F. G., & Freche, B. (2008). Bacterial pore-forming toxins: The (w)hole story? *Cellular and Molecular Life Sciences*, *65*(3), 493–507. https://doi.org/10.1007/s00018-007-7434-y.

Gouaux, E. (1997). Channel-forming toxins: Tales of transformation. *Current Opinion in Structural Biology*, 7(4), 566–573. https://doi.org/10.1016/s0959-440x(97)80123-6.

Gudbrandsson, J., Ahi, E. P., Franzdottir, S. R., Kapralova, K. H., Kristjansson, B. K., Steinhaeuser, S. S., et al. (2015). The developmental transcriptome of contrasting Arctic charr (Salvelinus alpinus) morphs. *F1000Res*, *4*, 136. https://doi.org/10.12688/f1000research.6402.2.

Hadders, M. A., Bubeck, D., Roversi, P., Hakobyan, S., Forneris, F., Morgan, B. P., et al. (2012). Assembly and regulation of the membrane attack complex based on structures of C5b6 and sC5b9. *Cell Reports*, *1*(3), 200–207. https://doi.org/10.1016/j.celrep.2012.02.003.

He, W. T., Wan, H., Hu, L., Chen, P., Wang, X., Huang, Z., et al. (2015). Gasdermin D is an executor of pyroptosis and required for interleukin-1beta secretion. *Cell Research*, *25*(12), 1285–1298. https://doi.org/10.1038/cr.2015.139.

Heuck, A. P., Tweten, R. K., & Johnson, A. E. (2001). Beta-barrel pore-forming toxins: Intriguing dimorphic proteins. *Biochemistry*, *40*(31), 9065–9073. https://doi.org/10.1021/bi0155394.

Hinds, M. G., Zhang, W., Anderluh, G., Hansen, P. E., & Norton, R. S. (2002). Solution structure of the eukaryotic pore-forming cytolysin equinatoxin II: Implications for pore formation. *Journal of Molecular Biology*, *315*(5), 1219–1229. https://doi.org/10.1006/jmbi.2001.5321.

Hong, Y., Ohishi, K., Inoue, N., Kang, J. Y., Shime, H., Horiguchi, Y., et al. (2002). Requirement of N-glycan on GPI-anchored proteins for efficient binding of aerolysin but not Clostridium septicum alpha-toxin. *The EMBO Journal*, *21*(19), 5047–5056. https://doi.org/10.1093/emboj/cdf508.

Hoover, D. L., Berger, M., Nacy, C. A., Hockmeyer, W. T., & Meltzer, M. S. (1984). Killing of Leishmania tropica amastigotes by factors in normal human serum. *Journal of Immunology*, *132*(2), 893–897. http://www.ncbi.nlm.nih.gov/pubmed/6690622.

Hotze, E. M., Le, H. M., Sieber, J. R., Bruxvoort, C., McInerney, M. J., & Tweten, R. K. (2013). Identification and characterization of the first cholesterol-dependent cytolysins from Gram-negative bacteria. *Infection and Immunity*, *81*(1), 216–225. https://doi.org/10.1128/IAI.00927-12.

Hotze, E. M., & Tweten, R. K. (2012). Membrane assembly of the cholesterol-dependent cytolysin pore complex. *Biochimica et Biophysica Acta*, *1818*(4), 1028–1038. https://doi.org/10.1016/j.bbamem.2011.07.036.

Howard, S. P., & Buckley, J. T. (1985). Activation of the hole-forming toxin aerolysin by extracellular processing. *Journal of Bacteriology*, *163*(1), 336–340. https://doi.org/10.1128/JB.163.1.336-340.1985.

Hunt, S., Green, J., & Artymiuk, P. J. (2010). Hemolysin E (HlyE, ClyA, SheA) and related toxins. *Advances in Experimental Medicine and Biology*, *677*, 116–126. https://doi.org/10.1007/978-1-4419-6327-7_10.

Iacovache, I., Bischofberger, M., & van der Goot, F. G. (2010). Structure and assembly of pore-forming proteins. *Current Opinion in Structural Biology*, *20*(2), 241–246. https://doi.org/10.1016/j.sbi.2010.01.013.

Iacovache, I., De Carlo, S., Cirauqui, N., Dal Peraro, M., van der Goot, F. G., & Zuber, B. (2016). Cryo-EM structure of aerolysin variants reveals a novel protein fold and the pore-formation process. *Nature Communications*, 7, 12062. https://doi.org/10.1038/ncomms12062.

Iacovache, I., Degiacomi, M. T., Pernot, L., Ho, S., Schiltz, M., Dal Peraro, M., et al. (2011). Dual chaperone role of the C-terminal propeptide in folding and oligomerization of the pore-forming toxin aerolysin. *PLoS Pathogens*, 7(7), e1002135. https://doi.org/10.1371/journal.ppat.1002135.

Iacovache, I., Paumard, P., Scheib, H., Lesieur, C., Sakai, N., Matile, S., et al. (2006). A rivet model for channel formation by aerolysin-like pore-forming toxins. *The EMBO Journal*, *25*(3), 457–466. https://doi.org/10.1038/sj.emboj.7600959.

Jansen, K. B., Inns, P. G., Housden, N. G., Hopper, J. T. S., Kaminska, R., Lee, S., et al. (2020). Bifurcated binding of the OmpF receptor underpins import of the bacteriocin colicin N into Escherichia coli. *The Journal of Biological Chemistry*, *295*(27), 9147–9156. https://doi.org/10.1074/jbc.RA120.013508.

Jia, N., Liu, N., Cheng, W., Jiang, Y. L., Sun, H., Chen, L. L., et al. (2016). Structural basis for receptor recognition and pore formation of a zebrafish aerolysin-like protein. *EMBO Reports*, *17*(2), 235–248. https://doi.org/10.15252/embr.201540851.

Jiao, F., Ruan, Y., & Scheuring, S. (2021). High-speed atomic force microscopy to study pore-forming proteins. *Methods in Enzymology*, *649*, 189–217. https://doi.org/10.1016/bs.mie.2021.01.033.

Kacprzyk-Stokowiec, A., Kulma, M., Traczyk, G., Kwiatkowska, K., Sobota, A., & Dadlez, M. (2014). Crucial role of perfringolysin O D1 domain in orchestrating structural transitions leading to membrane-perforating pores: A hydrogen-deuterium exchange study. *Journal of Cell Biology*, *289*(41), 28738–28752. https://doi.org/10.1074/jbc.M114.577981.

Kashimoto, T., Sugiyama, H., Kawamidori, K., Yamazaki, K., Kado, T., Matsuda, K., et al. (2020). Vibiro vulnificus hemolysin associates with gangliosides. *BMC Microbiology*, *20*(1), 69. https://doi.org/10.1186/s12866-020-01755-1.

Kathuria, R., & Chattopadhyay, K. (2018). Vibrio cholerae cytolysin: Multiple facets of the membrane interaction mechanism of a beta-barrel pore-forming toxin. *IUBMB Life*, *70*(4), 260–266. https://doi.org/10.1002/iub.1725.

Kathuria, R., Mondal, A. K., Sharma, R., Bhattacharyya, S., & Chattopadhyay, K. (2018). Revisiting the role of cholesterol in regulating the pore-formation mechanism of Vibrio cholerae cytolysin, a membrane-damaging beta-barrel pore-forming toxin. *The Biochemical Journal*, *475*(19), 3039–3055. https://doi.org/10.1042/BCJ20180387.

Kaus, K., Lary, J. W., Cole, J. L., & Olson, R. (2014). Glycan specificity of the Vibrio vulnificus hemolysin lectin outlines evolutionary history of membrane targeting by a toxin family. *Journal of Molecular Biology*, *426*(15), 2800–2812. https://doi.org/10.1016/j.jmb.2014.05.021.

Kayagaki, N., Stowe, I. B., Lee, B. L., O'Rourke, K., Anderson, K., Warming, S., et al. (2015). Caspase-11 cleaves gasdermin D for non-canonical inflammasome signalling. *Nature*, *526*(7575), 666–671. https://doi.org/10.1038/nature15541.

Khilwani, B., & Chattopadhyay, K. (2015). Signaling beyond punching holes: Modulation of cellular responses by Vibrio cholerae cytolysin. *Toxins (Basel)*, 7(8), 3344–3358. https://doi.org/10.3390/toxins7083344.

Khilwani, B., Mukhopadhaya, A., & Chattopadhyay, K. (2015). Transmembrane oligomeric form of Vibrio cholerae cytolysin triggers TLR2/TLR6-dependent proinflammatory responses in monocytes and macrophages. *The Biochemical Journal*, *466*(1), 147–161. https://doi.org/10.1042/BJ20140718.
Kienker, P. K., Qiu, X., Slatin, S. L., Finkelstein, A., & Jakes, K. S. (1997). Transmembrane insertion of the colicin Ia hydrophobic hairpin. *The Journal of Membrane Biology*, *157*(1), 27–37. https://doi.org/10.1007/s002329900213.
Koster, S., van Pee, K., Hudel, M., Leustik, M., Rhinow, D., Kuhlbrandt, W., et al. (2014). Crystal structure of listeriolysin O reveals molecular details of oligomerization and pore formation. *Nature Communications*, *5*, 3690. https://doi.org/10.1038/ncomms4690.
Kovacs, S. B., & Miao, E. A. (2017). Gasdermins: Effectors of pyroptosis. *Trends in Cell Biology*, *27*(9), 673–684. https://doi.org/10.1016/j.tcb.2017.05.005.
Kozorog, M., Sani, M. A., Separovic, F., & Anderluh, G. (2018). Listeriolysin O binding affects cholesterol and phospholipid acyl chain dynamics in fluid cholesterol-rich bilayers. *Chemistry*, *24*(53), 14220–14225. https://doi.org/10.1002/chem.201802575.
Kristan, K. C., Viero, G., Dalla Serra, M., Macek, P., & Anderluh, G. (2009). Molecular mechanism of pore formation by actinoporins. *Toxicon*, *54*(8), 1125–1134. https://doi.org/10.1016/j.toxicon.2009.02.026.
Kuang, S., Zheng, J., Yang, H., Li, S., Duan, S., Shen, Y., et al. (2017). Structure insight of GSDMD reveals the basis of GSDMD autoinhibition in cell pyroptosis. *Proceedings of the National Academy of Sciences of the United States of America*, *114*(40), 10642–10647. https://doi.org/10.1073/pnas.1708194114.
Kulma, M., Herec, M., Grudzinski, W., Anderluh, G., Gruszecki, W. I., Kwiatkowska, K., et al. (2010). Sphingomyelin-rich domains are sites of lysenin oligomerization: Implications for raft studies. *Biochimica et Biophysica Acta*, *1798*(3), 471–481. https://doi.org/10.1016/j.bbamem.2009.12.004.
Kundu, N., Tichkule, S., Pandit, S. B., & Chattopadhyay, K. (2017). Disulphide bond restrains the C-terminal region of thermostable direct hemolysin during folding to promote oligomerization. *The Biochemical Journal*, *474*(2), 317–331. https://doi.org/10.1042/BCJ20160728.
Kundu, N., Verma, P., Kumar, A., Dhar, V., Dutta, S., & Chattopadhyay, K. (2020). N-terminal region of Vibrio parahemolyticus thermostable direct hemolysin regulates the membrane-damaging action of the toxin. *Biochemistry*, *59*(4), 605–614. https://doi.org/10.1021/acs.biochem.9b00937.
Lakey, J. H., Massotte, D., Heitz, F., Dasseux, J. L., Faucon, J. F., Parker, M. W., et al. (1991). Membrane insertion of the pore-forming domain of colicin A. A spectroscopic study. *European Journal of Biochemistry*, *196*(3), 599–607. https://doi.org/10.1111/j.1432-1033.1991.tb15855.x.
Lamkanfi, M., & Dixit, V. M. (2014). Mechanisms and functions of inflammasomes. *Cell*, *157*(5), 1013–1022. https://doi.org/10.1016/j.cell.2014.04.007.
Law, R. H., Lukoyanova, N., Voskoboinik, I., Caradoc-Davies, T. T., Baran, K., Dunstone, M. A., et al. (2010). The structural basis for membrane binding and pore formation by lymphocyte perforin. *Nature*, *468*(7322), 447–451. https://doi.org/10.1038/nature09518.
Leung, C., Dudkina, N. V., Lukoyanova, N., Hodel, A. W., Farabella, I., Pandurangan, A. P., et al. (2014). Stepwise visualization of membrane pore formation by suilysin, a bacterial cholesterol-dependent cytolysin. *eLife*, *3*, e04247. https://doi.org/10.7554/eLife.04247.
Lieberman, J., Wu, H., & Kagan, J. C. (2019). Gasdermin D activity in inflammation and host defense. *Science Immunology*, *4*(39), eaav1447. https://doi.org/10.1126/sciimmunol.aav1447.
Lima, C., Falcao, M. A. P., Andrade-Barros, A. I., Seni-Silva, A. C., Grund, L. Z., Balogh, E., et al. (2021). Natterin an aerolysin-like fish toxin drives IL-1beta-dependent

neutrophilic inflammation mediated by caspase-1 and caspase-11 activated by the inflammasome sensor NLRP6. *International Immunopharmacology, 91*, 107287. https://doi.org/10.1016/j.intimp.2020.107287.

Liu, X., & Lieberman, J. (2020). Knocking 'em dead: Pore-forming proteins in immune defense. *Annual Review of Immunology, 38*, 455–485. https://doi.org/10.1146/annurev-immunol-111319-023800.

Liu, Z., Wang, C., Rathkey, J. K., Yang, J., Dubyak, G. R., Abbott, D. W., et al. (2018). Structures of the gasdermin D C-terminal domains reveal mechanisms of autoinhibition. *Structure, 26*(5), 778–784 e773. https://doi.org/10.1016/j.str.2018.03.002.

Liu, Z., Wang, C., Yang, J., Zhou, B., Yang, R., Ramachandran, R., et al. (2019). Crystal structures of the full-length murine and human gasdermin D reveal mechanisms of autoinhibition, lipid binding, and oligomerization. *Immunity, 51*(1), 43–49 e44. https://doi.org/10.1016/j.immuni.2019.04.017.

Liu, X., Zhang, Z., Ruan, J., Pan, Y., Magupalli, V. G., Wu, H., et al. (2016). Inflammasome-activated gasdermin D causes pyroptosis by forming membrane pores. *Nature, 535*(7610), 153–158. https://doi.org/10.1038/nature18629.

Los, F. C., Randis, T. M., Aroian, R. V., & Ratner, A. J. (2013). Role of pore-forming toxins in bacterial infectious diseases. *Microbiology and Molecular Biology Reviews*, 77(2), 173–207. https://doi.org/10.1128/MMBR.00052-12.

Lukoyanova, N., Hoogenboom, B. W., & Saibil, H. R. (2016). The membrane attack complex, perforin and cholesterol-dependent cytolysin superfamily of pore-forming proteins. *Journal of Cell Science, 129*(11), 2125–2133. https://doi.org/10.1242/jcs.182741.

Magalhaes, G. S., Lopes-Ferreira, M., Junqueira-de-Azevedo, I. L., Spencer, P. J., Araujo, M. S., Portaro, F. C., et al. (2005). Natterins, a new class of proteins with kininogenase activity characterized from Thalassophryne nattereri fish venom. *Biochimie, 87*(8), 687–699. https://doi.org/10.1016/j.biochi.2005.03.016.

Mancheno, J. M., Martin-Benito, J., Martinez-Ripoll, M., Gavilanes, J. G., & Hermoso, J. A. (2003). Crystal and electron microscopy structures of sticholysin II actinoporin reveal insights into the mechanism of membrane pore formation. *Structure, 11*(11), 1319–1328. https://doi.org/10.1016/j.str.2003.09.019.

Mancheno, J. M., Tateno, H., Goldstein, I. J., Martinez-Ripoll, M., & Hermoso, J. A. (2005). Structural analysis of the Laetiporus sulphureus hemolytic pore-forming lectin in complex with sugars. *The Journal of Biological Chemistry, 280*(17), 17251–17259. https://doi.org/10.1074/jbc.M413933200.

Mancheno, J. M., Tateno, H., Sher, D., & Goldstein, I. J. (2010). Laetiporus sulphureus lectin and aerolysin protein family. *Advances in Experimental Medicine and Biology, 677*, 67–80. https://doi.org/10.1007/978-1-4419-6327-7_6.

Marshall, P., Hasegawa, A., Davidson, E. A., Nussenzweig, V., & Whitlow, M. (1996). Interaction between complement proteins C5b-7 and erythrocyte membrane sialic acid. *The Journal of Experimental Medicine, 184*(4), 1225–1232. https://doi.org/10.1084/jem.184.4.1225.

McCormack, R. M., de Armas, L. R., Shiratsuchi, M., Fiorentino, D. G., Olsson, M. L., Lichtenheld, M. G., et al. (2015). Perforin-2 is essential for intracellular defense of parenchymal cells and phagocytes against pathogenic bacteria. *eLife, 4*, e06508. https://doi.org/10.7554/eLife.06508.

Mechaly, A. E., Bellomio, A., Gil-Carton, D., Morante, K., Valle, M., Gonzalez-Manas, J. M., et al. (2011). Structural insights into the oligomerization and architecture of eukaryotic membrane pore-forming toxins. *Structure, 19*(2), 181–191. https://doi.org/10.1016/j.str.2010.11.013.

Mechaly, A. E., Bellomio, A., Morante, K., Gonzalez-Manas, J. M., & Guerin, D. M. (2009). Crystallization and preliminary crystallographic analysis of fragaceatoxin C, a

pore-forming toxin from the sea anemone Actinia fragacea. *Acta Crystallographica. Section F, Structural Biology and Crystallization Communications*, *65*(Pt. 4), 357–360. https://doi.org/10.1107/S1744309109007064.
Menny, A., Serna, M., Boyd, C. M., Gardner, S., Joseph, A. P., Morgan, B. P., et al. (2018). CryoEM reveals how the complement membrane attack complex ruptures lipid bilayers. *Nature Communications*, *9*(1), 5316. https://doi.org/10.1038/s41467-018-07653-5.
Mesa-Galloso, H., Valiente, P. A., Valdes-Tresanco, M. E., Epand, R. F., Lanio, M. E., Epand, R. M., et al. (2019). Membrane remodeling by the lytic fragment of sticholysinII: Implications for the toroidal pore model. *Biophysical Journal*, *117*(9), 1563–1576. https://doi.org/10.1016/j.bpj.2019.09.018.
Metkar, S. S., Marchioretto, M., Antonini, V., Lunelli, L., Wang, B., Gilbert, R. J., et al. (2015). Perforin oligomers form arcs in cellular membranes: A locus for intracellular delivery of granzymes. *Cell Death and Differentiation*, *22*(1), 74–85. https://doi.org/10.1038/cdd.2014.110.
Mondal, A. K., & Chattopadhyay, K. (2020). Taking toll on membranes: Curious cases of bacterial beta-barrel pore-forming toxins. *Biochemistry*, *59*(2), 163–170. https://doi.org/10.1021/acs.biochem.9b00783.
Mondal, A. K., Sreekumar, A., Kundu, N., Kathuria, R., Verma, P., Gandhi, S., et al. (2018). Structural basis and functional implications of the membrane pore-formation mechanisms of bacterial pore-forming toxins. *Advances in Experimental Medicine and Biology*, *1112*, 281–291. https://doi.org/10.1007/978-981-13-3065-0_19.
Mondal, A. K., Verma, P., Lata, K., Singh, M., Chatterjee, S., & Chattopadhyay, K. (2020). Sequence diversity in the pore-forming motifs of the membrane-damaging protein toxins. *The Journal of Membrane Biology*, *253*(5), 469–478. https://doi.org/10.1007/s00232-020-00141-2.
Mondal, A. K., Verma, P., Sengupta, N., Dutta, S., Bhushan Pandit, S., & Chattopadhyay, K. (2021). Tyrosine in the hinge region of the pore-forming motif regulates oligomeric beta-barrel pore formation by Vibrio cholerae cytolysin. *Molecular Microbiology*, *115*(4), 508–525. https://doi.org/10.1111/mmi.14631.
Morton, C. J., Sani, M. A., Parker, M. W., & Separovic, F. (2019). Cholesterol-dependent cytolysins: Membrane and protein structural requirements for pore formation. *Chemical Reviews*, *119*(13), 7721–7736. https://doi.org/10.1021/acs.chemrev.9b00090.
Mueller, M., Grauschopf, U., Maier, T., Glockshuber, R., & Ban, N. (2009). The structure of a cytolytic alpha-helical toxin pore reveals its assembly mechanism. *Nature*, *459*(7247), 726–730. https://doi.org/10.1038/nature08026.
Nelson, L. D., Johnson, A. E., & London, E. (2008). How interaction of perfringolysin O with membranes is controlled by sterol structure, lipid structure, and physiological low pH: Insights into the origin of perfringolysin O-lipid raft interaction. *Journal of Biological Chemistry*, *283*(8), 4632–4642. https://doi.org/10.1074/jbc.M709483200.
Ni, T., Jiao, F., Yu, X., Aden, S., Ginger, L., Williams, S. I., et al. (2020). Structure and mechanism of bactericidal mammalian perforin-2, an ancient agent of innate immunity. *Science Advances*, *6*(5), eaax8286. https://doi.org/10.1126/sciadv.aax8286.
Nishimoto, A. T., Rosch, J. W., & Tuomanen, E. I. (2020). Pneumolysin: Pathogenesis and therapeutic target. *Frontiers in Microbiology*, *11*, 1543. https://doi.org/10.3389/fmicb.2020.01543.
Olson, R., & Gouaux, E. (2005). Crystal structure of the Vibrio cholerae cytolysin (VCC) pro-toxin and its assembly into a heptameric transmembrane pore. *Journal of Molecular Biology*, *350*(5), 997–1016. https://doi.org/10.1016/j.jmb.2005.05.045.
Opota, O., Vallet-Gely, I., Vincentelli, R., Kellenberger, C., Iacovache, I., Gonzalez, M. R., et al. (2011). Monalysin, a novel ss-pore-forming toxin from the Drosophila pathogen

Pseudomonas entomophila, contributes to host intestinal damage and lethality. *PLoS Pathogens*, 7(9), e1002259. https://doi.org/10.1371/journal.ppat.1002259.

Palacios-Ortega, J., Rivera-de-Torre, E., Garcia-Linares, S., Gavilanes, J. G., Martinez-Del-Pozo, A., & Slotte, J. P. (2021). Oligomerization of sticholysins from forster resonance energy transfer. *Biochemistry*, *60*(4), 314–323. https://doi.org/10.1021/acs.biochem.0c00840.

Pang, S. S., Bayly-Jones, C., Radjainia, M., Spicer, B. A., Law, R. H. P., Hodel, A. W., et al. (2019). The cryo-EM structure of the acid activatable pore-forming immune effector macrophage-expressed gene 1. *Nature Communications*, *10*(1), 4288. https://doi.org/10.1038/s41467-019-12279-2.

Parker, M. W., Buckley, J. T., Postma, J. P., Tucker, A. D., Leonard, K., Pattus, F., et al. (1994). Structure of the Aeromonas toxin proaerolysin in its water-soluble and membrane-channel states. *Nature*, *367*(6460), 292–295. https://doi.org/10.1038/367292a0.

Parker, M. W., Pattus, F., Tucker, A. D., & Tsernoglou, D. (1989). Structure of the membrane-pore-forming fragment of colicin A. *Nature*, *337*(6202), 93–96. https://doi.org/10.1038/337093a0.

Parker, M. W., Postma, J. P., Pattus, F., Tucker, A. D., & Tsernoglou, D. (1992). Refined structure of the pore-forming domain of colicin A at 2.4 A resolution. *Journal of Molecular Biology*, *224*(3), 639–657. https://doi.org/10.1016/0022-2836(92)90550-4.

Parsons, E. S., Stanley, G. J., Pyne, A. L. B., Hodel, A. W., Nievergelt, A. P., Menny, A., et al. (2019). Single-molecule kinetics of pore assembly by the membrane attack complex. *Nature Communications*, *10*(1), 2066. https://doi.org/10.1038/s41467-019-10058-7.

Paul, K., & Chattopadhyay, K. (2011). Unfolding distinguishes the Vibrio cholerae cytolysin precursor from the mature form of the toxin. *Biochemistry*, *50*(19), 3936–3945. https://doi.org/10.1021/bi200332g.

Podack, E. R., & Konigsberg, P. J. (1984). Cytolytic T cell granules. Isolation, structural, biochemical, and functional characterization. *The Journal of Experimental Medicine*, *160*(3), 695–710. https://doi.org/10.1084/jem.160.3.695.

Podobnik, M., Savory, P., Rojko, N., Kisovec, M., Wood, N., Hambley, R., et al. (2016). Crystal structure of an invertebrate cytolysin pore reveals unique properties and mechanism of assembly. *Nature Communications*, 7, 11598. https://doi.org/10.1038/ncomms11598.

Popoff, M. R. (2011). Epsilon toxin: A fascinating pore-forming toxin. *The FEBS Journal*, *278*(23), 4602–4615. https://doi.org/10.1111/j.1742-4658.2011.08145.x.

Potrich, C., Bastiani, H., Colin, D. A., Huck, S., Prevost, G., & Dalla Serra, M. (2009). The influence of membrane lipids in Staphylococcus aureus gamma-hemolysins pore formation. *The Journal of Membrane Biology*, *227*(1), 13–24. https://doi.org/10.1007/s00232-008-9140-6.

Praper, T., Sonnen, A., Viero, G., Kladnik, A., Froelich, C. J., Anderluh, G., et al. (2011). Human perforin employs different avenues to damage membranes. *The Journal of Biological Chemistry*, *286*(4), 2946–2955. https://doi.org/10.1074/jbc.M110.169417.

Preissner, K. T., Podack, E. R., & Muller-Eberhard, H. J. (1985). The membrane attack complex of complement: Relation of C7 to the metastable membrane binding site of the intermediate complex C5b-7. *Journal of Immunology*, *135*(1), 445–451. http://www.ncbi.nlm.nih.gov/pubmed/3998468.

Rai, A. K., & Chattopadhyay, K. (2015). Revisiting the membrane interaction mechanism of a membrane-damaging beta-barrel pore-forming toxin Vibrio cholerae cytolysin. *Molecular Microbiology*, *97*(6), 1051–1062. https://doi.org/10.1111/mmi.13084.

Rai, A. K., Paul, K., & Chattopadhyay, K. (2013). Functional mapping of the lectin activity site on the beta-prism domain of vibrio cholerae cytolysin: Implications for the membrane pore-formation mechanism of the toxin. *Journal of Biological Chemistry*, *288*(3), 1665–1673. https://doi.org/10.1074/jbc.M112.430181.

Rajan, B., Patel, D. M., Kitani, Y., Viswanath, K., & Brinchmann, M. F. (2017). Novel mannose binding natterin-like protein in the skin mucus of Atlantic cod (Gadus morhua). *Fish & Shellfish Immunology*, *68*, 452–457. https://doi.org/10.1016/j.fsi.2017.07.039.
Ram, S., Lewis, L. A., & Rice, P. A. (2010). Infections of people with complement deficiencies and patients who have undergone splenectomy. *Clinical Microbiology Reviews*, *23*(4), 740–780. https://doi.org/10.1128/CMR.00048-09.
Ramachandran, R., Tweten, R. K., & Johnson, A. E. (2005). The domains of a cholesterol-dependent cytolysin undergo a major FRET-detected rearrangement during pore formation. *Proceedings of the National Academy of Sciences of the United States of America*, *102*(20), 7139–7144. https://doi.org/10.1073/pnas.0500556102.
Reboul, C. F., Whisstock, J. C., & Dunstone, M. A. (2016). Giant MACPF/CDC pore forming toxins: A class of their own. *Biochimica et Biophysica Acta*, *1858*(3), 475–486. https://doi.org/10.1016/j.bbamem.2015.11.017.
Ridley, H., Johnson, C. L., & Lakey, J. H. (2010). Interfacial interactions of pore-forming colicins. *Advances in Experimental Medicine and Biology*, *677*, 81–90. https://doi.org/10.1007/978-1-4419-6327-7_7.
Roderer, D., Benke, S., Schuler, B., & Glockshuber, R. (2016). Soluble oligomers of the pore-forming toxin cytolysin A from Escherichia coli are off-pathway products of pore assembly. *The Journal of Biological Chemistry*, *291*(11), 5652–5663. https://doi.org/10.1074/jbc.M115.700757.
Rojko, N., Kristan, K. C., Viero, G., Zerovnik, E., Macek, P., Dalla Serra, M., et al. (2013). Membrane damage by an alpha-helical pore-forming protein, Equinatoxin II, proceeds through a succession of ordered steps. *The Journal of Biological Chemistry*, *288*(33), 23704–23715. https://doi.org/10.1074/jbc.M113.481572.
Ros, U., & Garcia-Saez, A. J. (2015). More than a pore: The interplay of pore-forming proteins and lipid membranes. *The Journal of Membrane Biology* *248*(3), 545–561. https://doi.org/10.1007/s00232-015-9820-y.
Rosado, C. J., Kondos, S., Bull, T. E., Kuiper, M. J., Law, R. H., Buckle, A. M., et al. (2008). The MACPF/CDC family of pore-forming toxins. *Cellular Microbiology*, *10*(9), 1765–1774. https://doi.org/10.1111/j.1462-5822.2008.01191.x.
Rossjohn, J., Feil, S. C., McKinstry, W. J., Tweten, R. K., & Parker, M. W. (1997). Structure of a cholesterol-binding, thiol-activated cytolysin and a model of its membrane form. *Cell*, *89*(5), 685–692. https://doi.org/10.1016/s0092-8674(00)80251-2.
Ruan, J., Xia, S., Liu, X., Lieberman, J., & Wu, H. (2018). Cryo-EM structure of the gasdermin A3 membrane pore. *Nature*, *557*(7703), 62–67. https://doi.org/10.1038/s41586-018-0058-6.
Sathyanarayana, P., Maurya, S., Behera, A., Ravichandran, M., Visweswariah, S. S., Ayappa, K. G., et al. (2018). Cholesterol promotes Cytolysin A activity by stabilizing the intermediates during pore formation. *Proceedings of the National Academy of Sciences of the United States of America*, *115*(31), E7323–E7330. https://doi.org/10.1073/pnas.1721228115.
Sathyanarayana, P., Visweswariah, S. S., & Ayappa, K. G. (2021). Mechanistic insights into pore formation by an alpha-pore forming toxin: Protein and lipid bilayer interactions of cytolysin A. *Accounts of Chemical Research*, *54*(1), 120–131. https://doi.org/10.1021/acs.accounts.0c00551.
Savva, C. G., Fernandes da Costa, S. P., Bokori-Brown, M., Naylor, C. E., Cole, A. R., Moss, D. S., et al. (2013). Molecular architecture and functional analysis of NetB, a pore-forming toxin from Clostridium perfringens. *The Journal of Biological Chemistry*, *288*(5), 3512–3522. https://doi.org/10.1074/jbc.M112.430223.
Schon, P., Garcia-Saez, A. J., Malovrh, P., Bacia, K., Anderluh, G., & Schwille, P. (2008). Equinatoxin II permeabilizing activity depends on the presence of sphingomyelin and lipid phase coexistence. *Biophysical Journal*, *95*(2), 691–698. https://doi.org/10.1529/biophysj.108.129981.

Schreiber, M. P., Chan, C. M., & Shorr, A. F. (2011). Bacteremia in Staphylococcus aureus pneumonia: Outcomes and epidemiology. *Journal of Critical Care*, *26*(4), 395–401. https://doi.org/10.1016/j.jcrc.2010.09.002.

Serna, M., Giles, J. L., Morgan, B. P., & Bubeck, D. (2016). Structural basis of complement membrane attack complex formation. *Nature Communications*, 7, 10587. https://doi.org/10.1038/ncomms10587.

Sher, D., Fishman, Y., Zhang, M., Lebendiker, M., Gaathon, A., Mancheno, J. M., et al. (2005). Hydralysins, a new category of beta-pore-forming toxins in cnidaria. *The Journal of Biological Chemistry*, *280*(24), 22847–22855. https://doi.org/10.1074/jbc.M503242200.

Shewell, L. K., Day, C. J., Jen, F. E., Haselhorst, T., Atack, J. M., Reijneveld, J. F., et al. (2020). All major cholesterol-dependent cytolysins use glycans as cellular receptors. *Science Advances*, *6*(21), eaaz4926. https://doi.org/10.1126/sciadv.aaz4926.

Soltani, C. E., Hotze, E. M., Johnson, A. E., & Tweten, R. K. (2007). Specific protein-membrane contacts are required for prepore and pore assembly by a cholesterol-dependent cytolysin. *The Journal of Biological Chemistry*, *282*(21), 15709–15716. https://doi.org/10.1074/jbc.M701173200.

Song, L., Hobaugh, M. R., Shustak, C., Cheley, S., Bayley, H., & Gouaux, J. E. (1996). Structure of staphylococcal alpha-hemolysin, a heptameric transmembrane pore. *Science*, *274*(5294), 1859–1866. https://doi.org/10.1126/science.274.5294.1859.

Soto, C., Del Valle, A., Valiente, P. A., Ros, U., Lanio, M. E., Hernandez, A. M., et al. (2017). Differential binding and activity of the pore-forming toxin sticholysin II in model membranes containing diverse ceramide-derived lipids. *Biochimie*, *138*, 20–31. https://doi.org/10.1016/j.biochi.2017.04.003.

Sousa, M. V., Richardson, M., Fontes, W., & Morhy, L. (1994). Homology between the seed cytolysin enterolobin and bacterial aerolysins. *Journal of Protein Chemistry*, *13*(8), 659–667. https://doi.org/10.1007/BF01886950.

Spicer, B. A., Law, R. H. P., Caradoc-Davies, T. T., Ekkel, S. M., Bayly-Jones, C., Pang, S. S., et al. (2018). The first transmembrane region of complement component-9 acts as a brake on its self-assembly. *Nature Communications*, *9*(1), 3266. https://doi.org/10.1038/s41467-018-05717-0.

Stewart, S. E., Kondos, S. C., Matthews, A. Y., D'Angelo, M. E., Dunstone, M. A., Whisstock, J. C., et al. (2014). The perforin pore facilitates the delivery of cationic cargos. *The Journal of Biological Chemistry*, *289*(13), 9172–9181. https://doi.org/10.1074/jbc.M113.544890.

Su, Z., Ho, D., Merrill, A. R., & Lipkowski, J. (2019). Situ electrochemical and PM-IRRAS studies of colicin E1 ion channels in the floating bilayer lipid membrane. *Langmuir*, *35*(25), 8452–8459. https://doi.org/10.1021/acs.langmuir.9b01251.

Tamura, M., Tanaka, S., Fujii, T., Aoki, A., Komiyama, H., Ezawa, K., et al. (2007). Members of a novel gene family, Gsdm, are expressed exclusively in the epithelium of the skin and gastrointestinal tract in a highly tissue-specific manner. *Genomics*, *89*(5), 618–629. https://doi.org/10.1016/j.ygeno.2007.01.003.

Tamura, S., Yamakawa, M., & Shiomi, K. (2011). Purification, characterization and cDNA cloning of two natterin-like toxins from the skin secretion of oriental catfish Plotosus lineatus. *Toxicon*, *58*(5), 430–438. https://doi.org/10.1016/j.toxicon.2011.08.001.

Tanaka, K., Caaveiro, J. M., Morante, K., Gonzalez-Manas, J. M., & Tsumoto, K. (2015). Structural basis for self-assembly of a cytolytic pore lined by protein and lipid. *Nature Communications*, *6*, 6337. https://doi.org/10.1038/ncomms7337.

Tomita, T., Noguchi, K., Mimuro, H., Ukaji, F., Ito, K., Sugawara-Tomita, N., et al. (2004). Pleurotolysin, a novel sphingomyelin-specific two-component cytolysin from the edible mushroom Pleurotus ostreatus, assembles into a transmembrane pore complex. *The Journal of Biological Chemistry*, *279*(26), 26975–26982. https://doi.org/10.1074/jbc.M402676200.

Tschopp, J. (1984). Ultrastructure of the membrane attack complex of complement. Heterogeneity of the complex caused by different degree of C9 polymerization. *The Journal of Biological Chemistry*, *259*(12), 7857–7863. http://www.ncbi.nlm.nih.gov/pubmed/6736027.

Unno, H., Matsuyama, K., Tsuji, Y., Goda, S., Hiemori, K., Tateno, H., et al. (2016). Identification, characterization, and X-ray crystallographic analysis of a novel type of mannose-specific lectin CGL1 from the Pacific oyster Crassostrea gigas. *Scientific Reports*, *6*, 29135. https://doi.org/10.1038/srep29135.

Vaidyanathan, M. S., Sathyanarayana, P., Maiti, P. K., Visweswariah, S. S., & Ayappa, K. G. (2014). Lysis dynamics and membrane oligomerization pathways for Cytolysin A (ClyA) pore-forming toxin. *RSC Advances*, *4*(10), 4930–4942. https://doi.org/10.1039/C3RA45159C.

van Pee, K., Neuhaus, A., D'Imprima, E., Mills, D. J., Kuhlbrandt, W., & Yildiz, O. (2017). CryoEM structures of membrane pore and prepore complex reveal cytolytic mechanism of Pneumolysin. *eLife*, *6*, e23644. https://doi.org/10.7554/eLife.23644.

Vanden Broeck, D., Horvath, C., & De Wolf, M. J. (2007). Vibrio cholerae: Cholera toxin. *The International Journal of Biochemistry & Cell Biology*, *39*(10), 1771–1775. https://doi.org/10.1016/j.biocel.2007.07.005.

Vasquez, M. T., Lubkin, A., Reyes-Robles, T., Day, C. J., Lacey, K. A., Jennings, M. P., et al. (2020). Identification of a domain critical for Staphylococcus aureus LukED receptor targeting and lysis of erythrocytes. *The Journal of Biological Chemistry*, *295*(50), 17241–17250. https://doi.org/10.1074/jbc.RA120.015757.

Verma, P., Gandhi, S., Lata, K., & Chattopadhyay, K. (2021). Pore-forming toxins in infection and immunity. *Biochemical Society Transactions*, *49*(1), 455–465. https://doi.org/10.1042/BST20200836.

Voskoboinik, I., Smyth, M. J., & Trapani, J. A. (2006). Perforin-mediated target-cell death and immune homeostasis. *Nature Reviews. Immunology*, *6*(12), 940–952. https://doi.org/10.1038/nri1983.

Voskoboinik, I., Thia, M. C., Fletcher, J., Ciccone, A., Browne, K., Smyth, M. J., et al. (2005). Calcium-dependent plasma membrane binding and cell lysis by perforin are mediated through its C2 domain: A critical role for aspartate residues 429, 435, 483, and 485 but not 491. *Journal of Biological Chemistry*, *280*(9), 8426–8434. https://doi.org/10.1074/jbc.M413303200.

Wade, K. R., Hotze, E. M., Kuiper, M. J., Morton, C. J., Parker, M. W., & Tweten, R. K. (2015). An intermolecular electrostatic interaction controls the prepore-to-pore transition in a cholesterol-dependent cytolysin. *Proceedings of the National Academy of Sciences of the United States of America*, *112*(7), 2204–2209. https://doi.org/10.1073/pnas.1423754112.

Wade, K. R., & Tweten, R. K. (2015). The Apicomplexan CDC/MACPF-like pore-forming proteins. *Current Opinion in Microbiology*, *26*, 48–52. https://doi.org/10.1016/j.mib.2015.05.001.

Wallace, A. J., Stillman, T. J., Atkins, A., Jamieson, S. J., Bullough, P. A., Green, J., et al. (2000). E. coli hemolysin E (HlyE, ClyA, SheA): X-ray crystal structure of the toxin and observation of membrane pores by electron microscopy. *Cell*, *100*(2), 265–276. https://doi.org/10.1016/s0092-8674(00)81564-0.

Westphal, D., Dewson, G., Czabotar, P. E., & Kluck, R. M. (2011). Molecular biology of Bax and Bak activation and action. *Biochimica et Biophysica Acta*, *1813*(4), 521–531. https://doi.org/10.1016/j.bbamcr.2010.12.019.

Wilke, G. A., & Bubeck Wardenburg, J. (2010). Role of a disintegrin and metalloprotease 10 in Staphylococcus aureus alpha-hemolysin-mediated cellular injury. *Proceedings of the National Academy of Sciences of the United States of America*, *107*(30), 13473–13478. https://doi.org/10.1073/pnas.1001815107.

Wu, F., Feng, B., Ren, Y., Wu, D., Chen, Y., Huang, S., et al. (2017). A pore-forming protein implements VLR-activated complement cytotoxicity in lamprey. *Cell Discovery*, *3*, 17033. https://doi.org/10.1038/celldisc.2017.33.

Xia, S., Zhang, Z., Magupalli, V. G., Pablo, J. L., Dong, Y., Vora, S. M., et al. (2021). Gasdermin D pore structure reveals preferential release of mature interleukin-1. *Nature*, *593*(7860), 607–611. https://doi.org/10.1038/s41586-021-03478-3.

Xu, C., Wang, B. C., Yu, Z., & Sun, M. (2014). Structural insights into Bacillus thuringiensis Cry, Cyt and parasporin toxins. *Toxins (Basel)*, *6*(9), 2732–2770. https://doi.org/10.3390/toxins6092732.

Xue, Z., Liu, X., Pang, Y., Yu, T., Xiao, R., Jin, M., et al. (2012). Characterization, phylogenetic analysis and cDNA cloning of natterin-like gene from the blood of lamprey, Lampetra japonica. *Immunology Letters*, *148*(1), 1–10. https://doi.org/10.1016/j.imlet.2012.08.005.

Yamaji, A., Sekizawa, Y., Emoto, K., Sakuraba, H., Inoue, K., Kobayashi, H., et al. (1998). Lysenin, a novel sphingomyelin-specific binding protein. *The Journal of Biological Chemistry*, *273*(9), 5300–5306. https://doi.org/10.1074/jbc.273.9.5300.

Yamaji-Hasegawa, A., Makino, A., Baba, T., Senoh, Y., Kimura-Suda, H., Sato, S. B., et al. (2003). Oligomerization and pore formation of a sphingomyelin-specific toxin, lysenin. *The Journal of Biological Chemistry*, *278*(25), 22762–22770. https://doi.org/10.1074/jbc.M213209200.

Yamashita, D., Sugawara, T., Takeshita, M., Kaneko, J., Kamio, Y., Tanaka, I., et al. (2014). Molecular basis of transmembrane beta-barrel formation of staphylococcal pore-forming toxins. *Nature Communications*, *5*, 4897. https://doi.org/10.1038/ncomms5897.

Yanagihara, I., Nakahira, K., Yamane, T., Kaieda, S., Mayanagi, K., Hamada, D., et al. (2010). Structure and functional characterization of Vibrio parahaemolyticus thermostable direct hemolysin. *The Journal of Biological Chemistry*, *285*(21), 16267–16274. https://doi.org/10.1074/jbc.M109.074526.

Yap, W. Y., Tan, K., & Hwang, J. S. (2019). Expansion of Hydra actinoporin-like toxin (HALT) gene family: Expression divergence and functional convergence evolved through gene duplication. *Toxicon*, *170*, 10–20. https://doi.org/10.1016/j.toxicon.2019.09.007.

Yilmaz, N., & Kobayashi, T. (2015). Visualization of lipid membrane reorganization induced by a pore-forming toxin using high-speed atomic force microscopy. *ACS Nano*, *9*(8), 7960–7967. https://doi.org/10.1021/acsnano.5b01041.

Yilmaz, N., Yamada, T., Greimel, P., Uchihashi, T., Ando, T., & Kobayashi, T. (2013). Real-time visualization of assembling of a sphingomyelin-specific toxin on planar lipid membranes. *Biophysical Journal*, *105*(6), 1397–1405. https://doi.org/10.1016/j.bpj.2013.07.052.

Yilmaz, N., Yamaji-Hasegawa, A., Hullin-Matsuda, F., & Kobayashi, T. (2018). Molecular mechanisms of action of sphingomyelin-specific pore-forming toxin, lysenin. *Seminars in Cell & Developmental Biology*, *73*, 188–198. https://doi.org/10.1016/j.semcdb.2017.07.036.

Zhao, F., Yan, C., Wang, X., Yang, Y., Wang, G., Lee, W., et al. (2014). Comprehensive transcriptome profiling and functional analysis of the frog (Bombina maxima) immune system. *DNA Research*, *21*(1), 1–13. https://doi.org/10.1093/dnares/dst035.

Zilnyte, M., Venclovas, C., Zvirbliene, A., & Pleckaityte, M. (2015). The cytolytic activity of vaginolysin strictly depends on cholesterol and is potentiated by human CD59. *Toxins (Basel)*, 7(1), 110–128. https://doi.org/10.3390/toxins7010110.

CHAPTER EIGHT

Ganglioside binding domains in proteins: Physiological and pathological mechanisms

Fodil Azzaz[a,b], Nouara Yahi[a,b], Coralie Di Scala[c], Henri Chahinian[a,b], and Jacques Fantini[a,b,*]
[a]INSERM UMR_S 1072, Marseille, France
[b]Aix-Marseille Université, Marseille, France
[c]Neuroscience Center-HiLIFE, University of Helsinki, Helsinki, Finland
*Corresponding author: e-mail address: jm.fantini@gmail.com

Contents

1. Structural features of gangliosides 290
2. Membrane localization of gangliosides 292
3. Molecular interactions: The "NH trick" 294
4. Flat surfaces and chalices: Same gangliosides, two topologies 300
5. Ganglioside binding domains (GBDs) 301
6. Helical GBDs 301
7. Loop shaped GBDs 303
8. Large flat GBDs 305
9. Gangliosides and synapses 306
10. Gangliosides and bacterial toxins 308
11. Gangliosides and virus infection 311
12. Gangliosides and neurodegenerative diseases 313
13. Conclusion and perspectives 315
References 317

Abstract

Gangliosides are anionic lipids that form condensed membrane clusters (lipid rafts) and exert major regulatory functions on a wide range of proteins. In this review, we propose a new view of the structural features of gangliosides with special emphasis on emerging properties associated with protein binding modes. We analyze the different possibilities of molecular associations of gangliosides in lipid rafts and the role of cholesterol in this organization. We are particularly interested in amide groups of N-acetylated sugars which make it possible to neutralize the negative charge of the carboxylate group of sialic acids. We refer to this effect as "NH trick" and we demonstrate that it is operative in GM1, GD1a, GD1b and GT1b gangliosides. The NH trick is key to understand the different topologies adopted by gangliosides (chalice-like at the edge of lipid rafts, condensed clusters in central areas) and their impact on protein binding. We define three

Advances in Protein Chemistry and Structural Biology, Volume 128
ISSN 1876-1623
https://doi.org/10.1016/bs.apcsb.2021.08.003

major types of ganglioside-binding domains (GBDs): α-helical, loop shaped, and large flat surface. We describe the mode of interaction of each GBD with typical reference proteins: synaptotagmin, 5HT1A receptor, cholera and botulinum toxins, HIV-1 surface envelope glycoprotein gp120, SARS-CoV-2 spike protein, cellular prion protein, Alzheimer's β-amyloid peptide and Parkinson's disease associated α-synuclein. We discuss the common mechanisms and peculiarities of protein binding to gangliosides in the light of physiological and pathological conditions. We anticipate that innovative ganglioside-based therapies will soon show an exponential growth for the treatment of cancer, microbial infections, and neurodegenerative diseases.

1. Structural features of gangliosides

Gangliosides belong to the family of sphingolipids, an important class of membrane lipids that share a common structural organization consisting of a fatty acid attached to sphingosine, a long chain amino-alcohol (Fantini, Garmy, Mahfoud, & Yahi, 2002). Based on this membrane-anchored structure called ceramide, sphingolipids are classified according to their polar part which protrudes in the extracellular milieu. Briefly, sphingomyelin has a phosphorylcholine head, and glycosphingolipids have an oligosaccharide moiety which, at physiological pH, can be either electrically neutral or charged depending on its sugar content. In this review we will focus on gangliosides (Schnaar, 2019), defined as anionic (electronegative) glycosphingolipids containing at least one sialic acid residue in their glycone part (Fig. 1A).

The biochemical diversity of gangliosides concerns all parts of the molecule, i.e., sphingosine, fatty acid and the oligosaccharide (Kolter, 2012). Several sphingosine derivatives with different chain length have been detected in mammal cells, but the C18:1 species is largely predominant (Sonnino & Chigorno, 2000). The fatty acid linked to sphingosine may also vary in chain length, insaturation, and hydroxylation (Fantini et al., 2002). Finally, the oligosaccharide part may consist of only two residues (galactose linked to a unique sialic acid in ganglioside GM4) or a sophisticated combination of sugars (glucose, galactose), amino-sugar derivatives (N-acetyl-galactosamine, GalNAc) and sialic acids. An additional level of biochemical diversity concerns the sialic acid itself, as two main species are expressed in mammals: N-acetyl-neuraminic acid (Neu5Ac) and N-glycolyl-neuraminic acid (Neu5Gc) which differ by a single oxygen atom (Fig. 1B). Most mammalian species are capable of synthesizing both Neu5Ac and Neu5Gc but humans express only Neu5Ac because of a deletion in the CMAH gene which codes for the enzyme that catalyzes the conversion

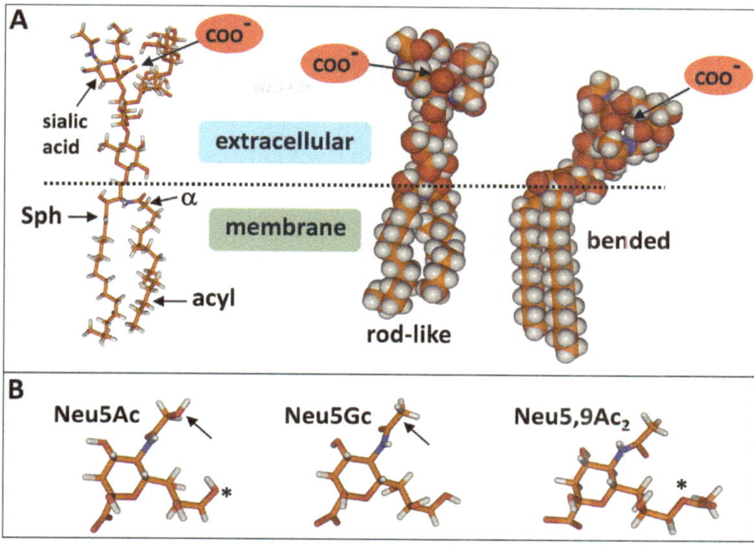

Fig. 1 Structural features of gangliosides. (A) The chemical structure of ganglioside GM1 is detailed on the left panel. The membrane-embedded part, ceramide, is built from an acyl chain condensed with sphingosine (Sph). The extracellular glycone part is an oligosaccharide containing one sialic acid whose carboxylate group (COO^-) is negatively charged at physiological pH. The C2 carbon of the acyl chain (α) can be hydroxylated. Gangliosides can adopt several distinct conformations in the plasma membrane, including linear rod-like structures (glycone part in the same axis as ceramide) to typical bended shapes as shown in the right panel. (B) Diversity of sialic acid residues found in gangliosides. The arrow indicates the difference between Neu5Ac and Neu5Gc and the asterisk the difference between Neu5Ac and Neu5,9Ac_2.

of CMP-Neu5Ac into CMP-Neu5Gc (Davies & Varki, 2015; Irie, Koyama, Kozutsumi, Kawasaki, & Suzuki, 1998). Apart from Neu5Ac and Neu5Gc, some sialic acid derivatives may show a specific tissue expression. Indeed, in the respiratory tract, the presence of ganglioside GM1 with a 9-*O*-acetyl derivative (Neu5,9Ac_2) (Fig. 1B) has a high biological significance since this sialic acid is specifically recognized by respiratory viruses (Hubert et al., 2020), including SARS-CoV-2, the etiological agent of Covid-19 (Hubert et al., 2020; Wu et al., 2020). Moreover, it has been shown that human cells expressing 9-*O*-acetylated sialic acid derivatives are more sensitive to *Salmonella Typhi* toxin (Nguyen et al., 2020), revealing a fine-tuning mechanism exploiting specific biochemical modifications of glycan receptor moieties for virulence (Bourdoulous & Lemichez, 2018).

The ganglioside nomenclature (Kolter, 2012) is based on the number of sialic residues combined with chromatographic properties. Gangliosides with one sialic acid are referred to as "mono-sialosyl" species which define the GM group (G for ganglioside, M for mono-sialosyl). Then a number indicates the distance of migration in thin layer chromatography. For mono-sialosyl gangliosides there are four main species: GM1, GM2, GM3 and GM4, ranged from the more polar (GM1) to the more apolar (GM4). This classification is interesting because it reflects merely the polarity of the glycone part of the ganglioside which has five residues for GM1 but only two for GM4. The same logic applies for di- and tri-sialosyl gangliosides (respectively GD and GT groups). Tetra-sialosyl gangliosides constitute the GQ family (Q for quadri).

Gangliosides have a selective expression pattern in human organs, tissues, and cells. In the brain, gangliosides represent 80% of all glycans and up to 75% of sialic acids whereas in peripheral organs and tissues those glycans are mostly linked to glycoproteins (Schnaar, Gerardy-Schahn, & Hildebrandt, 2014; Sipione, Monyror, Galleguillos, Steinberg, & Kadam, 2020). They are differently distributed in neurons and glial cells and this heterogeneous expression has a profound impact on synaptic functions (Schnaar, 2016). Gangliosides expressed on mucosal surfaces are key targets for pathogens whose adhesins and toxins have evolved to highjack their physiological functions (Fantini, Maresca, Hammache, Yahi, & Delézay, 2000). These different physiological and pathological aspects are developed in this chapter.

2. Membrane localization of gangliosides

Gangliosides, like other sphingolipids, are not stochastically distributed in the plasma membrane. On the one hand, these lipids are physiologically restricted to the outer leaflet. On the other hand, they are attracted in condensed molecular complexes that are stabilized by van der Waals interactions between their ceramide parts and by cholesterol. The presence of such cholesterol-sphingolipid complexes has been suspected in the years following the publication of the fluid mosaic model of the plasma membrane (Singer & Nicolson, 1972). Yet it is only in 1997 that Simons and Ikonen published a seminal article describing the functional organization of lipid rafts in biological membranes (Simons & Ikonen, 1997). After several years of controversy chiefly focused on detergent-based extraction procedures (Aureli, Grassi, Sonnino, & Prinetti, 2016; Munro, 2003), it is now widely accepted that lipid rafts do exist in the plasma membrane of mammalian cells

(Cai et al., 2012) (and incidentally even in prokaryotic membranes (LaRocca et al., 2013)) where they control key biological functions including membrane traffic (Brown & London, 1998; Diaz-Rohrer, Levental, & Levental, 2014), signaling (Staubach & Hanisch, 2011), exocytosis (Salaün, James, & Chamberlain, 2004), endocytosis (Nabi & Le, 2003), immune response (Varshney, Yadav, & Saini, 2016), and control of synaptic receptors (Allen, Halverson-Tamboli, & Rasenick, 2007; Fantini & Barrantes, 2009; Sonnino, Aureli, Mauri, Ciampa, & Prinetti, 2015). Lipid rafts (Fig. 2) are also involved in several pathologies such as cancer metastasis (Greenlee, Subramanian, Liu, & King, 2021), neurological disorders (Fantini & Yahi, 2010, 2015), and microbial infections (Mañes, del Real, & Martínez, 2003).

Apart from lipid rafts, gangliosides are also key components of caveolae, a subclass of plasma membrane microdomains involved in endocytosis (Kiss, 2012). It is important to note that the voluminous extracellular domain of gangliosides due to their large glycone part, together with the attraction of their ceramide moieties embedded in the plasma membrane have a marked effect on membrane curvature (Kabbani, Raghunathan, Lencer, Kenworthy, & Kelly, 2020; Kiss, 2012), consistent with endocytosis functions (Hurley & Wendland, 2002). Since lipid rafts are typical emerging properties (lipid rafts have properties that individual gangliosides do not have, just like the properties of liquid water cannot be anticipated from the study of the H_2O molecule), it is necessary to understand the molecular mechanisms controlling the coalescence of gangliosides in cholesterol-enriched plasma membrane microdomains.

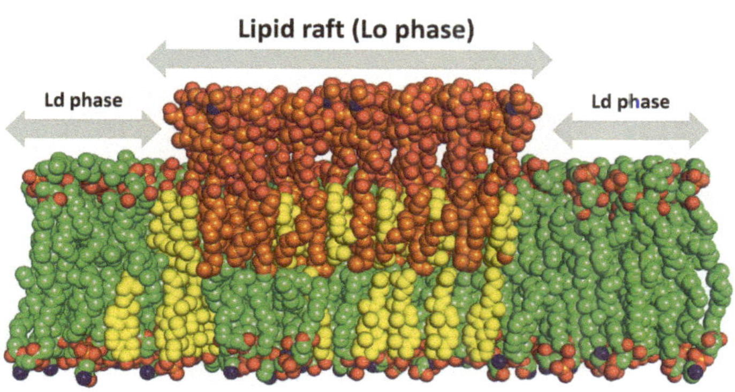

Fig. 2 Molecular model of a lipid raft. GM1 gangliosides are colored in orange, cholesterol in yellow, glycerophospholipids in green. Oxygen atoms are in red, nitrogen in blue and carbon in green. Lipid rafts are in the liquid ordered (Lo phase) whereas the bulk membrane is the liquid disordered (Ld) phase.

3. Molecular interactions: The "NH trick"

Twenty years ago, our group has adapted the "lipid shape" concept developed by Israelachvili (Israelachvili & Mitchell, 1975; Israelachvili, Mitchell, & Ninham, 1977) to model the molecular interactions between cholesterol and sphingolipids (Fantini et al., 2002). The basis of this concept is that cholesterol has a conic shape with a small polar head and a large apolar part, whereas sphingolipids have the opposite geometric features, i.e., a large polar part and a thin apolar domain (Fig. 3). Thus, there is an obvious geometric complementarity between cholesterol and sphingolipids (see the cholesterol molecule on the right side of the ganglioside-cholesterol complex in Fig. 3). Moreover, the junction between ceramide and the glycone part of the ganglioside forms a cavity into which cholesterol can easily find its room, as shown in Fig. 3 for the cholesterol molecule on the left side of the ganglioside-cholesterol complex.

Overall, these geometric features explain why sphingolipids, which do not display the classical cylindrical shape or glycerophospholipids can be part of a planar biological membrane (Fantini et al., 2002). Indeed, without cholesterol which serves as both molecular wedge and cement, sphingolipids would spontaneously adopt a micellar organization, which obviously is not consistent with a bilayer membrane leaflet (Fig. 4). The problem would

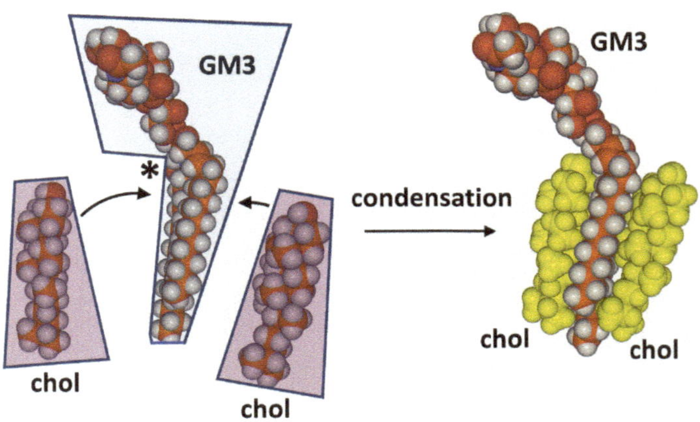

Fig. 3 The lipid shape concept and the condensing effect of cholesterol on gangliosides. Cholesterol and gangliosides have a cone/inverted cone geometric complementarity that facilitates the condensing effect of cholesterol, a motor of lipid raft formation. Gangliosides also display a cavity formed by the ceramide-oligosaccharide junction (on the left side of ganglioside GM3).

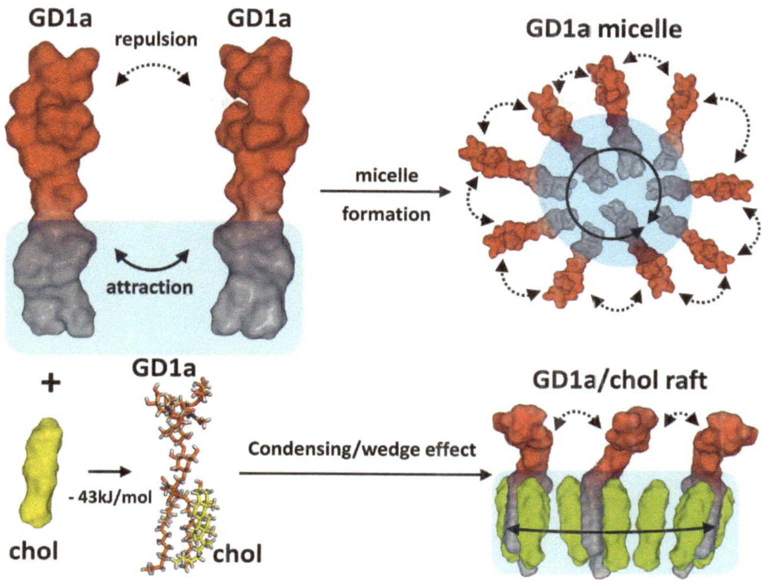

Fig. 4 Cholesterol prevents ganglioside micellization through a dual condensing/wedge mechanism. In absence of cholesterol, two vicinal GM1 molecules display both repulsive (electrostatic, dashed arrows) and attractive (van der Waals, full arrows) interactions which favor micelle formation (upper panel). At least two cholesterol molecules can bind to the ceramide part of GM1, which preserves van der Waals interactions between two vicinal GM1/cholesterol complexes and relaxes the electrostatic repulsion between the charged polar head groups of GM1 (lower panel). Thus, cholesterol has a condensing effect on GM1 and it also acts as a wedge preventing ganglioside micellization by maintaining a linear leaflet-compatible lipid organization.

be exacerbated with gangliosides because of their negative charge that would further repulse the polar head group of two vicinal gangliosides. The combination of polar repulsion with apolar attraction is the physicochemical basis of micelle formation. Cholesterol solves one problem (linearity). Yet, it remains to explain how ganglioside clusters in lipid rafts can accommodate such a concentration of electronegative charges.

In fact, gangliosides are equipped with a shield against electrostatic repulsion. This shield is conferred by the NH group of amino-sugar derivatives found in most gangliosides (the only exceptions being the two smallest gangliosides, GM4 and GM3). Structural NMR studies of ganglioside GM2 have revealed that the negative charge of the carboxylate can be neutralized by a hydrogen bond between the GalNAc amide proton and the Neu5Ac carboxylate group (Brocca, Acquotti, & Sonnino, 1993; Sonnino, Mauri,

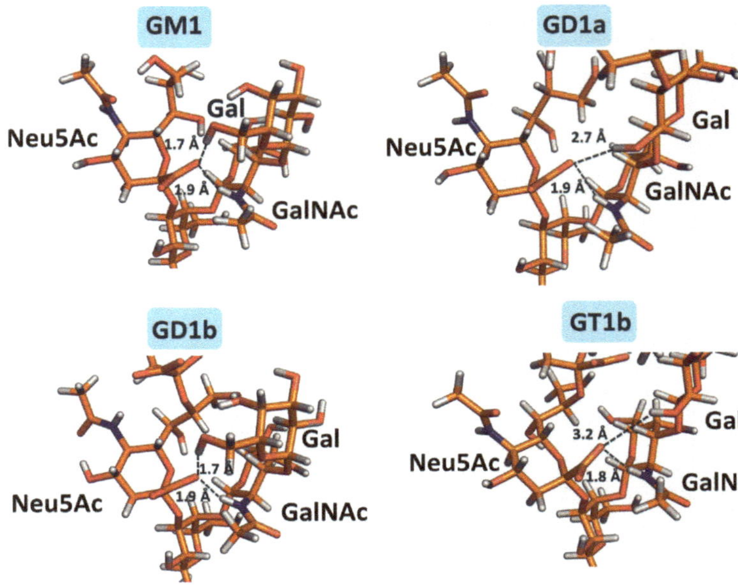

Fig. 5 An internal H-bond network neutralizes the carboxylate charge of sialic acid in gangliosides. For all gangliosides studied (GM1, GD1a, GD1b and GT1b), the carboxylate group of the first sialic acid (Neu5Ac) is neutralized by an internal hydrogen bond between the GalNAc amide proton and carboxylate (NH trick effect), which can be reinforced by the C5-OH group of the same Gal residue.

Chigorno, & Prinetti, 2007). In order to assess whether such a shield mechanism could be a general intrinsic property of gangliosides, we performed a series of molecular dynamics simulations of ganglioside dimers (Fig. 5). Regardless of the ganglioside considered (GM1, GD1a, GD1b and GT1b), the third residue of the main sugar branch is always GalNAc linked to galactose (Gal) via a β1–4 bond. During the biosynthetic pathway of gangliosides (van Echten & Sandhoff, 1993), some of them are assigned a second sugar branch which starts with a sialic acid linked to the Gal residue closest to ceramide, as it is indeed the case for GM1, GD1a, GD1b and GT1b. The minimized structures of these gangliosides (generated in CHARMM-GUI with the tool glycolipid modeler) showed that the NH group of GalNAc is optimally positioned near the acidic group of the neighbor sialic acid (Fig. 5). Thanks to this proximity, a hydrogen bond (H-bond) can be established between the hydrogen carried by the amide group and the carboxylate of the sialic acid, with an average distance of 1.9 Å. Additionally, there is a possibility of a second H-bond between carboxylate and the

C5-OH group of the same Gal residue. This secondary interaction may also contribute to carboxylate neutralization, although it is not as conserved as the H-bond between the NH and carboxylate. Overall, given the critical role of the NH of the amide group of GalNAc to "internalize" the negative charge of gangliosides inside their oligosaccharide part, this mechanism could be coined the "NH trick."

Interestingly, the crystal structures of the sugar moiety of various gangliosides bound to proteins seem to display a similar NH group position (Berntsson, Peng, Dong, & Stenmark, 2013; Buch et al., 2015; Davies, Masuyer, & Stenmark, 2020; Fotinou et al., 2001; Hamark et al., 2017; Holmner et al., 2011; Merritt et al., 1998; Neu, Woellner, Gauglitz, & Stehle, 2008; Ng et al., 2013; Reiss et al., 2012). Taken together, these data suggest that a key function of nitrogen atoms in gangliosides is to prevent the electrostatic repulsion between two vicinal gangliosides in lipid raft domains. Then we could ask why nature has not selected an oxygen atom instead of nitrogen since in this case the energy of interaction of the H-bond would be improved. The answer could be the valence. If an oxygen is placed in this position, it would most likely lead to a terminal OH group, forbidding any extension of the C2-linked group, depriving it of several H-bond and van der Waals possible interactions. Another possibility would be to link an acetyl group to this oxygen, as occasionally found in some sialic acid derivatives. However, this solution would add a second oxygen atom, so that the whole group would behave exclusively as H-bond acceptor. In the case of nitrogen, the NH group is both acceptor and donor of H-bonds, which increases the possibilities of interactions with vicinal oligosaccharide OH groups. Thus, as far as H-bonding is concerned, nitrogen appears clearly superior to oxygen, and this may explain the incorporation of N-acetyl sugars in the oligosaccharide part of most gangliosides. As shown in the Fig. 6, when two GM1 are placed in an orientation in which the sialic acids are as closest as possible for optimizing surface contact, the extension of the chain allows a molecular interaction with the GM1 partner in a way that the negative charges of both gangliosides are mitigated by the barrier highlighted in green. Overall, these data give a rational biochemical background explaining how gangliosides can form condensed microdomains despite the presence of a strong electronegative potential in their glycone parts.

The attraction and stabilization of gangliosides in lipid rafts is further facilitated by ceramide-ceramide and ceramide-cholesterol interactions. In this case, both van der Waals forces and H-bonds are involved (Fantini, Yahi, & Garmy, 2013; Yahi, Aulas, & Fantini, 2010). The smooth surfaces

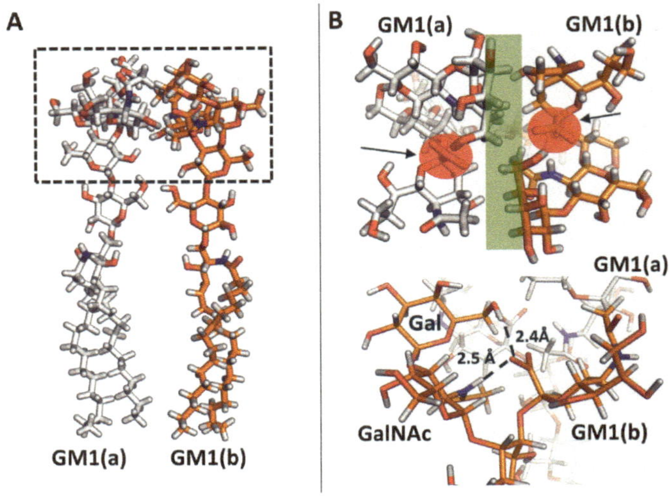

Fig. 6 An anti-repulsive shield allows close contacts between gangliosides in biological membranes. (A) Two GM1 molecules (GM1(a), carbon atoms in white, and GM1(b) with carbons in orange) in a membrane compatible orientation. The dashed frame is analyzed in (B). (B) Carboxylate groups are indicated by arrows with their negative charges highlighted in red. The anti-repulsive shield between these negative charges is highlighted in green (upper panel). The intramolecular H-bond network of GM1(b) is shown the lower panel. The acetyl group linked to the nitrogen atom of GalNac also contributes to the shield through the NH trick effect. A similar mechanism applies for GM1(a).

of ceramide and of the alpha face of cholesterol (Fig. 7) are particularly suited for establishing a coordinated network of van der Waals interactions over the apolar lipid chains. As sphingolipids have both acceptor and donor H-bond groups, cholesterol and sphingolipids are also stabilized in the membrane by H-bonds. Moreover, one should note that the acyl chain of ceramide may contain a hydroxyl group (noted α as it is linked to the C2 carbon, in α of the C1 carbonyl group) which has been shown to orient the sugar part of the ganglioside through a fine conformational tuning effect. In absence of such α-OH group, it is the 3-OH of cholesterol which fulfills this function (Fantini et al., 2013; Yahi et al., 2010).

Finally, lipid rafts are not only characterized by their lipid content, but also by their physical properties. The development of the lipid raft model of the plasma membrane led to new considerations on the phase separation between cholesterol-rich and cholesterol-poor domains (Brown, 1998; Fantini et al., 2002). The higher degree of freedom between glycerophospholipids confer liquid disorder properties of the phase which, for this

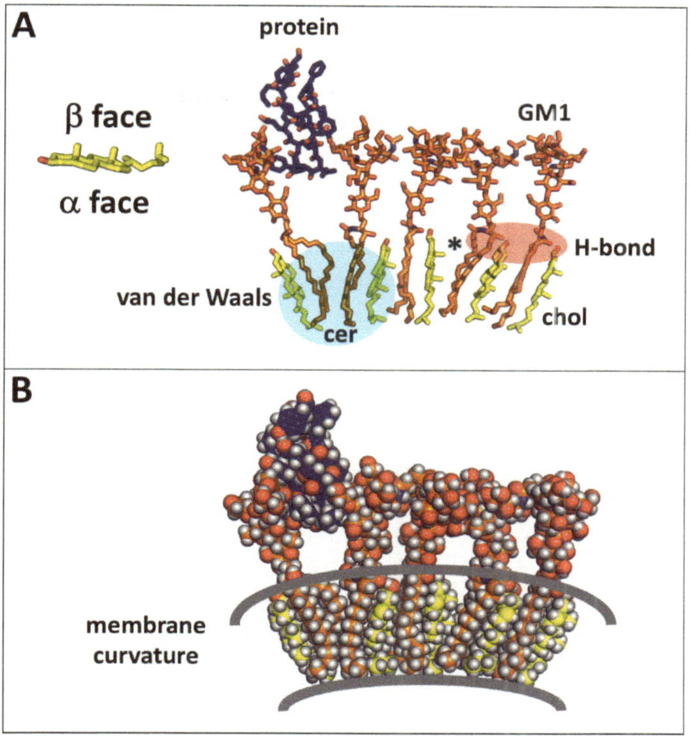

Fig. 7 Molecular organization of ganglioside-cholesterol complexes in lipid rafts. (A) Molecular mechanisms of lipid raft organization. The ceramide (cer) part of GM1 gangliosides interact with both the α (smooth) and β (rough) faces of cholesterol (chol) through van der Waals contacts. The β face of cholesterol generates some local disorder, which is compensated by the optimized interactions provided by the α face. A H-bond network stabilizes the binding of cholesterol to the ceramide part of GM1. The extracellular parts of GM1 are tightly bound in the central area of the raft, whereas there is more flexibility at the periphery of the raft where typical chalices can be penetrated by extracellular proteins. (B) Whole atom model of the raft shown in (A). Note that the raft organization induces a membrane curvature.

reason, is referred to as the Ld (liquid disorder phase). In contrast, lipid rafts are more cohesive and form tighter domains referred to as the Lo (liquid ordered) phase. This phase separation is also consistent with the higher melting temperature of gangliosides (207–230 °C for GM1) vs phosphatidylcholine (−2 °C for POPC). The enrichment of cholesterol (melting temperature of ca. 150 °C) in lipid rafts and its lower content in phosphatidylcholine-rich regions is also in line with the existence of two distinct physical states at physiological temperature (37 °C) (Fig. 2). Yet these physical parameters underline the overlooked notion that cholesterol increases membrane rigidity in the Ld phase whereas it fluidizes the Lo phase. In other words, it is not correct to assign a general rigidifying effect of cholesterol on biological membranes as it strongly depends on the biochemical content of the membrane and in particular the lipids with which cholesterol is in direct interaction (Fantini et al., 2002).

4. Flat surfaces and chalices: Same gangliosides, two topologies

The molecular organization of a typical lipid raft is shown in Fig. 7. It is obvious that the gangliosides located in the internal (central) region of the raft have a distinct topology compared with gangliosides at the edge of the microdomain. In central areas, there are numerous interactions between the glycone parts of neighbor gangliosides, which altogether form a high density network from which water molecules are excluded. Indeed, these zones form a coherent flat surface that is directly available for potential ligands without important conformational adjustments. In contrast, the edge of the raft is characterized by the presence of gangliosides with a higher degree of freedom. Such gangliosides can move their glycone moieties apart from the flat surface and have the possibility to form a chalice-like structure into which selected ligands can penetrate deeply (Fig. 7).

An intriguing aspect of the raft border is the systematic presence of cholesterol which may allow the coalescence of two distinct rafts (Fantini & Barrantes, 2013). In this case, the gangliosides previously located at the edge of both rafts now belong to the main flat surface that is characteristic of central raft areas. Thus, there is no predestination for an individual ganglioside to be part of a flat surface or to form a chalice with a partner at the edge of a raft. Again, this feature is typically an emerging property which may be switched (and potentially switched back) between two topologies, each being associated with a specific binding function. This specificity illustrates the high binding potential together with the versatility of gangliosides for proteins (Fantini, 2003). These considerations are of interest before thinking about the definition and characteristics of ganglioside binding domains in proteins. Indeed, in marked contrast with usual protein-ligand interactions, the way protein binds gangliosides depends not only on the biochemical structure of the ganglioside but also, and probably more importantly, on the way target gangliosides interact with their neighbors in specific areas of lipid raft domains.

5. Ganglioside binding domains (GBDs)

Schematically, two distinct classes of proteins may interact with gangliosides: host proteins (endogenous) and microbial proteins (exogenous). Endogenous proteins include either membrane or extracellular proteins. Microbial proteins (adhesins and toxins) are generally extracellular and most of them may use gangliosides as receptors. However, gangliosides are generally not the only molecular mechanism of interaction with the plasma membrane. Indeed, gangliosides are often referred to as "accessory" binding sites instead of true receptors for extracellular proteins (Hakomori, 1984). Nevertheless, at least the first contact of those proteins with the host cell membrane may rely on gangliosides exclusively. Extracellular proteins take advantage of the fact that their orientation is not limited by the network of molecular interactions which constrain host membrane proteins. Beyond these considerations, there are several parameters that control protein binding to gangliosides among which the geometric complementarity and the chemical compatibility play a critical role. However, as discussed above, the stoichiometry of the protein-ganglioside complex may strongly depend on the localization of the target gangliosides inside a raft or at its edge. When a protein binds to a single ganglioside, the energy of interaction of the resulting complex reflects the affinity of this protein for the ganglioside. When several gangliosides constitute the membrane binding site of the protein, the energy of interaction of the complex reflects the avidity of the protein for the cell surface.

This discussion illustrates the difficulty to precisely define the notion of ganglioside binding domain (GBD). Nevertheless, several lines can be drawn to circumvent the problem. Our strategy in this chapter is to characterize three distinct categories of GBDs instead of one generic GBD. At the molecular level, all these GBDs share common features and similar binding mechanisms. From a structural point of view, each GBD type has a characteristic topology, which can be a helical domain, a small loop, or a large flat surface.

6. Helical GBDs

The topology of transmembrane proteins allows simultaneous interactions with the ceramide part and the extracellular oligosaccharide domain of the same ganglioside (Fig. 8). A perfect example is given by synaptotagmin-2 (Syt-2), a protein associated with synaptic vesicles, and

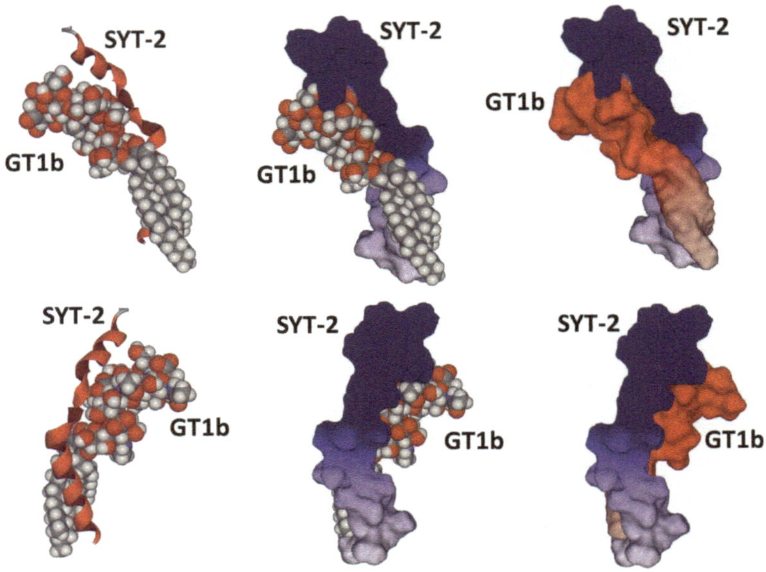

Fig. 8 Typical helical GBD: Synaptotagmin-2 (SYT-2). Front (upper panels) and rear (lower panels) views of SYT-2 bound to ganglioside GT1b. α-Helix backbones (red) are shown on the left panels. In the left and middle panels, GT1b is in atomic sphere rendition. Surface potential of GT1b and SYT-2 are shown in the middle and right panels (blue, positive; red, negative; white, neutral).

the tri-sialosyl ganglioside GT1b. The geometric complementarity of both partners in the complex results in part from the flexibility of the protein which can adapt its shape to wrap around the ganglioside. Interestingly, before interacting with GT1b, the extracellular domain of Syt-2 is intrinsically disordered (Fealey, Binder, Uversky, Hinderliter, & Thomas, 2018). Yet upon binding to GT1b, the protein adopts a helical configuration which forms a functional receptor for botulinum toxin (Flores et al., 2019). This phenomenon is a perfect illustration of the chaperone effect of gangliosides on proteins (Fantini & Barrantes, 2009, 2013; Sciacca et al., 2020; Yu, Bieberich, Xia, & Zeng, 2004). The electrostatic surface potential of Syt-2 is positive in the extracellular part, and neutral in the membrane-embedded portion of the protein. The extracellular part of GT1b is globally electronegative, whereas the ceramide is neutral. Thus, the geometric complementarity is associated with a chemical compatibility which facilitates the formation of the complex through a kinetic parameter. The combination of both kinetic and affinity parameters controls the protein-ganglioside binding reaction which, in this case, reached a ΔG value of $-168\,\text{kJ}\,\text{mol}^{-1}$ (Flores et al., 2019). Such high ΔG values are commonly measured for any ganglioside-protein complex.

The molecular mechanism of Syt-2-GT1b interaction involves the aromatic residues F54 and W66, the basic K53 and K60, and the acidic E57 (Flores et al., 2019). Despite the presence of three carboxylate groups in GT1b, the complex did not display any electrostatic interaction between those electronegative groups and basic amino acid residues of Syt-2. In fact, the complex was chiefly stabilized by hydrogen bonds, and—to a lesser extent—by sugar aromatic CH—π (Nishio, Umezawa, Fantini, Weiss, & Chakrabarti, 2014) and OH—π interactions (Fantini & Yahi, 2013).

7. Loop shaped GBDs

The finding of a structural homology between the V3 loop of HIV-1 surface envelope glycoprotein gp120 and amyloid proteins (Alzheimer's β-amyloid peptide and the cellular prion protein PrP^c) revealed the existence of a common sphingolipid-binding domain (SBD) in proteins with little or no amino acid sequence homology (Mahfoud et al., 2002). This universal structural domain has been found in a broad range of microbial (Fantini, 2003; Fantini, Garmy, & Yahi, 2006), amyloid (Di Scala, Yahi, Boutemeur, et al., 2016; Di Scala, Yahi, Flores, et al., 2016; Fantini & Yahi, 2010, 2011, 2013, 2015; Fantini et al., 2013), plasma membrane proteins and receptors (Aubert-Jousset et al., 2004; Chakrabandhu et al., 2008; Chattopadhyay et al., 2012; Hamel, Fantini, & Schweisguth, 2010; Heuss et al., 2013; Huang et al., 2018; Taïeb et al., 2009).

The SBD consist in a small loop (usually <30 amino acids) displaying a typical combination of basic, aromatic and turn-inducing residues. For instance, the tip of the V3 loop of HIV-1 gp120 has a consensus motif GPGRAF (Delézay, Hammache, Fantini, & Yahi, 1996; Yahi, Fantini, et al., 1995; Yahi et al., 1994; Yahi, Sabatier, Baghdiguian, Gonzalez-Scarano, & Fantini, 1995) which fulfills these criteria: GPG are turn-inducing amino acids, R is basic and F aromatic. Histidine, which can be considered as aromatic (rather than basic as the p*Ka* of the imidazolium group is <7) is also frequently found in these motifs. Several types of sphingolipids may interact with a SBD, including sphingomyelin, neutral glycosphingolipids, and gangliosides. The loop shaped structure of PrP^c, a typical ganglioside-binding domain (GBD) bound to GM1 (Fantini & Yahi, 2015), is shown in Fig. 9.

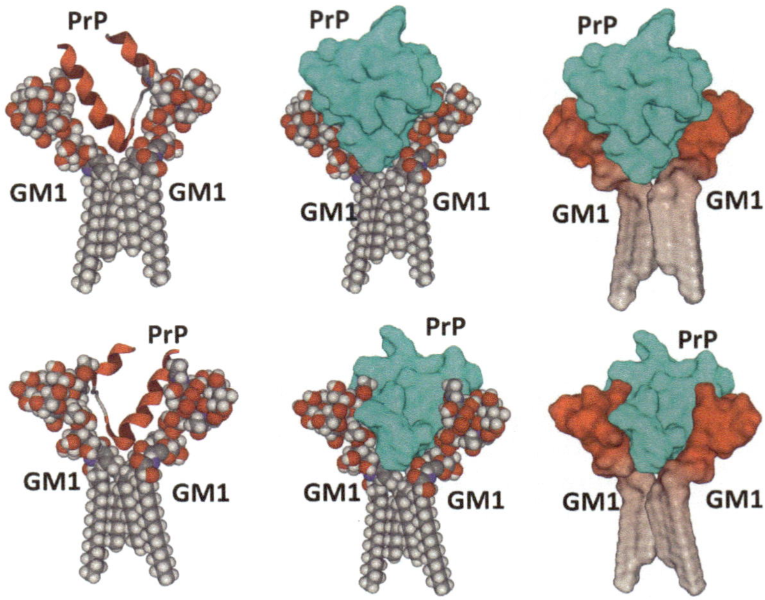

Fig. 9 A typical loop shaped GBD: the cellular prion protein PrP^C. The models represent the loop shaped GBD of human PrP^C (residues 198–208) bound to a dimer of gangliosides GM1 in a chalice-like configuration. Front views are shown in the upper panels, rear views in the lower panels. The GBD of PrP^C is represented in secondary structure (left panels) or in surface rendering (colored in cyan to improve clarity, middle and right panels). GM1 are represented in atomic sphere rendition (left and middle panel) or in electrostatic surface potential colors (middle and right panels), using the same color code as in Fig. 8.

Such loop shaped GBDs may preferentially interact with ganglioside dimers organized in chalices for which they have an optimized geometric complementarity. At the molecular level, the aromatic and basic residues of the motif generally interact with the oligosaccharide parts of the gangliosides, through H-bonds and sugar-aromatic CH—π stacking (Fantini, 2003; Fantini & Yahi, 2010). Since sialic acid carboxylate groups are generally neutralized by the NH of the amide group of GalNAc (NH trick effect) (Fig. 5), electrostatic bridges between the GBD and gangliosides are generally not observed.

For amyloid proteins (Aβ and α-synuclein), a consensus GBD motif has been defined: K/H/R/−X(1–4)−Y/F−X(4–5)−K/H/R, where at least one of the X(1–4) residues is glycine (Fantini & Yahi, 2011). This consensus motif can be used to predict the presence of functional GBDs from amino acid sequence data (Flores et al., 2019).

8. Large flat GBDs

A third type of GBD has been recently discovered in the spike protein of SARS-CoV-2 (Fantini, Di Scala, Chahinian, & Yahi, 2020), the coronavirus associated with the Covid-19 pandemic (Wu et al., 2020). The spike protein forms a trimeric structure which interacts with ACE-2, an enzymatic protein present in the plasma membrane of mucosal cells of the respiratory system (Fantini, Chahinian, & Yahi, 2020b; Fantini, Di Scala, et al., 2020) and, at higher levels, of the gastrointestinal tract (Devaux, Lagier, & Raoult, 2021; Zhang et al., 2020). However, ACE-2 recognition is not the first step of virus infection. In fact, the trimer displays three peripheral domains referred to as NTD (N-terminal domain) and a hidden receptor-binding domain (RBD) which interacts with the ACE-2 receptor after a conformational change which occurs in lipid raft under the control of gangliosides (Fantini, Chahinian, & Yahi, 2021). Interestingly, although both the NTD and the RBD cooperate to optimize the contacts between the virus and the host cell membrane, the strategy used by each domain to reach its respective target is totally different.

The NTD displays a large flat surface which can spread on a lipid raft as illustrated in Fig. 10. Thus, the whole NTD surface constitutes a new type of

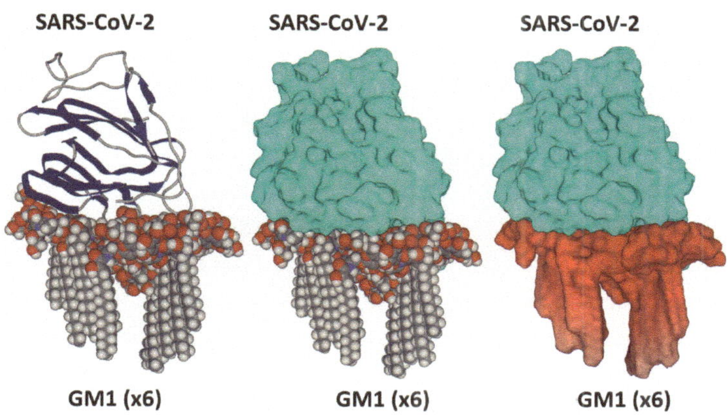

Fig. 10 The Flat GBD of the NTD of SARS-CoV-2 spike protein. The models represent the NTD bound to a cluster of 6 GM1 gangliosides in the central zone of a lipid raft. In the left panel the secondary structure of the NTD is shown. In the middle and right panels the surface of the NTD is represented in cyan. GM1 molecules are in atomic sphere rendition (left and middle panels) or in electrostatic surface potential colors (right panel), using the same color code as in Figs. 8–9.

GBD with more than 100 amino acid residues. The electrostatic surface potential of the NTD is globally positive, so that electrostatic interactions can mediate a rapid targeting to lipid rafts (Fantini, Yahi, Azzaz, & Chahinian, 2021). The mean distance between the NTDs is 120 Å (to be compared with 18 Å for the ganglioside-cholera toxin complex, as described below), so that each NTD may initially bind to a separate raft. By providing three stable contact points for the spike trimer, lipid rafts may then help the RBD to find ACE-2 on the host cell membrane (Fantini, Yahi, et al., 2021).

The binding process, which can be metaphorically compared to the docking of a spacecraft to an orbital station, can be broken down in four successive steps: (i) NTD binding to lipid raft gangliosides through electrostatic interactions, (ii) stabilization of the spike on the surface of the rafts following binding to three distinct rafts, one for each NTD, (iii) lipid raft dependent recruitment of ACE-2, and (iv) conformational adjustment of the central RBD allowing optimal positioning of ACE-2, ensuring the locking of the virus on the cell surface. This general mechanism is consistent with the absolute requirement of gangliosides as attachment cofactors of coronaviruses (Engin, Engin, & Engin, 2020; Hulswit et al., 2019; Li et al., 2017; Matrosovich, Herrler, & Klenk, 2015). In this respect, direct binding of cationic hydroxychloroquine to anionic cell surface gangliosides (Fantini, Di Scala, et al., 2020) is consistent with the antiviral activity of this compound against a broad range of viruses (Rolain, Colson, & Raoult, 2007), including SARS-CoV-2 (Lagier et al., 2020). From a methodological point of view, it could be interesting to use human intestinal cells (Galons, Fantini, Vion-Dury, Cozzone, & Canioni, 1989) as in vitro models for coronavirus infection, as these cells express a variety of neutral glycospingolipids and gangliosides able to promote the infection of several types of RNA viruses (Chan et al., 2004; Fantini, Cook, Nathanson, Spitalnik, & Gonzalez-Scarano, 1993; Hu et al., 2012; Huang, Chio, & Lin, 2018; Hubert et al., 2020; Qu et al., 2012; Yahi, Baghdiguian, & Fantini, 1995). Such models may help to identify and evaluate the efficiency of ganglioside-based therapeutic compounds as inhibitors of human RNA virus infection (Yager & Konan, 2019).

9. Gangliosides and synapses

Gangliosides have two distinct functions in synaptic regulation. On the one hand, most neurotransmitter receptors and ion channels segregate in lipid rafts, a location which facilitates physical contacts with gangliosides.

Extracellular juxtamembrane domains often display a ganglioside binding site. It is the case for the serotonin receptor 5HT1A which has been shown to specifically interact with the first extracellular loop (residues 99–109). This motif contains a characteristic combination of basic (K101), aromatic (W102) and turn-inducing residues (G105), which are often found in loop shaped GBDs (Mahfoud et al., 2002; Prasanna, Jafurulla, Sengupta, & Chattopadhyay, 2016).

On the other hand, ganglioside clusters in lipid rafts form an electronegative field on the post-synaptic neuronal surface which may attract cationic neurotransmitters (e.g., serotonin) and, on the contrary, repulse electronegative neurotransmitters (e.g., glutamate) (Fantini & Yahi, 2015). These functions are highly dependent on specific ganglioside expression in neurons, which synthesize all four major brain gangliosides (GM1, GD1a, GD1b, and GT1b) but with a different distribution in pre- and post-synaptic membranes, and in astrocytes, which express chiefly GM3, a mono-sialosyl ganglioside (Fantini & Yahi, 2015). Thus, neuronal membranes have a higher density of negative charges compared to astrocytes. Since glutamate is also negatively charged at physiological pH, it is repelled by neurons and directed to astrocytes which will uptake it and transform it into glutamine. The dissymmetric expression of gangliosides between astrocytes and neurons allows a rapid clearance of glutamate from the synaptic cleft. In the case of cationic neurotransmitters (e.g., serotonin), neuronal gangliosides are not repulsive but attractive, and their presence in the post-synaptic membrane facilitates the delivery of neurotransmitters to their receptors.

Rett syndrome, a neurodevelopmental disease affecting girls, illustrates the critical role of gangliosides in the control of neurotransmitters fluxes throughout the synapse. The root cause of Rett syndrome has not be elucidated, but increased levels of excitatory neurotransmitters such as glutamate have been postulated (Lappalainen & Riikonen, 1996). Children with the Rett syndrome also have a lower expression of di-sialosyl gangliosides (GD1a in the white matter and GD1b in temporal gray matter) (Lekman, Hagberg, & Svennerholm, 1991). Thus, it is possible that the electronegative field in this pathological condition is not strong enough to clear glutamate from the synapse. In this case, a therapeutic option would be to treat the patients with exogenous gangliosides, as suggested for Parkinson's disease patients (Ledeen & Wu, 2015; Schneider, Gollomp, et al., 2013; Schneider, Kean, & DiStefano, 1995; Wu, Lu, Kulkarni, & Ledeen, 2012).

10. Gangliosides and bacterial toxins

Cholera toxin is a bacterial toxin produced by *Vibrio cholerae* with an hexameric structure, a central A subunit and five peripheral B subunits (Merritt et al., 1998). Each B subunit displays a GM1 binding domain (Fig. 11). Analysis of the electrostatic surface potential of the pentamer clearly shows that one face (F1) is more electropositive than the other one (F2) (Fig. 11A). As expected, it is precisely this face of the toxin that interacts with lipid rafts. The GM1 binding site is organized in 3 distinct loops (L1, L2 and L3) harboring aromatic (Y12, H13, W88) cationic (K91) and H-bond forming (Q56) amino acid residues (Fig. 11A and B). Each of these loops can be interpreted as a single loop shaped GBD.

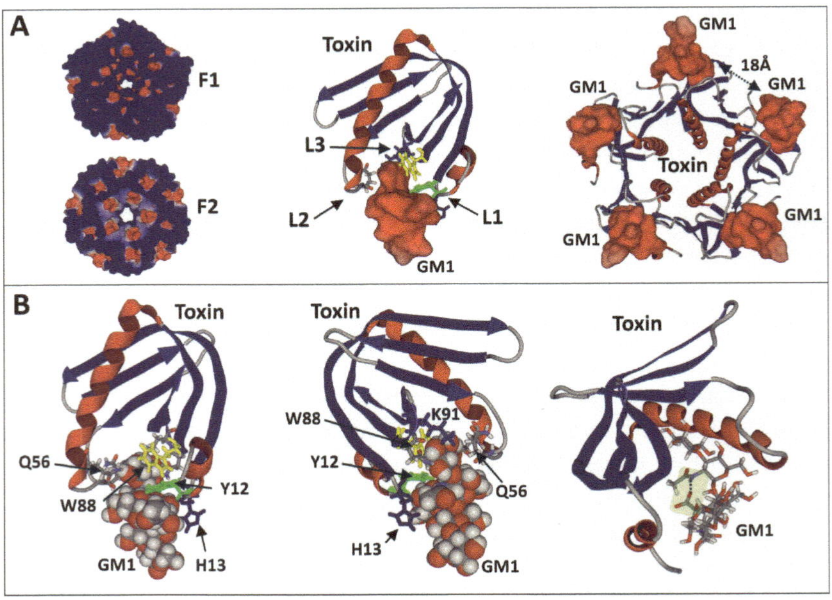

Fig. 11 The ganglioside binding site of cholera toxin is a cluster of 3 loop shaped GBDs. (A) Overall views of the structure of cholera toxin B subunits. Left panel, electrostatic surface potential of two opposite faces (F1 and F2) of the pentameric toxin; middle panel, structural organization of the ganglioside binding site in three distinct loop shaped GBDs (L1, L2 and L3). Amino acid residues highlighted are analyzed in (B); right panels, stoichiometry of GM1-toxin binding reaction. Note that two vicinal GM1 oligosaccharides at a mean distance of 18 Å. (B) Molecular mechanisms of GM1-toxin interactions (left and middle panels). The internal neutralizing shield conferred by the amide group of GalNAc is shown in the right panel.

The mean distance between two GM1 gangliosides in the pentameric complex is 18 Å (Fig. 11A), which suggests that a single pentamer may find all 5 GM1 gangliosides within the same lipid raft. As mentioned above, the carboxylate charge of Neu5Ac is neutralized by the NH of GalNAc (NH trick effect), so that the complex is stabilized by H-bonds and stacking interactions (CH—π and OH—π) rather than electrostatic bridges (Fig. 11B).

Botulinum toxin is another example of toxins that use gangliosides as initial binding sites on host cell membranes (Pirazzini, Rossetto, Eleopra, & Montecucco, 2017). The toxin consists of a light chain (L, 50 kDa) and a heavy chain (H, 100 kDa) linked by a disulfide bond. The C-terminal part of the H chain (HC) binds with high affinity to the presynaptic plasma membrane of skeletal and autonomic cholinergic nerve terminals. This unique nerve terminal binding property of the toxin is due to its capacity to recognize two independent receptors of the presynaptic membrane: a polysialosyl-ganglioside (e.g., GD1a or GT1b) and the extracellular part of synaptotagmin (Rossetto & Montecucco, 2007; Rossetto, Pirazzini, Fabris, & Montecucco, 2021), which also interacts with a ganglioside (Fig. 8). The recent demonstration that GT1b is an essential component of the synaptotagmin receptor suggested the interesting possibility that Botulinum toxin could form a tetrameric complex on the host cell surface, with two GT1b gangliosides and synaptotagmin bound to the toxin (Flores et al., 2019). Again, an analysis of the electrostatic surface potential of the HC domain of the botulinum toxin (serotype B, referred to as BoNT/B) gave some clues on the mode of toxin binding to the host cell surface (Fig. 12A). As previously noticed by Fogolari, Tosatto, Muraro, and Montecucco (2009), the surface of the toxin facing lipid raft gangliosides is clearly the electropositive one (F1). A part of the oligosaccharide of GT1b is inserted in a cavity of the toxin which delineates the ganglioside binding site (Berntsson et al., 2013). The interaction is stabilized by a network of H-bonds and stacking interactions involving Y117 ($-29.3\,kJ\,mol^{-1}$), W1266 ($-50.2\,kJ\,mol^{-1}$), and R1276 ($-18.6\,kJ\,mol^{-1}$), for a total energy of interaction of $-180\,kJ\,mol^{-1}$ (Fig. 12B). The ganglioside binding site consists of 4 distinct loop shaped GBDs. Electrostatic interactions are not involved in the complex, despite (i) the presence of two sialic acid residues in the GT1b analogue oligosaccharide used for crystallization and (ii) the proximity of several cationic residues of BoNT/B. Since the GBD and the synaptotagmin binding domains of the toxins do not overlap (Berntsson et al., 2013), it seems possible to create, in silico, a tetrameric complex in which the HC domain of BoNT/B interacts simultaneously with its two receptors, i.e., GT1b and a

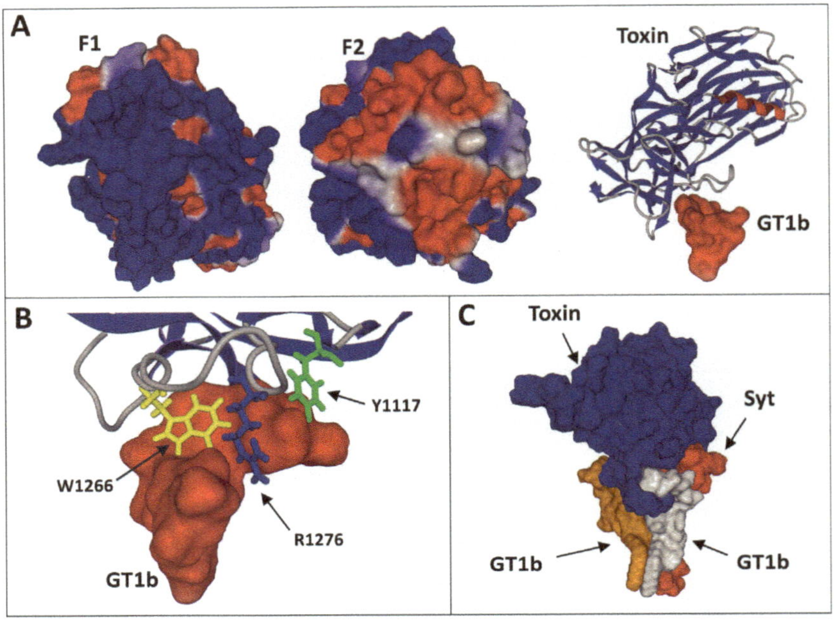

Fig. 12 The ganglioside binding site of botulinum toxin is a cluster of 4 loop shaped GBDs. (A) Overall views of the structure of botulinum toxin BoNT/B. Left and middle panels, electrostatic surface potential of two opposite faces (F1 and F2) of the HC domain of the toxin; right panel, structural organization of the ganglioside binding site in four distinct loop shaped GBDs that cover a large part of the GT1b oligosaccharide. (B) Localization of three amino acid residues that are critical for ganglioside recognition. (C) A tetrameric complex resulting from the molecular associations of the HC domain of BoNT/B (blue), Syt-2 (red) pe-bound to GT1b (white) and GT1b bound to the toxin (orange). Note how all components optimize the molecular contacts to obtain a high affinity complex.

preformed synaptotagmin-GT1b complex, both located in the same lipid raft. Preliminary data suggest that the topology of this tetrameric complex would be consistent with a reorganization of gangliosides whose ceramide portions might eventually converge and interact to further stabilize the tetramolecular assembly (Fig. 12C). In this case, the typical chalice-like shape of the ganglioside dimer is reinforced by a transmembrane protein partner (synaptotagmin). Such molecular mechanisms cannot be studied by classical crystallographic approaches since the crystals are obtained with the soluble parts of gangliosides (oligosaccharide) and synaptotagmin (extracellular α-helix) that totally ignore membrane constraints. In particular, the frequent use of oligosaccharides as surrogate for gangliosides in X-ray diffraction studies of crystals may generate artefactual molecular complexes. Therefore, in silico building of transmembrane domains of proteins and of the ceramide moiety of gangliosides is a prerequisite to molecular modeling studies (MD simulations or at least energy minimization routines) of such complex systems in a membrane compatible environment.

11. Gangliosides and virus infection

Gangliosides play several major roles in the mechanisms of viral infection. Some of these roles are linked to their presence in lipid rafts, as well as to their association with cholesterol, others are more precisely due to their chemical structure which offers multiple possibilities of interaction with viral proteins. As major components of lipid rafts, gangliosides form landing areas for most viruses which target these plasma membrane microdomains as privileged initial contact zones on the host cell surface. Then, the following step depends on the absence or presence of a protein receptor in the lipid raft which has caught the virus. Although most virus receptors are also located in lipid rafts, they may be absent from the specific raft selected by the virus. In this case, the virus will surf on the raft and drift in the plasma membrane until finding a raft with an appropriate receptor. This mechanism is used by HIV-1 which needs to find a fusion coreceptor (e.g., CCR5 or CXCR4) to fuse its envelope with the plasma membrane (Chandramouli et al., 2013). It is also used by coronaviruses such SARS-CoV-2 which seek for ACE-2 after its trimeric spike protein has been trapped in a lipid raft (Fantini, Yahi, et al., 2021). This step will be bypassed if the protein receptor is already present in the lipid raft onto which the virus is bound. The simultaneous presence of a ganglioside and a protein, both used as viral receptors, has been reported for HIV-1 bound to a preformed CD4-GM3 complex (Hammache, Yahi, Maresca, Piéroni, & Fantini, 1999; Hammache, Yahi, Piéroni, et al., 1998). This specific case is particularly interesting as it illustrates a second important mechanism of ganglioside-assisted virus infection. Once bound to CD4, HIV-1 gp120 undergoes a conformational change which demasks its V3 domain, a loop shaped region that binds to GM3 (Hammache et al., 1999). At this stage, the viral glycoprotein is ready to interact with a fusion co-receptor which has a high affinity for the V3 loop. In absence of GM3 (or other accessory glycosphingolipids), the conformational change does not occur, and the virus cannot infect the cell (Puri, Hug, Muñoz-Barroso, & Blumenthal, 1998). Mouse fibroblasts co-transfected with human CD4 and CCR5 can bind HIV-1, but are not infected by the virus unless human glycosphingolipids are fused with the cells (Puri, Hug, Jernigan, et al., 1998; Puri, Hug, Muñoz-Barroso, & Blumenthal, 1998). In this respect, there is an intriguing similarity between HIV-1 and SARS-CoV-2, as both viruses require both lipid raft gangliosides and a protein receptor to gain entry into host cells (Fantini, Yahi, et al., 2021).

Yet this is not the only lesson we can learn from the V3 loop of HIV-1. This domain is the principal neutralization determinant (PND) of HIV-1 and it is highly variable, which explains why the virus is so prone to escape from neutralizing antibodies. The variability of the V3 loop is also associated with a switch of coreceptor usage, starting from CCR5 in the early stages of infection to CXCR4 in later stages. This virus evolution allows the emergence of more aggressive strains in an infected individual if viral replication is not fully controlled by antiretrovirals. Once again, an analysis of the electrostatic surface potential of the coreceptors will help to understand the molecular basis of V3 loop controlled virus selection (Fig. 13). The surface of

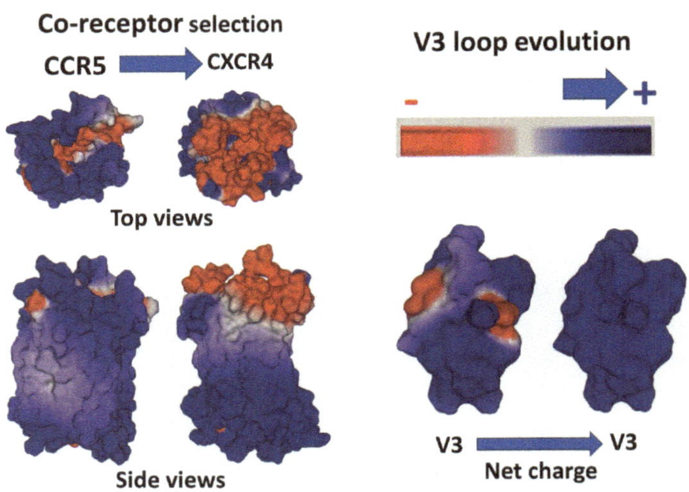

Fig. 13 Virus evolution driven by electrostatic fields. During the course of HIV-1 infection, there is a shift in co-receptor usage from CCR5 to CXCR4. Both co-receptors are G-protein coupled receptors with seven transmembrane domains, yet they markedly differ by their electrostatic surface potential (CXCR4 is more electronegative than CCR5). As these receptors are flush with the membrane, the differences are restricted to the parts that are accessible to extracellular proteins. Virus evolution follows an electrostatic logic, so that the net electric charge of the V3 loop increases stepwise by mutations, until it becomes sufficiently electropositive to use CCXCR4 instead of CCR5. The surface potential scale is indicated (red, negative, white, neutral, blue, positive).

CCR5 that is accessible to the V3 loop is globally positive, with two small electronegative spots. In contrast, the surface of CXCR4 is mostly electronegative. The evolution of the V3 loop will follow these criteria, as an increase of electropositive surface potential will render the virus compatible for CXCR4. Indeed, V3 loops evolve by acquiring a series of mutations which increase the number of cationic amino acid residues (K, R) (Chandramouli et al., 2013; Fantini, Yahi, et al., 2021). An example of such evolution is given in Fig. 13.

Now, can we compare HIV-1 and SARS-CoV-2? ACE-2, the main receptor for SARS-CoV-2 has a strong electronegative surface potential, and it is located in lipid rafts (Lu, Liu, & Tam, 2008). Some mutations of the SARS-CoV-2 spike protein (e.g., E484K) significantly increase the surface potential of the RBD, which may confer a kinetic advantage for reaching the raft and favor the selection of variants. The same mechanisms may also apply for the NTD, which does not interact with ACE-2 but has a high affinity for lipid raft gangliosides as illustrated in Fig. 10 (Fantini, Yahi, et al., 2021). To predict the transmissibility of SARS-CoV-2 variants, we recently proposed an index (T-index) which takes into account: (i) the affinity of the NTD for lipid raft gangliosides, (ii) the affinity of the RBD for ACE-2, and (iii) the electrostatic surface potential of both domains that controls the kinetics of virus binding to electronegative host cell membranes (Fantini, Yahi, et al., 2021). The T-index has correctly predicted the emergence of the delta variant of SARS-CoV-2 which became dominant in several countries at the beginning of summer 2021. Thus, this method has a high potential as a health monitoring strategy for anticipating Covid-19 outbreaks. Overall, these data underscore the critical role of the electrostatic surface potential in virus infection and, more globally, on the binding of microbial proteins to cell surface gangliosides.

12. Gangliosides and neurodegenerative diseases

Gangliosides play a critical role in neurodegenerative diseases, including Alzheimer's (AD) and Parkinson's disease (PD). The proteins responsible for these disorders (Aβ for AD and α-synuclein for PD) are referred to as amyloid proteins. Among their multiple characteristics, they behave as neurotoxins that bind to the plasma membrane of brain cells and form ion-permeable pores in the plasma membrane of brain cells. These neurotoxic structures, called oligomers or amyloid pores (Fantini, Chahinian, &

Yahi, 2020c) display a typical donut-like shape that is clearly visible by atomic force microscopy (Quist et al., 2005). Oligomers of Aβ and α-synuclein have been detected in the brain parenchyma and cerebrospinal fluid (CSF) of both AD (Bruggink et al., 2013) and PD (Emmanouilidou et al., 2011) patients, and there is a clear-cut relationship between the presence of these oligomers and the symptoms (Cline, Bicca, Viola, & Klein, 2018; Esparza et al., 2013). Permanent Ca^{2+} fluxes through amyloid pores trigger a cascade of neurotoxic events including Tau hyperphosphorylation, mitochondrial dysfunction, and finally cell death (Fantini, Chahinian, & Yahi, 2020c).

The very first step of amyloid pore formation is the binding of the amyloid protein to a dimer of gangliosides in a typical chalice-like configuration (Fig. 14). Although the repertory of amyloid proteins for ganglioside is degenerated, Aβ has a marked preference for GM1, which is expressed in neurons, whereas α-synuclein binds preferentially to GM3, which is expressed in astrocytes (Fantini & Yahi, 2011). In both cases, ganglioside recognition is mediated by a common structural loop shaped GBD, which corresponds to amino acids 5–16 in Aβ and 34–45 in α-syn (Di Scala, Yahi, Boutemeur, et al., 2016; Di Scala, Yahi, Flores, et al., 2016; Fantini & Yahi, 2011).

A unique computational approach allowed to simulate the whole process of amyloid pore formation, and to elucidate the biological code controlling ganglioside-protein recognition (Yahi & Fantini, 2014). By combining the respective ganglioside-binding domains of α-synuclein and Aβ, two of us (J. F. and N. Y.) designed a therapeutic synthetic peptide (Fantini, Chahinian, & Yahi, 2020c; Fantini & Yahi, 2011, 2013; Yahi & Fantini, 2014), which blocks neurotoxic pore formation at nanomolar concentrations without any undesirable side effect (Di Scala, Yahi, Boutemeur, et al., 2016; Di Scala, Yahi, Flores, et al., 2016; Yahi & Fantini, 2014). AmyP53 is a competitive inhibitor of Aβ and α-synuclein binding to gangliosides. Its activity is dependent on the presence of two histidine residues derived from the GBD of Aβ (Fantini, Chahinian, & Yahi, 2020c) (Fig. 14). AmyP53 is the first ever therapeutic compound that targets brain gangliosides for treating both AD and PD. It has been intentionally designed to tackle a common molecular mechanism of neurotoxicity that is considered as the actual root cause of both neurodegenerative diseases.

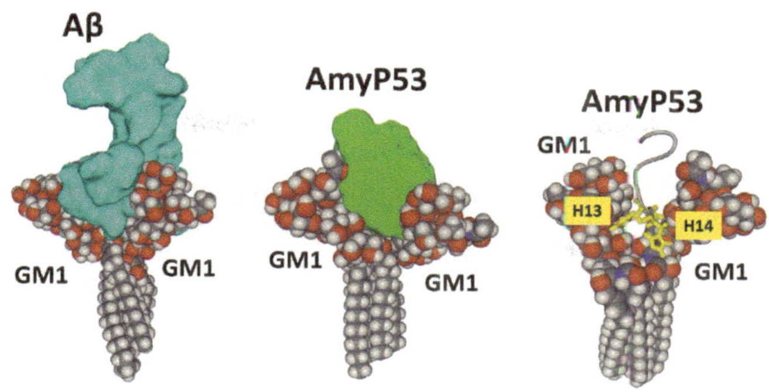

Fig. 14 Gangliosides and amyloid proteins. Ganglioside dimers are used by amyloid proteins (e.g., Aβ) as molecular factories that generate neurotoxic oligomers (Ca^{2+} permeable amyloid pores) in Alzheimer's and related neurodegenerative diseases. The very first step of this process is the insertion of Aβ within the chalice-like structure of a GM1 dimer (left panel). The therapeutic peptide AmyP53 has been designed to prevent this mechanism of neurotoxicity by competitive inhibition (middle panel). Two histidine residues (which correspond to H13 and H14 in Aβ) are critical for the biological activity of AmyP53 (right panel).

13. Conclusion and perspectives

Gangliosides play a pivotal role in a broad range of physiological processes and pathological conditions. These glycosphingolipids are rarely working alone so that their biological significance is typically an emergent property resulting from multiple molecular interactions involving several plasma membrane lipids. Like other sphingolipids, gangliosides have a high affinity for cholesterol whose conic shape shows a remarkable geometric complementarity for inverted cone lipids. For this reason, gangliosides do not systematically micellize in the outer leaflet of the plasma membrane but can form instead condensed microdomains referred to as lipid rafts. The surface potential of a ganglioside-enriched lipid raft is globally electronegative, due to the presence of a high number of oxygen atoms in the extracellular oligosaccharide portion of these glycolipids. Microbial proteins that interact with gangliosides have an electropositive surface that is oriented and attracted to the cell surface by ganglioside clusters. Virus evolution is often a race for cationic amino acid residues that increase the surface potential and give a kinetic advantage for reaching gangliosides on the plasma membrane of host cells. Such evolutionary pathways have been described for HIV-1 gp120 and more recently for SARS-CoV-2, the etiologic agent of the Covid-19 pandemic. Interestingly, viruses with a highly positive surface potential are not only more easily transmitted than other variants, but they tend to become resistant to neutralizing antibodies because their ganglioside binding site is also a major neutralization domain of the virus. In this respect, there is a striking functional similarity between the V3 loop of HIV-1 gp120 and the NTD of the SARS-CoV-2 spike protein. Both interact with gangliosides, both trigger the synthesis of neutralizing antibodies. The same rules apply for bacterial toxins, as illustrated in this review with cholera and botulinum toxins which display an electropositive surface particularly well adapted for cell surface gangliosides.

Common molecular mechanisms control protein-ganglioside interactions, which heavily rely on H-bonding and stacking interactions (CH—π, OH—π). However, despite the presence of at least one anionic charge in all gangliosides, electrostatic interactions are generally not operative in protein-ganglioside complexes. The reason for this intriguing observation is the "NH trick" that we define as an electrostatic shield generated by the neutralization of carboxylate by a H-bond formed by the amide group of GalNAc in the oligosaccharide portion of gangliosides GM1, GD1a, GD1b and GT1b.

Based on these biochemical features, one can predict that most GBDs fulfill a consensus amino acid sequence such as the consensus motif found in amyloid proteins (Fantini & Yahi, 2011). However, a major outcome of this survey is that GBDs may adopt several distinct topologies, including a α-helix (e.g., synaptotagmin), a single loop (e.g., the 5-HT1A receptor, and the V3 loop of HIV-1 gp120), a cluster of distinct loops (e.g., cholera and botulinum toxins) or a large flat surface (e.g., the NTD of the SARS-CoV-2 spike protein). So far, the consensus algorithm has successfully predicted α-helical and single loop GBDs in a broad range of physiological and pathological proteins (Aubert-Jousset et al., 2004; Chakrabandhu et al., 2008; Fantini & Yahi, 2015; Fantini et al., 2006; Flores et al., 2019; Taïeb et al., 2009; Yahi & Fantini, 2014), which is indeed remarkable considering the simplicity of the formula (K/H/R/−X(1–4)−Y/F−X (4–5)−K/H/R, where at least one of the X(1–4) residues is glycine). We are currently working on an improved algorithm aimed at detecting three-dimensional GBDs resulting from the clustering of distinct protein regions creating a functional ganglioside binding site.

One significant difficulty to predict such ganglioside binding sites is that the geometry of gangliosides is different according to their localization in the lipid raft. Chalice-like dimers are probably restricted to the edge of the raft, which is the privileged binding area of loop shaped GBDs (e.g., amyloid proteins Aβ and α-synuclein that penetrate deeply into the chalice and eventually separate the gangliosides to form a pore in the membrane). The central zone of the raft is more rigid and condensed and it can be used as a solid large and flat landing area for coronaviruses. Obviously, the structure of a GBD that binds to a ganglioside dimer is very different from a GBD that targets a large flat surface.

This versatile topology is consistent by the pleiotropic regulatory functions exerted by gangliosides. Apart from constituting a large repertory of binding sites for a broad range of proteins, gangliosides can also influence protein behavior by both indirect and direct effects. Indirect effects include membrane curvature, membrane fluidity, and optimized delivery to cholesterol. Direct effects are chiefly conformational, as gangliosides have a chaperone activity that controls or at least finely tunes protein conformation, oligomerization, and sorting through direct GBD-mediated interactions.

For all these reasons, gangliosides are now considered as major targets for innovative therapeutic avenues in cancer, microbial and neurodegenerative diseases. We anticipate that such therapies will grow exponentially in the near future as the scientific rationale and biochemical background are now well established. In this respect, the elucidation of the role of gangliosides in the formation of neurotoxic oligomers in AD and PD is clearly a breakthrough that will help to design therapeutic strategies that tackle for the first time the root cause of these diseases (Fantini, Chahinian, & Yahi, 2020c; Yahi, Di Scala, Chahinian, & Fantini, 2021).

References

Allen, J. A., Halverson-Tamboli, R. A., & Rasenick, M. M. (2007). Lipid raft microdomains and neurotransmitter signalling. *Nature Reviews. Neuroscience, 8*, 128–140.

Aubert-Jousset, E., Garmy, N., Sbarra, V., Fantini, J., Sadoulet, M. O., & Lombardo, D. (2004). The combinatorial extension method reveals a sphingolipid binding domain on pancreatic bile salt-dependent lipase: Role in secretion. *Structure, 12*, 1437–1447.

Aureli, M., Grassi, S., Sonnino, S., & Prinetti, A. (2016). Isolation and analysis of detergent-resistant membrane fractions. *Methods in Molecular Biology, 1376*, 107–131.

Berntsson, R. P., Peng, L., Dong, M., & Stenmark, P. (2013). Structure of dual receptor binding to botulinum neurotoxin B. *Nature Communications, 4*, 2058.

Bourdoulous, S., & Lemichez, E. (2018). Decoding glycan recognition by bacterial toxins. *Nature Microbiology, 3*, 124–126.

Brocca, P., Acquotti, D., & Sonnino, S. (1993). 1H-NMR study on ganglioside amide protons: Evidence that the deuterium exchange kinetics are affected by the preparation of samples. *Glycoconjugate Journal, 10*, 441–446.

Brown, R. E. (1998). Sphingolipid organization in biomembranes: What physical studies of model membranes reveal. *Journal of Cell Science, 111*(Pt. 1), 1–9.

Brown, D. A., & London, E. (1998). Functions of lipid rafts in biological membranes. *Annual Review of Cell and Developmental Biology*, *14*, 111–136.

Bruggink, K. A., Jongbloed, W., Biemans, E. A., Veerhuis, R., Claassen, J. A., Kuiperij, H. B., et al. (2013). Amyloid-β oligomer detection by ELISA in cerebrospinal fluid and brain tissue. *Analytical Biochemistry*, *433*, 112–120.

Buch, M. H., Liaci, A. M., O'Hara, S. D., Garcea, R. L., Neu, U., & Stehle, T. (2015). Structural and functional analysis of murine polyomavirus capsid proteins establish the determinants of ligand recognition and pathogenicity. *PLoS Pathogens*, *11*, e1005104.

Cai, M., Zhao, W., Shang, X., Jiang, J., Ji, H., Tang, Z., et al. (2012). Direct evidence of lipid rafts by in situ atomic force microscopy. *Small*, *8*, 1243–1250.

Chakrabandhu, K., Huault, S., Garmy, N., Fantini, J., Stebe, E., Mailfert, S., et al. (2008). The extracellular glycosphingolipid-binding motif of Fas defines its internalization route, mode and outcome of signals upon activation by ligand. *Cell Death and Differentiation*, *15*, 1824–1837.

Chan, P. K., To, K. F., Lo, A. W., Cheung, J. L., Chu, I., Au, F. W., et al. (2004). Persistent infection of SARS coronavirus in colonic cells in vitro. *Journal of Medical Virology*, *74*, 1–7.

Chandramouli, B., Chillemi, G., Giombini, E., Capobianchi, M. R., Rozera, G., & Desideri, A. (2013). Structural dynamics of V3 loop with different electrostatics: Implications on co-receptor recognition: A molecular dynamics study of HIV gp120. *Journal of Biomolecular Structure & Dynamics*, *31*, 403–413.

Chattopadhyay, A., Paila, Y. D., Shrivastava, S., Tiwari, S., Singh, P., & Fantini, J. (2012). Sphingolipid-binding domain in the serotonin(1A) receptor. *Advances in Experimental Medicine and Biology*, *749*, 279–293.

Cline, E. N., Bicca, M. A., Viola, K. L., & Klein, W. L. (2018). The amyloid-β oligomer hypothesis: Beginning of the third decade. *Journal of Alzheimer's Disease*, *64*, S567–s610.

Davies, J. R., Masuyer, G., & Stenmark, P. (2020). Structural and biochemical characterization of botulinum neurotoxin subtype B2 binding to its receptors. *Toxins (Basel)*, *12*, 603.

Davies, L. R., & Varki, A. (2015). Why is N-Glycolylneuraminic acid rare in the vertebrate brain? *Topics in Current Chemistry*, *366*, 31–54.

Delézay, O., Hammache, D., Fantini, J., & Yahi, N. (1996). SPC3, a V3 loop-derived synthetic peptide inhibitor of HIV-1 infection, binds to cell surface glycosphingolipids. *Biochemistry*, *35*, 15663–15671.

Devaux, C. A., Lagier, J. C., & Raoult, D. (2021). New insights into the physiopathology of COVID-19: SARS-CoV-2-associated gastrointestinal illness. *Frontiers in Medicine*, *8*, 640073.

Di Scala, C., Yahi, N., Boutemeur, S., Flores, A., Rodriguez, L., Chahinian, H., et al. (2016). Common molecular mechanism of amyloid pore formation by Alzheimer's β-amyloid peptide and α-synuclein. *Scientific Reports*, *6*, 28781.

Di Scala, C., Yahi, N., Flores, A., Boutemeur, S., Kourdougli, N., Chahinian, H., et al. (2016). Broad neutralization of calcium-permeable amyloid pore channels with a chimeric Alzheimer/Parkinson peptide targeting brain gangliosides. *Biochimica et Biophysica Acta*, *1862*, 213–222.

Diaz-Rohrer, B., Levental, K. R., & Levental, I. (2014). Rafting through traffic: Membrane domains in cellular logistics. *Biochimica et Biophysica Acta*, *1838*, 3003–3013.

Emmanouilidou, E., Elenis, D., Papasilekas, T., Stranjalis, G., Gerozissis, K., Ioannou, P. C., et al. (2011). Assessment of α-synuclein secretion in mouse and human brain parenchyma. *PLoS One*, *6*, e22225.

Engin, A. B., Engin, E. D., & Engin, A. (2020). Dual function of sialic acid in gastrointestinal SARS-CoV-2 infection. *Environmental Toxicology and Pharmacology*, *79*, 103436.

Esparza, T. J., Zhao, H., Cirrito, J. R., Cairns, N. J., Bateman, R. J., Holtzman, D. M., et al. (2013). Amyloid-β oligomerization in Alzheimer dementia versus high-pathology controls. *Annals of Neurology*, *73*, 104–119.

Fantini, J. (2003). How sphingolipids bind and shape proteins: Molecular basis of lipid-protein interactions in lipid shells, rafts and related biomembrane domains. *Cellular and Molecular Life Sciences, 60*, 1027–1032.

Fantini, J., & Barrantes, F. J. (2009). Sphingolipid/cholesterol regulation of neurotransmitter receptor conformation and function. *Biochimica et Biophysica Acta, 1788*, 2345–2361.

Fantini, J., & Barrantes, F. J. (2013). How cholesterol interacts with membrane proteins: An exploration of cholesterol-binding sites including CRAC, CARC, and tilted domains. *Frontiers in Physiology, 4*, 31.

Fantini, J., Chahinian, H., & Yahi, N. (2020b). Synergistic antiviral effect of hydroxychloroquine and azithromycin in combination against SARS-CoV-2: What molecular dynamics studies of virus-host interactions reveal. *International Journal of Antimicrobial Agents, 56*, 106020.

Fantini, J., Chahinian, H., & Yahi, N. (2020c). Progress toward Alzheimer's disease treatment: Leveraging the Achilles' heel of Aβ oligomers? *Protein Science, 29*, 1748–1759.

Fantini, J., Chahinian, H., & Yahi, N. (2021). Leveraging coronavirus binding to gangliosides for innovative vaccine and therapeutic strategies against COVID-19. *Biochemical and Biophysical Research Communications, 538*, 132–136.

Fantini, J., Cook, D. G., Nathanson, N., Spitalnik, S. L., & Gonzalez-Scarano, F. (1993). Infection of colonic epithelial cell lines by type 1 human immunodeficiency virus is associated with cell surface expression of galactosylceramide, a potential alternative gp120 receptor. *Proceedings of the National Academy of Sciences of the United States of America, 90*, 2700–2704.

Fantini, J., Di Scala, C., Chahinian, H., & Yahi, N. (2020). Structural and molecular modelling studies reveal a new mechanism of action of chloroquine and hydroxychloroquine against SARS-CoV-2 infection. *International Journal of Antimicrobial Agents, 55*, 105960.

Fantini, J., Garmy, N., Mahfoud, R., & Yahi, N. (2002). Lipid rafts: Structure, function and role in HIV, Alzheimer's and prion diseases. *Expert Reviews in Molecular Medicine, 4*, 1–22.

Fantini, J., Garmy, N., & Yahi, N. (2006). Prediction of glycolipid-binding domains from the amino acid sequence of lipid raft-associated proteins: Application to HpaA, a protein involved in the adhesion of helicobacter pylori to gastrointestinal cells. *Biochemistry, 45*, 10957–10962.

Fantini, J., Maresca, M., Hammache, D., Yahi, N., & Delézay, O. (2000). Glycosphingolipid (GSL) microdomains as attachment platforms for host pathogens and their toxins on intestinal epithelial cells: Activation of signal transduction pathways and perturbations of intestinal absorption and secretion. *Glycoconjugate Journal, 17*, 173–179.

Fantini, J., & Yahi, N. (2010). Molecular insights into amyloid regulation by membrane cholesterol and sphingolipids: Common mechanisms in neurodegenerative diseases. *Expert Reviews in Molecular Medicine, 12*, e27.

Fantini, J., & Yahi, N. (2011). Molecular basis for the glycosphingolipid-binding specificity of α-synuclein: Key role of tyrosine 39 in membrane insertion. *Journal of Molecular Biology, 408*, 654–669.

Fantini, J., & Yahi, N. (2013). The driving force of alpha-synuclein insertion and amyloid channel formation in the plasma membrane of neural cells: Key role of ganglioside- and cholesterol-binding domains. *Advances in Experimental Medicine and Biology, 991*, 15–26.

Fantini, J., & Yahi, N. (2015). *Brain lipids in synaptic function and neurological disease. Clues to innovative therapeutic strategies for brain disorders*. Elsevier Academic Press.

Fantini, J., Yahi, N., Azzaz, F., & Chahinian, H. (2021). Structural dynamics of SARS-CoV-2 variants: A health monitoring strategy for anticipating Covid-19 outbreaks. *The Journal of Infection*.

Fantini, J., Yahi, N., & Garmy, N. (2013). Cholesterol accelerates the binding of Alzheimer's β-amyloid peptide to ganglioside GM1 through a universal hydrogen-bond-dependent sterol tuning of glycolipid conformation. *Frontiers in Physiology, 4*, 120.

Fealey, M. E., Binder, B. P., Uversky, V. N., Hinderliter, A., & Thomas, D. D. (2018). Structural impact of phosphorylation and dielectric constant variation on Synaptotagmin's IDR. *Biophysical Journal*, *114*, 550–561.

Flores, A., Ramirez-Franco, J., Desplantes, R., Debreux, K., Ferracci, G., Wernert, F., et al. (2019). Gangliosides interact with synaptotagmin to form the high-affinity receptor complex for botulinum neurotoxin B. *Proceedings of the National Academy of Sciences of the United States of America*, *116*, 18098–18108.

Fogolari, F., Tosatto, S. C., Muraro, L., & Montecucco, C. (2009). Electric dipole reorientation in the interaction of botulinum neurotoxins with neuronal membranes. *FEBS Letters*, *583*, 2321–2325.

Fotinou, C., Emsley, P., Black, I., Ando, H., Ishida, H., Kiso, M., et al. (2001). The crystal structure of tetanus toxin Hc fragment complexed with a synthetic GT1b analogue suggests cross-linking between ganglioside receptors and the toxin. *The Journal of Biological Chemistry*, *276*, 32274–32281.

Galons, J. P., Fantini, J., Vion-Dury, J., Cozzone, P. J., & Canioni, P. (1989). Metabolic changes in undifferentiated and differentiated human colon adenocarcinoma cells studied by multinuclear magnetic resonance spectroscopy. *Biochimie*, *71*, 949–961.

Greenlee, J. D., Subramanian, T., Liu, K., & King, M. R. (2021). Rafting down the metastatic cascade: The role of lipid rafts in cancer metastasis, cell death, and clinical outcomes. *Cancer Research*, *81*, 5–17.

Hakomori, S. (1984). Ganglioside receptors: A brief overview and introductory remarks. *Advances in Experimental Medicine and Biology*, *174*, 333–339.

Hamark, C., Berntsson, R. P., Masuyer, G., Henriksson, L. M., Gustafsson, R., Stenmark, P., et al. (2017). Glycans confer specificity to the recognition of ganglioside receptors by botulinum neurotoxin A. *Journal of the American Chemical Society*, *139*, 218–230.

Hamel, S., Fantini, J., & Schweisguth, F. (2010). Notch ligand activity is modulated by glycosphingolipid membrane composition in Drosophila melanogaster. *The Journal of Cell Biology*, *188*, 581–594.

Hammache, D., Yahi, N., Maresca, M., Piéroni, G., & Fantini, J. (1999). Human erythrocyte glycosphingolipids as alternative cofactors for human immunodeficiency virus type 1 (HIV-1) entry: Evidence for CD4-induced interactions between HIV-1 gp120 and reconstituted membrane microdomains of glycosphingolipids (Gb3 and GM3). *Journal of Virology*, *73*, 5244–5248.

Hammache, D., Yahi, N., Piéroni, G., Ariasi, F., Tamalet, C., & Fantini, J. (1998). Sequential interaction of CD4 and HIV-1 gp120 with a reconstituted membrane patch of ganglioside GM3: Implications for the role of glycolipids as potential HIV-1 fusion cofactors. *Biochemical and Biophysical Research Communications*, *246*, 117–122.

Heuss, S. F., Tarantino, N., Fantini, J., Ndiaye-Lobry, D., Moretti, J., Israël, A., et al. (2013). A glycosphingolipid binding domain controls trafficking and activity of the mammalian notch ligand delta-like 1. *PLoS One*, *8*, e74392.

Holmner, A., Mackenzie, A., Okvist, M., Jansson, L., Lebens, M., Teneberg, S., et al. (2011). Crystal structures exploring the origins of the broader specificity of Escherichia coli heat-labile enterotoxin compared to cholera toxin. *Journal of Molecular Biology*, *406*, 387–402.

Hu, L., Crawford, S. E., Czako, R., Cortes-Penfield, N. W., Smith, D. F., Le Pendu, J., et al. (2012). Cell attachment protein VP8* of a human rotavirus specifically interacts with A-type histo-blood group antigen. *Nature*, *485*, 256–259.

Huang, H. I., Chio, C. C., & Lin, J. Y. (2018). Inhibition of EV71 by curcumin in intestinal epithelial cells. *PLoS One*, *13*, e0191617.

Huang, Y., Huang, S., Di Scala, C., Wang, Q., Wandall, H. H., Fantini, J., et al. (2018). The glycosphingolipid MacCer promotes synaptic Bouton formation in Drosophila by interacting with Wnt. *eLife*, *7*, e38183.

Hubert, M., Jeannin, P., Burlaud-Gaillard, J., Roingeard, P., Gessain, A., Ceccaldi, P. E., et al. (2020). Evidence that Zika virus is transmitted by breastfeeding to newborn A129 (Ifnar1 Knock-out) mice and is able to infect and cross a tight monolayer of human intestinal epithelial cells. *Frontiers in Microbiology*, *11*, 524678.

Hulswit, R. J. G., Lang, Y., Bakkers, M. J. G., Li, W., Li, Z., Schouten, A., et al. (2019). Human coronaviruses OC43 and HKU1 bind to 9-O-acetylated sialic acids via a conserved receptor-binding site in spike protein domain A. *Proceedings of the National Academy of Sciences of the United States of America*, *116*, 2681–2690.

Hurley, J. H., & Wendland, B. (2002). Endocytosis: Driving membranes around the bend. *Cell*, *111*, 143–146.

Irie, A., Koyama, S., Kozutsumi, Y., Kawasaki, T., & Suzuki, A. (1998). The molecular basis for the absence of N-glycolylneuraminic acid in humans. *The Journal of Biological Chemistry*, *273*, 15866–15871.

Israelachvili, J. N., & Mitchell, D. J. (1975). A model for the packing of lipids in bilayer membranes. *Biochimica et Biophysica Acta*, *389*, 13–19.

Israelachvili, J. N., Mitchell, D. J., & Ninham, B. W. (1977). Theory of self-assembly of lipid bilayers and vesicles. *Biochimica et Biophysica Acta*, *470*, 185–201.

Kabbani, A. M., Raghunathan, K., Lencer, W. I., Kenworthy, A. K., & Kelly, C. V. (2020). Structured clustering of the glycosphingolipid GM1 is required for membrane curvature induced by cholera toxin. *Proceedings of the National Academy of Sciences of the United States of America*, *117*, 14978–14986.

Kiss, A. L. (2012). Caveolae and the regulation of endocytosis. *Advances in Experimental Medicine and Biology*, *729*, 14–28.

Kolter, T. (2012). Ganglioside biochemistry. *ISRN Biochemistry*, *2012*, 506160.

Lagier, J. C., Million, M., Gautret, P., Colson, P., Cortaredona, S., Giraud-Gatineau, A., et al. (2020). Outcomes of 3,737 COVID-19 patients treated with hydroxychloroquine/azithromycin and other regimens in Marseille, France: A retrospective analysis. *Travel Medicine and Infectious Disease*, *36*, 101791.

Lappalainen, R., & Riikonen, R. S. (1996). High levels of cerebrospinal fluid glutamate in Rett syndrome. *Pediatric Neurology*, *15*, 213–216.

LaRocca, T. J., Pathak, P., Chiantia, S., Toledo, A., Silvius, J. R., Benach, J. L., et al. (2013). Proving lipid rafts exist: Membrane domains in the prokaryote Borrelia burgdorferi have the same properties as eukaryotic lipid rafts. *PLoS Pathogens*, *9*, e1003353.

Ledeen, R. W., & Wu, G. (2015). The multi-tasked life of GM1 ganglioside, a true factotum of nature. *Trends in Biochemical Sciences*, *40*, 407–418.

Lekman, A. Y., Hagberg, B. A., & Svennerholm, L. T. (1991). Membrane cerebral lipids in Rett syndrome. *Pediatric Neurology*, *7*, 186–190.

Li, W., Hulswit, R. J. G., Widjaja, I., Raj, V. S., McBride, R., Peng, W., et al. (2017). Identification of sialic acid-binding function for the Middle East respiratory syndrome coronavirus spike glycoprotein. *Proceedings of the National Academy of Sciences of the United States of America*, *114*, E8508–e8517.

Lu, Y., Liu, D. X., & Tam, J. P. (2008). Lipid rafts are involved in SARS-CoV entry into Vero E6 cells. *Biochemical and Biophysical Research Communications*, *369*, 344–349.

Mahfoud, R., Garmy, N., Maresca, M., Yahi, N., Puigserver, A., & Fantini, J. (2002). Identification of a common sphingolipid-binding domain in Alzheimer, prion, and HIV-1 proteins. *The Journal of Biological Chemistry*, *277*, 11292–11296.

Mañes, S., del Real, G., & Martínez, A. C. (2003). Pathogens: Raft hijackers. *Nature Reviews. Immunology*, *3*, 557–568.

Matrosovich, M., Herrler, G., & Klenk, H. D. (2015). Sialic acid receptors of viruses. *Topics in Current Chemistry*, *367*, 1–28.

Merritt, E. A., Kuhn, P., Sarfaty, S., Erbe, J. L., Holmes, R. K., & Hol, W. G. (1998). The 1.25 a resolution refinement of the cholera toxin B-pentamer: Evidence of peptide backbone strain at the receptor-binding site. *Journal of Molecular Biology*, *282*, 1043–1059.

Munro, S. (2003). Lipid rafts: Elusive or illusive? *Cell*, *115*, 377–388.

Nabi, I. R., & Le, P. U. (2003). Caveolae/raft-dependent endocytosis. *The Journal of Cell Biology*, *161*, 673–677.

Neu, U., Woellner, K., Gauglitz, G., & Stehle, T. (2008). Structural basis of GM1 ganglioside recognition by simian virus 40. *Proceedings of the National Academy of Sciences of the United States of America*, *105*, 5219–5224.

Ng, N. M., Littler, D. R., Paton, A. W., Le Nours, J., Rossjohn, J., Paton, J. C., et al. (2013). EcxAB is a founding member of a new family of metalloprotease AB5 toxins with a hybrid cholera-like B subunit. *Structure*, *21*, 2003–2013.

Nguyen, T., Lee, S., Yang, Y. A., Ahn, C., Sim, J. H., Kei, T. G., et al. (2020). The role of 9-O-acetylated glycan receptor moieties in the typhoid toxin binding and intoxication. *PLoS Pathogens*, *16*, e1008336.

Nishio, M., Umezawa, Y., Fantini, J., Weiss, M. S., & Chakrabarti, P. (2014). CH-π hydrogen bonds in biological macromolecules. *Physical Chemistry Chemical Physics*, *16*, 12648–12683.

Pirazzini, M., Rossetto, O., Eleopra, R., & Montecucco, C. (2017). Botulinum neurotoxins: Biology, pharmacology, and toxicology. *Pharmacological Reviews*, *69*, 200–235.

Prasanna, X., Jafurulla, M., Sengupta, D., & Chattopadhyay, A. (2016). The ganglioside GM1 interacts with the serotonin1A receptor via the sphingolipid binding domain. *Biochimica et Biophysica Acta (BBA) - Biomembranes*, *1858*, 2818–2826.

Puri, A., Hug, P., Jernigan, K., Barchi, J., Kim, H. Y., Hamilton, J., et al. (1998). The neutral glycosphingolipid globotriaosylceramide promotes fusion mediated by a CD4-dependent CXCR4-utilizing HIV type 1 envelope glycoprotein. *Proceedings of the National Academy of Sciences of the United States of America*, *95*, 14435–14440.

Puri, A., Hug, P., Muñoz-Barroso, I., & Blumenthal, R. (1998). Human erythrocyte glycolipids promote HIV-1 envelope glycoprotein-mediated fusion of CD4 + cells. *Biochemical and Biophysical Research Communications*, *242*, 219–225.

Qu, B., Li, X., Gao, W., Sun, W., Jin, Y., Cardona, C. J., et al. (2012). Human intestinal epithelial cells are susceptible to influenza virus subtype H9N2. *Virus Research*, *163*, 151–159.

Quist, A., Doudevski, I., Lin, H., Azimova, R., Ng, D., Frangione, B., et al. (2005). Amyloid ion channels: A common structural link for protein-misfolding disease. *Proceedings of the National Academy of Sciences of the United States of America*, *102*, 10427–10432.

Reiss, K., Stencel, J. E., Liu, Y., Blaum, B. S., Reiter, D. M., Feizi, T., et al. (2012). The GM2 glycan serves as a functional coreceptor for serotype 1 Reovirus. *PLoS Pathogens*, *8*, e1003078.

Rolain, J. M., Colson, P., & Raoult, D. (2007). Recycling of chloroquine and its hydroxyl analogue to face bacterial, fungal and viral infections in the 21st century. *International Journal of Antimicrobial Agents*, *30*, 297–308.

Rossetto, O., & Montecucco, C. (2007). Peculiar binding of botulinum neurotoxins. *ACS Chemical Biology*, *2*, 96–98.

Rossetto, O., Pirazzini, M., Fabris, F., & Montecucco, C. (2021). Botulinum neurotoxins: Mechanism of action. *Handbook of Experimental Pharmacology*, *263*, 35–47.

Salaün, C., James, D. J., & Chamberlain, L. H. (2004). Lipid rafts and the regulation of exocytosis. *Traffic*, *5*, 255–264.

Schnaar, R. L. (2016). Gangliosides of the vertebrate nervous system. *Journal of Molecular Biology*, *428*, 3325–3336.

Schnaar, R. L. (2019). The biology of gangliosides. *Advances in Carbohydrate Chemistry and Biochemistry*, *76*, 113–148.

Schnaar, R. L., Gerardy-Schahn, R., & Hildebrandt, H. (2014). Sialic acids in the brain: Gangliosides and polysialic acid in nervous system development, stability, disease, and regeneration. *Physiological Reviews*, *94*, 461–518.

Schneider, J. S., Gollomp, S. M., Sendek, S., Colcher, A., Cambi, F., & Du, W. (2013). A randomized, controlled, delayed start trial of GM1 ganglioside in treated Parkinson's disease patients. *Journal of the Neurological Sciences*, *324*, 140–148.

Schneider, J. S., Kean, A., & DiStefano, L. (1995). GM1 ganglioside rescues substantia nigra pars compacta neurons and increases dopamine synthesis in residual nigrostriatal dopaminergic neurons in MPTP-treated mice. *Journal of Neuroscience Research*, *42*, 117–123.

Sciacca, M. F., Lolicato, F., Tempra, C., Scollo, F., Sahoo, B. R., Watson, M. D., et al. (2020). Lipid-chaperone hypothesis: A common molecular mechanism of membrane disruption by intrinsically disordered proteins. *ACS Chemical Neuroscience*, *11*, 4336–4350.

Simons, K., & Ikonen, E. (1997). Functional rafts in cell membranes. *Nature*, *387*, 569–572.

Singer, S. J., & Nicolson, G. L. (1972). The fluid mosaic model of the structure of cell membranes. *Science*, *175*, 720–731.

Sipione, S., Monyror, J., Galleguillos, D., Steinberg, N., & Kadam, V. (2020). Gangliosides in the brain: Physiology, pathophysiology and therapeutic applications. *Frontiers in Neuroscience*, *14*, 572965.

Sonnino, S., Aureli, M., Mauri, L., Ciampa, M. G., & Prinetti, A. (2015). Membrane lipid domains in the nervous system. *Frontiers in Bioscience (Landmark Edition)*, *20*, 280–302.

Sonnino, S., & Chigorno, V. (2000). Ganglioside molecular species containing C18- and C20-sphingosine in mammalian nervous tissues and neuronal cell cultures. *Biochimica et Biophysica Acta*, *1469*, 63–77.

Sonnino, S., Mauri, L., Chigorno, V., & Prinetti, A. (2007). Gangliosides as components of lipid membrane domains. *Glycobiology*, **17**, 1r–13r.

Staubach, S., & Hanisch, F. G. (2011). Lipid rafts: Signaling and sorting platforms of cells and their roles in cancer. *Expert Review of Proteomics*, *8*, 263–277.

Taïeb, N., Maresca, M., Guo, X. J., Garmy, N., Fantini, J., & Yahi, N. (2009). The first extracellular domain of the tumour stem cell marker CD133 contains an antigenic ganglioside-binding motif. *Cancer Letters*, *278*, 164–173.

van Echten, G., & Sandhoff, K. (1993). Ganglioside metabolism. Enzymology, topology, and regulation. *The Journal of Biological Chemistry*, *268*, 5341–5344.

Varshney, P., Yadav, V., & Saini, N. (2016). Lipid rafts in immune signalling: Current progress and future perspective. *Immunology*, *149*, 13–24.

Wu, G., Lu, Z. H., Kulkarni, N., & Ledeen, R. W. (2012). Deficiency of ganglioside GM1 correlates with Parkinson's disease in mice and humans. *Journal of Neuroscience Research*, *90*, 1997–2008.

Wu, F., Zhao, S., Yu, B., Chen, Y. M., Wang, W., Song, Z. G., et al. (2020). A new coronavirus associated with human respiratory disease in China. *Nature*, *579*, 265–269.

Yager, E. J., & Konan, K. V. (2019). Sphingolipids as potential therapeutic targets against enveloped human RNA viruses. *Viruses*, *11*.

Yahi, N., Aulas, A., & Fantini, J. (2010). How cholesterol constrains glycolipid conformation for optimal recognition of Alzheimer's beta amyloid peptide (Abeta1-40). *PLoS One*, *5*, e9079.

Yahi, N., Baghdiguian, S., & Fantini, J. (1995). Production of a highly cytopathic HIV-1 isolate from a human mucosal epithelial cell line cultured on microcarrier beads in serum-free medium. *In Vitro Cellular & Developmental Biology Animal*, *31*, 62–66.

Yahi, N., Di Scala, C., Chahinian, H., & Fantini, J. (2021). Innovative treatment targeting gangliosides aimed at blocking the formation of neurotoxic α-synuclein oligomers in Parkinson's disease. *Glycoconjugate Journal*. https://doi.org/10.1007/s10719-021-10012-0. In press.

Yahi, N., & Fantini, J. (2014). Deciphering the glycolipid code of Alzheimer's and Parkinson's amyloid proteins allowed the creation of a universal ganglioside-binding peptide. *PLoS One*, *9*, e104751.

Yahi, N., Fantini, J., Baghdiguian, S., Mabrouk, K., Tamalet, C., Rochat, H., et al. (1995). SPC3, a synthetic peptide derived from the V3 domain of human immunodeficiency virus type 1 (HIV-1) gp120, inhibits HIV-1 entry into CD4+ and CD4- cells by two distinct mechanisms. *Proceedings of the National Academy of Sciences of the United States of America*, *92*, 4867–4871.

Yahi, N., Fantini, J., Mabrouk, K., Tamalet, C., de Micco, P., van Rietschoten, J., et al. (1994). Multibranched V3 peptides inhibit human immunodeficiency virus infection in human lymphocytes and macrophages. *Journal of Virology*, *68*, 5714–5720.

Yahi, N., Sabatier, J. M., Baghdiguian, S., Gonzalez-Scarano, F., & Fantini, J. (1995). Synthetic multimeric peptides derived from the principal neutralization domain (V3 loop) of human immunodeficiency virus type 1 (HIV-1) gp120 bind to galactosylceramide and block HIV-1 infection in a human CD4-negative mucosal epithelial cell line. *Journal of Virology*, *69*, 320–325.

Yu, R. K., Bieberich, E., Xia, T., & Zeng, G. (2004). Regulation of ganglioside biosynthesis in the nervous system. *Journal of Lipid Research*, *45*, 783–793.

Zhang, H., Li, H. B., Lyu, J. R., Lei, X. M., Li, W., Wu, G., et al. (2020). Specific ACE2 expression in small intestinal enterocytes may cause gastrointestinal symptoms and injury after 2019-nCoV infection. *International Journal of Infectious Diseases*, *96*, 19–24.

CHAPTER NINE

Development of enhanced conformational sampling methods to probe the activation landscape of GPCRs

Ravinder Abrol*, Erik Serrano, and Luis Jaimes Santiago
Department of Chemistry and Biochemistry, California State University, Northridge, CA, United States
*Corresponding author: e-mail address: abrol@csun.edu

Contents

1. Introduction 326
   1.1 Revolution in molecular biology and structural dynamics of GPCRs 327
   1.2 Complexity of GPCR signaling and pharmacology 329
   1.3 Sampling conformational landscapes of GPCRs 330
2. Molecular dynamics based methods 332
   2.1 Ultra-long classical MD 333
   2.2 Metadynamics and other biased MD methods 335
   2.3 Accelerated MD 338
   2.4 Markov state models and MD 340
   2.5 RMSD based clustering-guided MD 341
3. Discrete conformational sampling methods 343
   3.1 Ligand-independent discrete sampling 343
   3.2 Ligand-dependent discrete sampling 345
   3.3 Coarse-grained energy landscapes 346
4. Future of GPCR conformational sampling is hybrid approaches 347
Acknowledgments 348
References 348

Abstract

G protein-coupled receptors (GPCRs) make up the largest superfamily of integral membrane proteins and play critical signal transduction roles in many physiological processes. Developments in molecular biology, biophysical, biochemical, pharmacological, and computational techniques aimed at these important therapeutic targets are beginning to provide unprecedented details on the structural as well as functional basis of their pleiotropic signaling mediated by G proteins, β arrestins, and other transducers. This pleiotropy presents a pharmacological challenge as the same ligand-receptor interaction can cause a therapeutic effect as well as an undesirable on-target side-effect through different

Advances in Protein Chemistry and Structural Biology, Volume 128
ISSN 1876-1623
https://doi.org/10.1016/bs.apcsb.2021.11.001

downstream pathways. GPCRs don't function as simple binary on-off switches but as finely tuned shape-shifting machines described by conformational ensembles, where unique subsets of conformations may be responsible for specific signaling cascades. X-ray crystallography and more recently cryo-electron microscopy are providing snapshots of some of these functionally-important receptor conformations bound to ligands and/or transducers, which are being utilized by computational methods to describe the dynamic conformational energy landscape of GPCRs. In this chapter, we review the progress in computational conformational sampling methods based on molecular dynamics and discrete sampling approaches that have been successful in complementing biophysical and biochemical studies on these receptors in terms of their activation mechanisms, allosteric effects, actions of biased ligands, and effects of pathological mutations. Some of the sampled simulation time scales are beginning to approach receptor activation time scales. The list of conformational sampling methods and example uses discussed is not exhaustive but includes representative examples that have pushed the limits of classical molecular dynamics and discrete sampling methods to describe the activation energy landscape of GPCRs.

Abbreviations

β2AR β2 adrenergic receptor
CV collective variable
DS discrete sampling
GaMD Gaussian accelerated MD
GPCR G protein coupled receptor
MD molecular dynamics
MSM markov state model
REMD replica exchange MD
TM trans membrane

1. Introduction

G protein-coupled receptors (GPCRs) mediate transmembrane (TM) signal transduction by binding to a diverse set of extracellular signals and transducing that information into the cell through interaction with many transducers like G proteins, α/β arrestins, and receptor kinases (Galandrin, Oligny-Longpre, & Bouvier, 2007; Hill, 2006; Kenakin & Miller, 2010; Lagerstrom & Schioth, 2008; Lefkowitz, 2004). The diversity of extracellular signals sensed by these receptors is extraordinary as it covers photons (for rhodopsin), small molecules (like hormones, neurotransmitters, metabolites, odorants, tastants, etc.), and large oligopeptides (like chemokines, incretins, etc.). GPCRs are upstream of many intracellular signaling cascades, which makes them attractive therapeutic targets in a large number of diseases (Lagerstrom & Schioth, 2008; Lundstrom, 2006; Zalewska, Siara, & Sajewicz, 2014). Nonsynonymous single nucleotide polymorphisms (nsSNPs) in GPCRs have been linked with specific disease pathologies as they can cause many pathophysiological changes due to altered signaling via conformational changes in the receptor (Thompson et al., 2014). Evidence is also accumulating in plants for G protein coupled signaling and corresponding receptors (Liu et al., 2021).

1.1 Revolution in molecular biology and structural dynamics of GPCRs

Past couple of decades has seen exponential growth in our ability to connect a GPCR activation event at the cell membrane to multitudes of intracellular signaling events that mediate the physiological cell responses to that receptor. This has been predominantly driven by the development of resonance energy based biosensors (Wright & Bouvier, 2021; Zhou, Meng, Xu, & Liu, 2021) that are illuminating the complexity of GPCR signaling and pathway selectivity. Parallel and complementary to these developments in GPCR functional biology, there are *three* broad and highly successful efforts in GPCR structural biology that are shedding unprecedented insight into their structural dynamics and conformational heterogeneity that underlies the complex GPCR functional biology.

Static structure determination: There has been strong interest in the structural basis of GPCR function to understand their biology and to rationalize their dysfunction in diseases (Schoneberg, Schulz, & Gudermann, 2002). The first atomic structure of a GPCR was obtained in 2000 for the dark state of Bovine rhodopsin (Palczewski et al., 2000). It took 7 years to get the atomic structure for a second GPCR, when the human β2 adrenergic receptor (β2AR) structure was solved (Cherezov et al., 2007; Rasmussen et al., 2007). Since then, the number of experimentally determined GPCR structures has increased essentially exponentially every year and new structures are appearing almost every week. This has been due to many membrane protein focused methodological developments in protein X-ray diffraction (XRD) (Blois & Bowie, 2009; Ghosh, Kumari, Jaiman, & Shukla, 2015; Milic & Veprintsev, 2015) and protein cryo-electron microscopy (cryo-EM) (Garcia-Nafria & Tate, 2020; Peisley & Skiniotis, 2015; Skiniotis & Southworth, 2016). Experimental structures are now available for 110+ GPCRs (Kooistra et al., 2021) covering all major GPCR phylogenetic classes, bound to single or multiple ligands (inverse agonists, antagonists, agonists without or with allosteric ligands), and/or bound to different G proteins/β arrestins, including the first structure of a ternary receptor-Gprotein-βarrestin megaplex (Thomsen et al., 2016). These structures, that suggest the GPCRs to be conformationally malleable machines, are catalyzing the full spectrum of mechanistic studies into GPCR functional biology to understand the implications of their gene sequence and protein structure in the function and pharmacology of these receptors (Hanson & Stevens, 2009; Heifetz et al., 2015; Hilger, Masureel, & Kobilka, 2018; Rosenbaum, Rasmussen, & Kobilka, 2009).

Conformational ensemble and dynamics studies: Beyond static structures, many biophysical methods have been developed for GPCRs to probe and characterize their conformational dynamics (Gusach et al., 2020) that include NMR based methods (Casiraghi et al., 2019; Eddy et al., 2018; Huang et al., 2021; Picard & Prosser, 2021; Xu et al., 2019), single molecule FRET (Kauk & Hoffmann, 2018; Quast & Margeat, 2019), DEER spectroscopy (Elgeti & Hubbell, 2021), and mass spectroscopy (Du et al., 2019). These studies paint the picture of GPCRs as nature's finely tuned malleable machines (Hilger et al., 2018) capable of orchestrating signaling cascades behind their complex functional biology and that they reside in complex energy landscapes in which different ligands (Xu et al., 2019) and transducers (Hilger et al., 2018) can stabilize distinct sets of receptor conformations.

Computational studies of GPCR dynamics: Since the early days of GPCR structural biology, computational biophysical methods have complemented and supplemented experimental studies to bridge the gaps between GPCR sequence/structure space, GPCR signaling landscape, receptor function, and drug discovery (Abrol, Kim, Bray, Trzaskowski, & Goddard, 2013; Hilger, 2021; Miao & McCammon, 2016a; Paiva, Oliveira, Horn, Bywater, & Vriend, 2006; Schlyer & Horuk, 2006; Torrens-Fontanals, Stepniewski, Gloriam, & Selent, 2021; Vaidehi et al., 2002; Vohra et al., 2007). Two major computational approaches have been developed and applied to GPCRs to understand their conformational dynamics. (A) *Molecular Dynamics*: First approach is the use of standard molecular dynamics (MD) simulations to understand receptor dynamics in its lipid bilayer environment (Deganutti, Moro, & Reynolds, 2019; Grossfield, 2011; Johnston & Filizola, 2011; Vanni & Rothlisberger, 2012). These simulations are not practical for studying GPCR activation due to receptor activation timescales spanning milliseconds or even higher. Beyond long-timescale MD simulations that can observe some activation related changes in an unbiased manner, MD protocols have been adapted in many ways to enhance the conformational sampling of GPCRs (Harpole & Delemotte, 2018; Johnston & Filizola, 2014; McRobb, Negri, Beuming, & Sherman, 2016), and they continue to be adapted, which will be covered in detail in the next section. (B) *Discrete Sampling*: Second approach takes advantage of the two-dimensional constraints put on receptor structure by the lipid bilayer, which in principle enables a brute-force discrete sampling of conformations since the TM helices cannot tilt in the membrane beyond a certain angle (Abrol, Kim, Bray, Griffith, & Goddard, 2011). This provides ways to systematically attempt the sampling of GPCR conformational landscape. This situation is different from that for water-soluble proteins, whose enormous diversity of tertiary folds prevents such discrete sampling of conformations. In addition, the increase in static GPCR structures from experimental biophysical methods, which are complexed with and without different ligand types as well as with and without different transducers, is beginning to enable a thermodynamic analysis of at least the major valleys (stable conformations) in the GPCR conformational space. Some of these methods will be described in a later section below.

1.2 Complexity of GPCR signaling and pharmacology

The above mentioned path-breaking developments in GPCR dynamics and function have laid bare a conformational ensemble view of these receptors that is being linked to the intricately coupled patch work of downstream signaling cascades. Signal transduction by GPCRs is a result of the coupling of a receptor with one or more heterotrimeric G proteins, which activate several pathways through either the Gα monomeric subunit or the Gβγ heterodimeric subunit of the G proteins. After G protein coupling, GPCR receptor kinases (GRKs) and arrestins are recruited for receptor desensitization and trafficking, however, it has been shown that GRKs and arrestins can trigger their own signaling cascades independent of G protein pathways (Barki-Harrington & Rockman, 2008; Pavlos & Friedman, 2017; Reiter & Lefkowitz, 2006; Seyedabadi, Gharghabi, Gurevich, & Gurevich, 2021). This information has expanded the knowledge of mechanisms underlying GPCR function, where their pleiotropic signaling can be either balanced or biased towards specific signaling pathways. This has been termed as biased agonism or functional selectivity (Gurevich & Gurevich, 2020; Kenakin, 2011; Wisler, Xiao, Thomsen, & Lefkowitz, 2014), where a ligand (or a receptor mutation) can bias the signaling toward a subset of potential downstream pathways. The balanced signaling can cause

GPCR-targeted drugs to have on-target side-effects, for example, arrestin has been shown to mediate the flushing side-effect of niacin through its receptor GPR109A, whereas the beneficial anti-lipolytic effect is G protein mediated (Walters et al., 2009). Such observations are generating novel therapeutic approaches utilizing biased ligands aimed at minimizing on-target side-effects (Kenakin, 2012; Kenakin & Christopoulos, 2013; Whalen, Rajagopal, & Lefkowitz, 2011). A conceptual understanding of the molecular mechanisms behind biased signaling (or functional selectivity) is emerging about the conformational flexibility and dynamics of GPCRs. This understanding is supported by an ever increasing number of experimental and computational studies (Abrol et al., 2013; Ahn, Ham, & Chung, 2021; Deupi & Kobilka, 2010; Gurevich & Gurevich, 2017; Hilger, 2021; Hilger et al., 2018; Kenakin & Miller, 2010; Latorraca, Venkatakrishnan, & Dror, 2017; Ma, Nivedha, & Vaidehi, 2021; Manglik & Kobilka, 2014; Preininger, Meiler, & Hamm, 2013; Vaidehi & Kenakin, 2010) that point to different ensembles of receptor conformations behind pleiotropic signaling, ligand selectivity, and biased signaling of GPCRs. Fig. 1 captures this complexity semi-quantitatively in terms of a conformational ensemble view of GPCR function. This picture will look different for each GPCR, however, it captures how functionally specific conformations have been or can be stabilized under different receptor conditions (e.g., ligands, mutations, transducers). One major biophysical insight into receptor activation has been that the agonist alone may not be enough to activate a GPCR (Rosenbaum et al., 2011), even though it might bias the receptor conformational ensemble toward the active state through conformational selection (Fernandes et al., 2021). A transducer like G protein also utilizes conformational selection to stabilize the active state by further pushing the conformational ensemble toward that state (Dong, Goddard, & Abrol, 2016; Huang et al., 2021; Manglik & Kobilka, 2014; Rasmussen, Choi, et al., 2011; Rasmussen, DeVree, et al., 2011).

1.3 Sampling conformational landscapes of GPCRs

All the advances described in the last section are sampling (directly or indirectly) the conformational energy landscapes of GPCRs. Energy landscapes continue to provide great mechanistic insight into chemical and biochemical processes (Wales, 2003). These landscapes also hold the key to understanding the complex signaling profiles of GPCRs (Deupi & Kobilka, 2010) and to rationally design (McCorvy et al., 2018) the next generation of biased

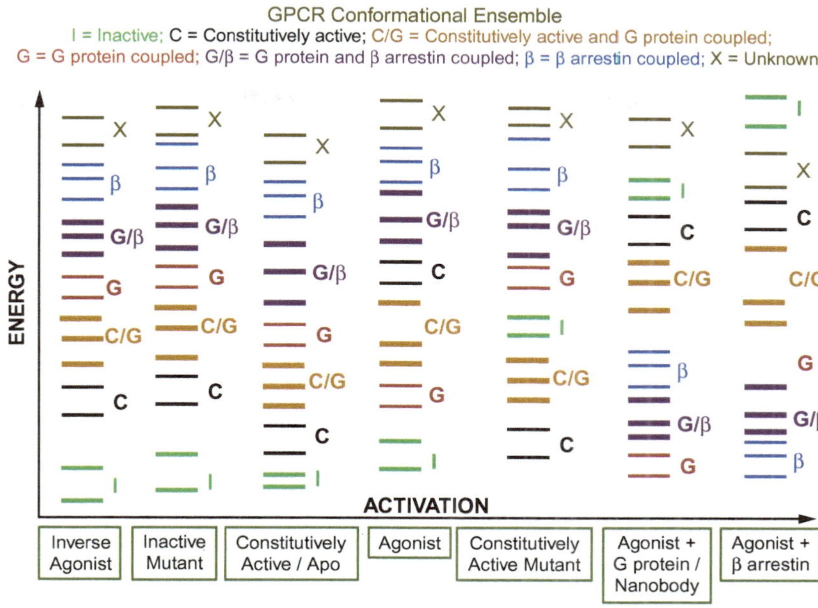

Fig. 1 An updated conformational ensemble view of GPCR activation, built upon the original version that appeared 10 years ago (Abrol et al., 2011). It highlights how different receptor conditions (ligands, mutations, transducers) can stabilize different receptor conformations. Some of the constitutively active (C) or G protein coupled (G) conformations may serve as intermediate conformations between the inactive and active receptor states that have been observed for some GPCRs so far.

ligands targeting these receptors (Smith, Lefkowitz, & Rajagopal, 2018). The conformational ensembles stabilized by these landscapes also provide a mechanistic framework to understand the molecular basis of GPCR signaling and to rationalize the biophysical basis for the evolution of the GPCR sequence space (Fig. 2).

In this chapter, we will highlight the progress in computational methods aimed at characterizing these conformational landscapes and understanding their roles in GPCR functions. The time scales of GPCR activation lie beyond milliseconds and higher, which are still beyond the reach of molecular dynamics and rigorous conformational sampling methods. However, enhanced sampling methods have been developed based on molecular dynamics (MD) and discrete sampling (DS) techniques to bridge the simulation timescales with the receptor activation timescales. Here, we will review some of these methods and their applications to showcase the mechanistic insight into GPCR signaling that is being provided by computational

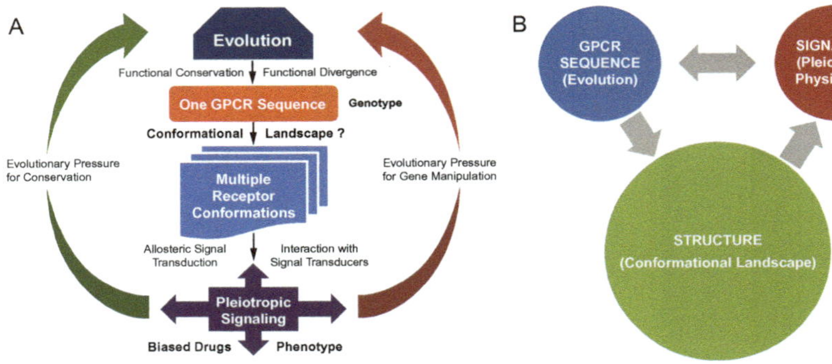

Fig. 2 (A) Conformational landscapes of receptors enable a mechanistic understanding of signaling and how it relates to evolution of GPCR sequences. (B) Sequence (evolution)–Structure (biophysics)–Signaling (biochemistry, physiology) nexus of GPCRs and the central role of receptor conformational ensembles.

methods. This perspective review is not meant to be exhaustive, but it demonstrates the progress on computational studies of GPCR dynamics using some of the methods that have pushed the envelope of MD and DS based approaches to understand the role of GPCR conformational ensembles in their signaling.

2. Molecular dynamics based methods

MD simulations of proteins (and nucleic acids) embedded in their native environments (ions, aqueous and/or lipid bilayer) have provided realistic dynamics (Karplus & Petsko, 1990; Levitt & Sharon, 1988) and "wigglings" of protein and solvent atoms (as envisioned by Richard Feynman) caused by the physiological temperature needed for protein function (Sengupta & Garrity, 2013). MD simulations solve Newton's equations of motion by taking into account the forces between all interacting atoms (protein–protein, protein–solvent, solvent–solvent), which are represented by force fields (Ponder & Case, 2003; Wang, Donini, Reyes, & Kollman, 2001). The length and accuracy of these simulations is controlled by the 1–2 fs time-steps needed for the force iterations to generate a realistic dynamics of the proteins as a function of time. To simulate physiological processes like GPCR activations that take several milliseconds or longer, the at least $\sim 10^{12}$ MD steps needed ($\sim$1 ms/$\sim$1 fs) to simulate those processes is a very real technological barrier. Advances in graphics processor units (GPUs) have been utilized for enabling longer MD simulations

(Friedrichs et al., 2009; Stone et al., 2007), however, only specialized supercomputing hardware dedicated to MD simulations like MDGRAPE (Komeiji et al., 1996; Ohmura, Morimoto, Ohno, Hasegawa, & Taiji, 2014) and Anton (Shaw et al., 2007, 2009) have enabled the protein simulations to attempt an approach towards physiological time scales (Dror, Dirks, Grossman, Xu, & Shaw, 2012; Dror, Jensen, Borhani, & Shaw, 2010; Kuroda, Suenaga, Sato, Kosuda, & Taiji, 2016). General access to these specialized supercomputers is very limited, but, independently of that, many methods have been developed based on classical MD that have enhanced the conformational sampling enabled by the classical simulations. First, we will describe some of the GPCR applications of ultra-long classical MD simulations and in the next few subsections we will briefly describe some of the MD based enhanced sampling methods along with their applications to GPCRs.

2.1 Ultra-long classical MD

Classical MD simulation method has been briefly described above, so here the focus is on their application to ultra-long simulation time scales ranging from tens of microseconds to hundreds of microseconds and even milliseconds for small protein systems. These ultra-long time scales are achievable due to the state of the art architectures of specialized MD-focused supercomputers like Anton-2 and MDGRAPE-4 that can achieve ~10 μs/day and ~1 μs/day simulation times, respectively, for a simulation system with ~100,000 atoms. Latest GPUs (e.g., RTX3090) are reaching ~250 ns/day in latest benchmarks, which is still ~1/4th the speed of MDGRAPE-4 and 1/40th the speed of Anton-2 supercomputers. Microseconds long MD simulations enable the exploration of molecular events like ligand binding and transducer binding to the receptors that shed direct light on the first molecular steps of receptor activation.

How drugs bind to receptors? One of the first exciting applications of ultra-long MD trajectories was the simulation of agonist and antagonist drug binding to β-adrenergic receptors (Dror et al., 2011). In these unbiased MD simulations, scores of 1-19 μs-long trajectories were carried out with multiple copies of the same ligand randomly distributed on the extracellular side of the receptor. Two major energetic barriers were identified that the ligands had to surmount on their way to the final orthosteric binding site present deep in the TM domain of the receptor. The larger of the two barriers occurred about 15 Å away from the final binding site in an extracellular vestibule region, where the drug molecules spend some time until the receptor changes shape and reveals a narrow passageway to the final binding pocket. Surprisingly, receptor conformational changes in the extracellular domain did not limit the rate of association of the ligands. Such simulations provide a front-row seat to the receptor-ligand association movie. A subsequent study looked at the binding of positive and negative allosteric modulators to the M2 muscarinic acetylcholine receptor (Dror et al., 2013) and found a similar vestibule region for the ligands as the earlier study. This study additionally revealed mechanisms that contribute to the positive and negative allosterism, which is a key insight useful for rationally designing GPCR drugs with unique therapeutic profiles.

How transducers bind to the receptors? Receptor activation involves interaction with the transducers like G proteins and arrestins. The time scales of these interactions are in the milliseconds or higher time scales so ultra-long MD simulations have tried to capture key events during these interactions. One study looked at the nucleotide exchange in heterotrimeric G proteins (Dror et al., 2015) upon engagement of the transducer with the receptor using an experimentally determined receptor-G protein interface. Steps of association were not determined as that will require longer time scales or a different sampling method. The 50 μs simulations of the receptor-G protein complex with and without the nucleotide GDP gave mechanistic insight into receptor mediated activation. The computational models found that the G protein's Ras and helical domains separate upon receptor binding exposing the nucleotide binding site, however, its α5 helix interaction in the receptor core was found to accelerate nucleotide release. Additionally, structural models were coupled alongside temperature-accelerated MD simulation to capture GDP release on a computationally accessible time scale. In a similar study, the interaction of the β-arrestin was simulated with the receptor (Latorraca et al., 2018) that showed multiple points of contact between arrestin and the receptor. Using 20 simulations ranging from 4 to 7 μs each, the study showed that the receptor's intracellular loops change conformations in response to arrestin. The simulations were able to simulate different mechanisms of arrestin activation. The receptor core and the phosphorylated cytoplasmic tail of the receptor each individually stabilized the active arrestin. Both receptor domains triggered transitions to the active state, and certain receptor core conformations favored arrestin stimulation over G protein stimulation. These and similar studies are laying the foundation for studying biased signaling in GPCRs.

The conformational sampling during long MD simulations can be enhanced by using replica-exchange MD (REMD) approaches, where multiple copies (replicas) of the system are run at different temperatures (most used REMD parameter) and the snapshots from neighboring temperature replicas are exchanged at regular intervals based on a Metropolis condition (Qi, Wei, Ma, & Nussinov, 2018). This approach helps overcome energetic barriers encountered in classical MD simulations to enhance the sampled conformational space as many of the replica temperatures sampled are higher than the physiological temperature. REMD approaches have been applied to different GPCRs (Cong, Fiorucci, & Golebiowski, 2018; Sena, Cong, Giorgetti, Kless, & Carloni, 2017) to enhance the conformational dynamics during receptor activation. Next, we will discuss several methods that have adapted classical MD in different ways but with the common goal to enhance sampling of receptor conformations not easily accessible during classical MD simulations.

2.2 Metadynamics and other biased MD methods

The time scales of physiological processes like GPCR activation are rarely approachable by unbiased long MD simulations, so biased simulations have been used to characterize the free energy landscapes along coordinates of functional interest. This requires a priori knowledge of a subset of the degrees of freedom (referred to as collective variables or CVs) present in the protein system that may have importance in terms of mimicking an(the) activation coordinate(s), or a ligand's motion relative to the final binding site, or morphing a protein's conformation A into conformation B. Four major approaches have emerged for biased MD simulations: metadynamics, adaptively-biased MD, string of swarms MD, and gently-biased MD.

2.2.1 Metadynamics

The methods that fall under the umbrella of metadynamics (Laio & Parrinello, 2002) essentially fill free energy wells encountered along functionally important collective variables (CVs) with positive Gaussian potentials (or "computational sand"), so that the molecular system is discouraged to visit the previously sampled points in the conformational space defined by the CVs. This enables the exploration of the full energy landscape in the subspace of CVs. Metadynamics has been applied to GPCRs to characterize ligand binding pathways, to determine the basis of ligand-receptor selectivity, and to estimate the thermodynamics/kinetics of receptor activation. The binding pathways explored by naloxone ligand, when it binds to the δ-opioid receptor, were studied by microsecond-timescale metadynamics (Provasi, Bortolato, & Filizola, 2009), using the CVs describing the ligand position and extracellular loop 2 position relative to the orthosteric binding site. The study identified metastable binding sites that usually act as a vestibule along the ligand binding pathway to the final binding site. Another study implemented a metadynamics-based protocol to determine ligand binding modes in GPCRs (Soldner, Horn, & Sticht, 2019) by clustering multiple "walker" metadynamics simulations that used ligand's distances to a binding site residue as CVs to "walk" the ligands towards their binding sites and in the process overcome barriers in the free energy landscape. This study used a funnel shaped bias potential to focus the ligand sampling toward the binding site residues and promises to be more accurate than docking to identify correct ligand binding modes. A recent study looked at the norepinephrine ligand's 10-fold selectivity for β1 vs β2 adrenergic receptors using metadynamics simulations biased along a single CV represented by the distance between ligand's center of mass and the bottom of the receptor binding pocket (Xu et al., 2021). It identified the role of an extracellular vestibule region's shape and electrostatic properties that are different for β1 vs β2 adrenergic receptors to be behind norepinephrine ligand's preference for the β1 adrenergic receptor.

2.2.2 Adaptively-biased MD

These methods were developed in response to speeding up metadynamics simulations and to control their accuracy (Babin, Roland, & Sagui, 2008). Metadynamics can require long MD simulations for convergence, needs many parameters, and its computing cost scales as $O(t^2)$. Even though the adaptively-biased simulations also require CV(s) and a judicious choice of these functionally important system variables is critical to capture the relevant parts of the system's energy landscape, the method is characterized by only two parameters and scales as $O(t)$ in computing time. This efficient method also allows extensions based on multiple walkers and replica exchange, where the different replicas can have different temperatures and/or CVs. It can enable up to two orders of magnitude speed up in conformational sampling depending on the implementation. A recent study applied this method to study the undesirable arrestin-biased μ-opioid receptor signaling of fentanyl using the distance between the ligand's rigid/flexible regions and the bottom of the receptor binding site as CVs (de Waal et al., 2020). The structural insight gained in the study was used to successfully design fentanyl analogs that biased the signaling towards the therapeutically desirable G protein coupling.

2.2.3 String of swarms MD

The original string method in CVs (Maragliano, Fischer, Vanden-Eijnden, & Ciccotti, 2006) developed a parametrized string (which is essentially a transition pathway along the CVs), represented by the protein system's snapshots from conformation A to conformation B, by estimating the mean force and the metric tensor at each snapshot using constrained MD simulations. This resulted in a minimum free energy path in the subspace of CVs. This method was adapted into string of swarms MD (Pan, Sezer, & Roux, 2008), which evolved the pathway string using a swarm of unbiased MD trajectories at each snapshot to refine a putative transition path in the CVs using the average dynamic drift of those variables observed in the swarms of trajectories. This method can provide reliable energy landscapes using ~1–2 orders of magnitude less simulation time than unbiased MD simulations. Recently, this method was applied to apo and holo states of β2AR (Fleetwood, Matricon, Carlsson, & Delemotte, 2020) to show how receptor microswitches (subsets of residues) control the conformational equilibrium during receptor activation. This study used five CVs, which were distances between TM helix pairs: TM2-TM7, TM6-TM4, TM7-TM4, TM3-TM6, and TM6-TM5. The free energy landscapes projected on to two CVs (TM2-TM7 and TM6-TM4 distances) clearly captured the outward TM6 motion followed by inward TM7 motion during the receptor activation process. These transitions were analyzed to show the communication between TM5 bulge, connector region ($P^{5.50}/I^{3.40}/F^{6.44}$ motif), and the NPxxY motif in TM7 during activation. The string of swarms method scales well with increased number of CVs, so one is not limited to choosing only one or two CVs as was demonstrated by this study on β2AR.

2.2.4 Gently-biased MD

These methods have been developed to enhance conformational sampling in realistic MD time scales by introducing macroscale biases in classical MD simulations that don't add any potentials to fill the energy wells. In a recent study, an MD protocol was developed to identify allosteric binding sites in GPCRs by applying a cylindrical harmonic wall potential and by preventing membrane distortions to enhance ligand sampling of the receptor-vicinity regions (Ciancetta et al., 2021). The study identified a novel allosteric site prospectively for D2 dopamine receptor and then validated that site by mutagenesis. Computationally investigating such putative allosteric sites on receptors can provide new avenues to target for biased ligand discovery. A method called supervised MD has been implemented to characterize receptor-ligand recognition pathways by supervising the unbiased MD simulated ligand entry at specific checkpoints (Sabbadin & Moro, 2014). Any MD trajectory that doesn't bring the ligand closer to the binding site is halted and restarted from the previous checkpoint with new sets of velocities. This process can be applied in reverse to look at ligand unbinding processes as was implemented in a study involving four GPCR ligands dissociating from their receptors (Deganutti, Moro, & Reynolds, 2020) without introducing any energy bias in the simulations.

2.3 Accelerated MD

Most methods described in the previous section require knowledge of functionally relevant collective variables (CVs). This information may not be available or known for the protein systems of interest. Even if some CVs are known, it may not be easy to preclude other CVs that may be important but may not have been included in the biased simulations. Two flavors of accelerated MD methods have been developed that don't require information on CVs, one based on applying a boost potential to enhance conformational sampling, and the other one based on applying random forces to accelerate the simulations.

2.3.1 Boost potential accelerated MD

As a step toward unbiased exploration of conformational spaces, boost potential based accelerated MD (aMD) methods were developed (Hamelberg, de Oliveira, & McCammon, 2007; Hamelberg, Mongan, & McCammon, 2004; Miao, Feher, & McCammon, 2015) that enable the enhanced sampling of distinct protein conformations and of rare conformational changes without requiring a priori knowledge of CVs. These methods add a non-negative boost potential to smoothen the protein's potential energy surface, which lowers energy barriers to enhance conformational sampling and to ease transitions between low-lying states. The harmonic boost potential version of aMD method is called GaMD (Miao, Nichols, & McCammon, 2014; Wang et al., 2021). These methods have now been extensively applied to GPCRs with some highlights being to accelerate ligand binding (Kappel, Miao, & McCammon, 2015), to enhance the exploration of receptor free energy landscape (Miao et al., 2014), to generate receptor conformational ensembles used in the discovery of novel allosteric modulators for the M2 muscarinic acetylcholine receptor (Miao et al., 2016), to gradually activate the same M2 receptor (Miao & McCammon, 2016b), to map allosteric binding sites of the A2a adenosine receptor (Caliman, Miao, & McCammon, 2018), to generate mechanistic insights in A2a receptor–G protein interactions (Wang & Miao, 2019), and to provide allosteric mechanisms behind adenosine A1 receptor mediated analgesia (Draper-Joyce et al., 2021). These different example applications have provided detailed insight into GPCR function and show the diversity of questions around receptor signaling that can be probed by this acceleration method. The conformation sampling in GaMD simulations can be further enhanced by using them in the context of replica exchange MD framework (Fajer, Hamelberg, & McCammon, 2008; Huang, McCammon, & Miao, 2018). Based on different implementations, a speed-up of 2–3 orders of magnitude in simulation time is within reach with these aMD methods.

2.3.2 Random acceleration MD

This method was developed as a random expulsion MD simulation protocol to accelerate the entry (and exit) of ligands into (from) their binding sites in proteins (Ludemann, Lounnas, & Wade, 2000). This enables the detailed exploration of binding pathways and egress pathways in protein–ligand interactions as these pathways invariably contain large barriers. Multiple simulations are run with a random force added to the normal force-field, but each simulation is then carried out in an unbiased manner. This method has been applied to study the ligand entry/exit pathways in the β2AR (Wang & Duan, 2009), which showed that: (a) an extracellular egress point in the loops is frequently visited by the ligands; (b) when the ligands escaped the binding site residues they didn't encounter any major barriers until their encounter with extracellular loops; and (c) ligand entry requires rewiring of the extracellular region interactions with the TM domains in the apo state enabling a transformation to the holo state. In another study, the egress route of retinal from the opsin state of rhodopsin was simulated and found to be through the lipid-tail facing TM5-TM6 opening (Wang & Duan, 2011). Both of these studies required 100–200 independent simulations with random forces to obtain statistically significant pathways preferred by the ligands during their interactions with the receptor.

2.4 Markov state models and MD

Markov state models (MSM) represent a master equation framework that attempts to uncover the underlying kinetics that exists within complex protein conformational landscapes (Husic & Pande, 2018). It has been gaining in popularity due to MSM's capability of returning interpretable results of conformational equilibrium (Kabir, Akhter, & Shehu, 2019) by providing the probability of transferring to different conformational states and also providing potential pathways. MSMs work by analyzing any molecular dynamics (MD) simulation data of a system and extracting its structural features (Kabir et al., 2019). These features can range from dihedral angles, root-mean-square deviation (RMSD) from a reference structure, relative positions of some Cα atoms, and residue-residue contacts. Extracted features are analyzed by a dimensionality reduction algorithm that converts the feature data into principal components which capture all the largest conformational variations in the system (Pande, Beauchamp, & Bowman, 2010). The principal components are ranked, where the first component captures the largest conformational motion that occurs within the system. The top principal components are used to cluster observed MD conformations into micro-states. These micro-states are used to build a transition matrix that measures the probability of transitioning between all micro-states (Kabir et al., 2019). However, when building a transition matrix, one must select a lag time that is Markovian. This means that the selected lag time should allow transitions independent of the previous state (Husic & Pande, 2018). Micro-states go through a coarse grain clustering to find "interesting" (some being novel) conformations (macro-states) in the constructed MSM model (Husic & Pande, 2018). MSM models have been applied to numerous protein systems as they interpret simulation data in terms of intuitive conformational states, and they enable the application of this information to guide any new rounds of MD simulations to especially enhance conformational spaces less traveled.

In the context of GPCR applications, the MSM framework was demonstrated using a 150 μs long unbiased MD simulation data (generated by a different research group) on the β2AR (Shukla, Lawrenz, & Pande, 2015). The simulation data were clustered into micro-states through a hybrid metric, where it looked at four regions of the β2AR: RMSD of ligand binding residues of carazol (pdbid: 2RH1) and BI-167107 (pdbid: 3P0G) bound, RMSD of connecter regions, RMSD of the NPxxY motif, and the Cα distances between helix 3 and 6. Through this hybrid metric, they generated almost 3000 micro-states that were used to build their

MSM model. Their model revealed the major difference between agonist-bound β2AR intermediate and inactive states, being the flipping of Ile121 and Phe282 residues in the connector region, while for the inverse agonist-bound conformation, it was the twisting of Helix 7 NPxxY motif region. MSMs were also used to analyze a 2 ms MD simulation of β2AR that was generated using cloud computing (Kohlhoff et al., 2014), which revealed multiple activation pathways as well as how agonists and inverse agonists interact in different ways with these pathways. Since then, MSMs have seen applications for many GPCRs to analyze long time scale MD simulation data related to receptor-ligand selectivities (Liao, Remington, May, & Li, 2021), receptor activation pathways (Lu et al., 2021), receptor oligomerization (Song, Duncan, & Sansom, 2021), and ligand modulation of receptor conformational landscapes (Taylor, Lee, & Amaro, 2019).

MSMs are also increasingly being combined with biased and accelerated MD methods to create hybrid approaches to enhance conformational sampling that will be discussed in a later section below.

2.5 RMSD based clustering-guided MD

Complementary to MSM models, we are developing a method based on Root Mean Squared Deviation (RMSD) analysis of MD simulation data called SeedPicker, that attempts to enhance the exploration of conformational landscapes by using deviations from multiple functionally distinct conformations as a clustering metric to drive subsequent rounds of unbiased MD simulations. Starting structures of functionally distinct conformations are used as a guide to determine if selected seeds have sampled novel conformational landscapes. Multiple functionally distinct conformations are used in parallel unbiased MD simulations to generate first round of MD simulation data. This data is then analyzed in the SeedPicker algorithm where the RMSD calculations are applied to all frames, thus forming a RMSD matrix. The algorithm then analyzes the RMSD matrix in search of the outliers by selecting a frame that contains the highest average RMSD value to all of the frames. The selected seeds are compared with the known experimental conformations to characterize how distinct they are and how much conformational space has been sampled so far in terms of an effective conformational surface area. The selected seeds are then used for the next round of unbiased MD simulations. The new MD round data is combined with data from the prior rounds and analyzed through the SeedPicker pipeline again to select

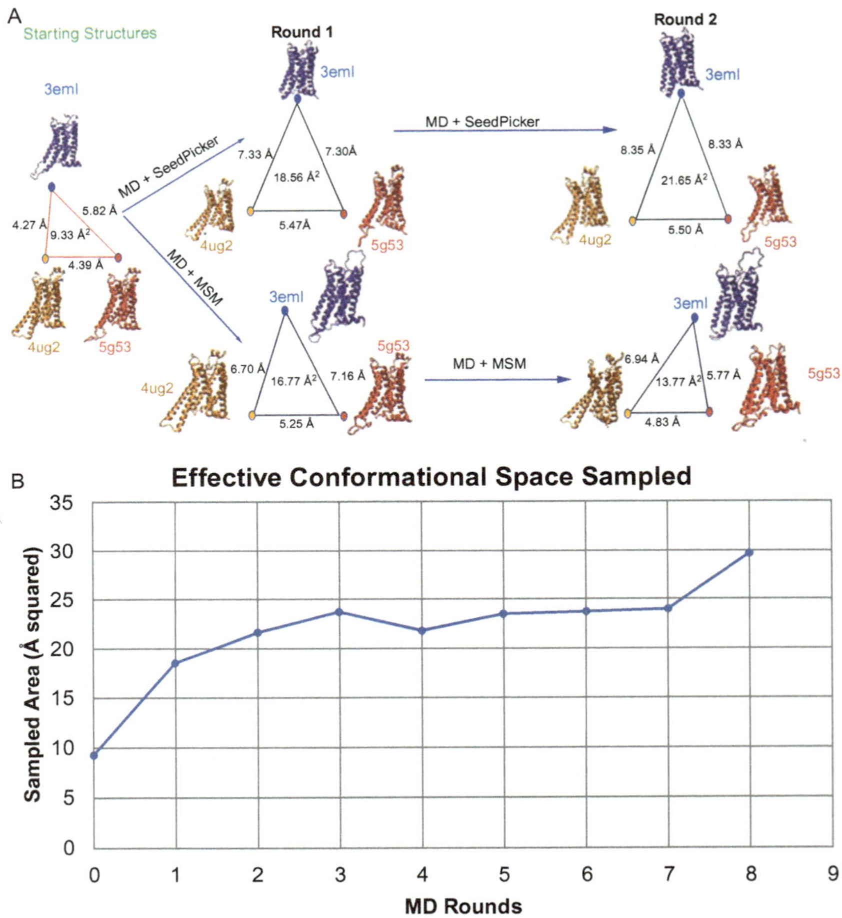

Fig. 3 (A) Comparison of SeedPicker and MSM based conformational space sampling after two rounds of MD starting with the same three conformations of the A2a adenosine receptor. The pdbids after each round represent which starting structure generated that conformation. The RMSDs between three seeds and the effective surface area of the enclosed triangle are also shown. (B) Effective conformational space in terms of three-conformation based surface area sampled by SeedPicker based seeds for eight consecutive rounds of unbiased MD simulations.

the new set of seed conformations for the next round of MD simulations. This process is repeated until the effective surface area of sampled conformational space converges. Fig. 3A shows a demonstration of this method based on the three conformational states of the A2a adenosine receptor (inactive, intermediate, and active) along with a comparison to the MSM based clustering using Cα atom positions as metrics that sampled a smaller conformational space (compared to SeedPicker) after two rounds of MD. Fig. 3B shows implementation of the SeedPicker algorithm to seed conformations for eight rounds of MD and the corresponding effective conformational surface area being sampled after each round of MD. The last MD round data shows that the conformational sampling is not converged yet and has potentially entered into a novel space not represented by starting conformations. The method is being developed with a goal to find new conformations not yet seen experimentally and which will then be used to guide experimental studies.

3. Discrete conformational sampling methods

The lipid bilayer membrane environment of α-helical membrane proteins provides a two-dimensional constraint on the protein topology that can be utilized to sample and characterize their *in membrano* dynamics. This stems from the fact that the α-helices in membrane proteins cannot tilt in the membrane beyond a certain angle. This enables, at least in principle, a grid-based conformational sampling of "all" protein conformations. This approach has been implemented in the absence and in the presence of ligands in different methods as described below. In addition, the availability of experimental structures of the same receptor bound to different ligand types and also bound to different intracellular transducers is allowing the use of coarse-graining of the free energy landscape where the different experimental structures represent different local minima or discrete conformations whose relative free energies can shed light on the receptor activation related processes.

3.1 Ligand-independent discrete sampling

The transmembrane (TM) helices exhibit very dynamic behavior in the lipid bilayer as has been observed by many biophysical approaches, so as a first approximation one can assume that extracellular and intracellular loops have only minor effects on TM helix dynamics. In this limit, we can "count" all TM helix conformations in the membrane. The coordinate system shown in Fig. 4 helps visualize helix orientations relative to the membrane, where one usually treats the middle plane of the membrane as the x-y plane (equivalent to $z = 0$ plane). Each TM helix is treated as a cylinder with the least moment of inertia vector going through the middle of the helix and describing the helical axis. This mathematical definition of the helix vector is uniquely

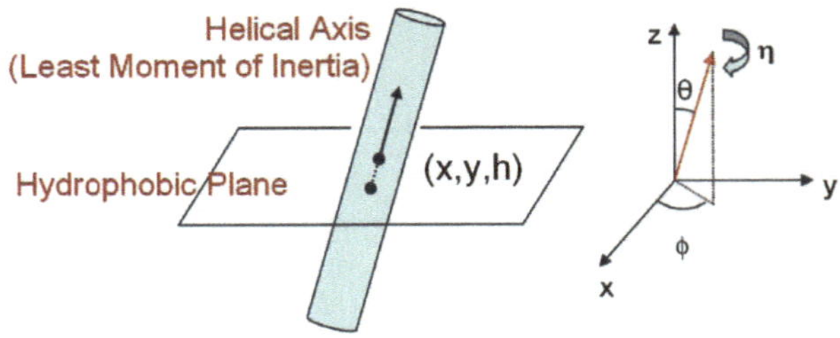

Fig. 4 TM helix coordinates in a lipid bilayer, where the hydrophobic plane goes through the middle of the membrane block (Abrol et al., 2011).

defined even for bent TM helices, which many GPCRs exhibit due to the presence of conserved proline residues in the middle of many TM helices. Each helical axis crosses the membrane middle (x-y plane) at a specific residue position called the hydrophobic center **h**. The helical axis helps define three orientation angles (helix tilt angle, helix sweep angle, helix rotation angle, labeled θ, ϕ, η, respectively) as displayed in Fig. 4 and described in detail elsewhere (Abrol et al., 2011).

This coordinate systems enables us to sample all TM conformations for a GPCR using an efficient two-helix decomposition method called BiHelix/SuperBiHelix (Abrol, Bray, & Goddard, 2012; Bray, Abrol, Goddard, Trzaskowski, & Scott, 2014) that deconstructs the seven-helix interaction problem into many two-helix problems. This brute-force approach for discrete conformational sampling is called GEnSeMBLE (GPCR Ensemble of Structures in Membrane BiLayer Environment) and has been applied to many GPCRs to predict their structures, their ligand binding sites (Abrol, Griffith, Bray, & Goddard, 2012; Dong, Abrol, & Goddard, 2015; Kim, Riley, Abrol, Jacobson, & Goddard, 2011; Tan, Abrol, Trzaskowski, & Goddard, 2012), and to study the effects of ligands and mutations on receptor conformations (Abrol et al., 2013, 2014; Ahn, Scott, Abrol, Goddard, & Kendall, 2013; Carpentier et al., 2012; Scott, Abrol, Ahn, Kendall, & Goddard, 2013). The method has been extended to predict active conformations (ActiveGEnSeMBLE) starting only from the inactive state (Dong et al., 2016; Dong, Goddard, & Abrol, 2017), one of the handful of methods capable of doing so in an unbiased manner. In this method, first a coarse-grained conformational sampling is carried out using SuperBiHelix to cast a wide net in the conformational space, which is followed by several fine-grained samplings to find local minima in different conformational spaces identified by the coarse-grained sampling. ActiveGEnSeMBLE was

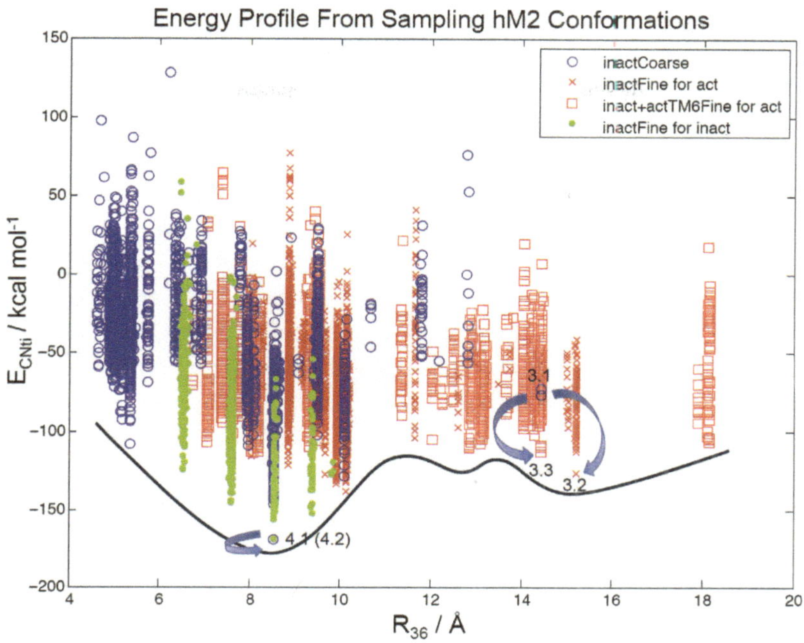

Fig. 5 Predicted energy landscape for human M2 muscarinic receptor. The red "x"s represent the results from fine-grain sampling after a round of coarse-grain sampled (blue circles) conformations (Dong et al., 2016).

applied to the β2AR and M2 muscarinic receptor cases to predict their active conformations correctly starting only from the inactive state (Dong et al., 2016). Fig. 5 shows the sampled conformational space for human M2 receptor using the intracellular TM3-TM6 distance as an effective coordinate to display the predicted conformations in the inactive and active energy wells.

The study also looked at the relative stability of the inactive, intermediate, and active conformations of the β2AR with and without the ligand as well as with and without the G protein. It provided a basis for the experimental observation (Manglik & Kobilka, 2014) that β2AR's active conformation cannot be stabilized by the agonist alone and that the G protein is necessary for the active conformation stabilization.

3.2 Ligand-dependent discrete sampling

The effect of the ligand on receptor conformations was not considered during conformational sampling steps of the methods described in the last section. Ideally, the effect of the ligand should be accounted for during conformational sampling; however, it can be potentially computationally expensive. This was made practical in a method called LITiCon (Ligand-induced transmembrane rotational conformational changes), where the conformational space of TM helices is systematically sampled in the presence of ligands combined with energy minimization to shed light on the ligand's effect on receptor conformations (Bhattacharya, Hall, Li, & Vaidehi, 2008; Bhattacharya, Hall, & Vaidehi, 2008; Bhattacharya & Vaidehi, 2012). The method was successful in characterizing the agonist induced conformational changes in rhodopsin activation (Bhattacharya, Hall, & Vaidehi, 2008) and ligand stabilized conformations in the β2AR (Bhattacharya, Hall, Li, et al., 2008). As the method accurately captures interhelical interactions, it has been utilized to predict thermostable receptor mutants (Bhattacharya, Lee, Grisshammer, Tate, & Vaidehi, 2014). This is an especially challenging problem as mutations can have dramatic as well as subtle effects on interhelical interactions and receptor stability.

3.3 Coarse-grained energy landscapes

In the introduction section, we had discussed the explosion in the number of structures/conformations that are currently available for many GPCRs. This situation is only getting better with time as we now have structures for the same receptor bound to inverse agonists, antagonists, or agonists, with or without allosteric modulators, and with or without intracellular transducers. If one imagines a free energy landscape for receptor activation, each of these structures will represent local minima in different regions of the landscape. Even if structures corresponding to some minima might be unavailable, there is enough structural information to construct and relax receptor conformations corresponding to some of those minima. The free energy landscape describing all these minima can help identify mechanisms of receptor activation and how different ligands/transducers stabilize or destabilize different receptor conformations, which provides biochemical insight into receptor signaling and pharmacology. This coarse-graining of the energy landscape has been implemented to explore the free energy landscape for β2AR activation (Alhadeff, Vorobyov, Yoon, & Warshel, 2018) and glucagon-like peptide-1 receptor (GLP-1R) activation (Alhadeff & Warshel, 2020). It has also been used to explore the activation process of the β2AR-Gs complex, especially in terms of the GDP/GTP exchange that occurs during G protein engagement (Bai et al., 2021). This study identified the location of the conformational barrier and the rate limiting step for the G protein to get to the GDP-free state. It also provided a rationale for the effect of G-protein mutations on its coupling to the receptor.

4. Future of GPCR conformational sampling is hybrid approaches

The scientific field of MD-based methods has been, in general, very open about the sharing of codes and protocols, so it is not a surprise that umbrella software interfaces like high-throughput MD (HTMD) (Doerr, Harvey, Noe, & De Fabritiis, 2016), OpenMM (Eastman et al., 2017), GROMACS (Pronk et al., 2013), and AMBER (Case et al., 2005) have pulled together MD pipelines, popular force-fields, and the best analysis tools to make it easier to execute enhanced conformational sampling for any protein with available structure and especially for membrane proteins (Doerr, Giorgino, Martinez-Rosell, Damas, & De Fabritiis, 2017). One can now manipulate structures to account for mutations (for example), build the system in its physiological environment (water and/or lipids) (e.g., CHARMM-GUI (Jo et al., 2017)), carry out MD simulations, and then perform extensive analysis on MD data through adaptive sampling, clustering, MSM model building, and free energy calculations.

Many studies have combined different MD approaches with specific analysis methods to enhance conformational sampling in GPCRs. In one study, adaptive sampling was used to explore ligand-specific conformational transitions in the μ-opioid receptor and the resulting simulation data was used to build MSM models of the receptor in the presence of two different ligands (Kapoor, Martinez-Rosell, Provasi, de Fabritiis, & Filizola, 2017). The study showed a different kinetic behavior of the receptor's conformational dynamics in the presence of a balanced agonist vs a G-protein biased agonist. In another study, metadynamics was combined with MSM models to look at the energetics of receptor activation by biasing the simulations along the path CVs chosen carefully to follow a low free energy activation pathway from the inactive state to the active state (Meral, Provasi, & Filizola, 2018). The resulting kinetics of receptor activation was consistent with that obtained from adaptive sampling. MSM models have been combined with adaptive sampling based MD simulation data using the HTMD pipelines to construct the activation landscape of A2a adenosine receptor (Lovera, Cuzzolin, Kelm, De Fabritiis, & Sands, 2019) that showed ligand-competent intermediate states. In this study, the adaptive sampling part was implemented using the inactive conformation with a goal-oriented scoring function (structural signatures of activation) as opposed to the actual active conformation. One study combined GaMD with multiple walker metadynamics and classical MD to explore agonist binding and G protein coupling in the histamine H2 receptor (Conrad, Soldner, Miao, & Sticht, 2020). The study showed that histamine ligand alone is not enough to stabilize the active conformation and the presence of the G protein coupling is important to stabilize the active state, as we had rationalized before for β2AR (Dong et al., 2016), and as has been shown before for other GPCRs as well. In addition, the hybrid study found evidence for a weak allosteric coupling between the intracellular and extracellular binding sites in the receptor.

We are in the middle of an exciting phase in understanding GPCR signaling through its conformational dynamics. The sustainable increase in receptor structures in different functional states, improvements in biophysical methods probing receptor dynamics, and maturing computational methods aimed at enhanced conformational sampling are providing unprecedented insight into receptor structure and function, which bodes well for rational structure-based drug discovery efforts targeting GPCRs and which in the light of our current understanding of GPCRs should be called conformation-based drug discovery. The stage is set for the design of finely tuned and highly efficacious drugs that exhibit biased signaling, if necessary, to minimize on-target side-effects.

Acknowledgments

Research in RA lab has been supported by the NIH SCORE grant SC2GM130480, CSUPERB New Investigator Award, and the institutional startup funds provided by the College of Science and Math at CSUN. L.J.S. was supported by NIH RISE grant GM063787 and E.S. was supported by NIH BUILD grant 5RL5GM118975.

References

Abrol, R., Bray, J. K., & Goddard, W. A. (2012). Bihelix: Towards de novo structure prediction of an ensemble of G-protein coupled receptor conformations. *Proteins: Structure, Function, and Bioinformatics*, *80*(2), 505–518. https://doi.org/10.1002/Prot.23216.

Abrol, R., Griffith, A. R., Bray, J. K., & Goddard, W. A., 3rd. (2012). Structure prediction of G protein-coupled receptors and their ensemble of functionally important conformations. In N. Vaidehi, & J. Klein-Seetharaman (Eds.), *Vol. 914. Membrane protein structure: Methods and protocols* (2012/09/15, pp. 237–254). Humana. https://doi.org/10.1007/978-1-62703-023-6_14.

Abrol, R., Kim, S. K., Bray, J. K., Griffith, A. R., & Goddard, W. A. (2011, Dec). Characterizing and predicting the functional and conformational diversity of seven-transmembrane proteins. *Methods*, *55*(4), 405–414. https://doi.org/10.1016/j.ymeth.2011.12.005.

Abrol, R., Kim, S. K., Bray, J. K., Trzaskowski, B., & Goddard, W. A., 3rd. (2013). Conformational ensemble view of G protein-coupled receptors and the effect of mutations and ligand binding. *Methods in Enzymology*, *520*, 31–48. https://doi.org/10.1016/B978-0-12-391861-1.00002-2.

Abrol, R., Trzaskowski, B., Goddard, W. A., 3rd, Nesterov, A., Olave, I., & Irons, C. (2014). Ligand- and mutation-induced conformational selection in the CCR5 chemokine G protein-coupled receptor. *Proceedings of the National Academy of Sciences of the United States of America, 111*(36), 13040–13045. https://doi.org/https://doi.org/10.1073/pnas.1413216111.

Ahn, D., Ham, D., & Chung, K. Y. (2021). The conformational transition during G protein-coupled receptor (GPCR) and G protein interaction. *Current Opinion in Structural Biology, 69*, 117–123. https://doi.org/10.1016/j.sbi.2021.03.013.

Ahn, K. H., Scott, C. E., Abrol, R., Goddard, W. A., 3rd, & Kendall, D. A. (2013). Computationally-predicted CB1 cannabinoid receptor mutants show distinct patterns of salt-bridges that correlate with their level of constitutive activity reflected in G protein coupling levels, thermal stability, and ligand binding. *Proteins: Structure, Function, and Bioinformatics, 81*(8), 1304–1317. https://doi.org/10.1002/prot.24264.

Alhadeff, R., Vorobyov, I., Yoon, H. W., & Warshel, A. (2018). Exploring the free-energy landscape of GPCR activation. *Proceedings of the National Academy of Sciences of the United States of America, 115*(41), 10327–10332. https://doi.org/10.1073/pnas.1810316115.

Alhadeff, R., & Warshel, A. (2020). A free-energy landscape for the glucagon-like peptide 1 receptor GLP1R. *Proteins: Structure, Function, and Bioinformatics, 88*(1), 127–134. https://doi.org/https://doi.org/10.1002/prot.25777.

Babin, V., Roland, C., & Sagui, C. (2008). Adaptively biased molecular dynamics for free energy calculations. *The Journal of Chemical Physics, 128*(13), 134101. https://doi.org/10.1063/1.2844595.

Bai, C., Wang, J., Mondal, D., Du, Y., Ye, R. D., & Warshel, A. (2021). Exploring the activation process of the beta2AR-Gs complex. *Journal of the American Chemical Society, 143*(29), 11044–11051. https://doi.org/https://doi.org/10.1021/jacs.1c03696.

Barki-Harrington, L., & Rockman, H. A. (2008). Beta-arrestins: Multifunctional cellular mediators. *Physiology (Bethesda), 23*, 17–22. https://doi.org/10.1152/physiol.00042.2007.

Bhattacharya, S., Hall, S. E., Li, H., & Vaidehi, N. (2008). Ligand-stabilized conformational states of human beta(2) adrenergic receptor: Insight into G-protein-coupled receptor activation. *Biophysical Journal, 94*(6), 2027–2042. https://doi.org/10.1529/biophysj.107.117648.

Bhattacharya, S., Hall, S. E., & Vaidehi, N. (2008). Agonist-induced conformational changes in bovine rhodopsin: Insight into activation of G-protein-coupled receptors. *Journal of Molecular Biology, 382*(2), 539–555. https://doi.org/10.1016/j.jmb.2008.06.084.

Bhattacharya, S., Lee, S., Grisshammer, R., Tate, C. G., & Vaidehi, N. (2014). Rapid computational prediction of Thermostabilizing mutations for G protein-coupled receptors. *Journal of Chemical Theory and Computation, 10*(11), 5149–5160. https://doi.org/10.1021/ct500616v.

Bhattacharya, S., & Vaidehi, N. (2012). LITiCon: A discrete conformational sampling computational method for mapping various functionally selective conformational states of transmembrane helical proteins. *Methods in Molecular Biology, 914*, 167–178. https://doi.org/10.1007/978-1-62703-023-6_10.

Blois, T. M., & Bowie, J. U. (2009). G-protein-coupled receptor structures were not built in a day. *Protein Science, 18*(7), 1335–1342. https://doi.org/10.1002/pro.165.

Bray, J. K., Abrol, R., Goddard, W. A., 3rd, Trzaskowski, B., & Scott, C. E. (2014). SuperBiHelix method for predicting the pleiotropic ensemble of G-protein-coupled receptor conformations. *Proceedings of the National Academy of Sciences of the United States of America, 111*(1), E72–E78. https://doi.org/10.1073/pnas.1321233111.

Caliman, A. D., Miao, Y., & McCammon, J. A. (2018). Mapping the allosteric sites of the A2A adenosine receptor. *Chemical Biology & Drug Design, 91*(1), 5–16. https://doi.org/10.1111/cbdd.13053.

Carpentier, E., Greenbaum, L. A., Rochdi, D., Abrol, R., Goddard, W. A., 3rd, Bichet, D. G., et al. (2012). Identification and characterization of an activating F229V substitution in the V2 vasopressin receptor in an infant with NSIAD. *Journal of the American Society of Nephrology*, *23*(10), 1635–1640. https://doi.org/10.1681/ASN.2012010077.

Case, D. A., Cheatham, T. E., 3rd, Darden, T., Gohlke, H., Luo, R., Merz, K. M., Jr., et al. (2005). The Amber biomolecular simulation programs. *Journal of Computational Chemistry*, *26*(16), 1668–1688. https://doi.org/10.1002/jcc.20290.

Casiraghi, M., Point, E., Pozza, A., Moncoq, K., Baneres, J. L., & Catoire, L. J. (2019). NMR analysis of GPCR conformational landscapes and dynamics. *Molecular and Cellular Endocrinology*, *484*, 69–77. https://doi.org/10.1016/j.mce.2018.12.019.

Cherezov, V., Rosenbaum, D. M., Hanson, M. A., Rasmussen, S. G., Thian, F. S., Kobilka, T. S., et al. (2007). High-resolution crystal structure of an engineered human beta2-adrenergic G protein-coupled receptor. *Science*, *318*(5854), 1258–1265. https://doi.org/https://doi.org/10.1126/science.1150577.

Ciancetta, A., Gill, A. K., Ding, T., Karlov, D. S., Chalhoub, G., McCormick, P. J., et al. (2021). Probe confined dynamic mapping for G protein-coupled receptor allosteric site prediction. *ACS Central Science*. https://doi.org/10.1021/acscentsci.1c00802.

Cong, X., Fiorucci, S., & Golebiowski, J. (2018). Activation dynamics of the Neurotensin G protein-coupled receptor 1. *Journal of Chemical Theory and Computation*, *14*(8), 4467–4473. https://doi.org/10.1021/acs.jctc.8b00216.

Conrad, M., Soldner, C. A., Miao, Y., & Sticht, H. (2020). Agonist binding and G protein coupling in histamine H2 receptor: A molecular dynamics study. *International Journal of Molecular Sciences*, *21*(18). https://doi.org/10.3390/ijms21186693.

de Waal, P. W., Shi, J., You, E., Wang, X., Melcher, K., Jiang, Y., et al. (2020). Molecular mechanisms of fentanyl mediated beta-arrestin biased signaling. *PLoS Computational Biology*, *16*(4), e1007394. https://doi.org/10.1371/journal.pcbi.1007394.

Deganutti, G., Moro, S., & Reynolds, C. A. (2019). Peeking at G-protein-coupled receptors through the molecular dynamics keyhole. *Future Medicinal Chemistry*, *11*(6), 599–615. https://doi.org/10.4155/fmc-2018-0393.

Deganutti, G., Moro, S., & Reynolds, C. A. (2020). A supervised molecular dynamics approach to unbiased ligand-protein unbinding. *Journal of Chemical Information and Modeling*, *60*(3), 1804–1817. https://doi.org/10.1021/acs.jcim.9b01094.

Deupi, X., & Kobilka, B. K. (2010). Energy landscapes as a tool to integrate GPCR structure, dynamics, and function. *Physiology (Bethesda)*, *25*(5), 293–303. https://doi.org/10.1152/physiol.00002.2010.

Doerr, S., Giorgino, T., Martinez-Rosell, G., Damas, J. M., & De Fabritiis, G. (2017). High-throughput automated preparation and simulation of membrane proteins with HTMD. *Journal of Chemical Theory and Computation*, *13*(9), 4003–4011. https://doi.org/https://doi.org/10.1021/acs.jctc.7b00480.

Doerr, S., Harvey, M. J., Noe, F., & De Fabritiis, G. (2016). HTMD: High-throughput molecular dynamics for molecular discovery. *Journal of Chemical Theory and Computation*, *12*(4), 1845–1852. https://doi.org/10.1021/acs.jctc.6b00049.

Dong, S. S., Abrol, R., & Goddard, W. A., 3rd. (2015). The predicted ensemble of low-energy conformations of human somatostatin receptor subtype 5 and the binding of antagonists. *ChemMedChem*, *10*(4), 650–661. https://doi.org/10.1002/cmdc.201500023.

Dong, S. S., Goddard, W. A., 3rd, & Abrol, R. (2016). Conformational and thermodynamic landscape of GPCR activation from theory and computation. *Biophysical Journal*, *110*(12), 2618–2629. https://doi.org/10.1016/j.bpj.2016.04.028.

Dong, S. S., Goddard, W. A., 3rd, & Abrol, R. (2017). Identifying multiple active conformations in the G protein-coupled receptor activation landscape using computational methods. *Methods in Cell Biology*, *142*, 173–186. https://doi.org/10.1016/bs.mcb.2017.07.009.

Draper-Joyce, C. J., Bhola, R., Wang, J., Bhattarai, A., Nguyen, A. T. N., Cowie-Kent, I., et al. (2021). Positive allosteric mechanisms of adenosine A1 receptor-mediated analgesia. *Nature, 597*(7877), 571–576. https://doi.org/10.1038/s41586-021-03897-2.

Dror, R. O., Dirks, R. M., Grossman, J. P., Xu, H., & Shaw, D. E. (2012). Biomolecular simulation: a computational microscope for molecular biology. *Annual Review of Biophysics, 41*, 429–452. https://doi.org/10.1146/annurev-biophys-042910-155245.

Dror, R. O., Green, H. F., Valant, C., Borhani, D. W., Valcourt, J. R., Pan, A. C., et al. (2013). Structural basis for modulation of a G-protein-coupled receptor by allosteric drugs. *Nature, 503*(7475), 295–299. https://doi.org/https://doi.org/10.1038/nature12595.

Dror, R. O., Jensen, M. O., Borhani, D. W., & Shaw, D. E. (2010). Exploring atomic resolution physiology on a femtosecond to millisecond timescale using molecular dynamics simulations. *The Journal of General Physiology, 135*(6), 555–562. https://doi.org/10.1085/jgp.200910373.

Dror, R. O., Mildorf, T. J., Hilger, D., Manglik, A., Borhani, D. W., Arlow, D. H., et al. (2015). SIGNAL TRANSDUCTION. Structural basis for nucleotide exchange in heterotrimeric G proteins. *Science, 348*(6241), 1361–1365. https://doi.org/10.1126/science.aaa5264.

Dror, R. O., Pan, A. C., Arlow, D. H., Borhani, D. W., Maragakis, P., Shan, Y., et al. (2011). Pathway and mechanism of drug binding to G-protein-coupled receptors. *Proceedings of the National Academy of Sciences of the United States of America, 108*(32), 13118–13123. https://doi.org/10.1073/pnas.1104614108.

Du, Y., Duc, N. M., Rasmussen, S. G. F., Hilger, D., Kubiak, X., Wang, L., et al. (2019). Assembly of a GPCR-G protein complex. *Cell, 177*(5), 1232–1242. e1211. https://doi.org/10.1016/j.cell.2019.04.022.

Eastman, P., Swails, J., Chodera, J. D., McGibbon, R. T., Zhao, Y., & Beauchamp, K. A. (2017). OpenMM 7: Rapid development of high performance algorithms for molecular dynamics. *PLoS Computational Biology, 13*(7), e1005659. https://doi.org/10.1371/journal.pcbi.1005659.

Eddy, M. T., Lee, M. Y., Gao, Z. G., White, K. L., Didenko, T., Horst, R., et al. (2018). Allosteric coupling of drug binding and intracellular signaling in the A2A adenosine receptor. *Cell, 172*(1–2), 68–80. e12. https://doi.org/10.1016/j.cell.2017.12.004.

Elgeti, M., & Hubbell, W. L. (2021). DEER analysis of GPCR conformational heterogeneity. *Biomolecules, 11*(6). https://doi.org/10.3390/biom11060778.

Fajer, M., Hamelberg, D., & McCammon, J. A. (2008). Replica-exchange accelerated molecular dynamics (REXAMD) applied to thermodynamic integration. *Journal of Chemical Theory and Computation, 4*(10), 1565–1569. https://doi.org/10.1021/ct800250m.

Fernandes, D. D., Neale, C., Gomes, G. W., Li, Y., Malik, A., Pandey, A., et al. (2021). Ligand modulation of the conformational dynamics of the A2A adenosine receptor revealed by single-molecule fluorescence. *Scientific Reports, 11*(1), 5910. https://doi.org/10.1038/s41598-021-84069-0.

Fleetwood, O., Matricon, P., Carlsson, J., & Delemotte, L. (2020). Energy landscapes reveal agonist control of G protein-coupled receptor activation via microswitches. *Biochemistry, 59*(7), 880–891. https://doi.org/10.1021/acs.biochem.9b00842.

Friedrichs, M. S., Eastman, P., Vaidyanathan, V., Houston, M., Legrand, S., Beberg, A. L., et al. (2009). Accelerating molecular dynamic simulation on graphics processing units. *Journal of Computational Chemistry, 30*(6), 864–872. https://doi.org/10.1002/jcc.21209.

Galandrin, S., Oligny-Longpre, G., & Bouvier, M. (2007). The evasive nature of drug efficacy: Implications for drug discovery. *Trends in Pharmacological Sciences, 28*(8), 423–430. https://doi.org/10.1016/j.tips.2007.06.005.

Garcia-Nafria, J., & Tate, C. G. (2020). Cryo-electron microscopy: Moving beyond X-ray crystal structures for drug receptors and drug development. *Annual Review of*

Pharmacology and Toxicology, *60*, 51–71. https://doi.org/10.1146/annurev-pharmtox-010919-023545.

Ghosh, E., Kumari, P., Jaiman, D., & Shukla, A. K. (2015). Methodological advances: The unsung heroes of the GPCR structural revolution. *Nature Reviews. Molecular Cell Biology*, *16*(2), 69–81. https://doi.org/10.1038/nrm3933.

Grossfield, A. (2011). Recent progress in the study of G protein-coupled receptors with molecular dynamics computer simulations. *Biochimica et Biophysica Acta*, *1808*(7), 1868–1878. https://doi.org/10.1016/j.bbamem.2011.03.010.

Gurevich, V. V., & Gurevich, E. V. (2017). Molecular mechanisms of GPCR signaling: A structural perspective. *International Journal of Molecular Sciences*, *18*(12). https://doi.org/10.3390/ijms18122519.

Gurevich, V. V., & Gurevich, E. V. (2020). Biased GPCR signaling: Possible mechanisms and inherent limitations. *Pharmacology & Therapeutics*, *211*, 107540. https://doi.org/10.1016/j.pharmthera.2020.107540.

Gusach, A., Maslov, I., Luginina, A., Borshchevskiy, V., Mishin, A., & Cherezov, V. (2020). Beyond structure: Emerging approaches to study GPCR dynamics. *Current Opinion in Structural Biology*, *63*, 18–25. https://doi.org/10.1016/j.sbi.2020.03.004.

Hamelberg, D., de Oliveira, C. A., & McCammon, J. A. (2007). Sampling of slow diffusive conformational transitions with accelerated molecular dynamics. *The Journal of Chemical Physics*, *127*(15), 155102. https://doi.org/10.1063/1.2789432.

Hamelberg, D., Mongan, J., & McCammon, J. A. (2004). Accelerated molecular dynamics: A promising and efficient simulation method for biomolecules. *The Journal of Chemical Physics*, *120*(24), 11919–11929. https://doi.org/10.1063/1.1755656.

Hanson, M. A., & Stevens, R. C. (2009). Discovery of new GPCR biology: One receptor structure at a time. *Structure*, *17*(1), 8–14. https://doi.org/10.1016/j.str.2008.12.003.

Harpole, T. J., & Delemotte, L. (2018). Conformational landscapes of membrane proteins delineated by enhanced sampling molecular dynamics simulations. *Biochimica et Biophysica Acta - Biomembranes*, *1860*(4), 909–926. https://doi.org/10.1016/j.bbamem.2017.10.033.

Heifetz, A., Schertler, G. F., Seifert, R., Tate, C. G., Sexton, P. M., Gurevich, V. V., et al. (2015). GPCR structure, function, drug discovery and crystallography: Report from academia-industry international conference (UK Royal Society) Chicheley Hall, 1-2 September 2014. *Naunyn-Schmiedeberg's Archives of Pharmacology*. https://doi.org/10.1007/s00210-015-1111-8.

Hilger, D. (2021). The role of structural dynamics in GPCR-mediated signaling. *The FEBS Journal*, *288*(8), 2461–2489. https://doi.org/10.1111/febs.15841.

Hilger, D., Masureel, M., & Kobilka, B. K. (2018). Structure and dynamics of GPCR signaling complexes. *Nature Structural & Molecular Biology*, *25*(1), 4–12. https://doi.org/10.1038/s41594-017-0011-7.

Hill, S. J. (2006). G-protein-coupled receptors: Past, present and future. *British Journal of Pharmacology*, *147 Suppl 1*, S27–S37. https://doi.org/10.1038/sj.bjp.0706455.

Huang, Y. M., McCammon, J. A., & Miao, Y. (2018). Replica exchange Gaussian accelerated molecular dynamics: Improved enhanced sampling and free energy calculation. *Journal of Chemical Theory and Computation*, *14*(4), 1853–1864. https://doi.org/10.1021/acs.jctc.7b01226.

Huang, S. K., Pandey, A., Tran, D. P., Villanueva, N. L., Kitao, A., Sunahara, R. K., et al. (2021). Delineating the conformational landscape of the adenosine A2A receptor during G protein coupling. *Cell*, *184*(7), 1884–1894 e1814. https://doi.org/10.1016/j.cell.2021.02.041.

Husic, B. E., & Pande, V. S. (2018). Markov state models: From an art to a science. *Journal of the American Chemical Society*, *140*(7), 2386–2396. https://doi.org/10.1021/jacs.7b12191.

Jo, S., Cheng, X., Lee, J., Kim, S., Park, S. J., & Patel, D. S. (2017). CHARMM-GUI 10 years for biomolecular modeling and simulation. *Journal of Computational Chemistry, 38*(15), 1114–1124. https://doi.org/10.1002/jcc.24660.
Johnston, J. M., & Filizola, M. (2011). Showcasing modern molecular dynamics simulations of membrane proteins through G protein-coupled receptors. *Current Opinion in Structural Biology, 21*(4), 552–558. https://doi.org/10.1016/j.sbi.2011.06.008.
Johnston, J. M., & Filizola, M. (2014). Beyond standard molecular dynamics: Investigating the molecular mechanisms of G protein-coupled receptors with enhanced molecular dynamics methods. *Advances in Experimental Medicine and Biology, 796*, 95–125. https://doi.org/10.1007/978-94-007-7423-0_6.
Kabir, K. L., Akhter, N., & Shehu, A. (2019). From molecular energy landscapes to equilibrium dynamics via landscape analysis and markov state models. *Journal of Bioinformatics and Computational Biology, 17*(6), 1940014. https://doi.org/10.1142/S0219720019400146.
Kapoor, A., Martinez-Rosell, G., Provasi, D., de Fabritiis, G., & Filizola, M. (2017). Dynamic and kinetic elements of micro-opioid receptor functional selectivity. *Scientific Reports*, 7(1), 11255. https://doi.org/10.1038/s41598-017-11483-8.
Kappel, K., Miao, Y., & McCammon, J. A. (2015). Accelerated molecular dynamics simulations of ligand binding to a muscarinic G-protein-coupled receptor. *Quarterly Reviews of Biophysics, 48*(4), 479–487. https://doi.org/10.1017/S0033583515000153.
Karplus, M., & Petsko, G. A. (1990). Molecular dynamics simulations in biology. *Nature, 347*(6294), 631–639. https://doi.org/10.1038/347631a0.
Kauk, M., & Hoffmann, C. (2018). Intramolecular and intermolecular FRET sensors for GPCRs - monitoring conformational changes and beyond. *Trends in Pharmacological Sciences, 39*(2), 123–135. https://doi.org/10.1016/j.tips.2017.10.011.
Kenakin, T. (2011). Functional selectivity and biased receptor signaling. *The Journal of Pharmacology and Experimental Therapeutics, 336*(2), 296–302. https://doi.org/10.1124/jpet.110.173948.
Kenakin, T. P. (2012). Biased signalling and allosteric machines: New vistas and challenges for drug discovery. *British Journal of Pharmacology, 165*(6), 1659–1669. https://doi.org/10.1111/j.1476-5381.2011.01749.x.
Kenakin, T., & Christopoulos, A. (2013). Signalling bias in new drug discovery: Detection, quantification and therapeutic impact. *Nature Reviews. Drug Discovery, 12*(3), 205–216. https://doi.org/10.1038/nrd3954.
Kenakin, T., & Miller, L. J. (2010). Seven transmembrane receptors as shapeshifting proteins: The impact of allosteric modulation and functional selectivity on new drug discovery. *Pharmacological Reviews, 62*(2), 265–304. https://doi.org/10.1124/pr.108.000992.
Kim, S.-K., Riley, L., Abrol, R., Jacobson, K. A., & Goddard, W. A., 3rd. (2011). Predicted structures of agonist and antagonist bound complexes of adenosine A_3 receptor. *Proteins: Structure, Function, and Bioinformatics, 79*(6), 1878–1897.
Kohlhoff, K. J., Shukla, D., Lawrenz, M., Bowman, G. R., Konerding, D. E., Belov, D., et al. (2014). Cloud-based simulations on Google Exacycle reveal ligand modulation of GPCR activation pathways. *Nature Chemistry, 6*(1), 15–21. https://doi.org/10.1038/nchem.1821.
Komeiji, Y., Yokoyama, H., Uebayasi, M., Taiji, M., Fukushige, T., Sugimoto, D., et al. (1996). A high performance system for molecular dynamics simulation of biomolecules using a special-purpose computer. *Pacific Symposium on Biocomputing*, 472–487. (https://www.ncbi.nlm.nih.gov/pubmed/9390252).
Kooistra, A. J., Mordalski, S., Pandy-Szekeres, G., Esguerra, M., Mamyrbekov, A., Munk, C., et al. (2021). GPCRdb in 2021: Integrating GPCR sequence, structure and function. *Nucleic Acids Research, 49*(D1), D335–D343. https://doi.org/10.1093/nar/gkaa1080.

Kuroda, Y., Suenaga, A., Sato, Y., Kosuda, S., & Taiji, M. (2016). All-atom molecular dynamics analysis of multi-peptide systems reproduces peptide solubility in line with experimental observations. *Scientific Reports, 6*, 19479. https://doi.org/10.1038/srep19479.

Lagerstrom, M. C., & Schioth, H. B. (2008). Structural diversity of G protein-coupled receptors and significance for drug discovery. *Nature Reviews. Drug Discovery*, 7(4), 339–357. https://doi.org/10.1038/nrd2518.

Laio, A., & Parrinello, M. (2002). Escaping free-energy minima. *Proceedings of the National Academy of Sciences of the United States of America, 99*(20), 12562–12566. https://doi.org/10.1073/pnas.202427399.

Latorraca, N. R., Venkatakrishnan, A. J., & Dror, R. O. (2017). GPCR dynamics: Structures in motion. *Chemical Reviews, 117*(1), 139–155. https://doi.org/10.1021/acs.chemrev.6b00177.

Latorraca, N. R., Wang, J. K., Bauer, B., Townshend, R. J. L., Hollingsworth, S. A., Olivieri, J. E., et al. (2018). Molecular mechanism of GPCR-mediated arrestin activation. *Nature, 557*(7705), 452–456. https://doi.org/10.1038/s41586-018-0077-3.

Lefkowitz, R. J. (2004). Historical review: A brief history and personal retrospective of seven-transmembrane receptors. *Trends in Pharmacological Sciences, 25*(8), 413–422. https://doi.org/10.1016/j.tips.2004.06.006.

Levitt, M., & Sharon, R. (1988). Accurate simulation of protein dynamics in solution. *Proceedings of the National Academy of Sciences of the United States of America, 85*(20), 7557–7561. https://doi.org/10.1073/pnas.85.20.7557.

Liao, C., Remington, J. M., May, V., & Li, J. (2021). Molecular basis of class B GPCR selectivity for the neuropeptides PACAP and VIP. *Frontiers in Molecular Biosciences, 8*, 644644. https://doi.org/10.3389/fmolb.2021.644644.

Liu, Y., Wang, X., Dong, D., Guo, L., Dong, X., Leng, J., et al. (2021). Research advances in heterotrimeric G-protein alpha subunits and uncanonical G-protein coupled receptors in plants. *International Journal of Molecular Sciences, 22*(16). https://doi.org/10.3390/ijms22168678.

Lovera, S., Cuzzolin, A., Kelm, S., De Fabritiis, G., & Sands, Z. A. (2019). Reconstruction of apo A2A receptor activation pathways reveal ligand-competent intermediates and state-dependent cholesterol hotspots. *Scientific Reports, 9*(1), 14199. https://doi.org/10.1038/s41598-019-50752-6.

Lu, S., He, X., Yang, Z., Chai, Z., Zhou, S., Wang, J., et al. (2021). Activation pathway of a G protein-coupled receptor uncovers conformational intermediates as targets for allosteric drug design. *Nature Communications, 12*(1), 4721. https://doi.org/10.1038/s41467-021-25020-9.

Ludemann, S. K., Lounnas, V., & Wade, R. C. (2000). How do substrates enter and products exit the buried active site of cytochrome P450cam? 1. Random expulsion molecular dynamics investigation of ligand access channels and mechanisms. *Journal of Molecular Biology, 303*(5), 797–811. https://doi.org/10.1006/jmbi.2000.4154.

Lundstrom, K. (2006). Latest development in drug discovery on G protein-coupled receptors. *Current Protein & Peptide Science*, 7(5), 465–470. http://www.ncbi.nlm.nih.gov/pubmed/17073697.

Ma, N., Nivedha, A. K., & Vaidehi, N. (2021). Allosteric communication regulates ligand-specific GPCR activity. *The FEBS Journal, 288*(8), 2502–2512. https://doi.org/10.1111/febs.15826.

Manglik, A., & Kobilka, B. (2014). The role of protein dynamics in GPCR function: Insights from the beta2AR and rhodopsin. *Current Opinion in Cell Biology, 27*, 136–143. https://doi.org/10.1016/j.ceb.2014.01.008.

Maragliano, L., Fischer, A., Vanden-Eijnden, E., & Ciccotti, G. (2006). String method in collective variables: Minimum free energy paths and isocommittor surfaces. *The Journal of Chemical Physics, 125*(2), 24106. https://doi.org/https://doi.org/10.1063/1.2212942.

McCorvy, J. D., Butler, K. V., Kelly, B., Rechsteiner, K., Karpiak, J., Betz, R. M., et al. (2018). Structure-inspired design of beta-arrestin-biased ligands for aminergic GPCRs. *Nature Chemical Biology, 14*(2), 126–134. https://doi.org/10.1038/nchembio.2527.

McRobb, F. M., Negri, A., Beuming, T., & Sherman, W. (2016). Molecular dynamics techniques for modeling G protein-coupled receptors. *Current Opinion in Pharmacology, 30*, 69–75. https://doi.org/10.1016/j.coph.2016.07.001.

Meral, D., Provasi, D., & Filizola, M. (2018). An efficient strategy to estimate thermodynamics and kinetics of G protein-coupled receptor activation using metadynamics and maximum caliber. *The Journal of Chemical Physics, 149*(22), 224101. https://doi.org/10.1063/1.5060960.

Miao, Y., Feher, V. A., & McCammon, J. A. (2015). Gaussian accelerated molecular dynamics: Unconstrained enhanced sampling and free energy calculation. *Journal of Chemical Theory and Computation, 11*(8), 3584–3595. https://doi.org/10.1021/acs.jctc.5b00436.

Miao, Y., Goldfeld, D. A., Moo, E. V., Sexton, P. M., Christopoulos, A., McCammon, J. A., et al. (2016). Accelerated structure-based design of chemically diverse allosteric modulators of a muscarinic G protein-coupled receptor. *Proceedings of the National Academy of Sciences of the United States of America, 113*(38), E5675–E5684. https://doi.org/10.1073/pnas.1612353113.

Miao, Y., & McCammon, J. A. (2016a). G-protein coupled receptors: Advances in simulation and drug discovery. *Current Opinion in Structural Biology, 41*, 83–89. https://doi.org/10.1016/j.sbi.2016.06.008.

Miao, Y., & McCammon, J. A. (2016b). Graded activation and free energy landscapes of a muscarinic G-protein-coupled receptor. *Proceedings of the National Academy of Sciences of the United States of America, 113*(43), 12162–12167. https://doi.org/https://doi.org/10.1073/pnas.1614538113.

Miao, Y., Nichols, S. E., & McCammon, J. A. (2014). Free energy landscape of G-protein coupled receptors, explored by accelerated molecular dynamics. *Physical Chemistry Chemical Physics, 16*(14), 6398–6406. https://doi.org/10.1039/c3cp53962h.

Milic, D., & Veprintsev, D. B. (2015). Large-scale production and protein engineering of G protein-coupled receptors for structural studies. *Frontiers in Pharmacology, 6*, 66. https://doi.org/10.3389/fphar.2015.00066.

Ohmura, I., Morimoto, G., Ohno, Y., Hasegawa, A., & Taiji, M. (2014). MDGRAPE-4: A special-purpose computer system for molecular dynamics simulations. *Philosophical Transactions. Series A, Mathematical, Physical, and Engineering Sciences, 372*(2021). https://doi.org/https://doi.org/10.1098/rsta.2013.0387.

Paiva, A. C., Oliveira, L., Horn, F., Bywater, R. P., & Vriend, G. (2006). Modeling GPCRs. *Ernst Schering Foundation Symposium Proceedings, 2*, 23–47. https://doi.org/10.1007/2789_2006_002.

Palczewski, K., Kumasaka, T., Hori, T., Behnke, C. A., Motoshima, H., Fox, B. A., et al. (2000). Crystal structure of rhodopsin: A G protein-coupled receptor. *Science, 289*(5480), 739–745. (http://www.ncbi.nlm.nih.gov/pubmed/10926528).

Pan, A. C., Sezer, D., & Roux, B. (2008). Finding transition pathways using the string method with swarms of trajectories. *The Journal of Physical Chemistry. B, 112*(11), 3432–3440. https://doi.org/10.1021/jp0777059.

Pande, V. S., Beauchamp, K., & Bowman, G. R. (2010). Everything you wanted to know about Markov state models but were afraid to ask. *Methods, 52*(1), 99–105. https://doi.org/10.1016/j.ymeth.2010.06.002.

Pavlos, N. J., & Friedman, P. A. (2017). GPCR signaling and trafficking: The long and short of it. *Trends in Endocrinology and Metabolism, 28*(3), 213–226. https://doi.org/10.1016/j.tem.2016.10.007.

Peisley, A., & Skiniotis, G. (2015). 2D projection analysis of GPCR complexes by negative stain electron microscopy. *Methods in Molecular Biology*, *1335*, 29–38. https://doi.org/10.1007/978-1-4939-2914-6_3.

Picard, L. P., & Prosser, R. S. (2021). Advances in the study of GPCRs by (19)F NMR. *Current Opinion in Structural Biology*, *69*, 169–176. https://doi.org/10.1016/j.sbi.2021.05.001.

Ponder, J. W., & Case, D. A. (2003). Force fields for protein simulations. *Advances in Protein Chemistry*, *66*, 27–85. https://doi.org/10.1016/s0065-3233(03)66002-x.

Preininger, A. M., Meiler, J., & Hamm, H. E. (2013). Conformational flexibility and structural dynamics in GPCR-mediated G protein activation: A perspective. *Journal of Molecular Biology*, *425*(13), 2288–2298. https://doi.org/https://doi.org/10.1016/j.jmb.2013.04.011.

Pronk, S., Pall, S., Schulz, R., Larsson, P., Bjelkmar, P., Apostolov, R., et al. (2013). GROMACS 4.5: A high-throughput and highly parallel open source molecular simulation toolkit. *Bioinformatics*, *29*(7), 845–854. https://doi.org/10.1093/bioinformatics/btt055.

Provasi, D., Bortolato, A., & Filizola, M. (2009). Exploring molecular mechanisms of ligand recognition by opioid receptors with metadynamics. *Biochemistry*, *48*(42), 10020–10029. https://doi.org/10.1021/bi901494n.

Qi, R., Wei, G., Ma, B., & Nussinov, R. (2018). Replica exchange molecular dynamics: A practical application protocol with solutions to common problems and a peptide aggregation and self-assembly example. *Methods in Molecular Biology*, *1777*, 101–119. https://doi.org/10.1007/978-1-4939-7811-3_5.

Quast, R. B., & Margeat, E. (2019). Studying GPCR conformational dynamics by single molecule fluorescence. *Molecular and Cellular Endocrinology*, *493*, 110469. https://doi.org/10.1016/j.mce.2019.110469.

Rasmussen, S. G., Choi, H. J., Fung, J. J., Pardon, E., Casarosa, P., Chae, P. S., et al. (2011). Structure of a nanobody-stabilized active state of the beta(2) adrenoceptor. *Nature*, *469*(7329), 175–180. https://doi.org/10.1038/nature09648.

Rasmussen, S. G., Choi, H. J., Rosenbaum, D. M., Kobilka, T. S., Thian, F. S., Edwards, P. C., et al. (2007). Crystal structure of the human beta2 adrenergic G-protein-coupled receptor. *Nature*, *450*(7168), 383–387. https://doi.org/https://doi.org/10.1038/nature06325.

Rasmussen, S. G., DeVree, B. T., Zou, Y., Kruse, A. C., Chung, K. Y., Kobilka, T. S., et al. (2011). Crystal structure of the beta2 adrenergic receptor-Gs protein complex (Research Support, N.I.H., Extramural Research Support, Non-U.S. Gov't). *Nature*, *477*(7366), 549–555. https://doi.org/10.1038/nature10361.

Reiter, E., & Lefkowitz, R. J. (2006). GRKs and beta-arrestins: Roles in receptor silencing, trafficking and signaling. *Trends in Endocrinology and Metabolism*, *17*(4), 159–165. https://doi.org/10.1016/j.tem.2006.03.008.

Rosenbaum, D. M., Rasmussen, S. G., & Kobilka, B. K. (2009). The structure and function of G-protein-coupled receptors. *Nature*, *459*(7245), 356–363. https://doi.org/10.1038/nature08144.

Rosenbaum, D. M., Zhang, C., Lyons, J. A., Holl, R., Aragao, D., Arlow, D. H., et al. (2011). Structure and function of an irreversible agonist-beta(2) adrenoceptor complex. *Nature*, *469*(7329), 236–240. https://doi.org/10.1038/nature09665.

Sabbadin, D., & Moro, S. (2014). Supervised molecular dynamics (SuMD) as a helpful tool to depict GPCR-ligand recognition pathway in a nanosecond time scale. *Journal of Chemical Information and Modeling*, *54*(2), 372–376. https://doi.org/10.1021/ci400766b.

Schlyer, S., & Horuk, R. (2006). I want a new drug: G-protein-coupled receptors in drug development. *Drug Discovery Today*, *11*(11 – 12), 481–493. https://doi.org/10.1016/j.drudis.2006.04.008.

Schoneberg, T., Schulz, A., & Gudermann, T. (2002). The structural basis of G-protein-coupled receptor function and dysfunction in human diseases. *Reviews of Physiology, Biochemistry and Pharmacology, 144*, 143–227. (http://www.ncbi.nlm.nih.gov/pubmed/11987825).

Scott, C. E., Abrol, R., Ahn, K. H., Kendall, D. A., & Goddard, W. A. (2013). Molecular basis for dramatic changes in cannabinoid CB1 G protein-coupled receptor activation upon single and double point mutations. *Protein Science, 22*(1), 101–113. https://doi.org/10.1002/Pro.2192.

Sena, D. M., Jr., Cong, X., Giorgetti, A., Kless, A., & Carloni, P. (2017). Structural heterogeneity of the mu-opioid receptor's conformational ensemble in the apo state. *Scientific Reports, 8*, 45761. https://doi.org/10.1038/srep45761.

Sengupta, P., & Garrity, P. (2013). Sensing temperature. *Current Biology, 23*(8), R304–R307. https://doi.org/10.1016/j.cub.2013.03.009.

Seyedabadi, M., Gharghabi, M., Gurevich, E. V., & Gurevich, V. V. (2021). Receptor-Arrestin interactions: The GPCR perspective. *Biomolecules, 11*(2). https://doi.org/10.3390/biom11020218.

Shaw, D. E., Deneroff, M. M., Dror, R. O., Kuskin, J. S., Larson, R. H., Salmon, J. K., et al. (2007). Anton, a special-purpose machine for molecular dynamics simulation. *SIGARCH Compututer Architecture News, 35*(2), 1–12. https://doi.org/10.1145/1273440.1250664.

Shaw, D. E., Dror, R. O., Salmon, J. K., Grossman, J. P., Mackenzie, K. M., Bank, J. A., et al. (2009). *Millisecond-scale molecular dynamics simulations on Anton* proceedings of the conference on high performance computing networking. *Storage and Analysis, Portland, Oregon*. https://doi.org/10.1145/1654059.1654126.

Shukla, D., Lawrenz, M., & Pande, V. S. (2015). Elucidating ligand-modulated conformational landscape of GPCRs using cloud-computing approaches. *Methods in Enzymology, 557*, 551–572. https://doi.org/10.1016/bs.mie.2014.12.007.

Skiniotis, G., & Southworth, D. R. (2016). Single-particle cryo-electron microscopy of macromolecular complexes. *Microscopy (Oxford), 65*(1), 9–22. https://doi.org/10.1093/jmicro/dfv366.

Smith, J. S., Lefkowitz, R. J., & Rajagopal, S. (2018). Biased signalling: From simple switches to allosteric microprocessors. *Nature Reviews. Drug Discovery, 17*(4), 243–260. https://doi.org/10.1038/nrd.2017.229.

Soldner, C. A., Horn, A. H. C., & Sticht, H. (2019). A Metadynamics-based protocol for the determination of GPCR-ligand binding modes. *International Journal of Molecular Sciences, 20*(8). https://doi.org/10.3390/ijms20081970.

Song, W., Duncan, A. L., & Sansom, M. S. P. (2021). Modulation of adenosine A2a receptor oligomerization by receptor activation and PIP2 interactions. *Structure*. https://doi.org/10.1016/j.str.2021.06.015.

Stone, J. E., Phillips, J. C., Freddolino, P. L., Hardy, D. J., Trabuco, L. G., & Schulten, K. (2007). Accelerating molecular modeling applications with graphics processors. *Journal of Computational Chemistry, 28*(16), 2618–2640. https://doi.org/10.1002/jcc.20829.

Tan, J., Abrol, R., Trzaskowski, B., & Goddard, W. A., 3rd. (2012). 3D structure prediction of TAS2R38 bitter receptors bound to agonists phenylthiocarbamide (PTC) and 6-n-Propylthiouracil (PROP). *Journal of Chemical Information and Modeling, 52*, 1875–1885. https://doi.ORG/10.1021/ci300133a.

Taylor, B. C., Lee, C. T., & Amaro, R. E. (2019). Structural basis for ligand modulation of the CCR2 conformational landscape. *Proceedings of the National Academy of Sciences of the United States of America, 116*(17), 8131–8136. https://doi.org/10.1073/pnas.1814131116.

Thompson, M. D., Cole, D. E., Capra, V., Siminovitch, K. A., Rovati, G. E., Burnham, W. M., et al. (2014). Pharmacogenetics of the G protein-coupled receptors. *Methods in Molecular Biology, 1175*, 189–242. https://doi.org/10.1007/978-1-4939-0956-8_9.

Thomsen, A. R. B., Plouffe, B., Cahill, T. J., 3rd, Shukla, A. K., Tarrasch, J. T., Dosey, A. M., et al. (2016). GPCR-G protein-beta-Arrestin super-complex mediates sustained G protein signaling. *Cell, 166*(4), 907–919. https://doi.org/https://doi.org/10.1016/j.cell.2016.07.004.

Torrens-Fontanals, M., Stepniewski, T. M., Gloriam, D. E., & Selent, J. (2021). Structural dynamics bridge the gap between the genetic and functional levels of GPCRs. *Current Opinion in Structural Biology, 69*, 150–159. https://doi.org/10.1016/j.sbi.2021.04.005.

Vaidehi, N., Floriano, W. B., Trabanino, R., Hall, S. E., Freddolino, P., Choi, E. J., et al. (2002). Prediction of structure and function of G protein-coupled receptors. *Proceedings of the National Academy of Sciences of the United States of America, 99*(20), 12622–12627. https://doi.org/10.1073/pnas.122357199.

Vaidehi, N., & Kenakin, T. (2010). The role of conformational ensembles of seven transmembrane receptors in functional selectivity. *Current Opinion in Pharmacology, 10*(6), 775–781. https://doi.org/10.1016/j.coph.2010.09.004.

Vanni, S., & Rothlisberger, U. (2012). A closer look into G protein coupled receptor activation: X-ray crystallography and long-scale molecular dynamics simulations. *Current Medicinal Chemistry, 19*(8), 1135–1145. https://doi.org/10.2174/092986712799320493.

Vohra, S., Chintapalli, S. V., Illingworth, C. J., Reeves, P. J., Mullineaux, P. M., Clark, H. S., et al. (2007). Computational studies of family a and family B GPCRs. *Biochemical Society Transactions, 35*(Pt 4), 749–754. https://doi.org/10.1042/BST0350749.

Wales, D. J. (2003). *Energy landscapes*. Cambridge University Press (Publisher description http://www.loc.gov/catdir/description/cam032/2003053071.html).

Walters, R. W., Shukla, A. K., Kovacs, J. J., Violin, J. D., DeWire, S. M., Lam, C. M., et al. (2009). Beta-Arrestin1 mediates nicotinic acid-induced flushing, but not its antilipolytic effect, in mice. *The Journal of Clinical Investigation, 119*(5), 1312–1321. https://doi.org/10.1172/JCI36806.

Wang, J., Arantes, P. R., Bhattarai, A., Hsu, R. V., Pawnikar, S., Huang, Y. M., et al. (2021). Gaussian accelerated molecular dynamics: Principles and applications. *WIREs Computational Molecular Science, e1521*. https://doi.org/10.1002/wcms.1521.

Wang, W., Donini, O., Reyes, C. M., & Kollman, P. A. (2001). Biomolecular simulations: Recent developments in force fields, simulations of enzyme catalysis, protein-ligand, protein-protein, and protein-nucleic acid noncovalent interactions. *Annual Review of Biophysics and Biomolecular Structure, 30*, 211–243. https://doi.org/10.1146/annurev.biophys.30.1.211.

Wang, T., & Duan, Y. (2009). Ligand entry and exit pathways in the beta2-adrenergic receptor. *Journal of Molecular Biology, 392*(4), 1102–1115. https://doi.org/10.1016/j.jmb.2009.07.093.

Wang, T., & Duan, Y. (2011). Retinal release from opsin in molecular dynamics simulations. *Journal of Molecular Recognition, 24*(2), 350–358. https://doi.org/10.1002/jmr.1087.

Wang, J., & Miao, Y. (2019). Mechanistic insights into specific G protein interactions with adenosine receptors. *The Journal of Physical Chemistry. B, 123*(30), 6462–6473. https://doi.org/10.1021/acs.jpcb.9b04867.

Whalen, E. J., Rajagopal, S., & Lefkowitz, R. J. (2011). Therapeutic potential of beta-arrestin- and G protein-biased agonists. *Trends in Molecular Medicine, 17*(3), 126–139. https://doi.org/10.1016/j.molmed.2010.11.004.

Wisler, J. W., Xiao, K., Thomsen, A. R., & Lefkowitz, R. J. (2014). Recent developments in biased agonism. *Current Opinion in Cell Biology, 27*, 18–24. https://doi.org/10.1016/j.ceb.2013.10.008.

Wright, S. C., & Bouvier, M. (2021). Illuminating the complexity of GPCR pathway selectivity - advances in biosensor development. *Current Opinion in Structural Biology*, *69*, 142–149. https://doi.org/10.1016/j.sbi.2021.04.006.
Xu, J., Hu, Y., Kaindl, J., Risel, P., Hubner, H., Maeda, S., et al. (2019). Conformational complexity and dynamics in a muscarinic receptor revealed by NMR spectroscopy. *Molecular Cell*, *75*(1), 53–65. e57. https://doi.org/https://doi.org/10.1016/j.molcel.2019.04.028.
Xu, X., Kaindl, J., Clark, M. J., Hubner, H., Hirata, K., Sunahara, R. K., et al. (2021). Binding pathway determines norepinephrine selectivity for the human beta1AR over beta2AR. *Cell Research*, *31*(5), 569–579. https://doi.org/10.1038/s41422-020-00424-2.
Zalewska, M., Siara, M., & Sajewicz, W. (2014). G protein-coupled receptors: Abnormalities in signal transmission, disease states and pharmacotherapy. *Acta Poloniae Pharmaceutica*, *71*(2), 229–243. http://www.ncbi.nlm.nih.gov/pubmed/25272642.
Zhou, Y., Meng, J., Xu, C., & Liu, J. (2021). Multiple GPCR functional assays based on resonance energy transfer sensors. *Frontiers in Cell and Development Biology*, *9*, 611443. https://doi.org/10.3389/fcell.2021.611443.

CHAPTER TEN

Molecular determinants of GPCR pharmacogenetics: Deconstructing the population variants in β_2-adrenergic receptor

Manali Joshi[a,*], Siddhanta V. Nikte[b,c], and Durba Sengupta[b,c,*]
[a]Bioinformatics Centre, Savitribai Phule Pune University, Pune, India
[b]CSIR-National Chemical Laboratory, Dr. Homi Bhabha Road, Pune, India
[c]Academy of Scientific and Innovative Research (AcSIR), Ghaziabad, India
*Corresponding authors: e-mail address: manalijoshi@unipune.ac.in; d.sengupta@ncl.res.in

Contents

1. Introduction 362
 1.1 β_2-Adrenergic receptors 364
 1.2 An overview of GPCR structures 365
 1.3 Structural domains of β_2-AR 367
 1.4 Activation mechanisms of GPCRs 371
 1.5 GPCR as drug targets 373
 1.6 Drugs targeting β_2-AR 375
 1.7 Genetic variations in GPCRs can lead to disease susceptibility and altered drug response 377
 1.8 Pharmacogenomics of GPCRs 377
 1.9 Overview of resources compiling SNPs in GPCRs 379
2. Computational tools to analyze SNPs in GPCRs 379
 2.1 Bioinformatics tools for predicting functional effects of nsSNPs 380
 2.2 Molecular dynamics and network analysis 381
3. An overview of the pharmacogenomics in β_2-AR 383
 3.1 Unraveling molecular mechanisms of clinical SNPs in β_2-AR 386
 3.2 N-terminal domain dynamics is modulated by variants at the 16th position 386
 3.3 Variation at the 27th position leads to a rewiring of the intra-protein network 387
 3.4 Water-mediated allosteric networks are disrupted by the 164th position variant 388
4. Conclusions 390
Acknowledgments 391
References 391

Advances in Protein Chemistry and Structural Biology, Volume 128
ISSN 1876-1623
https://doi.org/10.1016/bs.apcsb.2021.08.002

Abstract

G protein-coupled receptors (GPCRs) are membrane proteins that play a central role in cell signaling and constitute one of the largest classes of drug targets. The molecular mechanisms underlying GPCR function have been characterized by several experimental and computational methods and provide an understanding of their role in physiology and disease. Population variants arising from nsSNPs affect the native function of GPCRs and have been implicated in differential drug response. In this chapter, we provide an overview on GPCR structure and activation, with a special focus on the β_2-adrenergic receptor (β_2-AR). First, we discuss the current understanding of the structural and dynamic features of the wildtype receptor. Subsequently, the population variants identified in this receptor from clinical and large-scale genomic studies are described. We show how computational approaches such as bioinformatics tools and molecular dynamics simulations can be used to characterize the variant receptors in comparison to the wildtype receptor. In particular, we discuss three examples of clinically important variants and discuss how the structure and function of these variants differ from the wildtype receptor at a molecular level. Overall, the chapter provides an overview of structure and function of GPCR variants and is a step towards the study of inter-individual differences and personalized medicine.

1. Introduction

G protein-coupled receptors (GPCRs) are the largest family of transmembrane proteins that are expressed in various human cell types and act as receptors for extracellular signals (Marinissen & Gutkind, 2001). The receptors are composed of the characteristic seven transmembrane spanning helices that are connected by alternating intracellular and extracellular loops (Pierce, Premont, & Lefkowitz, 2002). A schematic representation of a cell membrane-embedded GPCR is shown in Fig. 1. As the name suggests, they generally couple to G proteins via the intracellular regions of the protein, although other effectors are well characterized. Since the GPCRs connect the extracellular to the intracellular region of cells and owing to their abundant expression in various cell types they are considered important cellular signaling hubs. There are around 800 GPCRs, around half of which play a role in sensory functions such as olfaction, taste and vision (Insel et al., 2019). The remaining play roles in various non-sensory processes. Using the same structural architecture, the GPCRs bind to signals ranging from photons, ions to neurotransmitters, hormones, peptides and even larger proteins. In fact, GPCRs are known to mediate the action of two-thirds of hormones and neurotransmitters (Foster et al., 2019). An overview of a canonical GPCR signaling step is shown in Fig. 1. As shown in the figure,

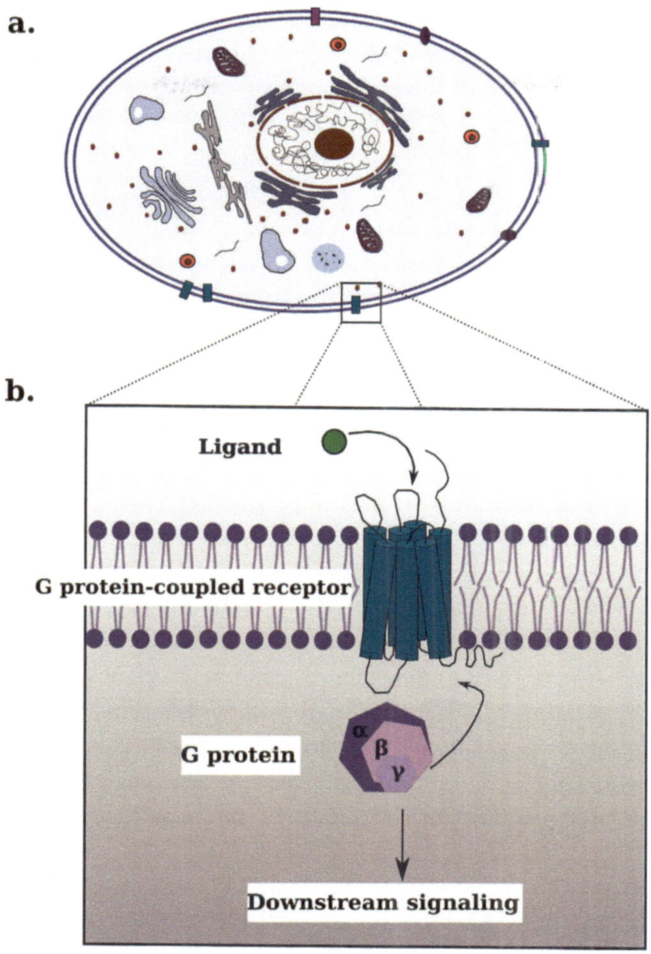

Fig. 1 An overview of canonical signaling mediated by membrane embedded GPCRs. (A) A simplified view of the cell with the (B) a zoom-in view of the GPCR embedded in the membrane. A schematic representation of the GPCR, the membrane, ligand and G protein is given.

the GPCR is embedded in the plasma membrane and upon ligand binding it couples to the effector, heterotrimeric G protein. Interestingly, the diversity of reported downstream effectors is less complex than the diversity of known GPCR ligands. The downstream effectors include proteins such as ion channels, or second messenger producing enzymes such as adenylyl cyclase which lead to generation of response in various cell types. Recent breakthroughs in GPCR biology suggest that several GPCRs could function mainly through non-canonical pathways such as β-arrestins (Luttrell & Gesty-Palmer, 2010). Similarly, some GPCRs have been reported to continue signaling from intracellular compartments, after endocytosis from the plasma membrane (Crilly & Puthenveedu, 2021). Recently, GPCR function has also been shown to be modulated by receptor dimerization (Chakraborty, Jafurulla, Clayton, & Chattopadhyay, 2018; Terrillon & Bouvier, 2004) and by membrane lipids (Oates & Watts, 2011; Prasanna, Mohole, Chattopadhyay, & Sengupta, 2020). Based on their critical role in cellular physiology and signaling mechanisms, the GPCRs are the largest family of drug targets with more than one-third of the FDA approved drugs targeting these proteins (Santos et al., 2017).

Despite the lack of high sequence similarity between the GPCRs, their overall structural topology and general signaling mechanisms share similar features. Based on these parameters, currently the GPCRs are sub-divided into various classes, namely class A (rhodopsin-like), class B1 (Secretin), class B2 (Adhesion), class C (Glutamate), class D1 (Ste-2 like fungal pheromone), class F (Frizzled), class T (Taste) and a separate class of "other GPCRs" (detailed at GPCRdb) (Kooistra et al., 2021). Of these, class A GPCRs are the largest class accounting for close to 85% of the GPCRs and are further divided into several subgroups such as aminergic, chemokine and melanocortin, depending on the type of ligand. For instance, the aminergic receptors bind to small molecules and the chemokine receptors bind to the chemokine proteins. Interestingly, the endogenous ligands of close to 20% GPCRs remain unidentified. Such GPCRs are referred to as orphan GPCRs (Tang, Wang, Li, Luo, & Liu, 2012). Although the class A GPCRs are one of the best studied, the IUPHAR reports close to 40 orphan members in this class (Alexander et al., 2019).

1.1 β_2-Adrenergic receptors

The β-adrenergic receptors are a member of the aminergic, class A family of GPCRs. These receptors are expressed on smooth muscles throughout the body, including bronchial, cardiac, vascular, gastrointestinal and uterine muscles (Johnson, 2006). The β-adrenergic receptors are divided into three sub-types, the β_1, β_2 and β_3 adrenergic receptors, which are expressed in cardiac, airway and adipose tissue, respectively (Wachter & Gilbert, 2012). Stimulation of these receptors causes smooth muscle relaxation. The receptor binds a range of agonists, such as epinephrine and norepinephrine, the natural agonists, antagonists (butaxamine) and inverse agonists such as carazolol. The natural agonist epinephrine binds tightest to β_2-AR followed by β_1-AR (around five times weaker) and β_3-AR (around 170 times weaker), similarly norepinephrine binds the β_1-AR and β_3-AR almost similarly and the β_2-AR six times weaker (Hoffmann, Leitz, Oberdorf-Maass, Lohse, & Klotz, 2004). β_2-AR is encoded by the ADRB2 gene which is localized to the chromosome region 5q31-q32 (Kobilka et al., 1987). The gene ADRB2 is a single exon which encodes for a 413 amino acid protein. Although the expression of β_2-AR is highest in airway smooth muscles it is also found on smooth muscle cells of other body tissues. In the airway muscles, the β_2-AR plays an important physiological role in breathing. The binding of the agonists epinephrine and norepinephrine leads to the activation of adenylyl cyclase via the Gs protein which leads to the relaxation of smooth muscles of lungs, aiding in respiration (Lefkowitz, Stadel, & Caron, 1983). This receptor is thus one of the important drug targets in asthma and chronic obstructive pulmonary disease (COPD).

1.2 An overview of GPCR structures

The overall structural architecture of GPCRs is conserved throughout the family and they are characterized by seven membrane spanning or transmembrane (TM) helices, and are often referred to as 7TM receptors (Rosenbaum, Rasmussen, & Kobilka, 2009). A representative snapshot of a GPCR is shown in Fig. 2 and highlights the different structural domains. Apart from the TM domain, GPCRs contain three extracellular and three intracellular loops connecting the helices, an extracellular N-terminal and an intracellular C-terminal domain of varying lengths. The C-terminal domain may be further subdivided into a surface aligned helix 8 and the more disordered C-terminal tail. The N-terminal region of class A GPCRs is usually disordered and flexible, on the contrary class C (and class B and F) GPCRs usually have larger, structured N-terminal domains. Despite the similarity of the transmembrane domain topology, GPCRs show high structural diversification in the loop regions, especially the N-terminal, C-terminal domain, extracellular loop 2 (ECL2) and intracellular loop 3 (ICL3) that connects transmembrane helices 5 and 6. The transmembrane helices are oriented like a bundle encompassing a central lumen that is the main orthosteric binding site for small molecules. Ligand binding in class C GPCRs occurs at the N-terminal domain, and the binding of larger proteins and peptides to other GPCRs occurs at the N-terminal domain in conjunction with the orthosteric ligand binding site. At the cytoplasmic end of the lumen,

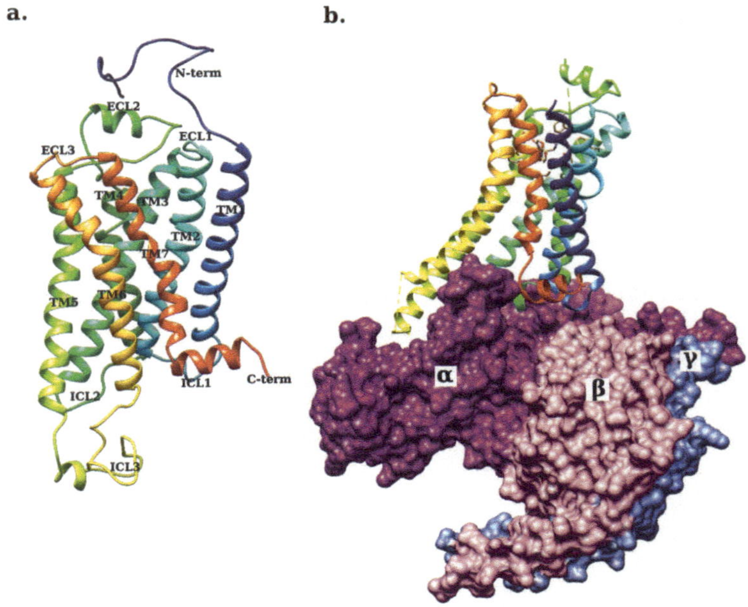

PDB ID: 3SN6

Fig. 2 A cartoon representation of (A) the inactive state structure of β_2-AR and (B) the active state structure bound to the trimeric G protein. The receptor is rendered in rainbow colors corresponding to the structural domains. The G protein is represented in the surface representation and the constituent $G_\alpha - G_\beta - G_\gamma$ subunits are colored light purple, pink and light blue, respectively. The figures were rendered using the visualization software, Chimera.

coupling of G protein and other effector occurs. As shown in Fig. 2 the coupling of G proteins usually occurs at the face formed by transmembrane helices 5 and 6. These interactions of GPCRs with other proteins including the ligand and G-protein can be quite dynamic (Kharche, Joshi, Chattopadhyay, & Sengupta, 2021; Kharche & Sengupta, 2020). Each of the structural domains is linked to multiple roles that are still emerging.

The structural characterization of GPCRs occurred in two main phases, with the first phase focused on rhodopsin (and archeal rhodopsin-like proteins) (Palczewski et al., 2000). In the second phase, a break-through occurred with the structural resolution of β_2-AR (Cherezov et al., 2007; Rasmussen et al., 2007; Rosenbaum et al., 2007). Since then over 500 structures of about 100 GPCRs have been resolved and the coordinates have been deposited to the protein data bank (PDB) (Berman et al., 2000). The second phase has been accelerated with the advent of cryo electron microscopy (cryo-EM) allowing the structural characterization of large GPCR complexes. The structural determination by crystallography of the hydrophobic transmembrane region in these receptors had been a hefty challenge, in part due to their high flexibility and the lack of membranes or membrane-mimetic matrices suitable for stabilization and crystallization. In addition, GPCRs occur in at least two states: active and inactive, rendering structural characterization more challenging. Consequently, GPCRs are heavily engineered to improve crystallization with several dynamic loops either removed or replaced and thermostabilizing mutations introduced at the transmembrane domain. In particular, innovative experimental techniques such as fusion chimeras with T4-lysozyme, and antibody/nanobody fragments have led to the exponential growth of GPCR crystal structures in the second phase. A careful inspection of the experimentally resolved structure is recommended to understand the effect of these protein engineering mutations before making further analysis based on these structures. Database resources such as GPCRdb (Kooistra et al., 2021) maintain a compilation of structural models for the wildtype receptor based on the core experimentally resolved (but engineered) structure.

1.3 Structural domains of β_2-AR

β_2-AR is one of the best structurally characterized GPCRs with about 37 structures (as of June 2021) since the first breakthrough structural resolution. Due to the extensive structural studies and important similarities to related receptors, β_2-AR acts as a representative member of the class A GPCRs binding to external ligands. The receptor has been crystallized in the *apo* inactive state, ligand-bound and nanobody-coupled intermediate states and ligand and G protein-coupled active states (Rasmussen, Choi, et al., 2011; Rasmussen, DeVree, et al., 2011; Rosenbaum et al., 2007). Overall, β_2-AR is observed to maintain the classical GPCR seven transmembrane topology but adopts an ensemble of inactive, intermediate and active states, perhaps leading toward different activation pathways (Nygaard et al., 2013). The two main inactive and active states are shown in Fig. 2. The striking difference between the structural states is the placement of transmembrane helices 5 and 6 that open outwards in the active state to allow interactions with the G-protein that inserts between the two helices (Fig. 3). However, subtle differences in structure are observed with varying agonists, antagonists, inverse agonists and the myriad of conformational states that confer β_2-AR a distinct conformational plasticity.

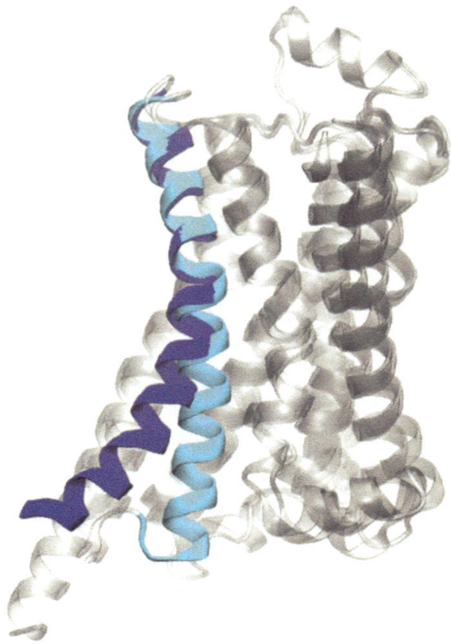

Fig. 3 Key difference in the inactive and active state of the GPCRs. As a representative GPCR, the structures of the active (PDB ID: 3SN6) and inactive (PDB ID: 2RH1) states of β_2-AR are superimposed. The protein is rendered as ribbons and colored gray, the helix 6 is colored light and dark blue in the inactive and active state, respectively. The figure was rendered using the visualization software, VMD.

The ligand binding site lies at the lumen, and in β_2-AR is positively charged. Different ligands such as agonists, antagonists and inverse agonists appear to bind in similar poses, although the size of the ligand dictates the coverage in the lumen. A zoom in to the main interacting residues from transmembrane helices 3, 5, 6 and 7 is shown in Fig. 4. Residues $D113^{3\times32}$, $S203^{5\times43}$, $S207^{5\times46}$, $F289^{6\times51}$, $N293^{6\times55}$ and $N312^{7\times38}$ define the topology of the binding pocket. Of these, residues $D113^{3\times32}$, $F289^{6\times51}$ and $N312^{7\times38}$ have been observed to make consensus contacts with various ligands in the class A GPCRs (Venkatakrishnan et al., 2013). The entry of the ligand to this site has been suggested to occur via two distinct entry sites (see Fig. 5). The two pathways are separated by a salt bridge between residues $D192^{45\times51}$ and $K305^{7\times31}$ at the top of the receptor. The first pathway is formed between residues from TM helices 5, 6 and 7 (referred to as vestibule 1) and the second pathway comprises residues from TM helices 2, 3 and 7

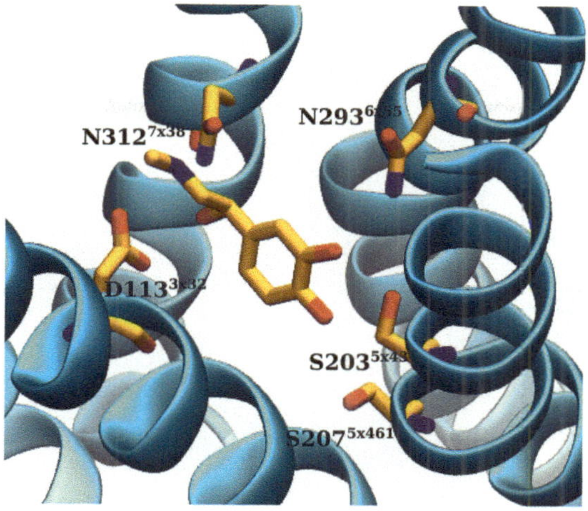

Fig. 4 A zoom in view of the orthosteric ligand binding site of β_2-AR with the ligand epinephrine. The protein (PDB ID: 4LDO) is rendered as ribbons and colored blue, the binding site residues and the ligand are shown in stick representation and colored by atom. The figure was rendered using the visualization software, VMD.

(referred to as vestibule 2). Previous computational studies (Wang & Duan, 2009) suggest that vestibule 1 is the dominant pathway and vestibule 2 may have a reduced frequency of ligand entry (or exit).

A large number of sequence and structural motifs in the transmembrane helices have been suggested (Venkatakrishnan et al., 2013). A few of these structural motifs are shown in Fig. 5. An important signature of the inactive conformational state of the receptor is the conserved NPxxY (N322$^{7\times49}$, P323$^{7\times50}$, Y326$^{7\times53}$) motif on transmembrane helix 7 of the receptor. It has been suggested to form contact with a tyrosine residue in transmembrane helix 5 in β_2-AR and mutating these residues appears to stabilize the inactive receptor conformations (Ragnarsson, Andersson, Thomas, & Lewis, 2019). Adjacent to this site lies the Na^+ pocket that is conserved amongst class A GPCRs (residues D79$^{2\times50}$, S120$^{3\times39}$, N318$^{7\times45}$ and N322$^{7\times49}$). Similarly, a highly conserved DRY motif is present at the cytoplasmic end of transmembrane helix 3. In the *apo* state of the receptor, the residues of the DRY motif (D130$^{3\times49}$, R131$^{3\times50}$, Y132$^{3\times51}$) forms salt bridges with surrounding residues and with transmembrane helix 6 stabilizing the receptor in an inactive conformation. Upon ligand binding, these interactions are broken and helps to stabilize the active conformation of the receptor. Another motif that was implicated in the stabilization of the inactive state

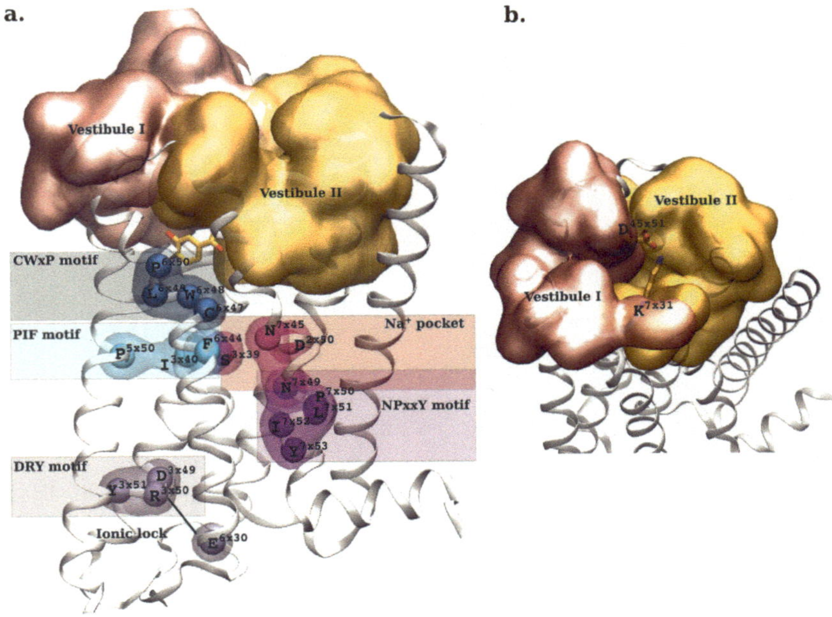

Fig. 5 Representation of sequence and structural motifs in β_2-AR. (A) Side view and (B) Top view of the receptor. The protein is rendered as ribbon and colored gray. Residues corresponding to the vestibules are rendered as surface and colored pink (Vestibule I) and orange (Vestibule II). Residues corresponding to different motifs are shown as spheres. The motifs are colored as CWxP: Blue, PIF: Cyan, Na+pocket: Magenta, NPxxY: Purple, DRY motif and ionic lock residues: Gray. The bound ligand is shown in the stick representation. The salt bridge between $D^{45\times51}$ and $K^{7\times31}$ that separates the ligand entry pathway in vestibule I and vestibule II is represented in the stick representation. The figures were rendered using the visualization software, VMD.

of GPCRs is the "ionic lock," a salt bridge formed between the residues R131$^{3\times50}$ and E268$^{6\times30}$, from adjacent transmembrane helices 3 and 6. The salt bridge was originally proposed based on rhodopsin crystal structures but is absent in several crystal structures of the *apo*-receptor and several computational studies confirm that the ionic lock forms only transiently in β_2-AR. A signature of receptor activation is the residue Trp286$^{6\times48}$ that was originally suggested to form the rotamer toggle switch and recently renamed as the transmission switch. This residue has been suggested to adopt multiple conformational states (like a toggle switch) and is an important link between the ligand-binding site and the G-protein coupling site. This site has now been extended to include the PIF or connector motif (P211$^{5\times50}$, I121$^{3\times40}$, F282$^{6\times44}$). Further, the activation of class A GPCRs has also been linked to the formation of a continuous water channel that is facilitated by the flipping of Tyr326$^{7\times53}$ (part of the NPxxY motif) and the opening of an adjacent hydrophobic layer of amino acids (Yuan, Filipek, Palczewski, & Vogel, 2014). Overall, a large number of conserved sequence motifs and residues have been mapped to functional roles but the conformational plasticity of the receptor suggests the involvement of only a few of these motifs/models in the different states.

1.4 Activation mechanisms of GPCRs

Multiple activation pathways interconnecting several conformational states have been suggested to accompany distinct intracellular signaling routes in response to different ligands. In the canonical pathway, ligand binding initiates receptor activation that involves a series of allosteric pathways interconnecting the structural and sequence motifs such as the DRY and NPxxY sites. The initial activation mechanism proposed was based on molecular dynamics simulations in conjunction with crystallography and spectroscopy (Kobilka, 2007). Since then the general mechanisms of GPCR activation have been suggested based on comparisons to related receptors.

The current consensus model of a common GPCR activation pathway involves four main layers of sites starting from the orthosteric ligand binding site and ending at the G-protein coupling site. This particular pathway was recently proposed by Zhou et al. by comparing all class A GPCR structures and highlights about 35 specific residues and residue pairs (Zhou et al., 2019) and is consistent with the mechanisms proposed from molecular dynamics simulations. The four layers identified correspond to (1) signal initiation, (2) signal propagation, (3) microswitch rewiring and (4) G-protein coupling. In the first step, binding of agonists at the orthosteric binding site leads to an allosteric signal to the Na^+ binding site and to the PIF motif/transmission switch (3×40, 5×50 and 6×44). In conjunction, the mainly hydrophobic interactions between the residues (6×40 and 6×41) on transmembrane helix 6 and residue at 3×43 on transmembrane helix 3 are lost. This leads to a loosening of the tight transmembrane helix 3 to transmembrane helix 6 packing, reminiscent of a hydrophobic lock. The third layer/step involves the conformational changes at the NPxxY motif, and the residue $Y^{7\times53}$ swivels to increase the interactions between transmembrane helix 7 and 3 that further reduces the contacts between transmembrane helix 6 and 3. And lastly, rearrangements occur at the DRY motif/ionic lock and the specific contact of $R^{3\times50}$ with both residues $D/E^{3\times49}$ and ionic lock with

D/E$^{6\times30}$ are lost and facilitates the outward motion of transmembrane helix 6. These rewired contacts of R$^{3\times50}$ and flanking and proximal residues (3×53, 3×54, 5×61 and 6×33) stabilize a receptor conformational state that is able to bind to G protein on the cytosolic side. The models of receptor activation validate that the intra-helical/inter-helical and switching contacts between residues aid the continuous and modular nature of the activation pathway and link the residue-level changes to large conformational changes in the receptor. Interestingly, although these common pathways allow mechanistic interpretation of several mutations, they are unable to reconcile the mutations that lie outside of these pathways. It should be emphasized that although the common activation pathway could be the main activation pathway in class A GPCRs, each receptor could have further ligand and effector-specific activation pathways.

In the specific case of β_2-AR, the common activation pathway appears to be robust and consistent with previous experimental and simulation data (Dror et al., 2011; Nygaard et al., 2013). However, the basal level activity of the receptor (Lamichhane et al., 2015) suggests a spontaneous transition between inactive and active-like conformations in the absence of rearrangements triggered by ligand binding. In addition, several pathways throughout the receptor and not just the direct pathway linking the ligand binding site and G-protein coupling site have been proposed. For instance, residues on the extracellular loops have been shown to modulate receptor activation. Allosteric pipelines have been reported to pass via these extracellular loops and have been implicated in secondary activation pathways. In addition, it has been shown that partial and full agonists differentially affect the rate at which GDP-bound β_2AR-G$_s$ complexes are formed (Gregorio et al., 2017). This was linked to a differential transmembrane helix 6 motion, with the maximal outward movement upon binding of the full agonist. This implies that based on the nature of the ligand, the residues involved that allow the outward motion of transmembrane helix 6 could vary, although it has not yet been shown. To test the conserved and unique activation signatures, a deep mutagenesis study of β_2-AR was performed (Jones et al., 2020) and based on mutational tolerance data of each residue, previously uncharacterized residues were shown to be functionally critical for this receptor. In particular, W99$^{23\times50}$ of extracellular loop 1 (ECL1) is the 13th most intolerant residue, which is unusual as mutationally intolerant residues are rare in the flexible loops. In addition, a subset of residues in extracellular loop 2 (ECL2), including C184 and C190 that form an intraloop disulfide bridge and proximal residues (A181, Y185) were found to be highly sensitive to mutation, suggesting this region is uniquely important to β_2-AR. In summary, there is a main conserved activation pathway across class A GPCRs, and specific additional pathways in β_2-AR.

1.5 GPCR as drug targets

GPCRs are the most successful drug target family owing to their critical role in cellular physiology and their druggability (see Fig. 6). As per recent data, close to ~600 drugs are approved for various GPCRs and a reasonable number are undergoing clinical trials (Sriram & Insel, 2018; Yang et al., 2021). On an average there are 10 drugs approved for a single GPCR (Hauser, Attwood, Rask-Andersen, Schioth, & Gloriam, 2017). Since the class A receptors form a large proportion of the GPCRs, close to 500 drugs target this class (see Fig. 6). The drugs are approved for diverse diseases ranging from pain relief, allergies, pulmonary and cardiovascular diseases, mental health related diseases and cancer to name a few. An analysis of the drugs approved for GPCRs reveal that a majority are synthetic organic molecules or peptides that work either as agonists or antagonists of the receptors. A few examples are metoprolol (beta blocker, an antagonist of the β_1-AR), oxycodone (against pain and interacting with μ and κ opioid receptors) and sumatriptan (used in migraine headaches and binds serotonin$_{1D}$ receptor). Very few inverse agonists or partial agonists are approved as drugs.

While most of the drugs bind at the orthosteric pocket encompassed in the TM helices, a few bind at extracellular allosteric pockets or the intracellular regions. Some successful examples of allosteric modulators include maraviroc, a modulator of the CCR5 for prevention of HIV-1 entry into cells (Woollard & Kanmogne, 2015) and cinacalcet, a modulator of the calcium sensing receptor for treatment of hyperparathyroidism (Riccardi & Martin, 2008). While most agonists work by activating the GPCR to recruit G protein, several drugs have been approved that recruit β-arrestins in receptors (Smith, Lefkowitz, & Rajagopal, 2018). One of the successful drugs that works as a biased agonist is oliceridine targeting the μ-opioid receptor (Lambert & Calo, 2020). By recruitment of β-arrestin and the downstream pathway activated by it, oliceridine leads to adequate analgesia with decreased risk of adverse effects such as constipation and respiratory depression caused by agonists that recruit G-proteins. Although modulating the single receptor has been the mainstay of drug design, recent studies have shown that several GPCRs are present as homo or heterodimers or even oligomers of higher order. It has been reported that several GPCRs form

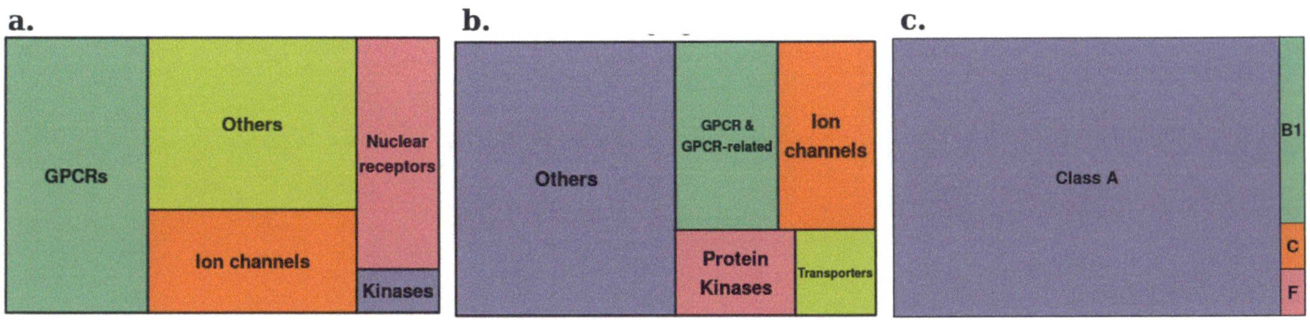

Fig. 6 An overview of the current statistics for the GPCRs as drug targets. Treemaps of the percentage of (A) FDA approved drugs targeting various protein families (B) protein targets (C) percentage of the drugs targeting to different classes of GPCRs. The values are taken from Sriram and Insel (2018). The treemaps were generated using R scripts.

functional heterodimers such as the DOR–CXCR4 heterodimer (Hereld & Jin, 2008) or the CCR5-MOR heterodimers (Zhang, Zhang, Hang, & Liu, 2020). These are potentially new drug targets for the design of bivalent ligands that will interact with both monomers.

Owing to the similarities between the structural scaffold of the GPCRs some drugs and even their natural ligands are known to crosstalk between the receptors. One of the classical cases of crosstalk between natural ligands is the chemokine system where certain chemokines bind to many receptors, while many receptors are observed to bind the same chemokine (Salanga, O'Hayre, & Handel, 2009). The crosstalk between chemokine and opioid GPCRs is also well documented, which has recently raised the possibility that the chemokine receptors can be manipulated to alleviate pain (Melik Parsadaniantz, Rivat, Rostene, & Reaux-Le Goazigo, 2015). Such polypharmacology is also observed with drugs like haloperidol which bind the dopamine as well as serotonin receptors. Similarly, antimuscarinics used in bladder dysfunction are observed to bind to all five receptor types, in spite of muscarinic acetylcholine M2 and M3 receptors being the primary receptors. Such cross talk between natural agonists and drugs is also observed in the β-adrenergic receptors. Interestingly, although the natural ligands are not identified for orphan GPCRs some are nonetheless being pursued as drug targets. For example, GPR119 is being pursued as a target for the treatment of diabetes (Ritter, Buning, Halland, Poverlein, & Schwink, 2016), GPR35 for the treatment of an allergic inflammatory condition (Divorty, Mackenzie, Nicklin, & Milligan, 2015) and GPR84 for the treatment of fibrotic disorders (Wojciechowicz & Ma'ayan, 2020).

1.6 Drugs targeting β_2-AR

One of the most successful targets of class A GPCRs is β_2-AR. Agonists of β_2-AR bind to the receptor and lead to muscle relaxation causing symptomatic relief from the bronchospasm associated with asthma and COPD. Depending on their action, the β_2-AR agonists used in respiratory distress are classified as short acting β_2-AR agonists (SABA), long acting (LABA) or ultra long acting. The most commonly used short acting agonists are salbutamol and terbutaline, the commonly used long acting agonists are salmeterol, formoterol, olodaterol and vilanterol (Waldeck, 2002) (see Fig. 7). The only widely used ultra long acting agonist is indacaterol. Most agonists possess a catecholamine group similar to the natural ligand.

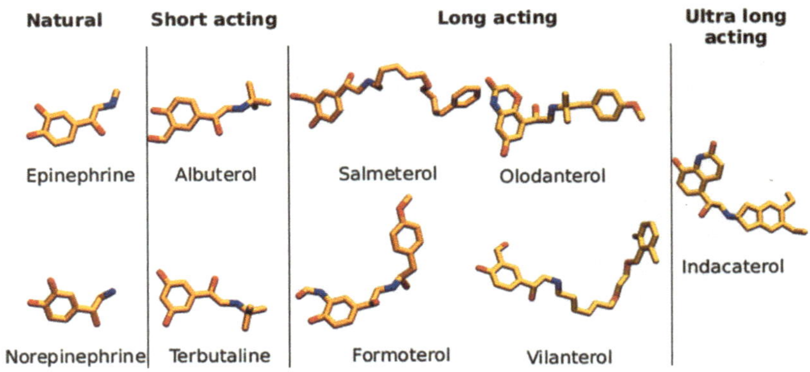

Fig. 7 Schematic representation of natural agonists and drug molecules targeting β_2-AR. The drug molecules are classified as short acting, long acting and ultra-long acting. The ligands are rendered as stick and colored by atom type. The drug molecules were retrieved from PubChem (Kim et al., 2019) and 3D conformers were generated using Open Babel (O'Boyle et al., 2011). The figure was rendered using the visualization software, VMD.

While all agonists activate the receptor by binding at the orthosteric pocket, it is thought that the LABAs are retained in the cell membranes and hence there is a continued presence of drug near the receptor leading to prolonged activation. Since the β_2-AR is also expressed on other smooth muscles of the body, β_2-AR agonists are also used to induce their relaxation. For instance, albuterol is used to relax uterine muscles preventing pre-term labour. The activation of β_2-AR in cardiac myocytes and vascular smooth muscle cells leads to increased rate and force of heart contractions. Thus, β_2-AR is also a drug target in conditions such as hypertension and congestive heart failure. In these disease conditions antagonists of β_2-AR such as carvedilol and propranolol are used.

Owing to the similarity between β_1-AR, β_2-AR and β_3-AR there is significant cross talk between the drugs. Isoproterenol, formoterol and fenoterol are all agonists of all three receptor subtypes. Salbutamol and terbutaline are partial agonists of the β_2-AR but antagonists of the β_1-AR. Their efficacy at the β_3-AR subtype ranges from 42% to 87% (Hoffmann et al., 2004). Salmeterol is the only β_2-AR agonist with no agonistic activity at the other receptor subtypes. Similarly, some beta-blockers targeted to β_1-AR overexpressed in cardiac cells also cross-talk with β_2-AR at higher concentrations, leading to a dual effect of smooth muscle relaxation by both β_1 and β_2-AR.

1.7 Genetic variations in GPCRs can lead to disease susceptibility and altered drug response

Genetic variations in GPCRs can lead to both disease susceptibility and differential drug response. The most common reason for such differences are Single Nucleotide Polymorphisms (SNPs). Clinical studies with limited patient data have been able to identify the most important SNPs. In fact, several large-scale genome sequencing projects such as the 1000 genome by US (Genomes Project, et al., 2012), 100,000 genome by UK (Turnbull et al., 2018), have enabled the detection of inter-individual genome variability, potentially uncovering rare and low frequency variants that have not been captured by clinical studies thus far. The SNPs in coding regions that cause change in amino acid are referred to as non-synonymous SNPs (nsSNPs) and lead to missense variants of proteins. Such nsSNPs involved in the pharmacokinetics of a drug or in the pharmacodynamics can lead to an adverse response to a particular drug. In addition to nsSNPs several other variations such as mutations introducing stop codon, frameshift mutation, loss-of-function (LoF) mutation can affect the structure function of GPCRs and cause its dysfunction (and therefore disease susceptibility) or altered response to drugs. For instance, the calcium sensing receptors, Class C GPCRs are involved in a variety of cellular processes. Variants of the calcium sensing receptor are associated with hyperparathyroidism and idiopathic calcium nephrolithiasis (Grzegorzewska et al., 2019). The receptor angiotensin II type 1 (AT1R) a class A GPCR is involved in the control of blood pressure and volume in the cardiovascular system. Variants of AT1R are positively correlated with higher risk of atrial fibrillation (Simoes et al., 2020).

1.8 Pharmacogenomics of GPCRs

When a drug enters the body, most commonly through the oral route, as a first step, it is absorbed in the intestine. From the intestine it enters hepatic circulation where the drug might undergo metabolism in the liver before being released in the blood for general circulation, distribution and eventual excretion. These processes are termed as the pharmacokinetics of the drug and a myriad of proteins are involved in them. Several recent surveys have indicated that the top selling drugs on the market benefit as little as 1 in 25 to as much as only 1 in 4 intended patients (Reardon, 2015). This is a worrisome statistic which indicates that there is a large population not benefited by the approved drugs. This is because the way in which a person processes a drug in their body could be very different from another. One of the most well studied examples is the case of the drug nortriptyline which is a tricyclic antidepressant used in the treatment of depression which acts via various targets including the serotonin$_{2A}$ receptor. The drug nortriptyline is metabolized by the cytochrome 2D6 (CYP2D6) in the liver and hence there is a reduced amount of the active drug in the blood. Since different people may have different number of CYP2D6 genes the standard dose administered may be subtherapeutic to some people while those who have no functional CYP2D6 might face toxicity due to the high plasma concentration. Interestingly, a significant number of drugs prescribed for psychiatry that target various GPCRs suffer from the issue of variable pharmacokinetics in patients (Kirchheiner et al., 2004). These patients are thus labeled as poor, intermediate rapid and ultra-rapid metabolizers and are dosed accordingly. A differential response can also arise due to inter-individual differences in the drug target that the drug is designed for (see Fig. 8). The relationship between the drug dose and effect upon interaction with the target is termed as pharmacodynamics. Pharmacodynamics can differ between patients if the target itself has differences between people. The study of the link between the effect of polymorphic expression of a single gene on a drugs response profile is pharmacogenetics. Most often the adverse drug response is not limited to a single gene, but several genes, the study of which is termed as pharmacogenomics.

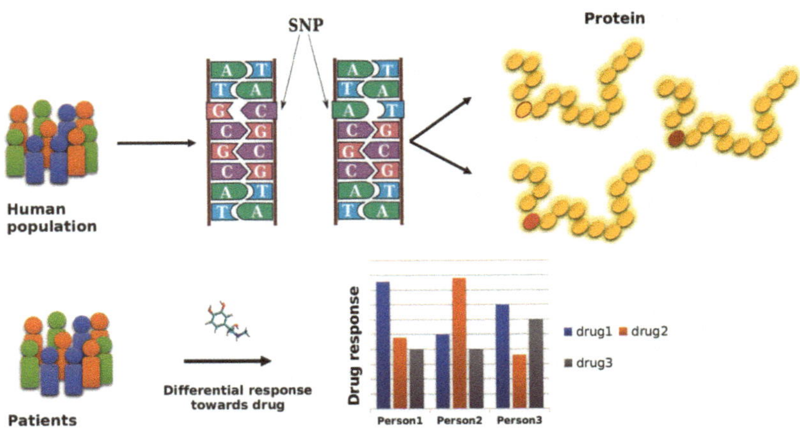

Fig. 8 Schematic representation of pharmacogenomics in GPCRs. The human population carries nsSNPs that encode variants of the same protein. Patients with such variants in drug targets can show differential response towards drugs.

1.9 Overview of resources compiling SNPs in GPCRs

Advances in sequencing technologies over the past 20 years have led to genome sequencing being cheaper and quicker. Owing to the large-scale genome sequencing projects, there is an increasing amount of genome (DNA level) and exome (mRNA level) information available leading to cataloging of the differences such as SNPs, indels and other structural variations.

Given below are resources that can be explored to identify variations in GPCRs. The dbSNP (https://www.ncbi.nlm.nih.gov/snp/) (Sherry et al., 2001) compiles genetic polymorphisms such as SNPs, small-scale multi base deletions or insertions, retroposable element insertions and microsatellite repeat variations. The GPCRdb (https://gpcrdb.org/) (Kooistra et al., 2021) is a database focussed on GPCRs only and included 65,539 missense variations (as of June 2021) in all GPCRs. As per the statistic of GPCRdb there are an average of 155 rare missense variations per receptor and on an average of 5 common missense variants per receptor. GPCRdb curates data from dbSNP for GPCRs specifically and annotates them based on experimental and clinical inputs. The PharmGKB (https://www.pharmgkb.org/) (Barbarino, Whirl-Carrillo, Altman, & Klein, 2018) is a database linking genetic variation with adverse drug response. It contains clinical information including dosing guidelines and drug labels and actionable gene-drug associations. As of now PharmGKB contains information for approximately 80 GPCRs.

2. Computational tools to analyze SNPs in GPCRs

With such large scale sequencing projects, experimental analysis of variant receptors is challenging for every new SNP reported. For each reported SNP, the physiological function of the receptor (and disease susceptibility) would need to be characterized. In addition, due to the role of GPCRs as drug targets, altered drug efficacy would need to be tested as well. In this regard, computational methodologies have been developed to help identify the modulation of function for each of these SNPs and prioritize them for experimental testing. A large number of computational tools have been developed that are mainly based on sequence, structure and/or reported experimental data. To account for the implicit limitation in each of these bioinformatics tools, several meta-analyzers have been reported that are based on multiple such tools. In addition, due to the inherent dynamics of these receptors, molecular dynamics simulations have been used in conjunction to capture the effect of the SNPs on receptor dynamics.

2.1 Bioinformatics tools for predicting functional effects of nsSNPs

Bioinformatics resources and tools have been proven invaluable for the analysis of nsSNPs and prioritizing the deleterious ones. Several tools have been developed such as SIFT (Vaser, Adusumalli, Leng, Sikic, & Ng, 2016), PolyPhen2 (Adzhubei, Jordan, & Sunyaev, 2013) and SNAP (Johnson et al., 2008). An overview of the tools and resources is given in Table 1. SIFT is one of the most common tools and takes a query sequence and uses

Table 1 A non-exhaustive list of bioinformatics tools available for predicting the functional effect of SNPs.

Name	Website/URL	Database/ training set	Methodology
LS-NSP	http://modbase.compbio.ucsf. edu/LS-SNP/	Human allele	Rule-based annotation
PANTHER	http://www.pantherdb.org/ tools/csnpScoreForm.jsp	Evolution/ human allele	HMM
Parepro	http://www.mobioinfor.cn/ parepro/	Human allele	ML-SVM
PMut	http://mmb2.pcb.ub.es:8080/ PMut/	Human allele	Neural network
PolyPhen	http://genetics.bwh.harvard. edu/pph/	Human allele	Empirical rules
Polyphen-2	http://genetics.bwh.harvard. edu/pph2/		ML-classification
SAPRED	http://sapred.cbi.pku.edu.cn/	Human allele	
SIFT	http://blocks.fhcrc.org/sift/ SIFT.html	Experimental	Empirical rules, PSSM
SNAP	http://cubic.bioc.columbia.edu/ servers/SNAP/	Experimental	Neural network
SNPs3D	http://www.snps3d.org/	Human allele	Empirical rules
PredictSNP	http://loschmidt.chemi.muni. cz/predictsnp		Consensus of multiple tools

multiple alignment information to predict tolerated and deleterious substitutions for every position of the query sequence. It calculates normalized probabilities for all possible substitutions from the alignment. Positions with normalized probabilities less than 0.05 are predicted to be deleterious, those greater than or equal to 0.05 are predicted to be tolerated. Another interesting tool is PolyPhen-2 that extracts structure-based information complemented by sequence features and using a probabilistic classifier identifies them as deleterious or not. An important meta-analysis tool is PredictSNP (Bendl et al., 2014) that has been suggested to provide an accurate and robust prediction ranging between 72% and 78% for various datasets. The individual tools comprising PredictSNP give predictions based on evolutionary, physico-chemical or structural characteristics and are supplemented by experimental annotations from Protein Mutant (Kawabata, Ota, & Nishikawa, 1999) and UniProtKB (Boutet et al., 2016) databases. More recently, machine learning approaches such as Parepro (Tian et al., 2007) and PMut (Lopez-Ferrando, Gazzo, de la Cruz, Orozco, & Gelpi, 2017) have been used to predict variant outcomes. Although these tools work well on large datasets, they miss out on several specific SNPs, especially those involving dynamic or allosteric effects.

2.2 Molecular dynamics and network analysis

Molecular dynamics simulations have been invaluable in analyzing molecular mechanisms in GPCRs and have provided key insights complementing experimental findings. In these simulations, the position and velocity of each atom in the receptor is estimated by assuming classical motion, described by Newtonian mechanics. From the simulation trajectories, the inherent dynamics of the receptor can be directly observed. As a very simplified example, the snapshots along the trajectory of the wildtype β_2-AR receptor are shown in Fig. 9. Loop dynamics that occurs at a nanosecond timescale can be easily distinguished from the figure. Transmembrane domain rearrangements such as those in the activation pathway occur at much longer timescales and are more challenging to sample with molecular dynamics simulations. From the simulations, further analysis can be performed to identify specific interactions such as protein-ligand and protein-lipid interactions. In addition, advanced approaches such as free energy calculations and metadynamics can be used to identify pathways and calculate energetics.

In the *apo*-receptor, the wildtype receptor and thermostabilizing mutations have been analyzed and the differential dynamics has been probed. Ligand binding has been analyzed for a large number of GPCRs using simple

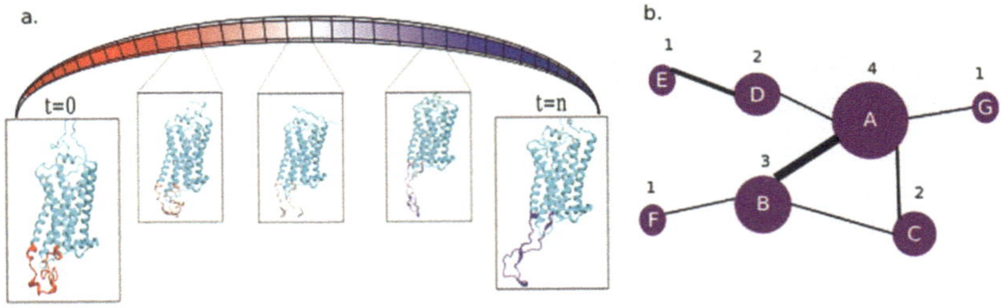

Fig. 9 Schematic representation of computational tools used to analyze SNPs. (A) An overview of a trajectory of a GPCR obtained from molecular dynamics simulations. (B) A general network graph depicting an inter-residue network. The nodes corresponding to the different residues are labeled (A)–(G), the thickness of edges interconnecting them are based on the strength of the interaction and the degree of the node (number of neighbors) is labeled above it.

docking approaches, free-energy calculations and long-time scale classical unbiased simulations. A variety of ligands have been probed, ranging from natural small molecule ligands, synthetic drugs, peptides, lipids, and proteins. In fact, molecular dynamics simulations were critical to identify the different GPCR conformational states including the vast repertoire of intermediate states. The molecular basis of basal-level activity can be easily identified from the *apo-* and ligand-bound receptor simulations. Molecular dynamics simulations were critical to identify that a ligand-bound state may not be enough to stabilize the active state in specific receptors and that effectors (and nanobodies) would be needed. The mechanism of activation was first proposed by using long time scale molecular dynamics and subsequently the activation of several GPCRs was probed by classical and enhanced sampling approaches. Using coarse-grain simulations to sample at even longer length and timescales, lipid interactions and organization of several GPCRs has been characterized. Since several aspects of the wildtype receptor are so well characterized and validated at the molecular level, the predictive accuracy of these approaches for the variant receptors could be considered to be high. However, the main limitation of these methods is that they are computationally expensive and very detailed analysis is required to identify the differences.

An interesting approach is to represent the receptor structure by using protein networks. In general, a network can be defined as a graph consisting of a set of nodes connected by lines, i.e., edges. The nodes in the network represent different entities (e.g., amino acid residues, proteins or genes in biological networks), and edges give the information about an interaction between the nodes. An example of a protein residue network is shown in Fig. 9. In most intra-protein networks, the nodes correspond to the protein residues and the edges are dependent on interactions between them. Several parameters such as distance, fluctuations, correlated motion and energetics have been used to build such networks. As an example, the electrostatic interactions between all-to-all residues can be calculated and edge weight can be determined from it. These individual networks can be quite difficult to interpret as the number of interactions are high and can be sensitive to cut-offs often used to simplify the network. To directly compare between receptor variants, a difference network has been proposed in which the difference of the interaction energies (between wildtype and variant) is used to determine the edges. As a result, the core interactions that are not affected by the variations are hidden from the network and only the perturbation arising due to the variations are observed on the network. These networks can be further simplified by distinguishing between interactions that are more or less favorable in the wildtype receptor compared to the variant receptor. Such difference networks provide a suitable handle to identify small but significant differences between receptor variants that could otherwise be difficult to tweeze out.

3. An overview of the pharmacogenomics in β_2-AR

Several SNPs have been recorded in the ADRB2 gene, in the untranslated as well as in the coding region. In the case of β_2-AR, dbSNP lists 319 missense variants in the coding region. In contrast, the curated database, GPCRdb reports 115 missense variants, 3 frameshift variants and one stop gained mutation (as of June 2021). Fig. 10 gives an overview of the nsSNPs/variants listed at GPCRdb. As can be observed from the figure, the nsSNPs are present all over the receptor, including each of the transmembrane helices. These nsSNPs have been characterized as deleterious (red), tolerated (green) or loss of function (violet/blue) based on the bioinformatics tools, SIFT and PolyPhen. A few of these nsSNPs corresponding to missense receptor variants have been reported to be correlated with a differential therapeutic response. In particular, the variants Arg16Gly, Gln27Glu and Thr164Ile have been associated with a differential response to asthma and COPD drugs. It was suggested that these variants may alter receptor dynamics affecting ligand binding and G-protein coupling through local and long-range effects in subtle but important ways (Green, Cole, Jacinto, Innis, & Liggett, 1993). It is thus critical to understand the molecular level details of how these variants could affect β_2-AR function and their implications in physiology.

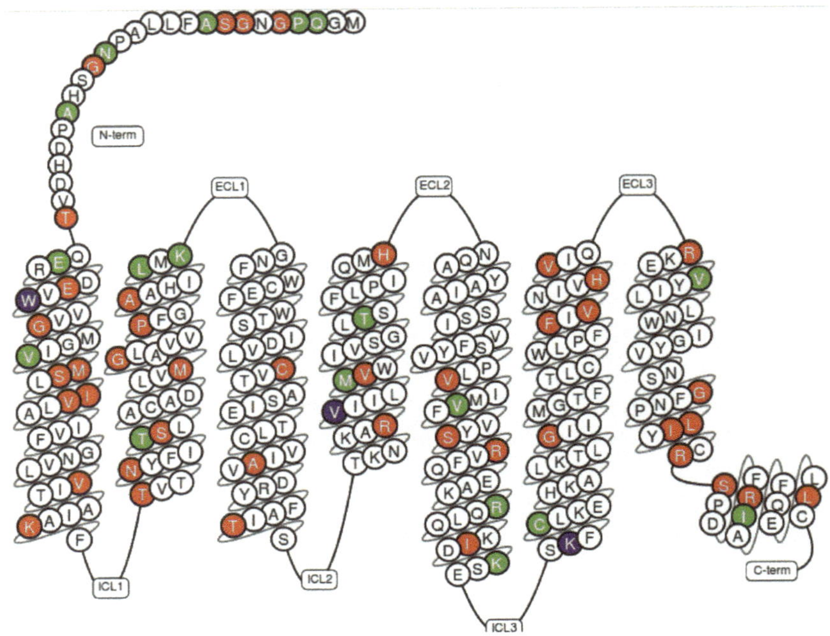

Fig. 10 Variants of β_2-AR marked on the snake plot of the receptor. The different positions of the variants are marked based on the structure and are colored based on predictions of their function. Variants that are predicted as deleterious are colored red, those that are predicted as tolerated are colored green and the loss of function variants are colored purple. The figure was generated using GPCRdb.

Clinical studies have most often reported three specific nsSNPs/variants that are shown in Fig. 11. The most common variant is the Arg16Gly that is present in the population with a minor allele frequency of 50% (Litonjua et al., 2010). Several clinical studies suggest that this variant leads to a differential response to albuterol treatment, although results vary between studies and across populations (Corvol & Burchard, 2008). In cell line experiments, the variants displayed dissimilar kinetics upon repeated agonist treatment although the expression levels of the receptors were similar (Ahles, Rochais, Frambach, Bunemann, & Engelhardt, 2011). Further, marginal differences were observed in the binding affinity of the natural ligand, epinephrine. One of the lesser studied variants in β_2-AR is the Gln27Glu in which the residue Gln is usually found at a higher frequency. Several clinical studies report that isoproterenol, albuterol and terbutaline bind better to the Glu27 variant as compared to the Gln variant (Dishy et al., 2001; Giubergia, Gravina, Castanos, Chertkoff, & Grenoville, 2008; Lee, 2011). A rarer

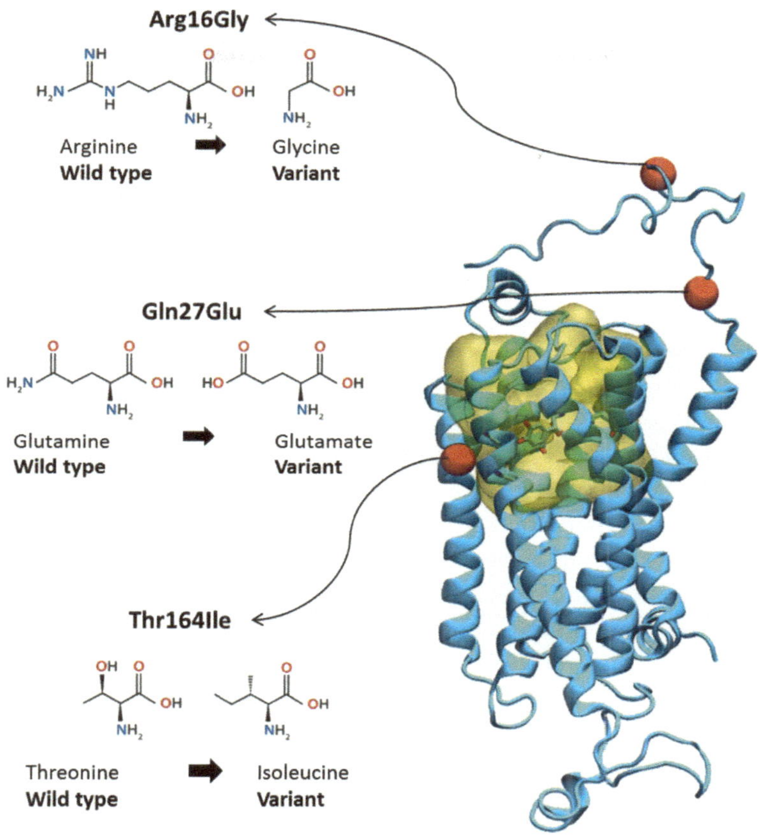

Fig. 11 Clinically important variants in β_2-AR. The protein is rendered as ribbon and colored blue. The variants positions are marked as red spheres. The orthosteric binding site is rendered as a transparent surface colored yellow. The wild type and variant residues are shown in 2D representations.

variant (heterozygous frequency of ~5%) that has also been reported in clinical studies is Thr164Ile. Interestingly, patients homozygous for the Ile164 variant have not yet been reported. Both clinical and laboratory-level experimental studies have shown that the Ile variant has a lower binding affinity for the agonists such as epinephrine, isoproterenol, albuterol and norepinephrine as compared to the wildtype receptor (Ahles & Engelhardt, 2014; Bandaru et al., 2016). In addition, studies on transgenic mice overexpressing Ile164 variant confirmed a lowered adenyl cyclase activity in response to isoproterenol along with lower heart rates based on hemodynamic parameters (Turki et al., 1996). Clinical studies have also identified a rare variant Val34Met that has a minor allele frequency less than 1% (Litonjua et al., 2010). A comparison to the predictions from bioinformatics tools (Fig. 10) reveals that these clinically-relevant nsSNPs may not be predicted to be deleterious.

3.1 Unraveling molecular mechanisms of clinical SNPs in β_2-AR

Surprisingly, the clinically identified variants of β_2-AR, Arg16Gly, Gln27Glu and Thr164Ile that have been correlated to a differential response to asthma drugs were not identified as deleterious by several bioinformatics tools (Tandale, Joshi, & Sengupta, 2016). This discrepancy could be due to the limited accuracy of these tools. For instance, one of the most robust prediction tools, PredictSNP, reports an accuracy range of just 72–78% for various datasets (Bendl et al., 2014). In addition, the variants at the 16th and 27th position lie at the highly variable and dynamic N-terminal domain that has not been shown to directly interact with small molecule ligands. Consequently, the tools that are based on structural and sequence similarity information assigned a low importance to these two variants. The wildtype residue, Thr164 is present on transmembrane helix 4 and is replaced by an aliphatic residue, isoleucine in the Thr164Ile variant. Since this position is present in the apolar part of the membrane, it would seem to be a favorable mutation in a transmembrane region to sequence and structure based tools. This necessitates more advanced computational approaches such as molecular dynamics simulations (Sengupta, Joshi, Athale, & Chattopadhyay, 2016; Sengupta, Sonar, & Joshi, 2017). We describe below the insights provided by atomistic simulations for these three variants.

3.2 N-terminal domain dynamics is modulated by variants at the 16th position

To identify the molecular origin of the differences between the two variants at the 16th position, long timescale molecular dynamics simulations were performed combined with docking approaches (Shahane, Parsania, Sengupta, & Joshi, 2014). Based on the simulations, several functional domains of the variant receptor were identified that deviated from the wildtype receptor. The work was made possible due to the detailed characterization of the wildtype receptor by previous computational studies. The main difference between the variants was the dynamics of the N-terminal domain. Fig. 12 shows an overview of the different conformations sampled in the molecular dynamics trajectories. The N-terminal domain containing

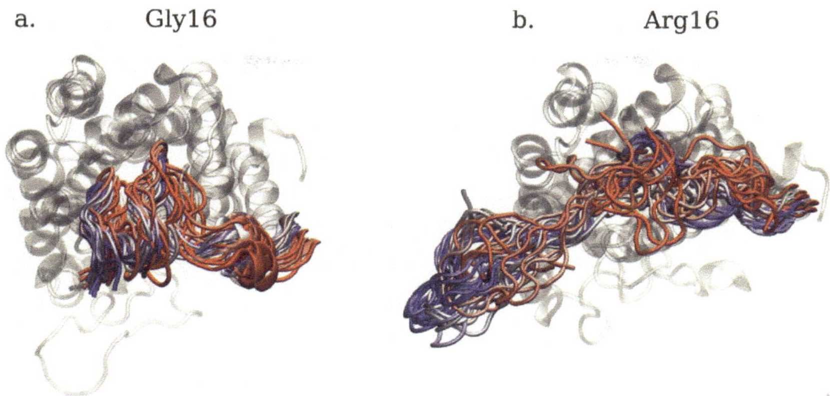

Fig. 12 Differential N-terminal dynamics in the 16th position variant of β_2-AR. Panel (A) represents the wildtype receptor with Gly at the 16th position, panel (B) represent the variant receptor with Arg at the 16th position. The receptor is rendered in ribbons and colored gray. Snapshots of the N terminal region corresponding to different time frames are superimposed and colored as red to blue.

the Arg variant was differentially positioned and sampled a large number of extended conformations interacting with transmembrane helices further away. In contrast the Gly variant N-terminal domain was less dynamic, predominantly positioned on top of the receptor. This led to cascading effects to both the ligand entry site and the ligand binding site. The results indicate that the Gly variants had a more open vestibule 1 than the Arg variants, while the opposite was observed at vestibule 2 and it was more accessible in Arg variants compared to the Gly variants. Interestingly, long-range effects were also seen at the ligand binding site, which was marginally larger in the Gly as compared to the Arg variant. Due to the increased binding site, a preferential docking of albuterol to the Gly variant was observed. Overall, molecular dynamics simulations suggest that due to the differential dynamics of the N-terminal domain, the ligand binding energetics and kinetics of the variants could be modulated.

3.3 Variation at the 27th position leads to a rewiring of the intra-protein network

A similar simulation study of the 27th variant position was performed in conjunction with network analysis (Bhosale, Nikte, Sengupta, & Joshi, 2019). In this case as well, it was shown that the residues Gln27 and Glu27 altered the dynamics of the N-terminal domain of the receptor. In turn, the differential dynamics of the N-terminal domain alters the chain of molecular level interactions towards the ligand binding site. To characterize these interactions, a network based on inter-residue electrostatics interactions and hydrogen bonds was constructed. An example of the network is shown in Fig. 13. At the variant position, there are local changes in the electrostatic interactions with Gln26 and Gln/Glu27 leading to differences in its placement with respect to other extracellular loops and the first residues of transmembrane helices 2, 3 and 7. In particular, it was observed that the salt bridge between residues D192$^{45\times51}$ and K305$^{7\times31}$ was broken in the Glu variant. This salt bridge lies just above the orthosteric ligand binding site and partitions the two ligand entry vestibules (see Fig. 5). It was hypothesized that the ligand entry would be altered in these variants. In addition, docking studies predicted that variant Glu27 has a more favorable binding free energy for ligands and preference to isoproterenol than Gln27, in line with experimental studies.

3.4 Water-mediated allosteric networks are disrupted by the 164th position variant

The polymorphism corresponding to the Thr164Ile variant is intriguing as an isoleucine residue I164$^{4\times56}$ in the variant should actually correspond to increased transmembrane domain stability compared to the wildtype residue, T164$^{4\times56}$. In fact, related adrenergic receptors (β_1-AR, β_3-AR)

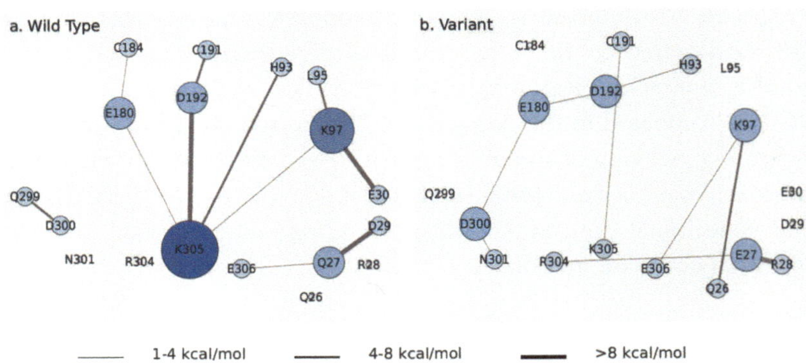

Fig. 13 Perturbed electrostatic network in the wildtype and variant receptor at the 27th position. A network of difference in the electrostatic interaction energies showing favorable interactions in (A) wildtype (Gln27) and (B) variant receptor (Glu27) is depicted. The residues (nodes) are represented as spheres in shades of blue, whose radius and color shade (light to dark) is proportional to number of connections (edges). The thickness of the connections (edges) are proportional to the difference in the electrostatic energies as per the scale given. The networks were drawn using the network visualization and analysis software, Cytoscape.

actually have a hydrophobic residue at the equivalent position (4×56). A computational study using mainly docking approaches predicted that the variant receptor would bind less favorably to agonists such as albuterol (Bandaru et al., 2017). However, they were unable to identify the origin of this discrepancy. To bridge this gap, an extensive molecular dynamics simulation study was performed and coupled with inter-residue network analysis (Nikte, Sonar, Tandale, Joshi, & Sengupta, 2021). Two specific hydration sites were identified in the transmembrane domain in the vicinity of the residue 164, one of which has been reported in several experimentally resolved crystal structures. The simulations indicate that the bound water molecules at these two hydration sites are necessary to maintain the essential conformational dynamics of the transmembrane domain. In the absence of these two sites in the variant, the transmembrane domain becomes more rigid and appears to be more compact. The loss of these hydration sites in the variant leads to perturbations in the intra-protein interaction network and rearrangements in the orthosteric ligand binding site (see Fig. 14). In conjunction, free-energy calculations validate an altered binding and reduced free energy of a series of agonists, in line with experimental trends. Overall, the computational study was able to identify a functional allosteric pathway connected by specific hydration sites in β_2-AR that has not been reported before and provides insight into water-mediated networks in GPCRs in general.

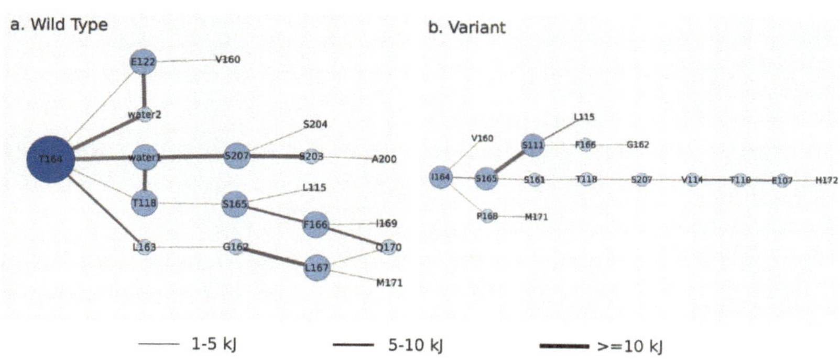

Fig. 14 Perturbed electrostatic network in the wildtype and variant receptor at the 164th position. A network of difference in the electrostatic interaction energies showing favorable interactions in (A) wildtype (Thr164) and (B) variant receptor (Ile164) is depicted. The residues (nodes) are represented as spheres in shades of blue, whose radius and color shade (light to dark) is proportional to number of connections (edges). The thickness of the connections (edges) are proportional to the difference in the electrostatic energies as per the scale given. The networks were drawn using the network visualization and analysis software, Cytoscape.

4. Conclusions

GPCRs are membrane embedded proteins that form central signaling hubs for ligands of various sizes. These receptors play a myriad of physiological roles such as muscle relaxation, immune response, olfactory/taste sensing, and cognition. Owing to their critical roles in cell function, GPCRs are important drug targets, and more than one-third of the FDA approved drugs target these receptors. Recent studies have revealed a worrisome statistic that shows that a small percent of patients is never benefited by the FDA approved drugs due to inter-individual differences. Advances in genome sequencing as well as clinical studies have revealed that the non-synonymous single nucleotide polymorphisms (nsSNPs) in genes encoding for GPCRs (with missense variants) are the single largest contributors for such inter-individual differences in drug response. While differences in the pharmacokinetics can be countered by dosage adjustments, differences due to variants in the drug targets might need dosage adjustment and or even a new drug. In this chapter we focus on the β_2-AR, a representative member of the class A GPCRs and provide an overview of its structure, dynamics, clinically identified variants and their pharmacogenetics.

The β_2-AR is an important drug target in asthma and COPD. Several short, long and ultralong acting agonists are approved drugs that are inhaled by patients to provide relief from bronchospasm. There are three clinically identified variants of the β_2-AR, the Arg16Gly, Gln27Glu and Thr164Ile that have been associated with an adverse response to the asthma/COPD drugs. In-vitro experimental studies have validated this observation for Thr164Ile and the binding with a series of agonists. We describe the use of bioinformatics tools in conjunction with molecular dynamics simulations, docking free energy calculations and network analysis to understand the molecular details of the adverse response of the receptor variants.

The β_2-AR has been extensively studied by crystallography, other experimental and computational methods and the activation mechanism and dynamics of the wildtype receptor is very well characterized. Due to this detailed characterization, it establishes a gold standard to compare the behavior of the variant receptors. In each of the variant receptors, we observe local changes in the immediate vicinity of the variant positions. In general, these small changes propagate via allosteric communication pipelines and lead to the binding site where it modulates ligand binding. We suggest that a fast screening can be performed by bioinformatics tools but detailed computational approaches are needed to identify receptor-specific and variant-specific mechanisms. A concerted effort between computational, experimental and clinical studies is needed to fully establish the effect of the variant receptors on drug response. Sequencing efforts have become mainstream in assessing disease susceptibility and we suggest that it should be an important part of assessing drug response as well.

Acknowledgments

We thank Chirag Parsania, Ganesh Shahane, Aditi Tandale, Krushna Sonar and Sumedha Bhosale for previous discussions. We thank members of our research group for critical feedback on the manuscript. D.S. thanks DST (National Supercomputing Mission) for the grant #2021/03.21. M.J. thanks Bioinformatics Centre, S. P. Pune University for infrastructure support. S.V.N. acknowledges the Department of Biotechnology, India, for the doctoral research fellowship.

References

Adzhubei, I., Jordan, D. M., & Sunyaev, S. R. (2013). Predicting functional effect of human missense mutations using PolyPhen-2. *Current Protocols in Human Genetics, 76*, 7.20.1–7.20.41.

Ahles, A., & Engelhardt, S. (2014). Polymorphic variants of adrenoceptors: Pharmacology, physiology, and role in disease. *Pharmacological Reviews, 66*(3), 598–637.

Ahles, A., Rochais, F., Frambach, T., Bunemann, M., & Engelhardt, S. (2011). A polymorphism-specific "memory" mechanism in the β_2-adrenergic receptor. *Science Signaling, 4*(185), ra53.

Alexander, S. P. H., Christopoulos, A., Davenport, A. P., Kelly, E., Mathie, A., Peters, J. A., et al. (2019). The concise guide to pharmacology 2019/20: G protein-coupled receptors. *British Journal of Pharmacology, 176*, S21–S141.

Bandaru, S., Alvala, M., Nayarisseri, A., Sharda, S., Goud, H., Mundluru, H. P., et al. (2017). Molecular dynamic simulations reveal suboptimal binding of salbutamol in T164I variant of β_2-adrenergic receptor. *PLoS One, 12*(10), e0186666.

Bandaru, S., Tarigopula, P., Akka, J., Marri, V. K., Kattamuri, R. K., Nayarisseri, A., et al. (2016). Association of β_2-adrenergic receptor (Thr164Ile) polymorphism with salbutamol refractoriness in severe asthmatics from Indian population. *Gene, 592*(1), 15–22.

Barbarino, J. M., Whirl-Carrillo, M., Altman, R. B., & Klein, T. E. (2018). PharmGKB: A worldwide resource for pharmacogenomic information. *Wiley Interdisciplinary Reviews. Systems Biology and Medicine, 10*(4), e1417.

Bendl, J., Stourac, J., Salanda, O., Pavelka, A., Wieben, E. D., Zendulka, J., et al. (2014). PredictSNP: Robust and accurate consensus classifier for prediction of disease-related mutations. *PLOS Computational Biology, 10*(1), e1003440.

Berman, H. M., Westbrook, J., Feng, Z., Gilliland, G., Bhat, T. N., Weissig, H., et al. (2000). The Protein Data Bank. *Nucleic Acids Research, 28*(1), 235–242.

Bhosale, S., Nikte, S. V., Sengupta, D., & Joshi, M. (2019). Differential dynamics underlying the Gln27Glu population variant of the β_2-adrenergic receptor. *The Journal of Membrane Biology, 252*(4–5), 499–507.

Boutet, E., Lieberherr, D., Tognolli, M., Schneider, M., Bansal, P., Bridge, A. J., et al. (2016). UniProtKB/Swiss-Prot, the manually annotated section of the UniProt KnowledgeBase: How to use the entry view. *Methods in Molecular Biology, 1374*, 23–54.

Chakraborty, H., Jafurulla, M., Clayton, A. H. A., & Chattopadhyay, A. (2018). Exploring oligomeric state of the serotonin$_{1A}$ receptor utilizing photobleaching image correlation spectroscopy: Implications for receptor function. *Faraday Discussions*, *207*, 409–421.
Cherezov, V., Rosenbaum, D. M., Hanson, M. A., Rasmussen, S. G., Thian, F. S., Kobilka, T. S., et al. (2007). High-resolution crystal structure of an engineered human β_2-adrenergic G protein-coupled receptor. *Science*, *318*(5854), 1258–1265.
Corvol, H., & Burchard, E. G. (2008). Pharmacogenetic response to albuterol among asthmatics. *Pharmacogenomics*, *9*(5), 505–510.
Crilly, S. E., & Puthenveedu, M. A. (2021). Compartmentalized GPCR signaling from intracellular membranes. *The Journal of Membrane Biology*, *254*(3), 259–271.
Dishy, V., Sofowora, G. G., Xie, H. G., Kim, R. B., Byrne, D. W., Stein, C. M., et al. (2001). The effect of common polymorphisms of the β_2-adrenergic receptor on agonist-mediated vascular desensitization. *The New England Journal of Medicine*, *345*(14), 1030–1035.
Divorty, N., Mackenzie, A. E., Nicklin, S. A., & Milligan, G. (2015). G protein-coupled receptor 35: An emerging target in inflammatory and cardiovascular disease. *Frontiers in Pharmacology*, *6*, 41.
Dror, R. O., Arlow, D. H., Maragakis, P., Mildorf, T. J., Pan, A. C., Xu, H., et al. (2011). Activation mechanism of the β_2-adrenergic receptor. *Proceedings of the National Academy of Sciences of the United States of America*, *108*(46), 18684–18689.
Foster, S. R., Hauser, A. S., Vedel, L., Strachan, R. T., Huang, X. P., Gavin, A. C., et al. (2019). Discovery of human signaling systems: Pairing peptides to G protein-coupled receptors. *Cell*, *179*(4), 895.
Genomes Project, C, Abecasis, G. R., Auton, A., Brooks, L. D., DePristo, M. A., Durbin, R. M., et al. (2012). An integrated map of genetic variation from 1,092 human genomes. *Nature*, *491*(7422), 56–65.
Giubergia, V., Gravina, L. P., Castanos, C., Chertkoff, L., & Grenoville, M. (2008). Influence of β_2-adrenoceptor polymorphisms on the response to chronic use of albuterol in asthmatic children. *Pediatric Pulmonology*, *43*(5), 421–425.
Green, S. A., Cole, G., Jacinto, M., Innis, M., & Liggett, S. B. (1993). A polymorphism of the human β_2-adrenergic receptor within the fourth transmembrane domain alters ligand binding and functional properties of the receptor. *The Journal of Biological Chemistry*, *268*(31), 23116–23121.
Gregorio, G. G., Masureel, M., Hilger, D., Terry, D. S., Juette, M., Zhao, H., et al. (2017). Single-molecule analysis of ligand efficacy in β2-AR-G-protein activation. *Nature*, *547*(7661), 68.
Grzegorzewska, A. E., Frycz, B. A., Swiderska, M., Niepolski, L., Mostowska, A., & Jagodzinski, P. P. (2019). Calcium-sensing receptor gene (CASR) polymorphisms and CASR transcript level concerning dyslipidemia in hemodialysis patients: A cross-sectional study. *BMC Nephrology*, *20*(1), 436.
Hauser, A. S., Attwood, M. M., Rask-Andersen, M., Schioth, H. B., & Gloriam, D. E. (2017). Trends in GPCR drug discovery: New agents, targets and indications. *Nature Reviews. Drug Discovery*, *16*(12), 829–842.
Hereld, D., & Jin, T. (2008). Slamming the DOR on chemokine receptor signaling: Heterodimerization silences ligand-occupied CXCR4 and delta-opioid receptors. *European Journal of Immunology*, *38*(2), 334–337.
Hoffmann, C., Leitz, M. R., Oberdorf-Maass, S., Lohse, M. J., & Klotz, K. N. (2004). Comparative pharmacology of human β-adrenergic receptor subtypes - characterization of stably transfected receptors in CHO cells. *Naunyn-Schmiedeberg's Archives of Pharmacology*, *369*(2), 151–159.
Insel, P. A., Sriram, K., Gorr, M. W., Wiley, S. Z., Michkov, A., Salmeron, C., et al. (2019). GPCRomics: An approach to discover GPCR drug targets. *Trends in Pharmacological Sciences*, *40*(6), 378–387.

Johnson, M. (2006). Molecular mechanisms of β_2-adrenergic receptor function, response, and regulation. *The Journal of Allergy and Clinical Immunology. In Practice*, *117*(1), 18–24.

Johnson, A. D., Handsaker, R. E., Pulit, S. L., Nizzari, M. M., O'Donnell, C. J., & de Bakker, P. I. (2008). SNAP: A web-based tool for identification and annotation of proxy SNPs using HapMap. *Bioinformatics*, *24*(24), 2938–2939.

Jones, E. M., Lubock, N. B., Venkatakrishnan, A., Wang, J., Tseng, A. M., Paggi, J. M., et al. (2020). Structural and functional characterization of G protein–coupled receptors with deep mutational scanning. *Elife*, 9.

Kawabata, T., Ota, M., & Nishikawa, K. (1999). The protein mutant database. *Nucleic Acids Research*, *27*(1), 355–357.

Kharche, S. A., Joshi, M., Chattopadhyay, A., & Sengupta, D. (2021). Conformational plasticity and dynamic interactions of the N-terminal domain of the chemokine receptor CXCR1. *PLoS Computational Biology*, *17*(5), e1008593.

Kharche, S. A., & Sengupta, D. (2020). Dynamic protein interfaces and conformational landscapes of membrane protein complexes. *Current Opinion in Structural Biology*, *61*, 191–197.

Kim, S., Chen, J., Cheng, T. J., Gindulyte, A., He, J., He, S. Q., et al. (2019). PubChem 2019 update: Improved access to chemical data. *Nucleic Acids Research*, *47*(D1), D1102–D1109.

Kirchheiner, J., Nickchen, K., Bauer, M., Wong, M. L., Licinio, J., Roots, I., et al. (2004). Pharmacogenetics of antidepressants and antipsychotics: The contribution of allelic variations to the phenotype of drug response. *Molecular Psychiatry*, *9*(5), 442–473.

Kobilka, B. K. (2007). G protein coupled receptor structure and activation. *Biochimica Biophysica Acta, Biomembranes*, *1768*(4), 794–807.

Kobilka, B. K., Dixon, R. A., Frielle, T., Dohlman, H. G., Bolanowski, M. A., Sigal, I. S., et al. (1987). cDNA for the human β_2-adrenergic receptor: A protein with multiple membrane-spanning domains and encoded by a gene whose chromosomal location is shared with that of the receptor for platelet-derived growth factor. *Proceedings of the National Academy of Sciences of the United States of America*, *84*(1), 46–50.

Kooistra, A. J., Mordalski, S., Pandy-Szekeres, G., Esguerra, M., Mamyrbekov, A., Munk, C., et al. (2021). GPCRdb in 2021: Integrating GPCR sequence, structure and function. *Nucleic Acids Research*, *49*(D1), D335–D343.

Lambert, D., & Calo, G. (2020). Approval of oliceridine (TRV130) for intravenous use in moderate to severe pain in adults. *British Journal of Anaesthesia*, *125*(6), e473–e474.

Lamichhane, R., Liu, J. J., Pljevaljcic, G., White, K. L., van der Schans, E., Katritch, V., et al. (2015). Single-molecule view of basal activity and activation mechanisms of the G protein-coupled receptor β_2-AR. *Proceedings of the National Academy of Sciences of the United States of America*, *112*(46), 14254–14259.

Lee, M. Y., Cheng, S. N., Chen, S. J., Huang, H. L., Wang, C. C., & Fan, H. C. (2011). Polymorphisms of the β2-adrenergic receptor correlated to nocturnal asthma and the response of terbutaline nebulizer. *Pediatrics and Neonatology*, *52*(1), 18–23.

Lefkowitz, R. J., Stadel, J. M., & Caron, M. G. (1983). Adenylate cyclase-coupled β-adrenergic receptors: Structure and mechanisms of activation and desensitization. *Annual Review of Biochemistry*, *52*, 159–186.

Litonjua, A. A., Gong, L., Duan, Q. L., Shin, J., Moore, M. J., Weiss, S. T., et al. (2010). Very important pharmacogene summary ADRB2. *Pharmacogenetics and Genomics*, *20*(1), 64–69.

Lopez-Ferrando, V., Gazzo, A., de la Cruz, X., Orozco, M., & Gelpi, J. L. (2017). PMut: A web-based tool for the annotation of pathological variants on proteins, 2017 update. *Nucleic Acids Research*, *45*(W1), W222–W228.

Luttrell, L. M., & Gesty-Palmer, D. (2010). Beyond desensitization: Physiological relevance of arrestin-dependent signaling. *Pharmacological Reviews*, *62*(2), 305–330.

Marinissen, M. J., & Gutkind, J. S. (2001). G-protein-coupled receptors and signaling networks: Emerging paradigms. *Trends in Pharmacological Sciences*, *22*(7), 368–376.

Melik Parsadaniantz, S., Rivat, C., Rostene, W., & Reaux-Le Goazigo, A. (2015). Opioid and chemokine receptor crosstalk: A promising target for pain therapy? *Nature Reviews. Neuroscience*, *16*(2), 69–78.

Nikte, S. V., Sonar, K., Tandale, A., Joshi, M., & Sengupta, D. (2021). Loss of a water-mediated network results in reduced agonist affinity in a β_2-adrenergic receptor clinical variant. *Biochimica et Biophysica Acta, Proteins and Proteomics*, *1869*(4), 140605.

Nygaard, R., Zou, Y., Dror, R. O., Mildorf, T. J., Arlow, D. H., Manglik, A., et al. (2013). The dynamic process of β_2-adrenergic receptor activation. *Cell*, *152*(3), 532–542.

Oates, J., & Watts, A. (2011). Uncovering the intimate relationship between lipids, cholesterol and GPCR activation. *Current Opinion in Structural Biology*, *21*(6), 802–807.

O'Boyle, N. M., Banck, M., James, C. A., Morley, C., Vandermeersch, T., & Hutchison, G. R. (2011). Open babel: An open chemical toolbox. *Journal of Cheminformatics*, *3*.

Palczewski, K., Kumasaka, T., Hori, T., Behnke, C. A., Motoshima, H., Fox, B. A., et al. (2000). Crystal structure of rhodopsin: A G protein-coupled receptor. *Science*, *289*(5480), 739–745.

Pierce, K. L., Premont, R. T., & Lefkowitz, R. J. (2002). Seven-transmembrane receptors. *Nature Reviews. Molecular Cell Biology*, *3*(9), 639–650.

Prasanna, X., Mohole, M., Chattopadhyay, A., & Sengupta, D. (2020). Role of cholesterol-mediated effects in GPCR heterodimers. *Chemistry and Physics of Lipids*, *227*, 104852.

Ragnarsson, L., Andersson, A., Thomas, W. G., & Lewis, R. J. (2019). Mutations in the NPxxY motif stabilize pharmacologically distinct conformational states of the alpha(1B)- and beta(2)-adrenoceptors. *Science Signaling*, *12*(572), eaas9485.

Rasmussen, S. G., Choi, H. J., Fung, J. J., Pardon, E., Casarosa, P., Chae, P. S., et al. (2011). Structure of a nanobody-stabilized active state of the β_2-adrenoceptor. *Nature*, *469*(7329), 175–180.

Rasmussen, S. G. F., Choi, H. J., Rosenbaum, D. M., Kobilka, T. S., Thian, F. S., Edwards, P. C., et al. (2007). Crystal structure of the human β_2-adrenergic G-protein-coupled receptor. *Nature*, *450*(7168), 383–U384.

Rasmussen, S. G., DeVree, B. T., Zou, Y., Kruse, A. C., Chung, K. Y., Kobilka, T. S., et al. (2011). Crystal structure of the β_2-adrenergic receptor-Gs protein complex. *Nature*, *477*(7366), 549–555.

Reardon, S. (2015). Personalized medicine: Health study set to decide data policy. *Nature*, *525*(7567), 16–17.

Riccardi, D., & Martin, D. (2008). The role of the calcium-sensing receptor in the pathophysiology of secondary hyperparathyroidism. *NDT Plus*, *1*(Suppl. 1), i7–i11.

Ritter, K., Buning, C., Halland, N., Poverlein, C., & Schwink, L. (2016). G protein-coupled receptor 119 (GPR119) agonists for the treatment of diabetes: Recent progress and prevailing challenges. *Journal of Medicinal Chemistry*, *59*(8), 3579–3592.

Rosenbaum, D. M., Cherezov, V., Hanson, M. A., Rasmussen, S. G. F., Thian, F. S., Kobilka, T. S., et al. (2007). GPCR engineering yields high-resolution structural insights into β_2-adrenergic receptor function. *Science*, *318*(5854), 1266–1273.

Rosenbaum, D. M., Rasmussen, S. G. F., & Kobilka, B. K. (2009). The structure and function of G-protein-coupled receptors. *Nature*, *459*(7245), 356–363.

Salanga, C. L., O'Hayre, M., & Handel, T. (2009). Modulation of chemokine receptor activity through dimerization and crosstalk. *Cellular and Molecular Life Sciences*, *66*(8), 1370–1386.

Santos, R., Ursu, O., Gaulton, A., Bento, A. P., Donadi, R. S., Bologa, C. G., et al. (2017). A comprehensive map of molecular drug targets. *Nature Reviews. Drug Discovery*, *16*(1), 19–34.

Sengupta, D., Joshi, M., Athale, C. A., & Chattopadhyay, A. (2016). What can simulations tell us about GPCRs: Integrating the scales. *Methods in Cell Biology*, *132*, 429–452.

Sengupta, D., Sonar, K., & Joshi, M. (2017). Characterizing clinically relevant natural variants of GPCRs using computational approaches. *Methods in Cell Biology*, *142*, 187–204.

Shahane, G., Parsania, C., Sengupta, D., & Joshi, M. (2014). Molecular insights into the dynamics of pharmacogenetically important N-terminal variants of the human β_2-adrenergic receptor. *PLoS Computational Biology*, *10*(12), e1004006.

Sherry, S. T., Ward, M. H., Kholodov, M., Baker, J., Phan, L., Smigielski, E. M., et al. (2001). dbSNP: The NCBI database of genetic variation. *Nucleic Acids Research*, *29*(1), 308–311.

Simoes, S. C., Balico-Silva, A. L., Parreiras-e-Silva, L. T., Bitencourt, A. L. B., Bouvier, M., & Costa-Neto, C. M. (2020). Signal transduction profiling of angiotensin II type 1 receptor with mutations associated to atrial fibrillation in humans. *Frontiers in Pharmacology*, *11*, 600132.

Smith, J. S., Lefkowitz, R. J., & Rajagopal, S. (2018). Biased signalling: From simple switches to allosteric microprocessors. *Nature Reviews. Drug Discovery*, *17*(4), 243–260.

Sriram, K., & Insel, P. A. (2018). G protein-coupled receptors as targets for approved drugs: How many targets and how many drugs? *Molecular Pharmacology*, *93*(4), 251–258.

Tandale, A., Joshi, M., & Sengupta, D. (2016). Structural insights and functional implications of inter-individual variability in β_2-adrenergic receptor. *Scientific Reports*, *6*, 24379.

Tang, X. L., Wang, Y., Li, D. L., Luo, J., & Liu, M. Y. (2012). Orphan G protein-coupled receptors (GPCRs): Biological functions and potential drug targets. *Acta Pharmacologica Sinica*, *33*(3), 363–371.

Terrillon, S., & Bouvier, M. (2004). Roles of G-protein-coupled receptor dimerization—From ontogeny to signalling regulation. *EMBO Reports*, *5*(1), 30–34.

Tian, J., Wu, N., Guo, X., Guo, J., Zhang, J., & Fan, Y. (2007). Predicting the phenotypic effects of non-synonymous single nucleotide polymorphisms based on support vector machines. *BMC Bioinformatics*, *8*, 450.

Turki, J., Lorenz, J. N., Green, S. A., Donnelly, E. T., Jacinto, M., & Liggett, S. B. (1996). Myocardial signaling defects and impaired cardiac function of a human β_2-adrenergic receptor polymorphism expressed in transgenic mice. *Proceedings of the National Academy of Sciences of the United States of America*, *93*(19), 10483–10488.

Turnbull, C., Scott, R. H., Thomas, E., Jones, L., Murugaesu, N., Pretty, F. B., et al. (2018). The 100 000 Genomes Project: Bringing whole genome sequencing to the NHS. *BMJ*, *361*, k1952.

Vaser, R., Adusumalli, S., Leng, S. N., Sikic, M., & Ng, P. C. (2016). SIFT missense predictions for genomes. *Nature Protocols*, *11*(1), 1–9.

Venkatakrishnan, A. J., Deupi, X., Lebon, G., Tate, C. G., Schertler, G. F., & Babu, M. M. (2013). Molecular signatures of G-protein-coupled receptors. *Nature*, *494*(7436), 185–194.

Wachter, S. B., & Gilbert, E. M. (2012). β-Adrenergic receptors, from their discovery and characterization through their manipulation to beneficial clinical application. *Cardiology*, *122*(2), 104–112.

Waldeck, B. (2002). β-Adrenoceptor agonists and asthma—100 years of development. *European Journal of Pharmacology*, *445*(1–2), 1–12.

Wang, T., & Duan, Y. (2009). Ligand entry and exit pathways in the β_2-adrenergic receptor. *Journal of Molecular Biology*, *392*(4), 1102–1115.

Wojciechowicz, M. L., & Ma'ayan, A. (2020). GPR84: An immune response dial? *Nature Reviews. Drug Discovery*, *19*(6), 374.

Woollard, S. M., & Kanmogne, G. D. (2015). Maraviroc: A review of its use in HIV infection and beyond. *Drug Design, Development and Therapy*, *9* 5447–5468.

Yang, D. H., Zhou, Q. T., Labroska, V., Qin, S. S., Darbalaei, S., Wu, Y. R., et al. (2021). G protein-coupled receptors: Structure- and function-based drug discovery. *Signal Transduction and Targeted Therapy*, *6*(1), 7.

Yuan, S., Filipek, S., Palczewski, K., & Vogel, H. (2014). Activation of G-protein-coupled receptors correlates with the formation of a continuous internal water pathway. *Nature Communications*, *5*, 4733.

Zhang, L., Zhang, J. T., Hang, L., & Liu, T. (2020). Mu opioid receptor heterodimers emerge as novel therapeutic targets: Recent progress and future perspective. *Frontiers in Pharmacology*, *11*, 1078.

Zhou, Q. T., Yang, D. H., Wu, M., Guo, Y., Guo, W. J., Zhong, L., et al. (2019). Common activation mechanism of class a GPCRs. *eLife*, *8*, e50279.

CHAPTER ELEVEN

The role of membranes in function and dysfunction of intrinsically disordered amyloidogenic proteins

Bani Kumar Pathak[a,†], Sandip Dey[a], Sukanya Mozumder[a,b], and Jayati Sengupta[a,b,*]

[a]Structural Biology and Bioinformatics Division, CSIR-Indian Institute of Chemical Biology, Jadavpur, Kolkata, India
[b]Academy of Scientific and Innovative Research (AcSIR), Ghaziabad, India
*Corresponding author: e-mail address: jayati@iicb.res.in

Contents

1. Introduction 398
2. Human disease-related, widely known IDPs 400
   2.1 Amyloid beta 400
   2.2 Alpha-synuclein 402
   2.3 Tau protein 402
   2.4 Amylin 403
   2.5 Prions 404
3. Association of IDPs with biological membranes 404
   3.1 Membrane lipid composition and effects of cholesterol 404
   3.2 Regulation of membrane-IDP interactions 406
   3.3 Impact on membrane permeability 409
4. Implications of IDP-membrane interactions in physiological functions 410
   4.1 Signal transduction 410
   4.2 Vesicle trafficking 411
5. Pathological implications of IDP-membrane interactions 412
   5.1 Nucleation in oligomer and amyloid fibril formation 412
   5.2 Structures of the cytotoxic oligomers and amyloid fibrils 414
   5.3 Mechanism of toxicity 415
   5.4 Small molecule and peptide based inhibitors 417
6. Outlook 419
Acknowledgments 420

† Present address: Department of Biotechnology, Maulana Abul Kalam Azad University of Technology, West Bengal, India.

Advances in Protein Chemistry and Structural Biology, Volume 128
ISSN 1876-1623
https://doi.org/10.1016/bs.apcsb.2021.08.001

Conflict of interest statement 420
Author contributions 420
References 421

Abstract

Membrane-protein interactions play a major role in human physiology as well as in diseases pathology. Interaction of a protein with the membrane was previously thought to be dependent on well-defined three-dimensional structure of the protein. In recent decades, however, it has become evident that a large fraction of the proteome, particularly in eukaryotes, stays disordered in solution and these proteins are termed as intrinsically disordered proteins (IDPs). Also, a vast majority of human proteomes have been reported to contain substantially long disordered regions, called intrinsically disordered regions (IDRs), in addition to the structurally ordered regions. IDPs exist in an ensemble of conformations and the conformational flexibility enables IDPs to achieve functional diversity. IDPs (and IDRs) are found to be important players in cell signaling, where biological membranes act as anchors for signaling cascades. Therefore, IDPs modulate the membrane architectures, at the same time membrane composition also affects the binding of IDPs. Because of intrinsic disorders, misfolding of IDPs often leads to formation of oligomers, protofibrils and mature fibrils through progressive self-association. Accumulation of amyloid-like aggregates of some of the IDPs is a known causative agent for numerous diseases. In this chapter we highlight recent advances in understanding membrane interactions of some of the intrinsically disordered proteins involved in the pathogenesis of human diseases.

1. Introduction

Proteins are traditionally described as ordered structures, dedicated for specialized functions. Nevertheless, intrinsically disordered proteins (IDPs) as well as structured proteins with intrinsically disordered regions (IDRs) are naturally abundant in all proteomes (Ward, Sodhi, McGuffin, Buxton, & Jones, 2004). The structure-function paradigm of protein is challenged by the structure-independent functions of IDPs. Structurally folded proteins attain minimum energy conformation. In contrast, IDPs rapidly interconvert within conformational states, closely spaced in energy landscape. Transient interactions (showing varied affinities) of IDPs with other molecules, often mediated by repeat sequences in IDP, are common in all living forms (Segrest, Jackson, Morrisett, & Gotto, 1974). The conformational dynamics allows IDPs/IDRs to achieve a wide range of functional capabilities (Uversky & Dunker, 2010). Interestingly, recent studies have suggested that conformational heterogeneity of IDPs steers phase separation in the cell (Brangwynne, Tompa, & Pappu, 2015).

The protein folding instructions are embedded in amino acid sequences in the classical foldable group of proteins (Uversky & Dunker, 2010). Interestingly, the amino acid sequence is also responsible for the production of unstructured biologically active IDPs. Protein sequencing data has unraveled the truth of being intrinsically unstructured under physiological conditions. The IDPs are enriched with Gly, Ala, Ser, Gln, Arg, Lys, Glu, and Pro residues which encourage the maintenance of disorder structure as well as high net charges at neutral pH (Gast et al., 1995; Uversky, Gillespie, & Fink, 2000; Weinreb, Zhen, Poon, Conway, & Lansbury, 1996). It has been reported that the net high charge, usually negative, and low content of hydrophobic amino acids, results in not having order structure in IDPs.

A large fraction of human proteomes consists of IDPs which are instrumental in crucial cellular signaling networks to regulate normal cellular activities (Dunker, Cortese, Romero, Iakoucheva, & Uversky, 2005). These disordered proteins are known to interact with a large array of cytosolic clients, including biological membranes (Yoon, Mitrea, Ou, & Kriwacki, 2012). Optimal activity of the membrane proteins having various functional roles, such as, receptors for multiple signals, electron transporters, ions and protein/peptide transporters, enzymes, are closely associated with membrane dynamics (Gould, 2018; Grecco, Schmick, & Bastiaens, 2011). Interestingly, a vast majority of such membrane proteins (>40%) have intrinsically disordered regions (IDRs) and some are IDPs itself (Tantos, Han, & Tompa, 2012; Sacks et al., 2018).

These membrane-bound IDPs are the hub for protein-protein interactions that mediates a wide range of downstream cell signaling. It has been reported that downstream signaling cascade carried out by the interaction of IDPs with intracellular and extracellular proteins which further triggers the growth and development of the neurons (Beraldo et al., 2011). Besides signaling, they are also involved in vesicular trafficking pathways. It has been observed that the IDPs bind the phospholipid bilayer and interacts with specific complexes (Lou, Kim, Hawk, & Shin, 2017).

These IDPs are also the causative agent for various cytotoxic protein aggregates related to human disorders (Abedini & Raleigh, 2009; Evangelisti et al., 2016; Tipping, Van Oosten-Hawle, Hewitt, & Radford, 2015). Protein aggregation, especially order amyloid type is formed by the errors occurring during synthesis and maturation of IDPs. Additionally, several intra and extra cellular factors are also responsible for the formation of order cytotoxic amyloid aggregates (Grant et al., 2007; Isaacson, Weeds, & Fersht, 1999;

Jesus, Almeida, Vaz, Faria, & Brito, 2016). Amyloidosis is the signature hallmark of IDP linked to human diseases. Familiar examples of some IDPs that amyloidosis are amyloid beta (Aβ) variants in Alzheimer's disease, alpha-synuclein (α-syn) in Parkinson's disease, Prion protein in Spongiform encephalopathies (mad cow disease), tau in frontotemporal dementia with Parkinsonism and Islet amyloid polypeptide (IAPP) or amylin in Type II diabetes (Abedini & Raleigh, 2009).

The multidimensional interactions between the IDPs and cellular membranes and other binding partners mark the mechanistic differences in amyloid aggregate formation. The disease-relation made the IDPs as good drug targets for several human disorders (Uversky, Oldfield, & Dunker, 2008). Particularly, highly specific membrane-IDP interactions are potential therapeutic targets and extensive research is being carried out to understand the implications of these interactions in disease pathogenesis.

Here, we have discussed general properties and significance of IDP interactions with biological membranes for selected IDPs, such as, Amyloid beta, Alpha-synuclein, tau protein, IAPP and prions, highlighting the molecular mechanisms and functional implications of IDP-membrane interactions in normal signaling pathways as well as in amyloid conditions.

2. Human disease-related, widely known IDPs

2.1 Amyloid beta

Fibrillar aggregates of Amyloid beta (Aβ) peptides (36–43 amino acids) are the key components of amyloid plaques found in the brain of the patients with Alzheimer's disease (AD) (Roberson & Mucke, 2006). The functional peptide is primarily derived from the amyloid precursor protein (APP), which is likely involved in synapse formation. APP is cleaved sequentially by the transmembrane enzymes, beta and gamma secretases, releasing mature Aβ peptide which has intracellular as well as extracellular functions. However, improper proteolytic cleavage of APP generates aggregation prone peptides, among which $A\beta_{1\text{-}40}$ $A\beta_{1\text{-}42}$ (generally termed as $A\beta_{40}$ and $A\beta_{42}$) are the predominant species. These peptides form amyloid fibrils which are primary constituents of neuritic plaques (Ross & Poirier, 2004). Solution NMR structure of the Aβ $(1-40)$ and Aβ $(1-42)$ peptides has been established (Hou et al., 2004). NMR structure of Aβ (1-42) shows two helical regions encompassing residues 8–25 and 28–38 which is connected by a regular type I beta-turn (Crescenzi et al., 2002) (Fig. 1A).

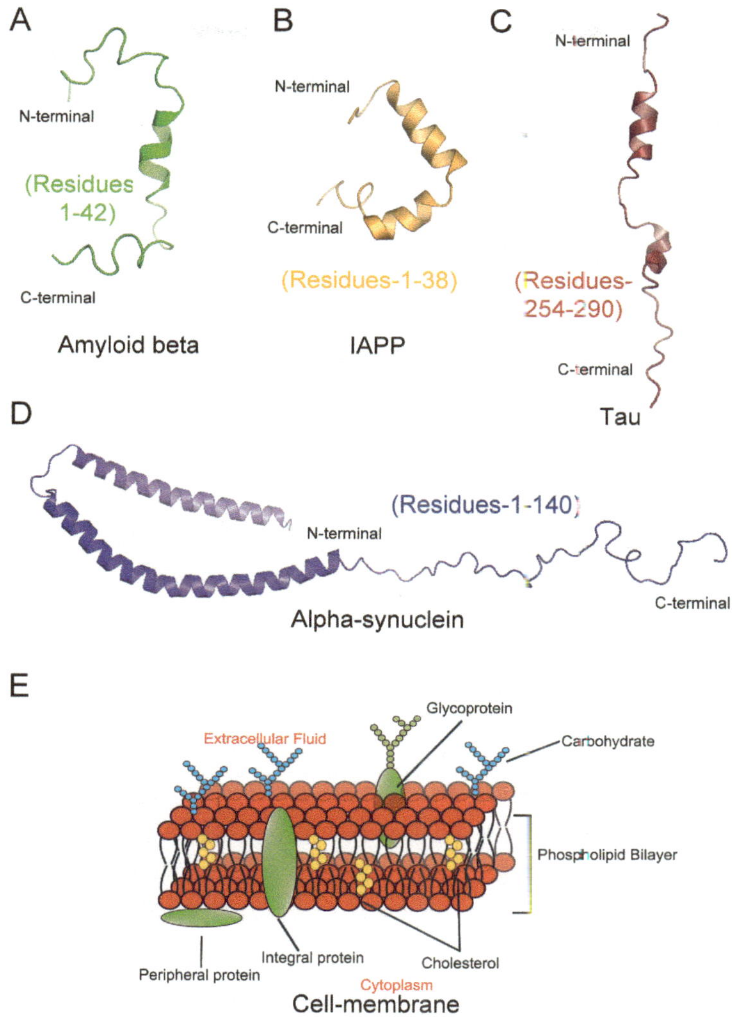

Fig. 1 Disease-related amyloidogenic proteins and biological membrane. Cartoon representations of the atomic structures of amyloid beta (PDB: 6SZF) (A); IAPP (PDB: 2L86) (B); tau (PDB: 5N5A) (C); and A-Syn (PDB: 1XQ8) (D). Schematic representation of a typical biological membrane (E) is shown.

It has been suggested that Aβ aggregation is promoted by the close association of Aβ and cell membranes in the brain (Straub & Thirumalai, 2014). Apparently, Aβ-membrane interaction plays a key role in the amyloid aggregation pathway (Hamley, 2012; Straub & Thirumalai, 2014). While many studies have been performed to investigate the neurotoxicity of the Aβ aggregates in the membrane environment, there are several difficulties for such experimentations and no conclusive model has been derived yet. Nonetheless, recent studies have provided substantial information on Aβ-membrane interactions (Hamley, 2012; Williams & Serpell, 2011).

2.2 Alpha-synuclein

Amyloid aggregation of Alpha-synuclein (α-syn) is a pathological hallmark of human neurodegenerative Parkinson's disease (PD). α-syn, a small peripheral membrane protein (140 amino acids long), is localized specifically at the axon terminal of neurons (Bendor, Logan, & Edwards, 2013; Iwai et al., 1995; Maroteaux, Campanelli, & Scheller, 1988). High resolution NMR structure of micelle-bound α-syn has been reported (Ulmer, Bax, Cole, & Nussbaum, 2005). The structure shows residues Val3-Val37 and Lys45-Thr92 form curved alpha-helices connected by a well-ordered extended linker in an unexpected anti-parallel arrangement which is further followed by another short-extended region of residues Gly93-Lys97 (Fig. 1D).

Although the exact cellular function of α-Syn is still unclear, emerging evidence suggests its involvement in neurotransmitter release and synaptic plasticity (Bendor et al., 2013). Recognition of curved membrane plays a vital role in α-syn function. It was found that conformational alteration of α-syn occurs in the course of membrane interactions (Eliezer, Kutluay, Bussell Jr., & Browne, 2001). In fact, it has been shown that interaction with lipid bilayers stabilizes α-syn majorly in α-helical conformation (Davidson, Jonas, Clayton, & George, 1998; Rovere, Sanderson, Fonseca-Ornelas, Patel, & Bartels, 2018).

2.3 Tau protein

The tau protein is primarily present in brain neurons and stabilizes the internal microtubules. The aberrant forms of tau protein are major constituent of the neurofibrillary tangles, a hallmark feature of Alzheimer disease (AD), and thus it is of key interest as an indicator of disease mechanisms (Jeganathan, Von Bergen, Mandelkow, & Mandelkow, 2008). Tau represents a group of six highly soluble protein isoforms which are derived by alternative splicing from the gene *MAPT* (microtubule-associated protein tau). Six isoforms of tau with a range of 352–441 amino acids are found in human brain. Tau is mainly expressed in the central nervous system and their significant function of promoting microtubule is carried out by six different isoforms which is mainly regulated by phosphorylation (Buée, Bussière, Buée-Scherrer,

Delacourte, & Hof, 2000; Weingarten, Lockwood, Hwo, & Kirschner, 1975). High resolution NMR structure of structure of tau (254–290) bound to filamentous actin (F-actin) has been reported (Cabrales Fontela et al., 2017). Tau adopts a helical structure on binding to F-actin. Tau uses several short helical segments to bind in a dynamic, multivalent process to the hydrophobic pocket between subdomains 1 and 3 of actin (Fig. 1C).

Several researches have shown that the plasma membrane modulates the tau protein aggregation. Tau interacts with the plasma membrane with its amino-terminal projection (Brandt, LÉger, & Lee, 1995) which is present in the detergent resistant membrane microdomain (Kawarabayashi et al., 2004). However, the exact mechanism of how tau mediates neurodegeneration remains unclear.

2.4 Amylin

Islet amyloid polypeptide (IAPP) or Amylin is a short peptide hormone (37 amino acids). It is co-secreted with insulin from the β-cells of pancreas in response to hyper blood glucose (Brereton, Vergari, Zhang, & Clark, 2015). Amyloid aggregation of IAPP is implicated in metabolic disorder such as type II diabetes (Haataja, Gurlo, Huang, & Butler, 2008; Westermark, Andersson, & Westermark, 2011). The newly synthesized prepro IAPP is an 89-residue long peptide. The maturation of IAPP is a multi-step process, associated with sequential peptide bond cleavage at both the N- and C-terminal regions of the prepro forms (Furuta et al., 1997). The amino group addition at the C-terminus tyrosine by the peptidyl amidating mono-oxygenase complex (PAM) and disulfide bond formation are said to be crucial steps to achieve final functional IAPP. High resolution NMR structures of hIAPP in SDS micelles at physiological pH was solved which shows the N-terminal helix from C7-V17 is separated from the other helix from N21-S28 by a turn comprising of residues H18-S20 and the C-terminus has a short 310-helix comprising of residues G33-N35 (Nanga, Brender, Vivekanandan, & Ramamoorthy, 2011) (Fig. 1B).

Interestingly, a study with INS-1β-cells showed that the negatively charged lipids of pancreatic islet β-cells promote human IAPP aggregation (Seeliger, Weise, Opitz, & Winter, 2012). When human IAPP is incubated with INS-1β-cells, IAPP crosses the plasma membrane and a significant proportion of the initial concentration of IAPP has been deposited in mitochondria (50%) and endoplasmic reticulum (5%) (Birol, Kumar, Rhoades, & Miranker, 2018).

2.5 Prions

Prions are glycoproteins and mostly localized at the surface of the plasma membrane. The central nervous system is crowded with cellular prion proteins (PrP^c) (Acevedo-Morantes & Wille, 2014). The cellular function of PrP^c is considered pleiotropic as it is likely associated with multifaceted activities such as cell death and proliferation, neurite outgrowth, myelin maintenance, synaptic transmission, β-amyloid toxicity, and several other functions (Linden et al., 2008; Laurén, Gimbel, Nygaard, Gilbert, & Strittmatter, 2009; Leighton, Nadolski, Morrill, Hamilton, & Allison, 2018; Roguski & Gill, 2017; Vassallo & Herms, 2003).

NMR structural study (Wulf, Senatore, & Aguzzi, 2017) revealed that PrP^c has a 23–121 residues long, largely flexible N-terminus whereas the same peptide adopts a highly structured 127–231 amino acids long globular C-terminal domain with three α-helices and two β-sheets (Wulf et al., 2017). These two domains are connected by a conserved hydrophobic core (HC) region (residues 105–126) that make contacts with the biological membrane. PrP^c can be glycosylated at two highly conserved amino acid positions (N180 and N196 in mice) and form diglycosylated, monoglycosylated at N180, monoglycosylated at N196, and unglycosylated PrP^c (Haraguchi et al., 1989).

Prions are attached to the membrane by a glycosyl-phosphatidyl inositol (GPI) anchor. The PrP^c is synthesized on the ER-bound ribosome like other membrane-bound and cellular secretory proteins. The pre PrP^c is modified inside the ER, significantly, the addition of a GPI anchor at the C-terminal segment. The 208 residue long mature PrP^c is targeted to the plasma membrane by Golgi derived systems (Orlean & Menon, 2007). Membrane bound cellular prion proteins are found in intracellular organelles. The two different forms of PrP^c, $^{Ntm}PrP^c$ and $^{Ctm}PrP^c$, were identified inside the ER membrane. These transmembrane forms have identical hydrophobic segments for membrane interactions but spotted in different orientation in the ER membrane (Yost, Lopez, Prusiner, Myers, & Lingappa, 1990).

3. Association of IDPs with biological membranes

3.1 Membrane lipid composition and effects of cholesterol

Phospholipids are the significant constituent of the membranes (Fig. 1E). The properties of the membrane, namely the fluidity, phase, charge, thickness, are solely controlled by the phospholipids (Yoda, Miura, & Takeuchi, 2008). The shape of the phospholipids, in addition their charge, plays an important role in determining the membrane curvature (Renne & De Kroon, 2018). For example, it has been reported that presence of an enhanced level of conical lipids, like phosphatidyl ethanolamine (PE), causes membrane packing defects which promotes membrane curving favorable for α-syn binding (Jo, Fuller, Rand, St George-Hyslop, & Fraser, 2002). Membrane lipids are major players in regulating IDP dynamics and thus membrane interaction has pivotal role in governing the physiological functions, as well as in promoting disease-causing amyloid aggregations of these proteins.

In order to investigate the Aβ dimerization on the membrane interface molecular dynamic simulation study using the model zwitterionic (DPPC) and the anionic (DOPC) membranes have been used (Davis & Berkowitz, 2010). The results of the thermodynamic cycle calculation in a recent study (Sahoo & Matysiak, 2019) has shown stronger peptide-peptide interaction as compared to peptide-lipid interaction occurs for DOPS, which in turn actively promotes dimerization. They have postulated that the effect is due to the formation of neutral Aβ peptide when the reduction in pH closer to the phosphatidylserine (PS) membrane occurs.

On the other hand, compared to the DOPS, peptides on DPPC generally show more parallel structure which further increases the peptide-membrane contacts, facilitating an added superficial and parallel orientation with the N-terminal hydrophobic heads of the embedded membrane bilayer (Sahoo & Matysiak, 2019), whereas, interaction with the PS membrane stabilizes the transient states with a high susceptibility towards aggregation. It has been observed that the aggregation propensity of $A\beta_{40}$ peptide increases rapidly by the negatively charged DMPG phospholipid (Niu et al., 2014). Tau proteins have been shown to favorably interact with the positively charged 1,2-dimyristoyl-3-trimethyl ammonium propane (DMTAP) membrane, mostly through its negatively charged projection N-terminus (Jones et al., 2012).

Cholesterol is the primary constituent of the membranes and is known to affect the Aβ-membrane interactions. It not only interacts with the Aβ monomer and fibrils but also binds to the APP and helps it to insert into the lipid bilayer (Barrett et al., 2012; Beel, Sakakura, Barrett, & Sanders, 2010; Harris, 2008). Furthermore, cholesterol maintains the structure and regulates the dynamics of the membranes and facilitates the Aβ-membrane interactions. Upon increasing the cholesterol content, the binding of Aβ to the membrane may increase as there is a change in the overall chemistry and structure of the bilayer (Kakio, Nishimoto, Yanagisawa, Kozutsumi, & Matsuzaki, 2001). On the other hand, if cholesterol is depleted Aβ nucleation has shown to increase and overall affect the aggregation of Aβ on the surface of the bilayer (Puglielli, Tanzi, & Kovacs, 2003; Qiu et al., 2009; Reiss & Voloshyna, 2012; Wakabayashi & Matsuzaki, 2007; Yu & Zheng, 2012).

Interestingly, IAPP fibril formation and their interactions with the membrane is strictly controlled by the cholesterol in all types of model membranes (Cho, Trikha, & Jeremic, 2009). The membrane-assisted IAPP amyloid formation was reduced significantly at low cholesterol respect to the physiological levels and eliminated the possibility of IAPP-mediated membrane leakage (Zhang, St Clair, London, & Raleigh, 2017). It has been reported that the cholesterol exerts its inhibitory effects on IAPP aggregation by increasing the activation energy of amyloid fibril formation (Cho, Jena, & Jeremic, 2008).

3.2 Regulation of membrane-IDP interactions

Membrane shows a lot of physico-chemical variations which depend on charge content, lipid headgroup chemistry, packing density and curvature (Bernardino De La Serna, Schutz, Eggeling, & Cebecauer, 2016; Sezgin, Levental, Mayor, & Eggeling, 2017). The highly transient and reversible nature of membrane-IDP interactions makes it difficult to capture using existing experimental methods. Different kinds of interactions between IDPs and the membrane are not fully explored yet. Nevertheless, IDP-membrane interactions have been investigated using different biophysical, biochemical and computational tools (Das & Eliezer, 2019; Theillet et al., 2014).

Through evolution, IDPs have got the capability to show a remarkable specificity to various types of membranes. The primary sequence of an IDP and the post translational modifications ensure the membrane specificity. Conformational change in IDP is often coupled with membrane binding (Fig. 2) (Hammes, Chang, & Oas, 2009; Zea et al., 2016). The membrane lipid chaperonin has been shown for human IAPP, Aβ peptide, and α-syn recently (Sciacca et al., 2020; Scollo et al., 2018). Conformational transition of N-terminal membrane-binding domain of α-syn induced by membrane interactions is a typical example of that.

Many membrane-bound helical conformations having amphipathic nature (hydrophobic and polar residues are clustered in two faces of

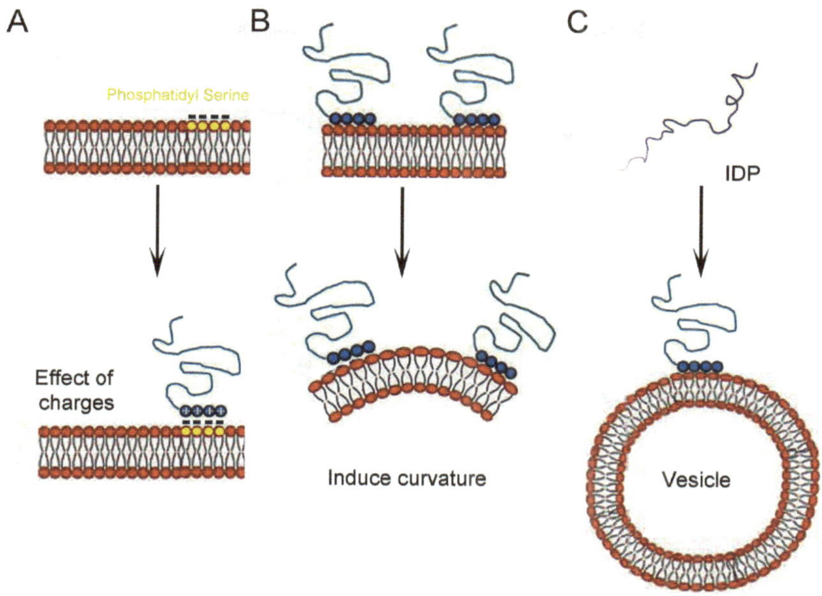

Fig. 2 Mutual perturbations due to IDP-membrane interactions. Negative phospholipids sometimes stabilize IDP-binding to the membrane through electrostatic interactions where charged face of the amphipathic helix formed at IDPs binding region is exposed to the lipid headgroups (A). IDP-binding may remodel the membrane and induce curvature by steric repulsion due to crowding on the membrane, or by lateral pressure created due to sliding of the bound protein over the surface (B). Membrane curvature may favour IDP binding and folding upon binding occurs (C).

the helix) have been reported. Such helices are placed along the apolar surface of the membrane. This event causes deep insertion at level of C3 or C4 acyl chain carbons of membranes (Bussell Jr., Ramlall, & Eliezer, 2005; Jao, Der-Sarkissian, Chen, & Langen, 2004; Jao, Hegde, Chen, Haworth, & Langen, 2008) and interfacial lysine amino acids interact with the negatively charged lipid head-group (Drin & Antonny, 2010; Segrest et al., 1992). Presence of seven imperfect repeats (having a hexapeptide motif (KTKEGV)) in the N-terminal membrane-binding region of alpha-synuclein promotes amphipathic helix formation.

Membrane curvature is another major factor which influence the membrane binding of particularly α-syn (Kjaer, Giehm, Heimburg, & Otzen, 2009; Middleton & Rhoades, 2010; Pranke et al., 2011; Rhoades, Ramlall, Webb, & Eliezer, 2006). Many studies have proposed that α-syn can sense the curvature of the membrane and shows high affinity to highly curved membrane surfaces (Fig. 2) (Jensen et al., 2011; Middleton & Rhoades, 2010). It has been found that the N-terminal amphipathic helix of α-syn has unusual helical twist that leads to formation of a curved helical structure matching well with the curved membrane surface (Bussell et al., 2005; Jao et al., 2004). Thus, α-syn binding specifically to highly curved membranes of synaptic vesicles is favored (Jao et al., 2008).

Mechanistically, extensive membrane curvature induces membrane packing defects (Cui, Lyman, & Voth, 2011; Jensen et al., 2011) which causes transiently exposure of the hydrophobic acyl regions of membrane. These transient exposure of hydrophobic patches play a key role in membrane interaction of α-syn (Jensen et al., 2011; Jo et al., 2002; Kamp & Beyer, 2006; Middleton & Rhoades, 2010; Nuscher et al., 2004; Pranke et al., 2011; Rhoades et al., 2006).

Increased amount of negatively charged lipids in the membrane facilitates electrostatic interaction of positively charged residues of α-syn (Fig. 2) (Davidson et al., 1998; Jo et al., 2002; Middleton & Rhoades, 2010; Nuscher et al., 2004; Perrin, Woods, Clayton, & George, 2000; Ramakrishnan, Jensen, & Marsh, 2003; Rhoades et al., 2006). This is another reason why a synaptic vesicle is an excellent target for α-syn binding; not only it is highly curved, but also it has negatively charged membrane surface (Takamori et al., 2006).

Changes in membrane environment lead to the formation of IDPs amyloid aggregates also. Aβ-membrane interactions are mutual. Not only the insertion or binding of Aβ peptide can affect the property and state of the membrane, but also membrane can regulate the fibrillation of Aβ peptides. It was observed that the electrostatic changes in lipid environment would induce the transformation of secondary structures of $A\beta_{40}$ (Bokvist, Lindström, Watts, & Gröbner, 2004). The overall binding of the Aβ peptide may compress the membrane and further it makes the membrane thinner. These interactions of Aβ oligomers can produce major membrane disruption (Gibson Wood, Eckert, Igbavboa, & Müller, 2003; Matsuzaki, 2007; Mclaurin & Chakrabartty, 1996; Murphy & Levine, 2010; Terzi, Hölzemann, & Seelig, 1995; Wong et al., 2009).

Few previous studies have proposed that the Aβ peptide directly interacts with the phospholipid. However, conclusive result is yet to come. Membrane integrity is mostly distorted by the amyloid fibril formation. Several reports have provided strong evidence that the mechanism of Aβ-membrane interactions is especially related to electrostatic and hydrophobic interactions (Sabaté, Espargaró, Barbosa-Barros, Ventura, & Estelrich, 2012). It has been reported that phosphatidyl ethanolamine (PE) lipid imposes a negative curvature on the membranes and upon decreasing the amount of PE in the bilayer, toxicity in the neuroblastoma cells decreases (Cazzaniga et al., 2011).

Paired helical filaments isolated from the AD brains have been shown to be associated with the membrane glycolipids (Gray, Paula-Barbosa, & Roher, 1987). It has been shown that membrane-association of chimeric tau facilitates the formation of paired helical filaments from the full-length tau (Campos-Peña, Tapia-Ramírez, Sánchez-Torres, & Meraz-Rios, 2009) and using anionic lipid vesicles it was further established that the aggregation of the microtubule-binding domain of tau (K18) at a sub molar concentration (Elbaum-Garfinkle, Ramlall, & Rhoades, 2010).

Furthermore, the interaction of IAPP with the membrane is regulated by the presence of cholesterol and the effect of cholesterol on IAPP polymerization has been investigated both on planar membranes and in solution (Cho et al., 2009). Time-lapse atomic force microscopy studies showed the on the cholesterol depleted plasma, IAPP formed symmetrical tetrameric and pentameric pore-like supramolecular oligomeric structures of intermediate-size. Whereas, on membranes with cholesterol, IAPP formed compact approximately large protein clusters that constituted seeds or nuclei for IAPP binding and aggregation (Cho et al., 2009).

3.3 Impact on membrane permeability

Permeability of the membrane is dependent on the fluidity of the bilayer as the fluidity decreases the membrane becomes substantially less permeable (Wong et al., 2009). Moreover, in some studies it has been reported that the process of amyloid aggregation is finely regulated by the thickness of the hydrophobic group; short-chain lipid 1,2-dilauroyl-sn-glycero-3-phosphatidylcholine (DLPC, 20.9 nm thick) inhibits the $A\beta_{40}$ aggregation compared to the 1,2-dioleoyl-sn-glycero-3-phosphocholine (DOPC, 26.8 nm thick) and 1-palmitoyl-2-oleoyl phosphatidylcholine (POPC, 27.1 nm thick) by a detergent like process (Korshavn et al., 2017). The Aβ peptide is often inserted into the membrane and subsequently forms a pore-like structure. This may stimulate membrane leakage and further leads to cell death.

Formation of ring like oligomer of α-Syn is occurred due to aggregation and it is believed that it is responsible for membrane porosity. A computational study on both wild type and mutant α-Syn (A53T) has shown the rapid penetration of α-Syn in membrane is responsible to change the conformation of α-helix of α-Syn to coiled structure. This conformational change facilitates binding of additional α-Syn monomers. In this study mutant A53T shows faster penetration and faster oligomerization than wild type. This computational work further has been proved using cell based biochemical assay (calcein efflux assay) (Tsigelny et al., 2012b).

The relocalization of mitochondrial membrane bound PrP^c is one of the significant event that promote the release of cytochrome-C and thus increase the mitochondrial permeability that triggers program cell death (Sorice et al., 2012). Human IAPP has multiple adverse effects on living cells including cell membrane permeabilization, endoplasmic reticulum (ER) damage and mitochondrial stress (Gurlo et al., 2010; Huang et al., 2007; Kayed et al., 2004).

4. Implications of IDP-membrane interactions in physiological functions

4.1 Signal transduction

The cellular scaffold proteins have multivalent faces to communicate with a wide variety of protein partners that fixes the downstream signal transmission (Good, Zalatan, & Lim, 2011; Langeberg & Scott, 2015; Morrison & Davis, 2003). The PrP^c interacts with both the intracellular and extracellular proteins and mediates signal transduction events (Budnik, Ruiz-Cañada, & Wendler, 2016; Vana, Zuber, Nikles, & Weiss, 2007). PrP^c can communicate with the neurotransmitter glutamate specific metabotropic receptors family mGluR1 and mGluR5 (Beraldo et al., 2011). The PrP^c is bound to the cell matrix protein laminin and induces phospholipase C regulated calcium signaling in hippocampal neurons that eventually triggers growth and development of the neurons in the same region (Beraldo et al., 2011).

It was reported that the PrP^c mediated pathways have a significant role for the development of AD. In AD, β- and γ-secretases are involved in the conversion of amyloid APP to toxic oligomer of Aβ peptide (AβO) into extracellular space (Hardy & Selkoe, 2002). The PrP^c interacts with the APP and reduces the neuronal damage, induces CNS development as well as maintains synaptic plasticity (Kaiser et al., 2012; Zheng & Koo, 2011). Moreover, Aβ mediates the neurodegenerative events through the post-translational changes of tau protein via transmembrane signaling. This involves the calcineurin mediated the glutamate receptor NMDAR and SIRT1/AMPK pathways (Smith, Pozueta, Gong, Arancio, & Shelanski, 2009; Tackenberg et al., 2013). An interconnected signaling pathway is involved in tau phosphorylation and mitochondrial functions in the neurons.

An active area of research is IDP-driven phase separation at the membrane surface. Increase in local concentration of IDPs at the membrane occurs due to phase separation. This may enhance the signals, selectively regulate signaling components, or even can act as controlling switch in a signaling pathway (reviewed in Banani, Lee, Hyman, & Rosen, 2017; Case, Ditlev, & Rosen, 2019). However, such phenomena have not been reported yet in case of the IDPs discussed here.

4.2 Vesicle trafficking

IDP-lipid bilayer interactions apparently regulate the release of synaptic vesicle's contents that eventually coordinate with systemic neuronal communications. In view of the high expression of α-Syn in neuronal cell, several in vivo functions of α-Syn has been suggested, such as, regulating vesicle transport (Jensen et al., 2011), controlling synaptic vesicle release (Snead & Eliezer, 2014), and even modulating secretory pathways (Wang & Hay, 2015). However, α-Syn is not present in yeasts, worms, and fruit flies, which indicates that it is not generally necessary for vesicle trafficking or synaptic transmission (Abeliovich et al., 2000; Burré et al., 2010; Jensen et al., 2011).

The role of α-Syn in synaptic vesicle trafficking has been explored. α-Syn has an N-terminal membrane binding region and a negatively charged C-terminal domain. N-terminal region of α-Syn was shown to bind to phospholipid bilayers, while C-terminal region interacts with N-terminus of VAMP2 of SNARE complexes (Burré et al., 2010; Lou et al., 2017). Another study has described that α-Syn has a role in assembly of SNARE complexes by cooperation with CSPa (Hou, Wang, Liu, Wang, & Long, 2017).

Effect of tau protein on the vesicle trafficking has been reported where tau inhibits kinesin-dependent transport of peroxisomes and Golgi-derived vesicles into neurites (Stamer, Vogel, Thies, Mandelkow, & Mandelkow, 2002). Particularly, tau inhibits transport of amyloid precursor protein (APP) into axons and dendrites which suggests a linkage between tau and APP trafficking, which may be significant in Alzheimer's disease.

The small guanosine triphosphate proteins, Rab, regulate the free movements of lipids/proteins between intracellular compartments and plasma membrane. They are considered as modulators for vesicular trafficking pathways (Zerial & McBride, 2001). It has been reported that the PrP^c drastically hampers the correct association between the membrane and Rab7 (GTPase) (Shim, Karri, Law, Schatzl, & Gilch, 2016). The prion, PrP^c can make contact directly with the Rab7a inside the cytosol. In addition, PrP^c is recruited by Rab for vesicle trafficking to endocytic compartments (Pereira-Leal & Seabra, 2001; Zerial & McBride, 2001).

5. Pathological implications of IDP-membrane interactions

5.1 Nucleation in oligomer and amyloid fibril formation

While IDP-membrane interaction plays a vital role in cellular signaling, abnormal IDP-membrane interplay leads to amyloid aggregation of IDPs. There are several intrinsic and extrinsic factors that influence the formation of amyloid aggregates. Most important intrinsic factors are protein sequences and dynamic conformational landscape of the amyloidogenic intrinsically unstructured proteins. Also, the pH, temperature, protein concentration and maturation steps have significant effects in amyloid aggregation. It has been recently reported that free membrane lipids interact with the IDPs in aqueous environment and then the same lipid-protein complex is inserted easily with the lipid bilayer, where free lipid molecules play a central role in membrane assisted amyloid fibril formations as well as membrane disruption (Sciacca et al., 2020; Scollo et al., 2018).

The IDPs form amyloid aggregates through a nucleation-dependent or a nucleation-independent process (Frieden, 2007). The nucleation-dependent amyloidosis is executed *via* three successive steps: lag phase, growth phase, and saturation phase (Morris, Watzky, & Finke, 2009). A pre-formed assembly or "nucleus" (kinetically the most unstable species) serves as an aggregation epicenter for other monomers and on-pathway oligomers in most of the amyloid prone proteins. These pre-formed fibrillar seeds impairs the formation of nucleation phase (Arosio, Knowles, & Linse, 2015; O'nuallain, Shivaprasad, Kheterpal, & Wetzel, 2005). In the growth phase, monomers, and oligomers are assembled to form stable protofibrils. The mature amyloid fibril is formed in the saturation phase. On the other hand, the nucleation-independent protein aggregation is considered as linear progressive process having an infinite number of steps to form mature amyloid fibril (Gosal et al., 2005). The main difference between the nucleation-dependent and independent amyloid formation is the absence of a multimeric nucleus in case of later aggregation process as it grows energetically favorable linear fashion (Gosal et al., 2005).

The Aβ amyloid fibrils are assembled through the nucleation-dependent pathway. There are mainly two different phases in the Aβ aggregation process: the nucleation or lag phase and the elongation phase. In the lag phase the nucleus plies up where the Aβ molecules persist in its unstructured monomer/monomeric from. Next in the elongation phase the Aβ accumulates into oligomers, protofibrils and mature fibrils all of which show different morphologies/polymorphism. Meanwhile APP secretes Aβ, the key component of amyloid plaques so Aβ is regarded as the crucial player of the AD (Ross & Poirier, 2004). This pathway is not unidirectional. It has been highlighted that the oligomeric intermediates are more neurotoxic than the mature Aβ amyloid fibrils (Chimon et al., 2007; Haass & Selkoe, 2007; Kayed et al., 2003). Thus, this interaction plays a key role in the formation of amyloid aggregates (Hamley, 2012; Straub & Thirumalai, 2014).

Tau undergoes conformational changes in order to aggregate into fibrils. Its aggregation into β-sheet rich fibrils is a nucleation-controlled polymerization pathway which requires the polyanionic cofactors addition like heparin and anionic micelles. The protein's local concentration is increased by the interaction of tau with the anionic inducers which improves its positive charges (Barghorn & Mandelkow, 2002; Chirita, Necula, & Kuret, 2003; Goedert et al., 1996; King, Gamblin, Kuret, & Binder, 2000). However the effect of the cofactors on the conformational changes is still unclear.

Processing errors assist IAPP linked amyloidosis that leads to the development of type II diabetes (Paulsson & Westermark, 2005; Wang & Raleigh, 2014). The ratio between molar concentration of IAPP and Insulin is 1:100 inside the storage granule. However, that amount of IAPP can promote the formation of toxic IAAP amyloid aggregates efficiently under certain conditions (Hull, Westermark, Westermark, & Kahn, 2004). The low pH value of the storage granule attenuates the IAPP aggregation process in healthy humans (Abedini & Raleigh, 2005; Tu, Serrano, Zanni, & Raleigh, 2014).

The abnormal, aggregation-prone conformation of prion protein is called PrP^{sc} as it can cause scrapie disease in animals (Prusiner, Groth, Bolton, Kent, & Hood, 1984). The transformation of PrP^{c} to PrP^{sc} is associated with progressive degenerative brain diseases. Such β-sheet rich infectious protein agents are responsible for the development and progression of Creutzfeldt–Jakob disease (CJD), Gerstmann–Sträussler–Scheinker disease (GSS), fatal familial insomnia (FFI), chronic wasting disease (CWD), bovine spongiform encephalopathy (BSE), and kuru in humans (Atkinson, Zhang, Munn, Wiegmans, & Wei, 2016; Prusiner et al., 1984).

5.2 Structures of the cytotoxic oligomers and amyloid fibrils

Aggregation of Aβ molecules can form flexible soluble oligomers that can exist in several different forms (Caughey & Lansbury, 2003; Chiti & Dobson, 2006; Mattson, 2004; Selkoe, 2001; Straub & Thirumalai, 2011). Amyloid fibrils are the major form of aggregates in the amyloid plaques. Significant advances have been made on deciphering the structures of amyloid fibrils but very little structural information on intermediate oligomers is available (because of their unstable and heterogeneous nature) despite the fact that they can be highly cytotoxic (Stefani & Dobson, 2003). Annular, pore-like structure formation of α-syn oligomers has been reported (Chen et al., 2015; Lashuel et al., 2002). While amyloid fibrils of Aβ, α-syn, tau share a common cross-β ladder-like core architecture where a steric zipper between the β-sheets stabilizes the core (Fitzpatrick et al., 2017; Li et al., 2018; Paravastu, Leapman, Yau, & Tycko, 2008), a "superpleated" β-sheet structure has been proposed for prion and IAPP fibrils (Kajava, Aebi, & Steven, 2005; Kajava, Baxa, Wickner, & Steven, 2004).

The aggregation prone IAPP monomers associate themselves to form dynamic oligomers with various shapes and sizes. High resolution microscopic data has been shown that the IAPP oligomers are the assembly of 10–20 monomers (Green, Goldsbury, Kistler, Cooper, & Aebi, 2004; Porat, Stepensky, Ding, Naider, & Gazit, 2003). There are few more highlights on the size of IAPP oligomers, one group showed that it has 20-40 monomers whereas another study showed a range of 25–500 monomers (Anguiano, Nowak, & Lansbury, 2002). Powerful spectroscopic and microscopic data revealed that β-sheet rich IAPP oligomers can organize themselves to form relatively stable straight or helical fibril with variable length and width (5–15 nm) (Luca, Yau, Leapman, & Tycko, 2007). The mature amyloid fibrils are connected through inter-chain hydrogen bonds and thus form higher order cross β structure. Two types of atomic models are considered for IAPP fibrils based on solid-state NMR spectroscopy or X-ray crystallographic examinations. The protofibril model proposed the involvement of two symmetric IAPP monomers for a single fibril. The unit monomer has two β-sheet segments, one in residues 8–17 and another spanning the 29–37 residues of IAPP. The side chain of amino acid residues 26–32 of each monomer interacts to form a single IAPP filament. Lateral associations between these unit filaments is prerequisite to form mature IAPP amyloid fibril (Luca et al., 2007). The second model for hIAPP fibrils states that the steric zippers of segment 20–27 of one monomer interact firmly with the 29–33 segment of the second monomer of IAPP. These unit peptide stacks with one another perpendicularly to the fibril's main axis that eventually form the mature fibril (Wiltzius et al., 2008).

5.3 Mechanism of toxicity

The membrane damage caused by the mature amyloid fibrils or by the on-pathway toxic oligomers has remarkable consequence in cell death. The membrane assisted amyloid fibril formation is associated with removal of free lipids from the membrane bilayers (Scalisi et al., 2010; Sciacca et al., 2012). The amyloidogenic IDPs travel from the aqueous environment to the membrane hydrophobic core and initiate global cytotoxic responses.

The extracellular amyloid fibrillar deposits have pathogenic consequence on various organs including brain and insulin secreting cells in the pancreas (Evangelisti et al., 2016; Tipping et al., 2015; Verma, Vats, & Taneja, 2015). In neurodegenerative amyloid disorders, cytotoxic, soluble protein oligomers are populated during the amyloid fibril formation or oligomeric units can be dislodge from the mature amyloid fibrils (Evangelisti et al., 2016).

The molecular basis of oligomer-mediated cytotoxicity is largely unknown. However, the favorable β-sheet structure in amyloid oligomers is not essentially linked with the toxicity of the oligomers. Toxic oligomers likely expose hydrophobic surfaces and perturb membranes exposing the cytoplasm to the extracellular space (Tsigelny et al., 2012a), as a consequence cause cellular dysfunctions (Mannini et al., 2014; Vivoli Vega et al., 2019). Thus, the toxicity confers on the intrinsic misfolded nature of the small aggregation-prone oligomers (Evangelisti et al., 2016). These abnormal structural designs allow the oligomers to interact with a wide range of cellular parts, like proteins, lipid bilayers, nucleic acids and metabolic intermediates (Benilova, Karran, & De Strooper, 2012; Goodchild et al., 2014; Olzscha et al., 2011; Roberts & Brown, 2015). Such unwanted interactions between amyloidogenic oligomers and cellular components drive the cytotoxic effects of the former.

A remarkable NMR-based study has identified two oligomeric species of α-syn recently (Fusco et al., 2017), one of them interacts strongly with the membrane and is capable of disrupting the membrane whereas the other one showing weaker membrane interaction is not. It was suggested that accessibility of lipophilic N-terminal region to interact with the lipid bilayer is a determining step for this protein to perturb cellular membranes.

Interestingly, human IAPP are prone to form amyloid type of aggregates whereas rodent IAPP does not. The residues 20–29 in IAPP are identified as crucial amino acids that can drive the amyloid fibrillation. Human and rodent IAPP differ in 6 amino acid residues in this said peptide region. Three proline substitutions among these 6 amino acids in human make human IAPP more amyloidogenic than the rodent variety. The amino acids positions at 25 (Ala), 28 (Ser) and 29 (Ser) are changed from proline residues in human IAPP through the course of evolution (Ryan, Jobe, & Martin, 2005). These substitutions make β-sheet structure more unstable which eventually promotes the amyloid fibril formation of human IAPP. These amyloid fibrils are considered to play the vital role in pancreatic islet β-cell toxicity and concomitant development of type 2 diabetes (Haataja et al., 2008; Westermark et al., 2011)

It has been confirmed that human IAPP fibrillation proceed through three successive phases: a lag phase (monomers and oligomers exist), an elongation phase (elongation of fibrils), and a saturation phase (mature amyloid fibrils) (Abedini et al., 2016). The lag phase oligomers are proposed to be toxic for pancreatic islets. In addition, human amylin fibrils in the elongation and saturation phases showed considerable cytotoxicity to INS-1β-cell line (Abedini et al., 2016; Butler, Janson, Soeller, & Butler, 2003).

It has been reported that the negative charge on plasma membrane's lipid foster the human IAPP amyloid fibril formation (Knight & Miranker, 2004; Terakawa et al., 2018; Zhang et al., 2017). In physiological pH (pH ~7), the net charge of human IAPP is positive because it contains a significant amount of basic amino acid residues and lacks acidic amino acids. Thus, positively charged N-terminus of human IAPP is inserted easily inside the membrane (Williamson, Loria, & Miranker, 2009). The membrane assists the amyloid nucleation of human IAPP on the surface of the membrane and promotes the assembly of IAPP oligomers as well as amyloid fibrils (Apostolidou, Jayasinghe, & Langen, 2008; Nanga et al., 2011; Williamson et al., 2009). On the other hand, the preformed IAPP amyloid fibrils interact with membrane firmly and induce their damage (Terakawa et al., 2018). Furthermore, human IAPP oligomers are enriched significantly inside the pancreatic islet β-cells of patients with T2D than in normal individuals (Gurlo et al., 2010).

Interestingly, the binding of PrP^c with mGluR5 is crucial for the induction of AβO-assisted neurotoxicity (Haas, Kostylev, & Strittmatter, 2014). In addition, the AβO/PrP^c complex mediates the phosphorylation events in both the GluN2B subunit of NMDARs and tau protein by activated Fyn kinase, leading to synaptic disruption (Larson et al., 2012; Um et al., 2012). The interactions between the N-terminal region of PrP^c (amino acids 93–109) and α-synuclein oligomers are required to activate NMDARs. Moreover, PrP^c–α-synuclein complex disrupts the phospholipase C mediated Ca^{2+} homeostasis and drives the neurodegeneration process (Ferreira et al., 2017).

5.4 Small molecule and peptide based inhibitors

Amyloid aggregation cascade is an attractive therapeutic target. The ubiquitin tagging with the abnormal proteins and subsequent proteasome mediated degradation of the target protein is one of the main cellular defense mechanisms against amyloidosis. Such defense is often bypassed by the cytotoxic amyloid aggregates that make the call for external therapeutic compounds to combat against protein aggregation diseases. A wide variety of amyloid inhibitors has been reported since the last two decades. However, only few have proven their anti-amyloid properties *in vivo*.

Most of the amyloid inhibitors are classified based on their mode of action which includes: prevention of toxic oligomer formation, ability to inhibit fibril growth, interference in membrane-amyloid interactions, and disassembly of preformed amyloid fibrils. In addition, there is another classification system which categorized such inhibitors into three groups based on the final outcome of the aggregation process: First one prevents amyloid oligomer formation, second group of inhibitors are effective against both the oligomers and final amyloid fibrils and the third category inhibitors can attenuate the formation of amyloid fibril only (Necula, Kayed, Milton, & Glabe, 2007).

A large number of natural or synthetic molecules have been reported as anti-amyloid agents (Table 1), such as small peptides (Griffin et al., 2012; Jarmuła & Stępkowski, 2013; Parthsarathy et al., 2013), antibodies (Legleiter et al., 2004), antibiotics (Costa, Speretta, Crowther, & Cardoso, 2011) organic substances (Abe et al., 2013; Alavez, Vantipalli, Zucker, Klang, & Lithgow, 2011; Huong et al., 2010; Mina, Lasagna-Reeves, Glabe, & Kayed, 2009; Yao et al. 2012), nanoparticles (Cabaleiro-Lago, Szczepankiewicz, & Linse, 2012), natural phenolic substances and many others. Small molecules and peptide/peptidomimetics have been successfully designed that specifically target steric zippers of amyloid fibrils (Jiang et al., 2013; Sievers et al., 2011).

There are only few compounds having real therapeutic effectiveness against amyloid aggregation mediated disorders and the number of clinically approved agents is almost nil. However, natural phenolic substances have been shown promising results against amyloidosis and have some potential

Table 1 Name of the naturally occurring small molecules acting as inhibitors and their mode of action on different IDPs.

Name of the compound	Target IDPs	Mode of action
Epigallocatechin-3-gallate	Aβ, tau, α-syn, IAPP (Cao & Raleigh, 2012; Ehrnhoefer et al., 2008; Wobst, Sharma, Diamond, Wanker, & Bieschke, 2015)	Prevention of seeding, attenuate amyloid fibril formation, and disaggregation of preformed aggregates
Curcumin	Aβ, α-syn, PrP (Daval et al., 2010; Ono & Yamada, 2006; Yang et al., 2005)	Inhibition and of disaggregation of amyloid fibrils
Resveratrol	Aβ, IAPP (Jiang, Li, Shea, & Mu, 2011; Ladiwala et al., 2010; Mishra, Sellin, Radovan, Gohlke, & Winter, 2009)	Induce off-pathway aggregates formation, inhibits fibril formation, disaggregation of preformed aggregates
Gallic acid	Aβ, α-syn, insulin, IAPP (Liu et al., 2013; Liu, Carver, Calabrese, & Pukala, 2014)	Inhibition of amyloid fibril formation, disaggregation of preformed fibrils
Kaempferol	Aβ, α-syn (Bhat, Bhat, & Bano, 2015; Ren et al., 2016; Yang et al., 2016)	Inhibition of conformational changes
Oleuropeinaglycon	Aβ, α-syn, IAPP, tau (Daccache et al., 2011; Palazzi et al., 2018; Rigacci et al., 2010, 2011)	Attenuation of amyloid fibril formation
Rosmarinic acid	Aβ, α-sy, IAPP (Hamaguchi, Ono, Murase, & Yamada, 2009; Rashidi, Mirhashemi, Taghizadeh, & Sarkhail, 2013; Takahashi et al., 2015)	Inhibition of amyloid fibril formation, disaggregation of preformed fibrils
Tannic acid	Aβ, α-syn, PrP (Caruana et al., 2011; Kocisko et al., 2005; Ono, Hasegawa, Naiki, & Yamada, 2004)	Antiaggregation, disaggregation of preformed fibrils
Lipoic acid and Ascorbic acid	IAPP (Azzam et al., 2018)	Inhibition of fibrillogenesis

advantage over others. They are natural small molecules commonly found in consumable substances; EGCG in green tea, curcumin in turmeric, and resveratrol in red wine and chocolate. It was reported that such small phenolic compounds can reduce and sometimes prevent the onset of both metabolic disorders (such as T2D) and neurodegenerative diseases (AD and PD) (Féart, Samieri, & Barberger-Gateau, 2010).

6. Outlook

The dynamic interplay between IDPs and membrane apparently is an integral part of many cellular processes (Fakhree, Blum, & Claessens, 2019), like vesicle trafficking, surface interactions, formations of signature domain and phase separation at the membrane (Fig. 3). The modulations occur in both ways: not only IDPs regulate membrane functioning, membrane lipids also assist IDP-binding. Aberrant IDP-membrane interactions, on the other hand, are implicated in many human diseases. Several models have been proposed for membrane disruption by cytotoxic oligomers and amyloid fibrils (Berthelot, Cullin, & Lecomte, 2013) (Fig. 3).

Due to the diverse and complex nature of membrane composition (Bernardino De La Serna et al., 2016) it is extremely challenging to mimic the cell membrane *in vitro*. Also, ensemble structures of IDPs and transient nature of IDP-membrane interactions make the studies difficult using any single, traditional experimental method. Despite these challenges, it has been established that membrane interactions play a pivotal role in functional regulation of IDPs as well as their abnormal amyloidogenic behavior. We have discussed here findings on membrane-protein cross-talks of some of the IDPs associated with neurodegenerative diseases and metabolic disorders. Further research using integrative approaches combining newer technologies is needed to illustrate specific mechanisms undertaken by individual proteins during such interactions.

The disease relation of IDP-membrane interactions has placed the event under spotlight for drug discovery research. A more holistic understanding of the mechanistic details of amyloid formation pathways and the role of collective properties of lipids, as well as a closer look into the intermediate oligomeric species (considered as the most attractive therapeutic targets), would provide important clues for designing therapeutics. Quantitative estimation of membrane-IDP interactions would render the possible routes for drug targeting against amyloidogenic IDPs in the metabolic and neurodegenerative disorders.

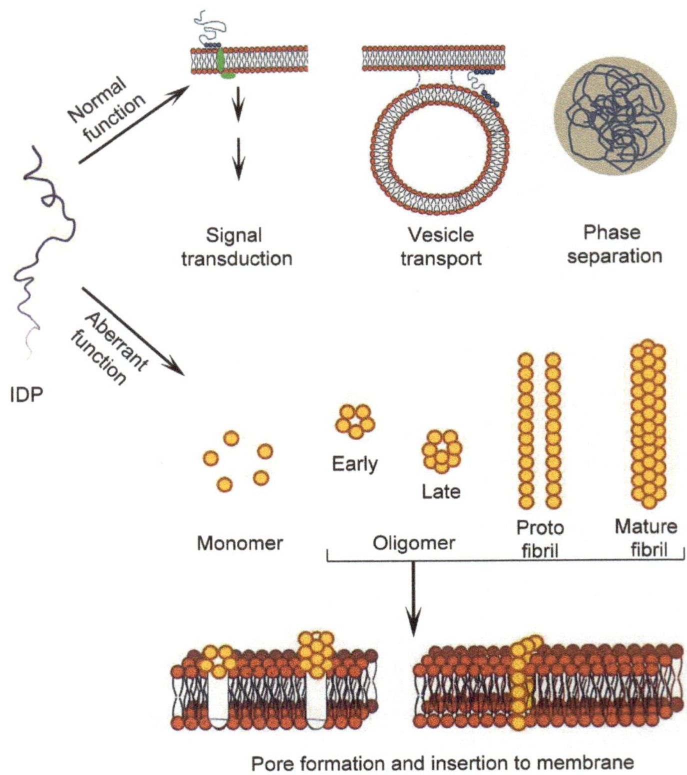

Fig. 3 Schematic display of the implications of IDP-membrane interactions. Membrane-associated cellular functions of IDPs, as well as abnormal fibrillation pathways of IDPs and consequent membrane damage are schematically depicted.

Acknowledgments

We acknowledge financial supports from SERB, DST (India) sponsored projects (SB/SO/BB-0025/2014 and CRG/2019/001788), and CSIR-Indian Institute of Chemical Biology, Kolkata, India. BKP, SD and SM thank DBT-RA, DST (SERB) (CRG/2019/001788) and DST-Inspire, India, respectively, for research fellowships. Financial support from the DBT-RA Program in Biotechnology and Life Sciences is gratefully acknowledged.

Conflict of interest statement

The authors declare no conflict of interest.

Author contributions

All authors searched the literature, contributed to data interpretation, preparation of the manuscript, and approved it for publication.

References

Abe, M., Abe, Y., Ohkuri, T., Mishima, T., Monji, A., Kanba, S., et al. (2013). Mechanism for retardation of amyloid fibril formation by sugars in Vλ6 protein. *Protein Science, 22*, 467–474.

Abedini, A., Plesner, A., Cao, P., Ridgway, Z., Zhang, J., Tu, L. H., et al. (2016). Time-resolved studies define the nature of toxic IAPP intermediates, providing insight for anti-amyloidosis therapeutics. *eLife, 5*, e12977. https://doi.org/10.7554/eLife.12977. PMID: 27213520.

Abedini, A., & Raleigh, D. P. (2005). The role of His-18 in amyloid formation by human islet amyloid polypeptide. *Biochemistry, 44*, 16284–16291.

Abedini, A., & Raleigh, D. P. (2009). A critical assessment of the role of helical intermediates in amyloid formation by natively unfolded proteins and polypeptides. *Protein Engineering, Design & Selection: PEDS, 22*, 453–459.

Abeliovich, A., Schmitz, Y., Fariñas, I., Choi-Lundberg, D., Ho, W. H., Castillo, P. E., et al. (2000). Mice lacking alpha-synuclein display functional deficits in the nigrostriatal dopamine system. *Neuron, 25*, 239–252.

Acevedo-Morantes, C. Y., & Wille, H. (2014). The structure of human prions: From biology to structural models—Considerations and pitfalls. *Viruses, 6*.

Alavez, S., Vantipalli, M. C., Zucker, D. J., Klang, I. M., & Lithgow, G. J. (2011). Amyloid-binding compounds maintain protein homeostasis during ageing and extend lifespan. *Nature, 472*, 226–229.

Anguiano, M., Nowak, R. J., & Lansbury, P. T. (2002). Protofibrillar islet amyloid polypeptide permeabilizes synthetic vesicles by a pore-like mechanism that may be relevant to type II diabetes. *Biochemistry, 41*, 11338–11343.

Apostolidou, M., Jayasinghe, S. A., & Langen, R. (2008). Structure of alpha-helical membrane-bound human islet amyloid polypeptide and its implications for membrane-mediated misfolding. *The Journal of Biological Chemistry, 283*, 17205–17210.

Arosio, P., Knowles, T. P., & Linse, S. (2015). On the lag phase in amyloid fibril formation. *Physical Chemistry Chemical Physics, 17*, 7606–7618.

Atkinson, C. J., Zhang, K., Munn, A. L., Wiegmans, A., & Wei, M. Q. (2016). Prion protein scrapie and the normal cellular prion protein. *Prion, 10*, 63–82.

Azzam, S. K., Jang, H., Choi, M. C., Alsafar, H., Lukman, S., & Lee, S. (2018). Inhibition of human amylin aggregation and cellular toxicity by lipoic acid and ascorbic acid. *Molecular Pharmaceutics, 15*, 2098–2106.

Banani, S. F., Lee, H. O., Hyman, A. A., & Rosen, M. K. (2017). Biomolecular condensates: Organizers of cellular biochemistry. *Nature Reviews. Molecular Cell Biology, 18*, 285–298.

Barghorn, S., & Mandelkow, E. (2002). Toward a unified scheme for the aggregation of tau into Alzheimer paired helical filaments. *Biochemistry, 41*, 14885–14896.

Barrett, P. J., Song, Y., Van Horn, W. D., Hustedt, E. J., Schafer, J. M., Hadziselimovic, A., et al. (2012). The amyloid precursor protein has a flexible transmembrane domain and binds cholesterol. *Science, 336*, 1168–1171.

Beel, A. J., Sakakura, M., Barrett, P. J., & Sanders, C. R. (2010). Direct binding of cholesterol to the amyloid precursor protein: An important interaction in lipid-Alzheimer's disease relationships? *Biochimica et Biophysica Acta, 1801*, 975–982.

Bendor, J. T., Logan, T. P., & Edwards, R. H. (2013). The function of alpha-synuclein. *Neuron, 79*, 1044–1066.

Benilova, I., Karran, E., & De Strooper, B. (2012). The toxic Aβ oligomer and Alzheimer's disease: An emperor in need of clothes. *Nature Neuroscience, 15*, 349–357.

Beraldo, F. H., Arantes, C. P., Santos, T. G., Machado, C. F., Roffe, M., Hajj, G. N., et al. (2011). Metabotropic glutamate receptors transduce signals for neurite outgrowth after binding of the prion protein to laminin gamma1 chain. *The FASEB Journal, 25*, 265–279.

Bernardino De La Serna, J., Schutz, G. J., Eggeling, C., & Cebecauer, M. (2016). There is no simple model of the plasma membrane organization. *Frontiers in Cell and Development Biology*, *4*, 106.

Berthelot, K., Cullin, C., & Lecomte, S. (2013). What does make an amyloid toxic: Morphology, structure or interaction with membrane? *Biochimie*, *95*, 12–19.

Bhat, W. F., Bhat, S. A., & Bano, B. (2015). Evaluation of polyphenols as possible therapeutics for amyloidoses: Comparative analysis of kaempferol and catechin. *International Journal of Biological Macromolecules*, *81*, 60–68.

Birol, M., Kumar, S., Rhoades, E., & Miranker, A. D. (2018). Conformational switching within dynamic oligomers underpins toxic gain-of-function by diabetes-associated amyloid. *Nature Communications*, *9*, 1312.

Bokvist, M., Lindström, F., Watts, A., & Gröbner, G. (2004). Two types of Alzheimer's beta-amyloid (1-40) peptide membrane interactions: Aggregation preventing transmembrane anchoring versus accelerated surface fibril formation. *Journal of Molecular Biology*, *335*, 1039–1049.

Brandt, R., Léger, J., & Lee, G. (1995). Interaction of tau with the neural plasma membrane mediated by tau's amino-terminal projection domain. *Journal of Cell Biology*, *131*, 1327–1340.

Brangwynne, C. P., Tompa, P., & Pappu, R. V. (2015). Polymer physics of intracellular phase transitions. *Nature Physics*, *11*, 899–904.

Brereton, M. F., Vergari, E., Zhang, Q., & Clark, A. (2015). Alpha-, delta- and Pp-cells: Are they the architectural cornerstones of islet structure and co-ordination? *The Journal of Histochemistry and Cytochemistry*, *63*, 575–591.

Budnik, V., Ruiz-Cañada, C., & Wendler, F. (2016). Extracellular vesicles round off communication in the nervous system. *Nature Reviews. Neuroscience*, *17*, 160–172.

Buée, L., Bussière, T., Buée-Scherrer, V., Delacourte, A., & Hof, P. R. (2000). Tau protein isoforms, phosphorylation and role in neurodegenerative disorders11These authors contributed equally to this work. *Brain Research Reviews*, *33*, 95–130.

Burré, J., Sharma, M., Tsetsenis, T., Buchman, V., Etherton, M. R., & SÜdhof, T. C. (2010). α-Synuclein promotes SNARE-complex assembly in vivo and in vitro. *Science*, *329*, 1663–1667.

Bussell, R., Jr., Ramlall, T. F., & Eliezer, D. (2005). Helix periodicity, topology, and dynamics of membrane-associated alpha-synuclein. *Protein Science*, *14*, 862–872.

Butler, A. E., Janson, J., Soeller, W. C., & Butler, P. C. (2003). Increased beta-cell apoptosis prevents adaptive increase in beta-cell mass in mouse model of type 2 diabetes: Evidence for role of islet amyloid formation rather than direct action of amyloid. *Diabetes*, *52*, 2304–2314.

Cabaleiro-Lago, C., Szczepankiewicz, O., & Linse, S. (2012). The effect of nanoparticles on amyloid aggregation depends on the protein stability and intrinsic aggregation rate. *Langmuir*, *28*, 1852–1857.

Cabrales Fontela, Y., Kadavath, H., Biernat, J., Riedel, D., Mandelkow, E., & Zweckstetter, M. (2017). Multivalent cross-linking of actin filaments and microtubules through the microtubule-associated protein Tau. *Nature Communications*, *8*, 1981.

Campos-Peña, V., Tapia-Ramírez, J., Sánchez-Torres, C., & Meraz-Rios, M. A. (2009). Pathological-like assembly of tau induced by a paired helical filament core expressed at the plasma membrane. *Journal of Alzheimer's Disease*, *18*, 919–933.

Cao, P., & Raleigh, D. P. (2012). Analysis of the inhibition and remodeling of islet amyloid polypeptide amyloid fibers by flavanols. *Biochemistry*, *51*, 2670–2683.

Caruana, M., Hogen, T., Levin, J., Hillmer, A., Giese, A., & Vassallo, N. (2011). Inhibition and disaggregation of alpha-synuclein oligomers by natural polyphenolic compounds. *FEBS Letters*, *585*, 1113–1120.

Case, L. B., Ditlev, J. A., & Rosen, M. K. (2019). Regulation of transmembrane signaling by phase separation. *Annual Review of Biophysics*, *48*, 465–494.

Caughey, B., & Lansbury, P. T. (2003). Protofibrils, pores, fibrils, and neurodegeneration: Separating the responsible protein aggregates from the innocent bystanders. *Annual Review of Neuroscience*, *26*, 267–298.

Cazzaniga, E., Bulbarelli, A., Lonati, E., Orlando, A., Re, F., Gregori, M., et al. (2011). Abeta peptide toxicity is reduced after treatments decreasing phosphatidylethanolamine content in differentiated neuroblastoma cells. *Neurochemical Research*, *36*, 863–869.

Chen, S. W., Drakulic, S., Deas, E., Ouberai, M., Aprile, F. A., Arranz, R., et al. (2015). Structural characterization of toxic oligomers that are kinetically trapped during alpha-synuclein fibril formation. *Proceedings of the National Academy of Sciences of the United States of America*, *112*, E1994–E2003.

Chimon, S., Shaibat, M. A., Jones, C. R., Calero, D. C., Aizezi, B., & Ishii, Y. (2007). Evidence of fibril-like β-sheet structures in a neurotoxic amyloid intermediate of Alzheimer's β-amyloid. *Nature Structural & Molecular Biology*, *14*, 1157–1164.

Chirita, C. N., Necula, M., & Kuret, J. (2003). Anionic micelles and vesicles induce tau fibrillization in vitro. *The Journal of Biological Chemistry*, *278*, 25644–25650.

Chiti, F., & Dobson, C. M. (2006). Protein misfolding, functional amyloid, and human disease. *Annual Review of Biochemistry*, *75*, 333–366.

Cho, W., Jena, B. P., & Jeremic, A. M. (2008). Chapter 13 nano-scale imaging and dynamics of amylin-membrane interactions and its implication in type II diabetes mellitus. *Methods in Cell Biology*, *90*, 267–286.

Cho, W. J., Trikha, S., & Jeremic, A. M. (2009). Cholesterol regulates assembly of human islet amyloid polypeptide on model membranes. *Journal of Molecular Biology*, *393*, 765–775.

Costa, R., Speretta, E., Crowther, D. C., & Cardoso, I. (2011). Testing the therapeutic potential of doxycycline in a Drosophila melanogaster model of Alzheimer disease. *The Journal of Biological Chemistry*, *286*, 41647–41655.

Crescenzi, O., Tomaselli, S., Guerrini, R., Salvadori, S., D'ursi, A. M., Temussi, P. A., et al. (2002). Solution structure of the Alzheimer amyloid beta-peptide (1-42) in an apolar microenvironment. Similarity with a virus fusion domain. *European Journal of Biochemistry*, *269*, 5642–5648.

Cui, H., Lyman, E., & Voth, G. A. (2011). Mechanism of membrane curvature sensing by amphipathic helix containing proteins. *Biophysical Journal*, *100*, 1271–1279.

Daccache, A., Lion, C., Sibille, N., Gerard, M., Slomianny, C., Lippens, G., et al. (2011). Oleuropein and derivatives from olives as Tau aggregation inhibitors. *Neurochemistry International*, *58*, 700–707.

Das, T., & Eliezer, D. (2019). Membrane interactions of intrinsically disordered proteins: The example of alpha-synuclein. *Biochimica et Biophysica Acta, Proteins and Proteomics*, *1867*, 879–889.

Daval, M., Bedrood, S., Gurlo, T., Huang, C. J., Costes, S., Butler, P. C., et al. (2010). The effect of curcumin on human islet amyloid polypeptide misfolding and toxicity. *Amyloid*, *17*, 118–128.

Davidson, W. S., Jonas, A., Clayton, D. F., & George, J. M. (1998). Stabilization of alpha-synuclein secondary structure upon binding to synthetic membranes. *The Journal of Biological Chemistry*, *273*, 9443–9449.

Davis, C. H., & Berkowitz, M. L. (2010). A molecular dynamics study of the early stages of amyloid-beta(1-42) oligomerization: The role of lipid membranes. *Proteins*, *78*, 2533–2545.

Drin, G., & Antonny, B. (2010). Amphipathic helices and membrane curvature. *FEBS Letters*, *584*, 1840–1847.

Dunker, A. K., Cortese, M. S., Romero, P., Iakoucheva, L. M., & Uversky, V. N. (2005). Flexible nets. *The FEBS Journal*, *272*, 5129–5148.

Ehrnhoefer, D. E., Bieschke, J., Boeddrich, A., Herbst, M., Masino, L., Lurz, R., et al. (2008). EGCG redirects amyloidogenic polypeptides into unstructured, off-pathway oligomers. *Nature Structural & Molecular Biology, 15*, 558–566.

Elbaum-Garfinkle, S., Ramlall, T., & Rhoades, E. (2010). The role of the lipid bilayer in tau aggregation. *Biophysical Journal, 98*, 2722–2730.

Eliezer, D., Kutluay, E., Bussell, R., Jr., & Browne, G. (2001). Conformational properties of alpha-synuclein in its free and lipid-associated states. *Journal of Molecular Biology, 307*, 1061–1073.

Evangelisti, E., Cascella, R., Becatti, M., Marrazza, G., Dobson, C. M., Chiti, F., et al. (2016). Binding affinity of amyloid oligomers to cellular membranes is a generic indicator of cellular dysfunction in protein misfolding diseases. *Scientific Reports, 6*, 32721.

Fakhree, M. A. A., Blum, C., & Claessens, M. (2019). Shaping membranes with disordered proteins. *Archives of Biochemistry and Biophysics, 677*, 108163.

FÉart, C., Samieri, C., & Barberger-Gateau, P. (2010). Mediterranean diet and cognitive function in older adults. *Current Opinion in Clinical Nutrition and Metabolic Care, 13*, 14–18.

Ferreira, D. G., Temido-Ferreira, M., Vicente Miranda, H., Batalha, V. L., Coelho, J. E., Szegö, É. M., et al. (2017). α-Synuclein interacts with PrP(C) to induce cognitive impairment through mGluR5 and Nmdar2B. *Nature Neuroscience, 20*, 1569–1579.

Fitzpatrick, A. W. P., Falcon, B., He, S., Murzin, A. G., Murshudov, G., Garringer, H. J., et al. (2017). Cryo-EM structures of tau filaments from Alzheimer's disease. *Nature, 547*, 185–190.

Frieden, C. (2007). Protein aggregation processes: In search of the mechanism. *Protein Science, 16*, 2334–2344.

Furuta, M., Yano, H., Zhou, A., Rouillé, Y., Holst, J. J., Carroll, R., et al. (1997). Defective prohormone processing and altered pancreatic islet morphology in mice lacking active SPC2. *Proceedings of the National Academy of Sciences of the United States of America, 94*, 6646–6651.

Fusco, G., Chen, S. W., Williamson, P. T. F., Cascella, R., Perni, M., Jarvis, J. A., et al. (2017). Structural basis of membrane disruption and cellular toxicity by alpha-synuclein oligomers. *Science, 358*, 1440–1443.

Gast, K., Damaschun, H., Eckert, K., Schulze-Forster, K., Maurer, H. R., Mueller-Frohne, M., et al. (1995). Prothymosin alpha: A biologically active protein with random coil conformation. *Biochemistry, 34*, 13211–13218.

Gibson Wood, W., Eckert, G. P., Igbavboa, U., & Müller, W. E. (2003). Amyloid beta-protein interactions with membranes and cholesterol: Causes or casualties of Alzheimer's disease. *Biochimica et Biophysica Acta, 1610*, 281–290.

Goedert, M., Jakes, R., Spillantini, M. G., Hasegawa, M., Smith, M. J., & Crowther, R. A. (1996). Assembly of microtubule-associated protein tau into Alzheimer-like filaments induced by sulphated glycosaminoglycans. *Nature, 383*, 550–553.

Good, M. C., Zalatan, J. G., & Lim, W. A. (2011). Scaffold proteins: Hubs for controlling the flow of cellular information. *Science, 332*, 680–686.

Goodchild, S. C., Sheynis, T., Thompson, R., Tipping, K. W., Xue, W. F., Ranson, N. A., et al. (2014). β2-Microglobulin amyloid fibril-induced membrane disruption is enhanced by endosomal lipids and acidic pH. *PLoS One, 9*, e104492.

Gosal, W. S., Morten, I. J., Hewitt, E. W., Smith, D. A., Thomson, N. H., & Radford, S. E. (2005). Competing pathways determine fibril morphology in the self-assembly of beta2-microglobulin into amyloid. *Journal of Molecular Biology, 351*, 850–864.

Gould, S. B. (2018). Membranes and evolution. *Current Biology: CB, 28*, R381–R385.

Grant, M. A., Lazo, N. D., Lomakin, A., Condron, M. M., Arai, H., Yamin, G., et al. (2007). Familial Alzheimer's disease mutations alter the stability of the amyloid β-protein

monomer folding nucleus. *Proceedings of the National Academy of Sciences of the United States of America, 104*, 16522–16527.
Gray, E. G., Paula-Barbosa, M., & Roher, A. (1987). Alzheimer's disease: Paired helical filaments and cytomembranes. *Neuropathology and Applied Neurobiology, 13*, 91–110.
Grecco, H. E., Schmick, M., & Bastiaens, P. I. (2011). Signaling from the living plasma membrane. *Cell, 144*, 897–909.
Green, J. D., Goldsbury, C., Kistler, J., Cooper, G. J., & Aebi, U. (2004). Human amylin oligomer growth and fibril elongation define two distinct phases in amyloid formation. *The Journal of Biological Chemistry, 279*, 12206–12212.
Griffin, M. D., Yeung, L., Hung, A., Todorova, N., Mok, Y. F., Karas, J. A., et al. (2012). A cyclic peptide inhibitor of apoC-II peptide fibril formation: Mechanistic insight from NMR and molecular dynamics analysis. *Journal of Molecular Biology, 416*, 642–655.
Gurlo, T., Ryazantsev, S., Huang, C. J., Yeh, M. W., Reber, H. A., Hines, O. J., et al. (2010). Evidence for proteotoxicity in beta cells in type 2 diabetes: Toxic islet amyloid polypeptide oligomers form intracellularly in the secretory pathway. *The American Journal of Pathology, 176*, 861–869.
Haas, L. T., Kostylev, M. A., & Strittmatter, S. M. (2014). Therapeutic molecules and endogenous ligands regulate the interaction between brain cellular prion protein (Prpc) and metabotropic glutamate receptor 5 (mGluR5). *The Journal of Biological Chemistry, 289*, 28460–28477.
Haass, C., & Selkoe, D. J. (2007). Soluble protein oligomers in neurodegeneration: Lessons from the Alzheimer's amyloid β-peptide. *Nature Reviews Molecular Cell Biology, 8*, 101–112.
Haataja, L., Gurlo, T., Huang, C. J., & Butler, P. C. (2008). Islet amyloid in type 2 diabetes, and the toxic oligomer hypothesis. *Endocrine Reviews, 29*, 303–316.
Hamaguchi, T., Ono, K., Murase, A., & Yamada, M. (2009). Phenolic compounds prevent Alzheimer's pathology through different effects on the amyloid-beta aggregation pathway. *The American Journal of Pathology, 175*, 2557–2565.
Hamley, I. W. (2012). The amyloid beta peptide: A chemist's perspective. Role in Alzheimer's and Fibrillization. *Chemical Reviews, 112*, 5147–5192.
Hammes, G. G., Chang, Y. C., & Oas, T. G. (2009). Conformational selection or induced fit: A flux description of reaction mechanism. *Proceedings of the National Academy of Sciences of the United States of America, 106*, 13737–13741.
Haraguchi, T., Fisher, S., Olofsson, S., Endo, T., Groth, D., Tarentino, A., et al. (1989). Asparagine-linked glycosylation of the scrapie and cellular prion proteins. *Archives of Biochemistry and Biophysics, 274*, 1–13.
Hardy, J., & Selkoe, D. J. (2002). The amyloid hypothesis of Alzheimer's disease: Progress and problems on the road to therapeutics. *Science, 297*, 353–356.
Harris, J. R. (2008). Cholesterol binding to amyloid-beta fibrils: a TEM study. *Micron, 39*, 1192–1196.
Hou, L., Shao, H., Zhang, Y., Li, H., Menon, N. K., Neuhaus, E. B., et al. (2004). Solution NMR studies of the A beta(1-40) and A beta(1-42) peptides establish that the Met35 oxidation state affects the mechanism of amyloid formation. *Journal of the American Chemical Society, 126*, 1992–2005.
Hou, C., Wang, Y., Liu, J., Wang, C., & Long, J. (2017). Neurodegenerative disease related proteins have negative effects on snare-mediated membrane fusion in pathological confirmation. *Frontiers in Molecular Neuroscience, 10*, 66.
Huang, C. J., Lin, C. Y., Haataja, L., Gurlo, T., Butler, A. E., Rizza, R. A., et al. (2007). High expression rates of human islet amyloid polypeptide induce endoplasmic reticulum stress mediated beta-cell apoptosis, a characteristic of humans with type 2 but not type 1 diabetes. *Diabetes, 56*, 2016–2027.

Hull, R. L., Westermark, G. T., Westermark, P., & Kahn, S. E. (2004). Islet amyloid: A critical entity in the pathogenesis of type 2 diabetes. *The Journal of Clinical Endocrinology & Metabolism*, *89*, 3629–3643.

Huong, V. T., Shimanouchi, T., Shimauchi, N., Yagi, H., Umakoshi, H., Goto, Y., et al. (2010). Catechol derivatives inhibit the fibril formation of amyloid-beta peptides. *Journal of Bioscience and Bioengineering*, *109*, 629–634.

Isaacson, R. L., Weeds, A. G., & Fersht, A. R. (1999). Equilibria and kinetics of folding of gelsolin domain 2 and mutants involved in familial amyloidosis-Finnish type. *Proceedings of the National Academy of Sciences of the United States of America*, *96*, 11247–11252.

Iwai, A., Masliah, E., Yoshimoto, M., Ge, N., Flanagan, L., De Silva, H. A., et al. (1995). The precursor protein of non-A beta component of Alzheimer's disease amyloid is a presynaptic protein of the central nervous system. *Neuron*, *14*, 467–475.

Jao, C. C., Der-Sarkissian, A., Chen, J., & Langen, R. (2004). Structure of membrane-bound alpha-synuclein studied by site-directed spin labeling. *Proceedings of the National Academy of Sciences of the United States of America*, *101*, 8331–8336.

Jao, C. C., Hegde, B. G., Chen, J., Haworth, I. S., & Langen, R. (2008). Structure of membrane-bound alpha-synuclein from site-directed spin labeling and computational refinement. *Proceedings of the National Academy of Sciences of the United States of America*, *105*, 19666–19671.

JarmułA, A., & Stępkowski, D. (2013). The β-sheet breakers and π-stacking. *Journal of Peptide Science*, *19*, 345–349.

Jeganathan, S., Von Bergen, M., Mandelkow, E. M., & Mandelkow, E. (2008). The natively unfolded character of tau and its aggregation to Alzheimer-like paired helical filaments. *Biochemistry*, *47*, 10526–10539.

Jensen, M. B., Bhatia, V. K., Jao, C. C., Rasmussen, J. E., Pedersen, S. L., Jensen, K. J., et al. (2011). Membrane curvature sensing by amphipathic helices: A single liposome study using α-synuclein and annexin B12. *The Journal of Biological Chemistry*, *286*, 42603–42614.

Jesus, C. S. H., Almeida, Z. L., Vaz, D. C., Faria, T. Q., & Brito, R. M. M. (2016). A new folding kinetic mechanism for human transthyretin and the influence of the amyloidogenic V30M mutation. *International Journal of Molecular Sciences*, *17*, 1428.

Jiang, P., Li, W., Shea, J. E., & Mu, Y. (2011). Resveratrol inhibits the formation of multiple-layered beta-sheet oligomers of the human islet amyloid polypeptide segment 22-27. *Biophysical Journal*, *100*, 1550–1558.

Jiang, L., Liu, C., Leibly, D., Landau, M., Zhao, M., Hughes, M. P., et al. (2013). Structure-based discovery of fiber-binding compounds that reduce the cytotoxicity of amyloid beta. *eLife*, *2*, e00857.

Jo, E., Fuller, N., Rand, R. P., St George-Hyslop, P., & Fraser, P. E. (2002). Defective membrane interactions of familial Parkinson's disease mutant A30P alpha-synuclein. *Journal of Molecular Biology*, *315*, 799–807.

Jones, E. M., Dubey, M., Camp, P. J., Vernon, B. C., Biernat, J., Mandelkow, E., et al. (2012). Interaction of tau protein with model lipid membranes induces tau structural compaction and membrane disruption. *Biochemistry*, *51*, 2539–2550.

Kaiser, D. M., Acharya, M., Leighton, P. L., Wang, H., Daude, N., Wohlgemuth, S., et al. (2012). Amyloid beta precursor protein and prion protein have a conserved interaction affecting cell adhesion and Cns development. *PLoS One*, *7*, e51305.

Kajava, A. V., Aebi, U., & Steven, A. C. (2005). The parallel superpleated beta-structure as a model for amyloid fibrils of human amylin. *Journal of Molecular Biology*, *348*, 247–252.

Kajava, A. V., Baxa, U., Wickner, R. B., & Steven, A. C. (2004). A model for Ure2p prion filaments and other amyloids: the parallel superpleated beta-structure. *Proceedings of the National Academy of Sciences of the United States of America*, *101*, 7885–7890.

Kakio, A., Nishimoto, S. I., Yanagisawa, K., Kozutsumi, Y., & Matsuzaki, K. (2001). Cholesterol-dependent formation of Gm1 ganglioside-bound amyloid beta-protein, an endogenous seed for Alzheimer amyloid. *The Journal of Biological Chemistry, 276*, 24985–24990.

Kamp, F., & Beyer, K. (2006). Binding of alpha-synuclein affects the lipid packing in bilayers of small vesicles. *The Journal of Biological Chemistry, 281*, 9251–9259.

Kawarabayashi, T., Shoji, M., Younkin, L. H., Wen-Lang, L., Dickson, D. W., Murakami, T., et al. (2004). Dimeric amyloid beta protein rapidly accumulates in lipid rafts followed by apolipoprotein E and phosphorylated tau accumulation in the Tg2576 mouse model of Alzheimer's disease. *The Journal of Neuroscience, 24*, 3801–3809.

Kayed, R., Head, E., Thompson, J. L., McIntire, T. M., Milton, S. C., Cotman, C. W., et al. (2003). Common structure of soluble amyloid oligomers implies common mechanism of pathogenesis. *Science, 300*, 486–489.

Kayed, R., Sokolov, Y., Edmonds, B., McIntire, T. M., Milton, S. C., Hall, J. E., et al. (2004). Permeabilization of lipid bilayers is a common conformation-dependent activity of soluble amyloid oligomers in protein misfolding diseases. *The Journal of Biological Chemistry, 279*, 46363–46366.

King, M. E., Gamblin, T. C., Kuret, J., & Binder, L. I. (2000). Differential assembly of human tau isoforms in the presence of arachidonic acid. *Journal of Neurochemistry, 74*, 1749–1757.

Kjaer, L., Giehm, L., Heimburg, T., & Otzen, D. (2009). The influence of vesicle size and composition on alpha-synuclein structure and stability. *Biophysical Journal, 96*, 2857–2870.

Knight, J. D., & Miranker, A. D. (2004). Phospholipid catalysis of diabetic amyloid assembly. *Journal of Molecular Biology, 341*, 1175–1187.

Kocisko, D. A., Engel, A. L., Harbuck, K., Arnold, K. M., Olsen, E. A., Raymond, L. D., et al. (2005). Comparison of protease-resistant prion protein inhibitors in cell cultures infected with two strains of mouse and sheep scrapie. *Neuroscience Letters, 388*, 106–111.

Korshavn, K. J., Satriano, C., Lin, Y., Zhang, R., Dulchavsky, M., Bhunia, A., et al. (2017). Reduced lipid bilayer thickness regulates the aggregation and cytotoxicity of amyloid-β. *The Journal of Biological Chemistry, 292*, 4638–4650.

Ladiwala, A. R., Lin, J. C., Bale, S. S., Marcelino-Cruz, A. M., Bhattacharya, M., Dordick, J. S., et al. (2010). Resveratrol selectively remodels soluble oligomers and fibrils of amyloid abeta into off-pathway conformers. *The Journal of Biological Chemistry, 285*, 24228–24237.

Langeberg, L. K., & Scott, J. D. (2015). Signalling scaffolds and local organization of cellular behaviour. *Nature Reviews. Molecular Cell Biology, 16*, 232–244.

Larson, M., Sherman, M. A., Amar, F., Nuvolone, M., Schneider, J. A., Bennett, D. A., et al. (2012). The complex PrP(c)-Fyn couples human oligomeric Aβ with pathological tau changes in Alzheimer's disease. *The Journal of Neuroscience, 32* 16857-71a.

Lashuel, H. A., Petre, B. M., Wall, J., Simon, M., Nowak, R. J., Walz, T., et al. (2002). Alpha-synuclein, especially the Parkinson's disease-associated mutants, forms pore-like annular and tubular protofibrils. *Journal of Molecular Biology, 322*, 1089–1102.

Laurén, J., Gimbel, D. A., Nygaard, H. B., Gilbert, J. W., & Strittmatter, S. M. (2009). Cellular prion protein mediates impairment of synaptic plasticity by amyloid-β oligomers. *Nature, 457*, 1128–1132.

Legleiter, J., Czilli, D. L., Gitter, B., Demattos, R. B., Holtzman, D. M., & Kowalewski, T. (2004). Effect of different anti-Abeta antibodies on Abeta fibrillogenesis as assessed by atomic force microscopy. *Journal of Molecular Biology, 335*, 997–1006.

Leighton, P. L. A., Nadolski, N. J., Morrill, A., Hamilton, T. J., & Allison, W. T. (2018). An ancient conserved role for prion protein in learning and memory. *Biology Open, 7*.

Li, B., Ge, P., Murray, K. A., Sheth, P., Zhang, M., Nair, G., et al. (2018). Cryo-Em of full-length alpha-synuclein reveals fibril polymorphs with a common structural kernel. *Nature Communications*, *9*, 3609.

Linden, R., Martins, V. R., Prado, M. A., Cammarota, M., Izquierdo, I., & Brentani, R. R. (2008). Physiology of the prion protein. *Physiological Reviews*, *88*, 673–728.

Liu, Y., Carver, J. A., Calabrese, A. N., & Pukala, T. L. (2014). Gallic acid interacts with alpha-synuclein to prevent the structural collapse necessary for its aggregation. *Biochimica et Biophysica Acta*, *1844*, 1481–1485.

Liu, Y., Pukala, T. L., Musgrave, I. F., Williams, D. M., Dehle, F. C., & Carver, J. A. (2013). Gallic acid is the major component of grape seed extract that inhibits amyloid fibril formation. *Bioorganic & Medicinal Chemistry Letters*, *23*, 6336–6340.

Lou, X., Kim, J., Hawk, B. J., & Shin, Y. K. (2017). α-Synuclein may cross-bridge v-Snare and acidic phospholipids to facilitate snare-dependent vesicle docking. *The Biochemical Journal*, *474*, 2039–2049.

Luca, S., Yau, W.-M., Leapman, R., & Tycko, R. (2007). Peptide conformation and supramolecular organization in amylin fibrils: Constraints from solid-state NMR. *Biochemistry*, *46*, 13505–13522.

Mannini, B., Mulvihill, E., Sgromo, C., Cascella, R., Khodarahmi, R., Ramazzotti, M., et al. (2014). Toxicity of protein oligomers is rationalized by a function combining size and surface hydrophobicity. *ACS Chemical Biology*, *9*, 2309–2317.

Maroteaux, L., Campanelli, J. T., & Scheller, R. H. (1988). Synuclein: A neuron-specific protein localized to the nucleus and presynaptic nerve terminal. *The Journal of Neuroscience*, *8*, 2804–2815.

Matsuzaki, K. (2007). Physicochemical interactions of amyloid beta-peptide with lipid bilayers. *Biochimica et Biophysica Acta*, *1768*, 1935–1942.

Mattson, M. P. (2004). Pathways towards and away from Alzheimer's disease. *Nature*, *430*, 631–639.

Mclaurin, J., & Chakrabartty, A. (1996). Membrane disruption by Alzheimer beta-amyloid peptides mediated through specific binding to either phospholipids or gangliosides. Implications for neurotoxicity. *The Journal of Biological Chemistry*, *271*, 26482–26489.

Middleton, E. R., & Rhoades, E. (2010). Effects of curvature and composition on α-synuclein binding to lipid vesicles. *Biophysical Journal*, *99*, 2279–2288.

Mina, E. W., Lasagna-Reeves, C., Glabe, C. G., & Kayed, R. (2009). Poloxamer 188 copolymer membrane sealant rescues toxicity of amyloid oligomers in vitro. *Journal of Molecular Biology*, *391*, 577–585.

Mishra, R., Sellin, D., Radovan, D., Gohlke, A., & Winter, R. (2009). Inhibiting islet amyloid polypeptide fibril formation by the red wine compound resveratrol. *ChemBioChem*, *10*, 445–449.

Morris, A. M., Watzky, M. A., & Finke, R. G. (2009). Protein aggregation kinetics, mechanism, and curve-fitting: A review of the literature. *Biochimica et Biophysica Acta*, *1794*, 375–397.

Morrison, D. K., & Davis, R. J. (2003). Regulation of map kinase signaling modules by scaffold proteins in mammals. *Annual Review of Cell and Developmental Biology*, *19*, 91–118.

Murphy, M. P., & Levine, H., 3rd. (2010). Alzheimer's disease and the amyloid-beta peptide. *Journal of Alzheimer's Disease*, *19*, 311–323.

Nanga, R. P., Brender, J. R., Vivekanandan, S., & Ramamoorthy, A. (2011). Structure and membrane orientation of IAPP in its natively amidated form at physiological pH in a membrane environment. *Biochimica et Biophysica Acta*, *1808*, 2337–2342.

Necula, M., Kayed, R., Milton, S., & Glabe, C. G. (2007). Small molecule inhibitors of aggregation indicate that amyloid beta oligomerization and fibrillization pathways are independent and distinct. *The Journal of Biological Chemistry*, *282*, 10311–10324.

Niu, Z., Zhao, W., Zhang, Z., Xiao, F., Tang, X., & Yang, J. (2014). The molecular structure of Alzheimer β-amyloid fibrils formed in the presence of phospholipid vesicles. *Angewandte Chemie (International Ed. in English), 53*, 9294–9297.

Nuscher, B., Kamp, F., Mehnert, T., Odoy, S., Haass, C., Kahle, P. J., et al. (2004). Alpha-synuclein has a high affinity for packing defects in a bilayer membrane: a thermodynamics study. *The Journal of Biological Chemistry, 279*, 21966–21975.

Olzscha, H., Schermann, S. M., Woerner, A. C., Pinkert, S., Hecht, M. H., Tartaglia, G. G., et al. (2011). Amyloid-like aggregates sequester numerous metastable proteins with essential cellular functions. *Cell, 144*, 67–78.

Ono, K., Hasegawa, K., Naiki, H., & Yamada, M. (2004). Anti-amyloidogenic activity of tannic acid and its activity to destabilize Alzheimer's beta-amyloid fibrils in vitro. *Biochimica et Biophysica Acta, 1690*, 193–202.

Ono, K., & Yamada, M. (2006). Antioxidant compounds have potent anti-fibrillogenic and fibril-destabilizing effects for alpha-synuclein fibrils in vitro. *Journal of Neurochemistry, 97*, 105–115.

O'nuallain, B., Shivaprasad, S., Kheterpal, I., & Wetzel, R. (2005). Thermodynamics of A beta(1-40) amyloid fibril elongation. *Biochemistry, 44*, 12709–12718.

Orlean, P., & Menon, A. K. (2007). Thematic review series: lipid posttranslational modifications. GPI anchoring of protein in yeast and mammalian cells, or: how we learned to stop worrying and love glycophospholipids. *Journal of Lipid Research, 48*, 993–1011.

Palazzi, L., Bruzzone, E., Bisello, G., Leri, M., Stefani, M., Bucciantini, M., et al. (2018). Oleuropein aglycone stabilizes the monomeric alpha-synuclein and favours the growth of non-toxic aggregates. *Scientific Reports, 8*, 8337.

Paravastu, A. K., Leapman, R. D., Yau, W. M., & Tycko, R. (2008). Molecular structural basis for polymorphism in Alzheimer's beta-amyloid fibrils. *Proceedings of the National Academy of Sciences of the United States of America, 105*, 18349–18354.

Parthsarathy, V., McClean, P. L., HÖlscher, C., Taylor, M., Tinker, C., Jones, G., et al. (2013). A novel retro-inverso peptide inhibitor reduces amyloid deposition, oxidation and inflammation and stimulates neurogenesis in the Appswe/PS1ΔE9 mouse model of Alzheimer's disease. *PLoS One, 8*, e54769.

Paulsson, J. F., & Westermark, G. T. (2005). Aberrant processing of human proislet amyloid polypeptide results in increased amyloid formation. *Diabetes, 54*, 2117–2125.

Pereira-Leal, J. B., & Seabra, M. C. (2001). Evolution of the Rab family of small GTP-binding proteins. *Journal of Molecular Biology, 313*, 889–901.

Perrin, R. J., Woods, W. S., Clayton, D. F., & George, J. M. (2000). Interaction of human alpha-Synuclein and Parkinson's disease variants with phospholipids. Structural analysis using site-directed mutagenesis. *The Journal of Biological Chemistry, 275*, 34393–34398.

Porat, Y., Stepensky, A., Ding, F. X., Naider, F., & Gazit, E. (2003). Completely different amyloidogenic potential of nearly identical peptide fragments. *Biopolymers, 69*, 161–164.

Pranke, I. M., Morello, V., Bigay, J., Gibson, K., Verbavatz, J. M., Antonny, B., et al. (2011). α-Synuclein and Alps motifs are membrane curvature sensors whose contrasting chemistry mediates selective vesicle binding. *The Journal of Cell Biology, 194*, 89–103.

Prusiner, S. B., Groth, D. F., Bolton, D. C., Kent, S. B., & Hood, L. E. (1984). Purification and structural studies of a major scrapie prion protein. *Cell, 38*, 127–134.

Puglielli, L., Tanzi, R. E., & Kovacs, D. M. (2003). Alzheimer's disease: the cholesterol connection. *Nature Neuroscience, 6*, 345–351.

Qiu, L., Lewis, A., Como, J., Vaughn, M. W., Huang, J., Somerharju, P., et al. (2009). Cholesterol modulates the interaction of beta-amyloid peptide with lipid bilayers. *Biophysical Journal, 96*, 4299–4307.

Ramakrishnan, M., Jensen, P. H., & Marsh, D. (2003). Alpha-synuclein association with phosphatidylglycerol probed by lipid spin labels. *Biochemistry 42*, 12919–12926.

Rashidi, A. A., Mirhashemi, S. M., Taghizadeh, M., & Sarkhail, P. (2013). Iranian medicinal plants for diabetes mellitus: a systematic review. *Pakistan Journal of Biological Sciences*, *16*, 401–411.

Reiss, A. B., & Voloshyna, I. (2012). Regulation of cerebral cholesterol metabolism in Alzheimer disease. *Journal of Investigative Medicine*, *60*, 576–582.

Ren, R., Shi, C., Cao, J., Sun, Y., Zhao, X., Guo, Y., et al. (2016). Neuroprotective effects of a standardized flavonoid extract of safflower against neurotoxin-induced cellular and animal models of Parkinson's disease. *Scientific Reports*, *6*, 22135.

Renne, M. F., & De Kroon, A. (2018). The role of phospholipid molecular species in determining the physical properties of yeast membranes. *FEBS Letters*, *592*, 1330–1345.

Rhoades, E., Ramlall, T. F., Webb, W. W., & Eliezer, D. (2006). Quantification of alpha-synuclein binding to lipid vesicles using fluorescence correlation spectroscopy. *Biophysical Journal*, *90*, 4692–4700.

Rigacci, S., Guidotti, V., Bucciantini, M., Nichino, D., Relini, A., Berti, A., et al. (2011). Abeta(1-42) aggregates into non-toxic amyloid assemblies in the presence of the natural polyphenol oleuropein aglycon. *Current Alzheimer Research*, *8*, 841–852.

Rigacci, S., Guidotti, V., Bucciantini, M., Parri, M., Nediani, C., Cerbai, E., et al. (2010). Oleuropein aglycon prevents cytotoxic amyloid aggregation of human amylin. *The Journal of Nutritional Biochemistry*, *21*, 726–735.

Roberson, E. D., & Mucke, L. (2006). 100 years and counting: Prospects for defeating Alzheimer's disease. *Science*, *314*, 781–784.

Roberts, H. L., & Brown, D. R. (2015). Seeking a mechanism for the toxicity of oligomeric α-synuclein. *Biomolecules*, *5*, 282–305.

Roguski, A., & Gill, A. C. (2017). The role of the mammalian prion protein in the control of sleep. *Pathogens*, *6*.

Ross, C. A., & Poirier, M. A. (2004). Protein aggregation and neurodegenerative disease. *Nature Medicine*, *10*, S10–S17.

Rovere, M., Sanderson, J. B., Fonseca-Ornelas, L., Patel, D. S., & Bartels, T. (2018). Refolding of helical soluble alpha-synuclein through transient interaction with lipid interfaces. *FEBS Letters*, *592*, 1464–1472.

Ryan, G. J., Jobe, L. J., & Martin, R. (2005). Pramlintide in the treatment of type 1 and type 2 diabetes mellitus. *Clinical Therapeutics*, *27*, 1500–1512.

Sabaté, R., Espargaró, A., Barbosa-Barros, L., Ventura, S., & Estelrich, J. (2012). Effect of the surface charge of artificial model membranes on the aggregation of amyloid β-peptide. *Biochimie*, *94*, 1730–1738.

Sacks, D., Baxter, B., Campbell, B. C. V., Carpenter, J. S., Cognard, C., Dippel, D., et al. (2018). Multisociety consensus quality improvement revised consensus statement for endovascular therapy of acute ischemic stroke. *International Journal of Stroke*, *13*, 612–632.

Sahoo, A., & Matysiak, S. (2019). Computational insights into lipid assisted peptide misfolding and aggregation in neurodegeneration. *Physical Chemistry Chemical Physics*, *21*, 22679–22694.

Scalisi, S., Sciacca, M. F., Zhavnerko, G., Grasso, D. M., Marletta, G., & La Rosa, C. (2010). Self-assembling pathway of HiApp fibrils within lipid bilayers. *ChemBioChem*, *11*, 1856–1859.

Sciacca, M. F., Kotler, S. A., Brender, J. R., Chen, J., Lee, D. K., & Ramamoorthy, A. (2012). Two-step mechanism of membrane disruption by Abeta through membrane fragmentation and pore formation. *Biophysical Journal*, *103*, 702–710.

Sciacca, M. F., Lolicato, F., Tempra, C., Scollo, F., Sahoo, B. R., Watson, M. D., et al. (2020). Lipid-chaperone hypothesis: A common molecular mechanism of membrane disruption by intrinsically disordered proteins. *ACS Chemical Neuroscience*, *11*, 4336–4350.

Scollo, F., Tempra, C., Lolicato, F., Sciacca, M. F. M., Raudino, A., Milardi, D., et al. (2018). Phospholipids critical micellar concentrations trigger different mechanisms of intrinsically disordered proteins interaction with model membranes. *Journal of Physical Chemistry Letters*, *9*, 5125–5129.

Seeliger, J., Weise, K., Opitz, N., & Winter, R. (2012). The effect of Aβ on Iapp aggregation in the presence of an isolated β-cell membrane. *Journal of Molecular Biology*, *421*, 348–363.

Segrest, J. P., Jackson, R. L., Morrisett, J. D., & Gotto, A. M., Jr. (1974). A molecular theory of lipid-protein interactions in the plasma lipoproteins. *FEBS Letters*, *38*, 247–258.

Segrest, J. P., Jones, M. K., De Loof, H., Brouillette, C. G., Venkatachalapathi, Y. V., & Anantharamaiah, G. M. (1992). The amphipathic helix in the exchangeable apolipoproteins: a review of secondary structure and function. *Journal of Lipid Research*, *33*, 141–166.

Selkoe, D. J. (2001). Alzheimer's disease: Genes, proteins, and therapy. *Physiological Reviews*, *81*, 741–766.

Sezgin, E., Levental, I., Mayor, S., & Eggeling, C. (2017). The mystery of membrane organization: Composition, regulation and roles of lipid rafts. *Nature Reviews. Molecular Cell Biology*, *18*, 361–374.

Shim, S. Y., Karri, S., Law, S., Schatzl, H. M., & Gilch, S. (2016). Prion infection impairs lysosomal degradation capacity by interfering with rab7 membrane attachment in neuronal cells. *Scientific Reports*, *6*, 21658.

Sievers, S. A., Karanicolas, J., Chang, H. W., Zhao, A., Jiang, L., Zirafi, O., et al. (2011). Structure-based design of non-natural amino-acid inhibitors of amyloid fibril formation. *Nature*, *475*, 96–100.

Smith, D. L., Pozueta, J., Gong, B., Arancio, O., & Shelanski, M. (2009). Reversal of long-term dendritic spine alterations in Alzheimer disease models. *Proceedings of the National Academy of Sciences of the United States of America*, *106*, 16877–16882.

Snead, D., & Eliezer, D. (2014). Alpha-synuclein function and dysfunction on cellular membranes. *Experimental Neurobiology*, *23*, 292–313.

Sorice, M., Mattei, V., Tasciotti, V., Manganelli, V., Garofalo, T., & Misasi, R. (2012). Trafficking of PrPc to mitochondrial raft-like microdomains during cell apoptosis. *Prion*, *6*, 354–358.

Stamer, K., Vogel, R., Thies, E., Mandelkow, E., & Mandelkow, E. M. (2002). Tau blocks traffic of organelles, neurofilaments, and APP vesicles in neurons and enhances oxidative stress. *The Journal of Cell Biology*, *156*, 1051–1063.

Stefani, M., & Dobson, C. M. (2003). Protein aggregation and aggregate toxicity: New insights into protein folding, misfolding diseases and biological evolution. *Journal of Molecular Medicine (Berlin, Germany)*, *81*, 678–699.

Straub, J. E., & Thirumalai, D. (2011). Toward a molecular theory of early and late events in monomer to amyloid fibril formation. *Annual Review of Physical Chemistry*, *62*, 437–463.

Straub, J. E., & Thirumalai, D. (2014). Membrane–protein interactions are key to understanding amyloid formation. *The Journal of Physical Chemistry Letters*, *5*, 633–635.

Tackenberg, C., Grinschgl, S., Trutzel, A., Santuccione, A. C., Frey, M. C., Konietzko, U., et al. (2013). NMDA receptor subunit composition determines beta-amyloid-induced neurodegeneration and synaptic loss. *Cell Death & Disease*, *4*, e608.

Takahashi, R., Ono, K., Takamura, Y., Mizuguchi, M., Ikeda, T., Nishijo, H., et al. (2015). Phenolic compounds prevent the oligomerization of alpha-synuclein and reduce synaptic toxicity. *Journal of Neurochemistry*, *134*, 943–955.

Takamori, S., Holt, M., Stenius, K., Lemke, E. A., Grønborg, M., Riedel, D., et al. (2006). Molecular anatomy of a trafficking organelle. *Cell*, *127*, 831–846.

Tantos, A., Han, K.-H., & Tompa, P. (2012). Intrinsic disorder in cell signaling and gene transcription. *Molecular and Cellular Endocrinology*, *348*, 457–465.

Terakawa, M. S., Lin, Y., Kinoshita, M., Kanemura, S., Itoh, D., Sugiki, T., et al. (2018). Impact of membrane curvature on amyloid aggregation. *Biochimica et Biophysica Acta - Biomembranes*, *1860*, 1741–1764.

Terzi, E., HÖlzemann, G., & Seelig, J. (1995). Self-association ofbeta-amyloid peptide (1-40) in solution and binding to lipid membranes. *Journal of Molecular Biology*, *252*, 633–642.

Theillet, F. X., Binolfi, A., Frembgen-Kesner, T., Hingorani, K., Sarkar, M., Kyne, C., et al. (2014). Physicochemical properties of cells and their effects on intrinsically disordered proteins (IDPs). *Chemical Reviews*, *114*, 6661–6714.

Tipping, K. W., Van Oosten-Hawle, P., Hewitt, E. W., & Radford, S. E. (2015). Amyloid fibres: Inert end-stage aggregates or key players in disease? *Trends in Biochemical Sciences*, *40*, 719–727.

Tsigelny, I. F., Sharikov, Y., Wrasidlo, W., Gonzalez, T., Desplats, P. A., Crews, L., et al. (2012a). Role of alpha-synuclein penetration into the membrane in the mechanisms of oligomer pore formation. *The FEBS Journal*, *279*, 1000–1013.

Tsigelny, I. F., Sharikov, Y., Wrasidlo, W., Gonzalez, T., Desplats, P. A., Crews, L., et al. (2012b). Role of α-synuclein penetration into the membrane in the mechanisms of oligomer pore formation. *The FEBS Journal*, *279*, 1000–1013.

Tu, L.-H., Serrano, A. L., Zanni, M. T., & Raleigh, D. P. (2014). Mutational analysis of preamyloid intermediates: the role of his-tyr interactions in islet amyloid formation. *Biophysical Journal*, *106*, 1520–1527.

Ulmer, T. S., Bax, A., Cole, N. B., & Nussbaum, R. L. (2005). Structure and dynamics of micelle-bound human alpha-synuclein. *The Journal of Biological Chemistry*, *280*, 9595–9603.

Um, J. W., Nygaard, H. B., Heiss, J. K., Kostylev, M. A., Stagi, M., Vortmeyer, A., et al. (2012). Alzheimer amyloid-β oligomer bound to postsynaptic prion protein activates Fyn to impair neurons. *Nature Neuroscience*, *15*, 1227–1235.

Uversky, V. N., & Dunker, A. K. (2010). Understanding protein non-folding. *Biochimica et Biophysica Acta*, *1804*, 1231–1264.

Uversky, V. N., Gillespie, J. R., & Fink, A. L. (2000). Why are "natively unfolded" proteins unstructured under physiologic conditions? *Proteins*, *41*, 415–427.

Uversky, V. N., Oldfield, C. J., & Dunker, A. K. (2008). Intrinsically disordered proteins in human diseases: introducing the D2 concept. *Annual Review of Biophysics*, *37*, 215–246.

Vana, K., Zuber, C., Nikles, D., & Weiss, S. (2007). Novel aspects of prions, their receptor molecules, and innovative approaches for TSE therapy. *Cellular and Molecular Neurobiology*, *27*, 107–128.

Vassallo, N., & Herms, J. (2003). Cellular prion protein function in copper homeostasis and redox signalling at the synapse. *Journal of Neurochemistry*, *86*, 538–544.

Verma, M., Vats, A., & Taneja, V. (2015). Toxic species in amyloid disorders: Oligomers or mature fibrils. *Annals of Indian Academy of Neurology*, *18*, 138–145.

Vivoli Vega, M., Cascella, R., Chen, S. W., Fusco, G., De Simone, A., Dobson, C. M., et al. (2019). The toxicity of misfolded protein oligomers is independent of their secondary structure. *ACS Chemical Biology*, *14*, 1593–1600.

Wakabayashi, M., & Matsuzaki, K. (2007). Formation of amyloids by abeta-(1-42) on NGF-differentiated PC12 cells: Roles of gangliosides and cholesterol. *Journal of Molecular Biology*, *371*, 924–933.

Wang, H., & Raleigh, D. P. (2014). The ability of insulin to inhibit the formation of amyloid by pro-islet amyloid polypeptide processing intermediates is significantly reduced in the presence of sulfated glycosaminoglycans. *Biochemistry*, *53*, 2605–2614.

Wang, T., & Hay, J. C. (2015). Alpha-synuclein toxicity in the early secretory pathway: How it drives neurodegeneration in Parkinsons disease. *Frontiers in Neuroscience*, *9*, 433. https://doi.org/10.3389/fnins.2015.00433.

Ward, J. J., Sodhi, J. S., McGuffin, L. J., Buxton, B. F., & Jones, D. T. (2004). Prediction and functional analysis of native disorder in proteins from the three kingdoms of life. *Journal of Molecular Biology*, *337*, 635–645.

Weingarten, M. D., Lockwood, A. H., Hwo, S. Y., & Kirschner, M. W. (1975). A protein factor essential for microtubule assembly. *Proceedings of the National Academy of Sciences of the United States of America*, *72*, 1858–1862.

Weinreb, P. H., Zhen, W., Poon, A. W., Conway, K. A., & Lansbury, P. T. (1996). NACP, a protein implicated in Alzheimer's disease and learning, is natively unfolded. *Biochemistry*, *35*, 13709–13715.

Westermark, P., Andersson, A., & Westermark, G. T. (2011). Islet amyloid polypeptide, islet amyloid, and diabetes mellitus. *Physiological Reviews*, *91*, 795–826.

Williams, T. L., & Serpell, L. C. (2011). Membrane and surface interactions of Alzheimer's Aβ peptide—Insights into the mechanism of cytotoxicity. *The FEBS Journal*, *278*, 3905–3917.

Williamson, J. A., Loria, J. P., & Miranker, A. D. (2009). Helix stabilization precedes aqueous and bilayer-catalyzed fiber formation in islet amyloid polypeptide. *Journal of Molecular Biology*, *393*, 383–396.

Wiltzius, J. J. W., Sievers, S. A., Sawaya, M. R., Cascio, D., Popov, D., Riekel, C., et al. (2008). Atomic structure of the cross-β spine of islet amyloid polypeptide (amylin). *Protein Science*, *17*, 1467–1474.

Wobst, H. J., Sharma, A., Diamond, M. I., Wanker, E. E., & Bieschke, J. (2015). The green tea polyphenol (-)-epigallocatechin gallate prevents the aggregation of tau protein into toxic oligomers at substoichiometric ratios. *FEBS Letters*, *589*, 77–83.

Wong, P. T., Schauerte, J. A., Wisser, K. C., Ding, H., Lee, E. L., Steel, D. G., et al. (2009). Amyloid-beta membrane binding and permeabilization are distinct processes influenced separately by membrane charge and fluidity. *Journal of Molecular Biology*, *386*, 81–96.

Wulf, M.-A., Senatore, A., & Aguzzi, A. (2017). The biological function of the cellular prion protein: An update. *BMC Biology*, *15*, 34.

Yang, F., Lim, G. P., Begum, A. N., Ubeda, O. J., Simmons, M. R., Ambegaokar, S. S., et al. (2005). Curcumin inhibits formation of amyloid beta oligomers and fibrils, binds plaques, and reduces amyloid in vivo. *The Journal of Biological Chemistry*, *280*, 5892–5901.

Yang, S., Liu, W., Lu, S., Tian, Y. Z., Wang, W. Y., Ling, T. J., et al. (2016). A novel multifunctional compound camellikaempferoside B decreases abeta production, interferes with abeta aggregation, and prohibits abeta-mediated neurotoxicity and neuroinflammation. *ACS Chemical Neuroscience*, 7, 505–518.

Yao, J., Ho, D., Calingasan, N. Y., Pipalia, N. H., Lin, M. T., & Beal, M. F. (2012). Neuroprotection by cyclodextrin in cell and mouse models of Alzheimer disease. *The Journal of Experimental Medicine*, *209*, 2501–2513.

Yoda, M., Miura, T., & Takeuchi, H. (2008). Non-electrostatic binding and self-association of amyloid beta-peptide on the surface of tightly packed phosphatidylcholine membranes. *Biochemical and Biophysical Research Communications*, *376*, 56–59.

Yoon, M.-K., Mitrea, D. M., Ou, L., & Kriwacki, R. W. (2012). Cell cycle regulation by the intrinsically disordered proteins p21 and p27. *Biochemical Society Transactions*, *40*, 981–988.

Yost, C. S., Lopez, C. D., Prusiner, S. B., Myers, R. M., & Lingappa, V. R. (1990). Non-hydrophobic extracytoplasmic determinant of stop transfer in the prion protein. *Nature*, *343*, 669–672.

Yu, X., & Zheng, J. (2012). Cholesterol promotes the interaction of Alzheimer β-amyloid monomer with lipid bilayer. *Journal of Molecular Biology*, *421*, 561–571.

Zea, D. J., Monzon, A. M., Gonzalez, C., Fornasari, M. S., Tosatto, S. C., & Parisi, G. (2016). Disorder transitions and conformational diversity cooperatively modulate biological function in proteins. *Protein Science*, *25*, 1138–1146.
Zerial, M., & McBride, H. (2001). Rab proteins as membrane organizers. *Nature Reviews. Molecular Cell Biology*, *2*, 107–117.
Zhang, X., St Clair, J. R., London, E., & Raleigh, D. P. (2017). Islet amyloid polypeptide membrane interactions: Effects of membrane composition. *Biochemistry*, *56*, 376–390.
Zheng, H., & Koo, E. H. (2011). Biology and pathophysiology of the amyloid precursor protein. *Molecular Neurodegeneration*, *6*, 27.

CHAPTER TWELVE

Dendritic spine membrane proteome and its alterations in autistic spectrum disorder

Ana Sofía Vallés[a] and Francisco J. Barrantes[b,*]
[a]Instituto de Investigaciones Bioquímicas de Bahía Blanca (UNS-CONICET), Bahía Blanca, Argentina
[b]Instituto de Investigaciones Biomédicas (BIOMED), UCA-CONICET, Buenos Aires, Argentina
*Corresponding author: e-mail address: francisco_barrantes@uca.edu.ar

Contents

1. Introduction 436
2. Molecular architecture of the dendritic spine 439
 2.1 The spine proteome: Receptors, scaffolding proteins, the PSD and the actin cytoskeletal meshwork 439
 2.2 The spine lipidome 444
3. Cholinergic neurotransmission and nicotinic receptors in dendritic spines 448
4. Spine alterations in neurological and neuropsychiatric disease 452
5. Spine dysfunction in autism spectrum disorder 453
6. Dysfunctional neurotransmission in autistic spectrum disorder 454
7. Concluding remarks 458
References 459

Abstract

Dendritic spines are small protrusions stemming from the dendritic shaft that constitute the primary specialization for receiving and processing excitatory neurotransmission in brain synapses. The disruption of dendritic spine function in several neurological and neuropsychiatric diseases leads to severe information-processing deficits with impairments in neuronal connectivity and plasticity. Spine dysregulation is usually accompanied by morphological alterations to spine shape, size and/or number that may occur at early pathophysiological stages and not necessarily be reflected in clinical manifestations. Autism spectrum disorder (ASD) is one such group of diseases involving changes in neuronal connectivity and abnormal morphology of dendritic spines on postsynaptic neurons. These alterations at the subcellular level correlate with molecular changes in the spine proteome, with alterations in the copy number, topography, or in severe cases in the phenotype of the molecular components, predominantly of those proteins involved in spine recognition and adhesion, reflected in abnormally short lifetimes of the synapse and compensatory increases in synaptic connections. Since cholinergic neurotransmission participates in the regulation of cognitive function (attention, memory, learning processes, cognitive flexibility, social interactions) brain acetylcholine

ISSN 1876-1623
https://doi.org/10.1016/bs.apcsb.2021.09.003

receptors are likely to play an important role in the dysfunctional synapses in ASD, either directly or indirectly via the modulatory functions exerted on other neurotransmitter receptor proteins and spine-resident proteins.

1. Introduction

Central nervous system (CNS) synapses are specialized structures where signal transmission between neurons takes place. Synapses comprise a presynaptic terminal and a postsynaptic specialization, which can occur in the soma, the axon or the dendritic arborization of the neuron, resulting in axo-somatic, axo-axonic or axodendritic synapses, respectively. In the latter case, the presynaptic terminal establishes contact with dendritic spines, very small protrusions of the neuronal dendritic tree that receive almost exclusively excitatory synaptic inputs and compartmentalize postsynaptic responses. From a physiological viewpoint, spines are the smallest unit at the receiving end of the vectorial information transfer machinery between neurons in CNS synapses.

Spines have been arbitrarily classified into five categories based on the relative length of their neck and the diameter of their head (Hering & Sheng, 2001). Mushroom, thin, stubby, filopodia, and bifurcation- or cup-shaped spines comprise the basic categories (Fig. 1). The remarkably dynamic nature of dendritic spines, which can alter their size and shape within seconds to minutes and undergo more lasting changes in the time scale of hours to days, adds complexity to this static picture. Spine motility is developmentally regulated and declines in mature neurons (Dunaevsky, Tashiro, Majewska, Mason, & Yuste, 1999).

During early stages of synaptogenesis, particularly in the first postnatal week, filopodia-shaped spines are most abundant. Dendritic filopodia are widely considered to be the precursors of mature spines and are believed to actively search for presynaptic partners to initiate the formation of neuronal connections during synaptogenesis (Ziv & Smith, 1996). Filopodia are subsequently replaced by stubby spines. Along development, the latter decrease in number and thin- and mushroom-shaped spines predominate (Fiala, Feinberg, Popov, & Harris, 1998). Some authors suggest that such a simple developmental relationship between filopodia and spines cannot be assumed since the transformation of spines back to filopodia has also been observed (Parnass, Tashiro, & Yuste, 2000).

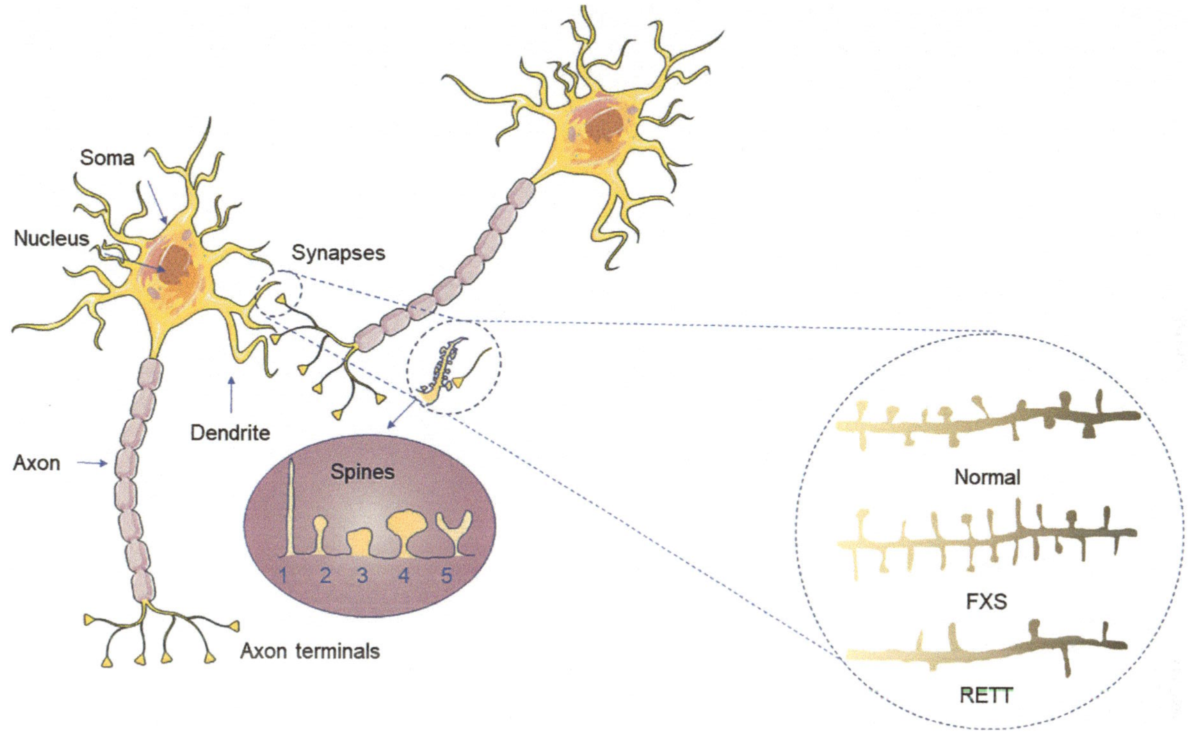

Fig. 1 Schematic illustration of a mature neuron with its basal and apical dendritic arborization, its axon, and incoming nerve endings establishing synapses. Enlarged is a dendrite containing dendritic spines that form synapses with other neurons. Based on their morphological characteristics, dendritic spines are classified as: (1) filopodia, (2) thin, (3) stubby, (4) mushroom, and (5) bifurcation- or cup-shaped. Mushroom-shaped spines predominate in subjects without ASD. In ASD-related monogenetic disorders such as fragile X syndrome (FXS) and Rett Syndrome (RETT) dendrites present an overabundance of dendritic spines with immature morphologies in the former case and reduced dendritic arborization and spine dysgenesis in RETT. *This illustration was created using Servier Medical Art Commons Attribution 3.0 Unported License (http://smart.servier.com). Servier Medical Art by Servier is licensed under a Creative Commons Attribution 3.0 Unported License.*

As neurons mature, the mushroom spines become more abundant and finally replace the immature filopodia (Fiala et al., 1998). These changes in spine morphology are crucial for synaptic function and plasticity since they are required for higher cognitive functions such as learning and memory formation. The width of the spine neck has been associated with the degree of compartmentalization of synaptic signals (Tønnesen, Katona, Rózsa, & Nägerl, 2014), and its length with the modulation of synaptic membrane tension, which in turn regulates the insertion of e.g. glutamate receptors (Vanderklish & Edelman, 2002).

Glutamatergic inputs from presynaptic terminals convey excitatory signals to these large spines in the CNS (McKinney, 2010; Okamura et al., 2004). Having the largest heads, mushroom spines probably constitute the most efficient structure for the vehiculization of excitatory neurotransmission since they harbor a greater number of postsynaptic glutamate receptors in an extended postsynaptic density (PSD) (Arellano, Benavides-Piccione, DeFelipe, & Yuste, 2007). Precisely for this reason they have been hypothesized to be the locus of long-term memories (Kasai, Matsuzaki, Noguchi, Yasumatsu, & Nakahara, 2003). In addition, mushroom-shaped spines are considered to be highly stable and less vulnerable to aging (Dumitriu et al., 2010), whereas thinner spines are generally assumed to be functionally weaker and have been associated with learning mechanisms (Dumitriu et al., 2010; Matsuzaki et al., 2001). Since they possess smaller heads, thinner spines contain lower amounts of the AMPA subtype glutamate receptors (Matsuzaki et al., 2001).

Anatomical evidence shows that aging-related cognitive impairment is associated with selective changes in spine density and morphology (Dumitriu et al., 2010). Spine enlargement and shrinkage are associated with long-term potentiation (LTP) and long-term depression (LTD), respectively (Tada & Sheng, 2006). Also, dendritic beading and focal or varicose swellings of dendritic arbors have been suggested to be morphological hallmarks of neuronal injury and dendrotoxicity (Ahlgren et al., 2014; Al-Noori & Swann, 2000).

As analyzed in this review, morphological changes in the spatial organization and/or morphology of the spine are observed in several neurological and neuropsychiatric diseases. Alterations in the copy number, topography, or in severe cases in the phenotype of the molecular constituents underlie the above morphological differences. Since the number of some synaptic proteins is rather low, minimal variations in their absolute figures can exert quite drastic changes in synaptic function. For instance, the presynaptic vesicle proton pump is only present in 1-2 copies per vesicle, thus constituting a rate-limiting factor in vesicle recycling, whereas proteins involved in the exocytic process are much more abundant and unlikely to pose a limiting problem (Denker et al., 2011).

2. Molecular architecture of the dendritic spine

2.1 The spine proteome: Receptors, scaffolding proteins, the PSD and the actin cytoskeletal meshwork

The spine harbors a constellation of molecular constituents in a femtoliter volume and the spine proteome is being actively explored. The diversity of synaptic subtypes is reflected at the molecular level in the wide spectrum of proteins present in dendritic spines, comprising various neurotransmitter receptors, scaffolding proteins, cytoskeletal and motor proteins, cell adhesion molecules and enzymes involved in the local biosynthesis and metabolism of proteins, lipids, and carbohydrates of the spine. The extracellular proteome -i.e. the proteins in the synaptic cleft- is also being studied using a variety of approaches e.g. peroxidase-based proximity labeling using the excitatory synaptic cell adhesion protein SynCAM 1 fused to horseradish peroxidase (HRP) as a reporter in cultured cortical neurons (Cijsouw et al., 2018).

A number of studies have addressed the quantitative assessment of the copy number of synaptic proteins. Traditionally, this information was obtained through biochemical methods ranging from quantitative Western blotting to biophysical approaches like mass spectrometry. In both cases, copy numbers resulting from these approaches are the outcome of averaging over thousands of cells. A semi-quantitative estimation of the synaptic protein copy number in synaptosomes, the biochemical subcellular fraction enriched in nerve terminals, has been obtained by combining the average protein copy number obtained *in vitro* through biochemical procedures (Wilhelm et al., 2014) with parallel measurements of fluorescence intensities in immunofluorescence specimens from the same samples, expressed as numbers of antibodies per synaptosome (Richter et al., 2018).

Although many neuronal proteins have been identified to date, most synapses share identical proteins, particularly those involved in physiological processes common to all synapses, like cell adhesion molecules, cytoskeletal proteins, motor proteins, enzymes and regulatory proteins, all of which can undergo post-translational modifications, thus contributing to the molecular diversity of synapses. Neurons are intimately connected with other cell types, such as astrocytes, oligodendrocytes and microglia, making unraveling of the neuron-exclusive proteome technically very difficult. Novel methodological developments are needed to overcome these limitations in the future (Dieterich & Kreutz, 2016).

The spine proteome is dynamically modulated by synaptic activity, which influences biochemical states and protein-protein interactions that regulate protein synthesis and reorganize cytoskeletal architecture (Nakahata & Yasuda, 2018). Dendritic spine morphology is maintained by and relies upon the actin cytoskeletal network (Penzes & Rafalovich, 2012). This network plays a pivotal role in the formation, elimination, motility, stability, size, and shape of dendritic spines (Tada & Sheng, 2006). Actin cytoskeleton dynamics and structure can be regulated by extracellular signals through its regulatory proteins (see below). Furthermore, the actin cytoskeleton and its regulatory proteins are necessary for memory formation and extinction. Moreover, the actin cytoskeleton has been suggested to intervene in synaptic transmission during learning and long-term cellular alterations mandatory for memory formation (Lamprecht, 2014). The actin cytoskeleton acts as a scaffold for stably positioning and anchoring ion-channels, cell-adhesion proteins and other macromolecules at the cell surface and also aids in the trafficking of subcellular structures such as endosomes and contributes to the architectural integrity of the PSD (Spence & Soderling, 2015), probably the most complex spine organelle.

The PSD is a protein network adjacent to the postsynaptic membrane, usually located at the head of the spine, where it occupies ~10% of its surface. The size of the PSD depends on the spine head volume (Arellano et al., 2007). The PSD harbors hundreds of molecular constituents, most of which are involved in a number of signaling pathways that control synaptic plasticity (Kennedy, 2000; Sheng & Pak, 2000). The PSD spatially and functionally organizes the neurotransmitter receptors at the synapse, and hence the number of postsynaptic receptors will also depend on the spine head volume (Arellano et al., 2007). The prototypical vertebrate excitatory CNS synapse is densely populated by ionotropic glutamatergic receptors organized and anchored at the PSD by specialized scaffolding molecules which are also involved in the regulation of their expression and trafficking (Dosemeci, Weinberg, Reese, & Tao-Cheng, 2016). The PSD is also connected to the subsynaptic actin-based cytoskeleton involved in spine motility (Hering, Lin, & Sheng, 2003). Lipid-protein interactions also appear to be important, as suggested by the links between cholesterol-rich membrane domains and F-actin and its local regulation (Föger, Marhaba, & Zöller,

2001; Itoh et al., 2002). Biological complexes are formed at the PSD through a series of protein–protein interactions. The PDZ domain-containing scaffold proteins provide a molecular interface between GluRs in the synaptic membrane and the spine cytoskeleton (Sheng & Pak, 2000; Sheng & Sala, 2001) (see Fig. 2). Genetic studies have shown that many mutations associated with neurodevelopmental disorders involve genes encoding regulators of the spine actin cytoskeleton (Borovac, Bosch, & Okamoto, 2018). Both PSD95 and SHANK (SH3 and multiple ankyrin repeat domains), two proteins that contain PDZ domains, bind through multiple interactions to NMDA receptors and to the metabotropic glutamate receptor (mGluR) (Naisbitt et al., 1999; Sala et al., 2001). In addition, SHANK cooperates with the protein HOMER that can also bind to mGluRs and inositol-1,4,5-trisphosphate receptors (InsP3R). Using cultured hippocampal neurons, an increase in the number and size of spines was found upon overexpression of PSD95 (El-Husseini, Schnell, Chetkovich, Nicoll, & Bredt, 2000). The authors of this study also found that overexpression of PSD-95 can drive maturation of glutamatergic synapses and enhanced clustering and activity of postsynaptic glutamate receptors. Expression of PSD-95 also enhances maturation of the presynaptic terminal (El-Husseini et al., 2000). Recent studies have reported that the alkaloid cotinine, the main metabolite of nicotine, acts as a positive allosteric modulator of the α7 nAChR, activating Akt, inhibiting GSK3b and increasing the expression of the PSD95 (Echeverria, Yarkov, & Aliev, 2016; Grizzell, Patel, Barreto, & Echeverria, 2017). The therapeutical possibilities of cotinine are currently being assayed in Alzheimer disease (AD), Parkinson disease, post-traumatic stress syndrome, and schizophrenia spectrum disorders. Another study showed that the overexpression of SHANK in developing hippocampal neurons promotes the maturation of mushroom-shaped spines and increases the size of spine heads in mature neurons without affecting spine number (Sala et al., 2001). In addition, the postsynaptic expression of SHANK enhances presynaptic function (Sala et al., 2001), thus suggesting a central role for the SHANK scaffold in the structural and functional organization of the dendritic spine and synaptic junction. All in all, synaptic plasticity appears to be governed by the strict regulation of the spine proteome.

Actin filaments interact with a variety of actin-binding proteins (ABPs) that provide the necessary structure for the formation of nascent protrusions and modifications of spine morphology (Runge, Cardoso, & de Chevigny, 2020). These ABPs play a role in many different aspects of actin dynamics

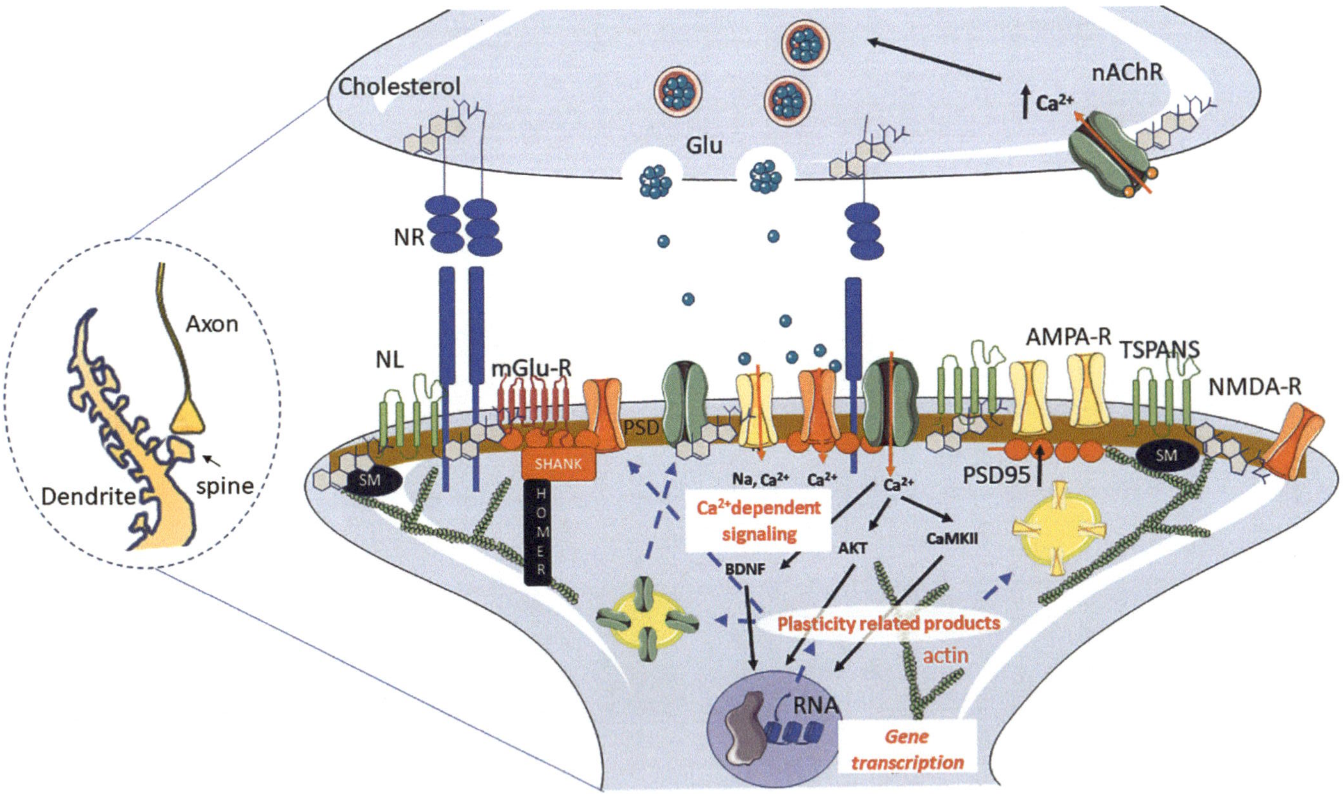

Fig. 2 See figure legend on opposite page.

such as actin polymerization-depolymerization, nucleation, branching, capping, cross-linking and trafficking (Nakahata & Yasuda, 2018). Although the precise nature of actin dynamics during synaptic plasticity remains elusive, it has been suggested that the actin network becomes fluid in the initial phase of structural changes, perhaps due to the dissociation of ABPs, subsequently allowing actin to be reorganized (Nakahata & Yasuda, 2018). After this reorganization has taken place, stabilization of the actin meshwork is mediated by the re-binding of ABPs (Kim et al., 2015; Okamoto, Nagai, Miyawaki, & Hayashi, 2004).

In the activated spine, Ca^{2+} enters through postsynaptic receptors such as α7AChR (Dani & Bertrand, 2007) or NMDA receptors and/or voltage-gated Ca^{2+} channels (Sabatini, Oertner, & Svoboda, 2002). This entry promotes the activation of multiple signaling pathways which have specific spatio-temporal patterns that orchestrate and regulate different aspects of cytoskeletal dynamics in the stimulated spines. Internally, the spine Ca^{2+} binds to calmodulin (CaM), a Ca^{2+}-binding protein which subsequently activates Ca^{2+}/CaM-dependent kinases and phosphatases such as CaMKII and calcineurin (Chang et al., 2017; Fujii et al., 2013). Both proteins have been reported to be required for inducing LTD as well as LTP (Fujii et al., 2013; Goodell, Zaegel, Coultrap, Hell, & Bayer, 2017) and their activity

Fig. 2 Schematic representation of the microanatomy of the excitatory synapse targeting on a mature mushroom-shaped spine. Cell adhesion molecules hold pre- and postsynaptic membranes together, separated only by the synaptic cleft. Several of these cell adhesion molecules, including neuroligins (NL) and neurexins (NR) have been implicated in ASD. The postsynaptic density (PSD) acts primarily as an organizing structure, and PDZ domain-containing proteins are found in the PSD. They interact with postsynaptic receptors anchored at the plasma membrane and connect them to the actin network that maintains the overall morphology of the dendritic spine. LTP elicited by Ca^{2+} entry to the postsynaptic spine activates CaMKIIα and its downstream signaling cascade, propagating from the spine to the nucleus to phosphorylate/dephosphorylate activity-dependent transcription factors. In turn, these factors induce the gene expression of plasticity-related products. TSPAN, tetraspanins; CaMKII, Ca^{2+}/calmodulin-dependent kinase II; AMPA-R, amino-3-hydroxy-5-methyl-4-isoazolepropionate receptor; GABA, γ-aminobutyric acid; GABA-R GABA receptor; NMDA-R, *N*-methyl-D-aspartate receptor; mGlu-R, metabotropic glutamate receptor; nAChR, nicotinic acetylcholine receptor; SM, sphingomyelin. BDNF, brain-derived neurotrophic factor; AKT, protein kinase B; Glu, glutamate; HOMER- Homer protein homolog 1; SHANK, SH3 and multiple ankyrin repeat domains proteins. *This illustration was created using Servier Medical Art Commons Attribution 3.0 Unported License (http://smart.servier.com). Servier Medical Art by Servier is licensed under a Creative Commons Attribution 3.0 Unported License.*

rapidly decays over a few seconds (Chang et al., 2017; Fujii et al., 2013). Down-regulation of CaMKII reduces the formation and maturation of dendritic spines (Okamoto, Narayanan, Lee, Murata, & Hayashi, 2007), highlighting the role of the enzyme in synaptic plasticity and memory formation (van Woerden et al., 2009). CaMKII activates small GTPases (Fig. 3) and these in turn modulate several kinases (Murakoshi, Wang, & Yasuda, 2011).These kinases have the ability to activate many ABPs, including Cofilin and Arp2/3, that play essential roles in actin reorganization (Borovac et al., 2018; Costa, Dines, & Lamprecht, 2020). Thus, the activation of multiple signaling pathways through the activation of small GTPases and associated ABPs via CaMKII activation after calcium entry, induces structural changes at the dendritic spine. In addition, calcium entry induces a cascade of kinase/phosphatase signaling pathways that propagate towards the nucleus to phosphorylate/dephosphorylate activity-dependent transcription factors (Flavell et al., 2006; Runge et al., 2020). These factors then induce the gene expression (Fig. 2) of plasticity-related products such as proteins or mRNAs in the cytoplasm (Benito & Barco, 2015; Hagenston & Bading, 2011).

Importantly, alterations of actin dynamics and dysregulation of associated activity-dependent signaling networks have been suggested to be among the major underlying causes of ASD and other ASD-related disorders (Lo & Lai, 2020; Stessman et al., 2017; Yap & Greenberg, 2018).

2.2 The spine lipidome

The spine lipidome does not fall short of diversity in comparison to the repertoire of molecular species of the spine proteome. A wide range of lipid classes are present in the spine, reflecting the importance of lipid-protein interactions for the correct functioning of the postsynapse.

Because of its role in the maintenance of spines and surface AMPA receptor stability, cholesterol has been qualified as a core component of spines and the synapse at large (Hering et al., 2003). The sterol plays a crucial role in synaptogenesis during brain development and has an important role in signal transduction, membrane trafficking and maturation of the CNS (Orth & Bellosta, 2012). Since the blood brain barrier (BBB) prevents the entry of lipoproteins into the brain parenchyma, brain cells must regulate their cholesterol levels independently from the rest of the body (Dietschy & Turley, 2004) and thus dysregulation of cholesterol homeostasis can perturb CNS development and function (Cartocci et al., 2016; Linetti et al., 2010; Segatto, Leboffe, Trapani, & Pallottini, 2014; Wang, 2014).

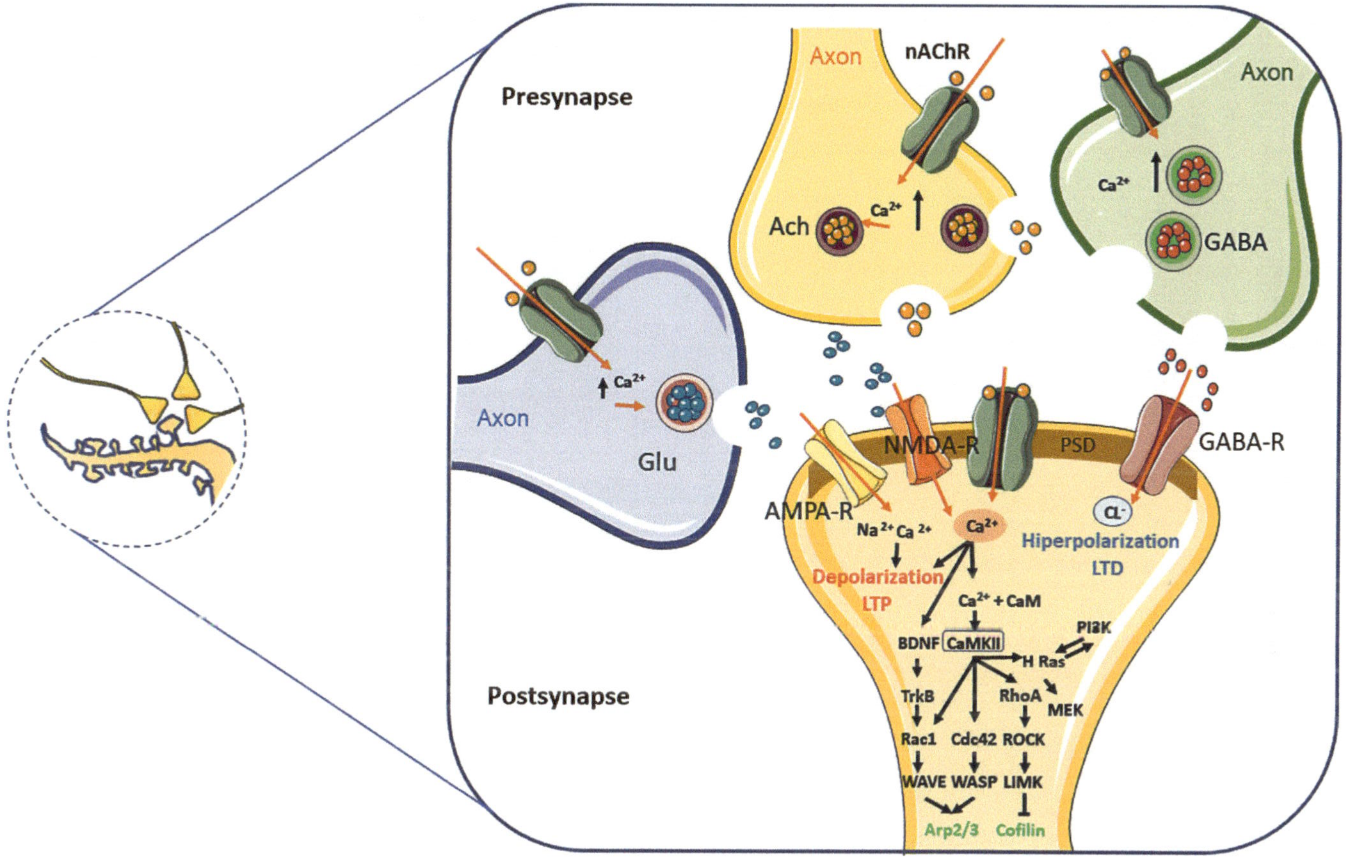

Fig. 3 See figure legend on next page.

Cholesterol and sphingolipids form highly dynamic and heterogeneous membrane nanodomains termed lipid rafts (Pike, 2006; Simons & Ikonen, 1997). These nanodomains compartmentalize cellular processes and serve as signaling platforms for the various neurotransmitter receptor systems harbored in the postsynaptic membrane. Their presence in dendritic spines has been reported to play a key role in the stability of AMPA receptors at the spine plasma membrane (Hering et al., 2003) (see review in ref. (Allen, Halverson-Tamboli, & Rasenick, 2007)).

Another important role played by lipid domains in dendritic spines is the recruitment of receptors from extrasynaptic sites. The migration of these receptor to the synaptic area may also involve lipid platforms, and once receptors reach the active zone of the spine, lipid nanodomains contribute to reduce their lateral mobility. This ultimately results in the anchoring of receptors at cholesterol- and sphingolipid-rich nanodomains at the spine head (Fernandes, Berg, & Gómez-Varela, 2010; Nagappan & Lu, 2005). In addition, cholesterol modulates the actin subcortical cytoskeletal meshwork that shapes spine structure and is a key element in receptor clustering and compartmentalization (Lozada et al., 2012a,b). Scaffolding proteins like the postsynaptic PSD95 also converge onto these areas of increased receptor

Fig. 3 Calcium influx at the presynaptic terminal promotes vesicle fusion and neurotransmitter release. Cholinergic agonists activating nAChR at these presynaptic terminal can induce acetylcholine, glutamate and GABA release. These neurotransmitters are liberated into the synaptic cleft activate NMDA-Rs, AMPA-Rs, mGlu-Rs, nAChRs and GABA-Rs. While glutamate and nAChR receptors induce depolarization and LTP of the postsynaptic spine, activation of GABA-R induces hyperpolarization and LTD. Ca^{2+} entry into the dendritic spine activates CaMKII which in turn activates small GTPases including Cdc42, Rac1, RhoA, and H-Ras. These small GTPases activate actin binding proteins (ABP) such as cofilin and Arp2/3. ABP regulate structural LTP via actin remodeling: e.g. Arp2/3 contributes to the increment in the number of branched actin filaments and targets dendritic spines into a more stable configuration. CaMKII, Ca^{2+}/calmodulin-dependent kinase II; AMPA-R, amino-3-hydroxy-5-methyl-4-isoazolepropionate receptor; GABA, γ-aminobutyric acid; GABA-R, GABA receptor; NMDA-R, *N*-methyl-D-aspartate receptor; nAChR, nicotinic acetylcholine receptor, BDNF, brain-derived neurotrophic factor; Glu; glutamate, ACh, acetylcholine; HOMER, Homer protein homolog 1; SHANK, SH3 and multiple ankyrin repeat domains proteins; ROCK, Rho-associated coiled-coil containing protein kinases; LIMK; LIM-kinase 1; WASP, Wiskott–Aldrich syndrome protein; WAVE; WASP-family verprolin-homologous protein; MEK, mitogen-activated protein kinase kinase; PI3K; phosphoinositide 3-kinases. *This illustrations was created using Servier Medical Art Commons Attribution 3.0 Unported License (http:/smart.servier.com). Servier Medical Art by Servier is licensed under a Creative Commons Attribution 3.0 Unported License.*

density (Hering et al., 2003; Perez & Bredt, 1998; Suzuki, 2002; Suzuki et al., 2011). Disruption of the cholesterol-rich lipid-ordered nanodomains in hippocampal neurons using acute cyclodextrin treatment blocks LTP (Frank et al., 2008; Koudinov & Koudinova, 2001). Cognitive deficits that appear with aging have also been associated with irreversible loss of membrane cholesterol (Martin et al., 2014b; Sodero, Weissmann, Ledesma, & Dotti, 2011). Conversely, replenishment of cholesterol in hippocampal slices from aged mice improves their learning and memory abilities (Martin, et al., 2014b). Cholesterol depletion by pharmacological treatments has been reported to produce a rapid collapse of spine morphological integrity associated with redistribution of F-actin from the spine proper to the dendritic shaft (Hering et al., 2003).

Dysregulation of cholesterol homeostasis leads to alterations in the biophysical properties of the membrane bilayer that affect the mobility and protein-protein interactions of neurotransmitter receptors (Fantini & Barrantes, 2013). Disruption of cholesterol metabolism has been implicated in the pathophysiology of ASD (Gillberg et al., 2017; Petrov, Kasimov, & Zefirov, 2017; Schengrund, Ali-Rahmani, & Ramer, 2012; Wang, 2014) and several neurological disorders related to ASD, like Rett, Asperger and Smith-Lemli-Opitz syndromes, all of which exhibit altered brain cholesterol homeostasis (Dziobek, Gold, Wolf, & Convit, 2007; Gillberg et al., 2017; Wang, 2014).

Sphingomyelin (SM) plays a relevant role in the spine membrane-actin cytoskeleton crosstalk. This sphingolipid modulates membrane binding and activity of the Rho GTPases, a family of small signaling G proteins and subfamily of the Ras superfamily, that are the key regulators of the actin cytoskeleton in the synapse. Low SM levels in postsynaptic membranes results in increased actin polymerization and augmented dendritic spine size in mice lacking the actin-related Wiskott–Aldrich syndrome protein (WASP)-interacting protein (WIP) (Franco-Villanueva et al., 2014). This pathological condition is due to the transcriptional upregulation of neutral sphingomyelinase with intervention of RhoA and partitioning of cholesterol-rich lipid domains. Thus, WIP is the nexus between the spine subcortical actin meshwork and postsynaptic plasma membrane lipid domains co-regulating dendritic spine size. In a Niemann-Pick disease type A mouse model defective in acid sphingomyelinase, SM accumulation decreases metabotropic glutamate receptors, impairing the membrane attachment of RhoA and its effectors ROCK and profilin IIa, and also decreases F-actin content, ultimately reducing spine number and size (Arroyo et al., 2014).

Ceramides, another major class of sphingolipids that intervenes in the regulation of cellular differentiation, activation, survival and apoptosis (Krönke, 1999), also contribute to spine plasticity by promoting receptor clustering and to spine maturation by boosting the transformation of dendritic filopodia into mature spines (Carrasco et al., 2012). The brain specific isoform of carnitine palmitoyl-transferase that modulates ceramide levels in hippocampal neurons is responsible for this transformation, since its deficiency increases immature filopodia and reduces mature mushroom and stubby spines, while not affecting total spine number or spine head area. In addition, mice lacking this enzyme have been reported to show defects in hippocampus-dependent learning abilities (Carrasco et al., 2012).

The occurrence of enzymes involved in phosphoinositide (PIP) interconversion supports a role for these lipids in the dendritic spine. PIPs have a high rate of metabolic turnover and their compartmentalization makes them key players in postsynaptic excitability (Hammond & Schiavo, 2007). Under basal conditions phosphatidylinositol (3,4,5)-triphosphate (PIP3) accumulates in spines rather than in dendritic shafts. PIP participates in spinule formation upon glutamate stimulation (Richards et al., 2005). PIP3 redistributes (Ueda & Hayashi, 2013) in these filopodia-like protrusions that project from spines (Richards et al., 2005). PIP3 has also been suggested to mediate membrane-cytoskeleton crosstalk at spines (Yin & Janmey, 2003) and to participate in dendritic and spine morphogenesis (Kumar, Zhang, Swank, Kunz, & Wu, 2005). Diacylglycerol (DAG) and 1,4,5-triphosphate (IP3) are PIP-derived second messengers of importance in dendritic spine organization and function. The former is reported to participate in spine maintenance (Brose, Betz, & Wegmeyer, 2004). Furthermore, phosphatidic acid (PA) and the α-p-21-activated kinase PAK1 also regulate spines by stabilizing actin filaments through myosin phosphorylation (Zhang, Webb, Asmussen, Niu, & Horwitz, 2005). Deeper insight into the role of lipids in dendritic spines, their interactions with postsynaptic receptors, and how they are regulated will contribute to our understanding of neurodegenerative disorders where brain lipid imbalances have been reported.

3. Cholinergic neurotransmission and nicotinic receptors in dendritic spines

Cholinergic signaling involving nicotinic acetylcholine receptors (nAChRs) modulates the excitatory–inhibitory homeostatic balance in brain. Cholinergic neurotransmission in cortical and hippocampal brain anatomical regions (Bentley, Driver, & Dolan, 2011) is intimately linked to the regulation of many important cognitive functions such as attention, working memory, and learning processes (Maurer & Williams, 2017), cognitive flexibility (Prado, Janickova, Al-Onaizi, & Prado, 2017) and social interactions (Wang et al., 2015).

Synaptic plasticity of excitatory synapses is modulated by the activation of presynaptic and postsynaptic neuronal nAChRs (McKinney, 2010; Okamura et al., 2004). Seventeen different subunits (α1–α10, β1–β4, γ, ε, and δ) encoded by 17 genes have been identified to date in the family of neuronal nAChRs in vertebrates. These subunits can assemble into homo- or hetero-pentameric structures (Jones & Sattelle, 2010). In the human brain, heteromeric α4β2 and homomeric α7 nAChRs constitute the most abundant subtypes of nAChRs, whereas other combinations are less common, are present in smaller numbers, and usually exhibit a more limited anatomical distribution, restricted to specific brain regions (Dani & Bertrand, 2007; Taly, Corringer, Guedin, Lestage, & Changeux, 2009). Activation of nAChRs can promote the release of various neurotransmitters, including the main excitatory transmitter, glutamate (Glu), in a Ca^{2+}-dependent manner (Fig. 3). For instance, activation of presynaptic α4β2 or α7 nAChRs depolarizes hippocampal interneurons and indirectly affects the release of neurotransmitters like Glu and ϒ-aminobutyric acid (GABA) by activating voltage-gated Ca^{2+} channels (McQuiston, 2014).

Presynaptic α7 nAChR in the rat auditory cortex has been shown to stimulate glutamate release and selectively potentiate NMDA receptor-mediated synaptic transmission (Aramakis & Metherate, 1998) albeit only in early postnatal life. α7 nAChR may thus contribute to the activity-dependent plasticity of the developing sensory cortex. Additionally, α7 nAChRs localized on human and rat brain glutamatergic terminals were shown to produce enhancement of glutamate release upon activation (Marchi, Risso, Viola, Cavazzani, & Raiteri, 2002).

At birth, most glutamatergic synapses in the hippocampus are immature and functionally "silent," either because the neurotransmitter is released in insufficient amounts to activate AMPA receptors or because the receptor proper is still missing or nonfunctional. At this critical period of postnatal development, acetylcholine acts presynaptically to strongly enhance neurotransmitter release and convert silent synapses into conductive ones. Nicotine mimics this role. It is thus clear that nAChR contributes to the maturation of functional synaptic contacts and the wiring of adult hippocampal circuitry (Maggi, Le Magueresse, Changeux, & Cherubini, 2003).

Similarly, Campbell and coworkers (Campbell, Fernandes, Halff, & Berg, 2010) demonstrated that in the dentate gyrus α7, nAChRs have a major role in adult neurogenesis. They also indicated that loss of α7 nAChRs can lead to the progressive impairment of hippocampal circuitry and function over time, as fewer neurons are added to the anatomical formation of the CNS, and those that are added integrate less well. Synaptic activation by nicotine has been reported to induce spine enlargement, as documented using structured illumination microscopy, a variant of super-resolution optical microscopy (Oda et al., 2014). The endogenous natural neurotransmitter, acetylcholine, as well as nAChR agonists have been suggested to participate in spine dendritic morphogenesis (Ballesteros-Yáñez, Benavides-Piccione, Bourgeois, Changeux, & DeFelipe, 2010). These authors found that in the absence of nAChRs containing the β2 subunit, the density of dendritic spines was significantly reduced. Lozada and collaborators reported that nicotine induces acute *de novo* glutamatergic synaptogenesis via β2 nAChRs both *in vivo* and *in vitro* (Lozada et al., 2012a,b). These observations reveal that activation of α4β2 nAChRs modulates synaptic architecture, the abundance of dendritic spines, and heterologous synaptogenesis.

Bailey and coworkers (Bailey, Alves, Nashmi, De Biasi, & Lambe, 2012) reported that normal developmental changes in nicotinic signaling and dendritic morphology in prefrontal cortex depend on nAChRs containing the α5 subunit. This latter study provides new insights into the possible cellular mechanisms underlying the known attention deficits in $\alpha 5^{-/-}$ mice and hypothesizes on the role of the α5 nAChR subunit in the pathophysiology of developmental neuropsychiatric disorders such as attention-deficit/hyperactivity disorder (ADHD) and ASD.

Perturbations in cholinergic signaling alter attention, cognition, anxiety and social behavior, all of which involve the integration of complex information by the prefrontal cortex (Kang, Tian, Bailey, & Lambe, 2015). *In vivo* chronic nicotine exposure during development significantly modified apical dendrite morphology and nAChR currents, compared with vehicle control in corticothalamic neurons within layer 6 of the medial prefrontal cortex (Bailey, Tian, Kang, O'Reilly, & Lambe, 2014). Among cortical neurons, prefrontal L6 pyramidal neurons receive unusually strong excitation by nicotinic acetylcholine receptors (Bailey, De Biasi, Fletcher, & Lambe, 2010; Kang et al., 2015). Thus, nicotinic perturbations may disrupt or distort the integration of information by prefrontal L6 neurons.

There is also evidence that the α7 subtype of neuronal nAChRs contributes to changes in synaptic architecture, promoting glutamatergic synaptogenesis (Lozada et al., 2012a,b) much like that resulting from α4β2 nAChR activation. A significant reduction in the number of glutamatergic synapses has been observed in α7 nAChR-knockout mice (Lozada et al., 2012a,b).

α7 nAChRs are also found postsynaptically in various brain regions including the neocortex and hippocampus where they cause fast inward cationic currents. At this location α7 nAChRs have also been proposed to modulate synaptic potentiation independently of fast excitatory transmission (Halff, Gómez-Varela, John, & Berg, 2014). These authors reported that stimulation of α7 nAChRs recruits GluA1 receptors from the surface pool of mobile extrasynaptic receptors, subsequently contributing to stabilizing them in the spine and thus increasing the density of this subtype of glutamatergic receptors. This stabilization requires functional PSD-95 scaffold protein family members that help reduce the local mobility of GluA1 macromolecules. As a result, the synapse enhances its signaling capacity, as reflected in larger miniature excitatory postsynaptic currents (mEPSCs). This mechanism of GluA1 activation via α7 nAChRs is independent of fast excitatory transmission mediated by AMPA or NMDA receptors.

Prenatal exposure to nicotine has been shown to induce significant long-term alterations in neuronal morphology (increased apical and basal spine density; decreased dendritic length) in parietal cortex of both male and female offspring (Mychasiuk, Muhammad, Gibb, & Kolb, 2013). The myriad of responses elicited by prenatal nicotine exposure on the various subtypes of nAChRs in different brain regions can lead to multiple pathophysiological outcomes. A recent study found altered neurite morphology and cholinergic function in a patient with autism and Kleefstra syndrome, a condition associated with intellectual disability, severely reduced or absent speech, and muscle hypotonia (Nagy et al., 2017).

ASD genome-wide association studies (GWAS) have provided estimates of the overlap of autism-related gene sets with genes involved in sensory processing and in cholinergic signal transmission. Martin and coworkers investigated the overlap of pathological structural variants in ADHD and ASD and found a significant overlap in genes including the α7 nAChR signaling pathway and genes involved in cell division (Martin et al., 2014a).

Brain-derived neurotrophic factor (BDNF) can regulate the level of α7 nAChRs in the hippocampus as well as in other brain regions (Zhou et al., 2004). Besides regulating the development of neuronal structure and function in the CNS, BDNF also serves as an activity-dependent regulator of synaptic plasticity and participates in synaptic transmission (Bramham &

Messaoudi, 2005; Massey, Zago, & Berg, 2006). Indeed, BDNF has been reported to increase both cell-surface and intracellular α7 nAChR pools in dissociated rat hippocampal neurons (Vallés & Barrantes, 2012). This upregulation of α7 nAChRs subunits depends on glutamatergic activity and is restricted to distinct neuronal subtypes, thus suggesting the existence of cell-type specific regulatory mechanisms (see Fig. 2). Since long exposure to BDNF is required to detect the stated increase in α7 nAChRs the suggestion was made that *de novo* synthesis is taking place (Massey et al., 2006). Activation of α7 nAChRs (Bertrand, Galzi, Devillers-Thiéry, Bertrand, & Changeux, 1993; Séguéla, Wadiche, Dineley-Miller, Dani, & Patrick, 1993) leads to calcium entry and cell depolarization and influences a variety of calcium-dependent events (Berg & Conroy, 2002) including changes in gene expression (Massey et al., 2006). Therefore, the distribution and regulation of these receptors can profoundly impact network function.

Neuronal nAChRs are modulated by cholesterol. The α7 nAChR localizes in cholesterol-rich liquid-ordered lipid domains at the surface of the somatic spines in chick ciliary ganglion sympathetic neurons (Brusés, Chauvet, & Rutishauser, 2001; Marchand, Devillers-Thiéry, Pons, Changeux, & Cartaud, 2002; Oshikawa et al., 2003; Zhu, Xiong, & Mei, 2006). Long-term inhibition of cholesterol biosynthesis by the statin lovastatin differentially augments cell-surface levels of α4β2 and α7 nAChRs in neurites and soma of rat hippocampal neurons (Borroni, Kamerbeek, Pediconi, & Barrantes, 2020). Disbalances in brain cholesterol homeostasis may therefore impact on the number of glutamate receptors at dendritic spines through cholinergic signaling involving α4β2 and α7 nAChRs.

4. Spine alterations in neurological and neuropsychiatric disease

Most neurological and neuropsychiatric diseases exhibit characteristic histopathological or gross anatomical pathological features in the necropsy (Taber, Hurley, & Yudofsky, 2010). The World Health Organization underscores that neuropsychiatric diseases are more common than almost any other type of disease, and are often accompanied by various comorbidities and shortened life expectancy (WHO | World Health Statistics, 2017: Monitoring health for the SDGs, 2017). Genetic studies indicate that excitatory postsynaptic mechanisms may be particularly important in the etiology of a wide range of neuropsychiatric disorders, although dysfunctional structural plasticity of neuronal circuits involving alterations in both presynaptic and postsynaptic sites has also been invoked (Sanders et al., 2015; Wefelmeyer, Puhl, & Burrone, 2016).

Distinct patterns of dendritic spine pathology have been associated with specific neuropsychiatric disorders. There is abundant literature analyzing these synaptopathies, which fall beyond the subject of this review. Briefly, disruption of spine shape, size or number that accompanies many neurological and neuropsychiatric disorders strongly suggests that dendritic spines are an important target in the etiopathogenesis of such diseases (Glantz & Lewis, 2000; Hutsler & Zhang, 2010; Selkoe, 2002; Sweet, Henteleff, Zhang, Sampson, & Lewis, 2009; Tackenberg, Ghori, & Brandt, 2009; Toro et al., 2010), if not the initial site of attack, as strongly suspected for AD (Selkoe, 2002; Tackenberg et al., 2009). We will now focus on the alterations of the spine specifically related to ASD.

5. Spine dysfunction in autism spectrum disorder

Alterations in synaptic homeostasis have been reported in ASD, a condition characterized by atypicalities in both social behavior and cognitive functions (Bonnet-Brilhault, 2017; Lai, Lombardo, & Baron-Cohen, 2014; Lord, Elsabbagh, Baird, & Veenstra-Vanderweele, 2018). Changes in the former can be identified in the first few months of life (Lord, 1995), suggesting the occurrence of neuroanatomical and/or neurochemical disbalances early on in the development of the CNS. ASD patients have severe information-processing deficits along with impairments in neuronal connectivity and plasticity (Penzes, Cahill, Jones, VanLeeuwen, & Woolfrey, 2011). A generalized reduction in the size and number of the dendritic spines, as well as an alteration of the morphology of dendrites and an increase in spine density with immature morphology has been reported in human and animal models of ASD (Lo & Lai, 2020; Martínez-Cerdeño, 2017) (see Fig. 1). The morphological alterations are suggestive of the immaturity of dendritic spines in ASD. In line with this hypothesis, it has been proposed that ASD may stem, at least in part, from loss of proteins that intervene in the stability and adhesivity of synapses and are crucial for spine development, structural organization and function, such as neuroligins, SHANK proteins, fragile X mental retardation protein (FMRP), methyl-CpG-binding protein 2 (MECP2) and cell adhesion molecules (Forrest, Parnell, & Penzes, 2018; Martínez-Cerdeño, 2017). Given the prominent roles these proteins play in the synapse, their loss effectively impacts on the tenacity and persistence of individual synaptic junctions.

6. Dysfunctional neurotransmission in autistic spectrum disorder

There is evidence of abnormal neurotransmission in ASD (Marotta et al., 2020). The alteration of glutamatergic neurotransmission has been suggested to play a central role in ASD, and both hyper- and hypo-glutamate models have been proposed for ASD phenotypes. Data supporting the hyper-glutamate model include increased levels of serum and plasma Glu in children and adults with ASD (Rojas, 2014; Zheng, Zhu, Qu, & Mu, 2016). Furthermore, upregulation of the GluN2A and GluN2B subunits of the NMDARs accompanied by enhancement of LTP are observed in the valproic acid (VPA)-induced ASD animal model (Rinaldi, Kulangara, Antoniello, & Markram, 2007). Interestingly, pharmacological modulation of AMPA receptors by antagonists rescue the social impairments in the VPA-induced ASD animal model (Kim et al., 2019). In patients with ASD the therapeutic efficacy of GluR antagonists has also been reported (Doyle & McDougle, 2012; Erickson et al., 2007, 2011, 2013; Rezaei et al., 2010). In support of the contrasting hypo-glutamate hypothesis, several animal models of ASD provide evidence of an altered NMDAR signaling pathway as a result of the downregulation of the principal NMDAR subunit, GluN1 (Uzunova, Hollander, & Shepherd, 2014). The beneficial pharmacological effects of glutamatergic agonists in rescuing some ASD symptoms in patients (Akhondzadeh et al., 2008; Urbano, Okwara, Manser, Hartmann, & Deutsch, 2015) and in mouse models of ASD-like phenotype (Eltokhi, Rappold, & Sprengel, 2018; Um et al., 2018) provide further support to the hypo-glutamate theory.

In agreement with the premise that ASD is at least in part the consequence of a developmental synaptopathy, evidence from rodent studies and postmortem human brains has revealed heterogeneous synaptic ultrastructural alterations related to this disorder (Eltokhi, Santuy, Merchan-Perez, & Sprengel, 2021). Changes in prefrontal axons likely contribute to the disruption of synaptic function in ASD (Zikopoulos & Barbas, 2010).

A genetic-associated rodent model of ASD carrying a human 15q11-13 chromosomal duplication was shown to have poor maturation of dendritic spines. The murine model also exhibits reduced PSD size and spine- head volume, and diminished spine head and spine neck widths in layer II/III of the somatosensory cortex. An altered balance between excitatory and inhibitory synapses has been suggested to be the functional correlate of the spine structural alterations observed (Sato & Okabe, 2019).

Integration of excitatory and inhibitory synaptic inputs are essential for neuronal signal processing. Accurate regulatory mechanisms provide excitatory/inhibitory balance in synaptic transmission. Thus, dysregulation of this balance, associated with mutations in synaptic adhesion and scaffolding molecules, has also been suggested to be related to ASD (Südhof, 2008).

Maintenance of cognition and normal behaviors is strongly dependent on precise formation and elimination of neuronal circuits. Firstly, during the initial postnatal brain development, synaptogenesis predominates over synaptic pruning, resulting in synapses in excess of functional needs. Subsequently, activity-dependent elimination of redundant synapses takes place, decreasing the number of synapses. A decade ago, Penzes and coworkers (Penzes et al., 2011) suggested that the increase in dendritic spine numbers in ASD may be the result of exaggerated spine formation or incomplete pruning during childhood. Activation upon neuronal depolarization of the myocyte enhancer factor 2 (MEF2) family of transcription factors is needed to reduce excitatory synapse numbers (Wilkerson et al., 2014). The *MEF2* family of transcription factors in vertebrates comprises four members: *MEF2A*, *MEF2B*, *MEF2C*, and *MEF2D*. The MEF2s are known to regulate ASD-associated genes important for neural development and synaptogenesis (Chaudhary, Agarwal, Kaushik, & Rehman, 2021; Flavell et al., 2006, 2008). Deletion of the key family member *MEF2C* in mouse brains was shown to increase synapse numbers and dendritic spines in both cortical and hippocampal neurons. The latter alteration of dendritic spine number may account for the learning and memory deficits and autistic phenotypes observed (Barbosa et al., 2008; Flavell et al., 2006).

As discussed in the previous section, dysfunction of the brain cholinergic system (Eissa et al., 2018; Hardan & Handen, 2002) as well as association with abnormal cholesterol metabolism (Cartocci et al., 2018; Segatto, Tonini, Pfrieger, Trezza, & Pallottini, 2019; Wang, 2014) has been reported in ASD. It should be noted that in addition to being the loci of ionotropic glutamate receptors of the AMPA- and NMDA- types and the metabotropic glutamate receptors (mGluR), dendritic spines also contain several other receptor species, including nicotinic receptors, which are regulated by neuroligins (NLs), among other endogenous ligands. NLs, brain-specific postsynaptic cell-adhesion molecules that are mediators of synaptogenesis, are altered in ASD (Bemben, Shipman, Nicoll, & Roche, 2015; Forrest et al., 2018). There is also evidence that mutations in NL1, NL3 and NL4 occur in patients with ASD (Blundell et al., 2010; Jamain et al., 2003; Laumonnier et al., 2004). NL1 upregulates the expression of the α7 nAChR in immortalized microglial cell lines (Mencel, Nash, &

Jacobson, 2013) and is implicated in the recruitment of α3-containing nAChR to the PSD (Conroy, Nai, Ross, Naughton, & Berg, 2007; Ross & Conroy, 2008).

Another family of transmembrane proteins enriched in spines is that of the tetraspanins (TSPANs). TSPANs play an important role as synaptic regulators (Murru, Moretto, Martano, & Passafaro, 2018). TSPAN7, for instance, is recognized to have a role in synapse formation, function, and plasticity as well as in learning and memory (Murru et al., 2017). TSPAN5 subtype is particularly abundant in rodent pyramidal neurons, where it modulates spine maturation by regulating NL1 clustering (Moretto et al., 2019). Some ASD patients with intellectual disabilities present mutations in TSPAN7 subtypes (Oeseburg, Dijkstra, Groothoff, Reijneveld, & Jansen, 2011; Piton et al., 2011; Schwartz & Neri, 2012). TSPANs self-organize in specialized membrane lipid domains enriched in cholesterol (Charrin et al., 2003), favoring the concentration of other transmembrane proteins (Moretto et al., 2019), an effect that further highlights the role of cholesterol in dendritic spines.

As referred to in the section on the spine proteome, SHANK are a group of postsynaptic scaffold proteins with PDZ domains present in the excitatory PSD, where they play a vital role in the formation, organization, and signaling of glutamatergic synapses (Eltokhi et al., 2018). Mutations in the proteins that are encoded by *SHANK* genes (*SHANK1*, *SHANK2*, and *SHANK3*) are found in ASD (Sala, Vicidomini, Bigi, Mossa, & Verpelli, 2015). Mutant mice with a deletion of *Shank1* display an ASD-like phenotype, altered PSD protein composition, smaller and thinner PSDs, reduced dendritic spine size, and a weaker basal synaptic transmission in the hippocampal CA1 region (Hung et al., 2008). SHANK1 deletions in a male patient with high functioning ASD have also been reported (Sato et al., 2012). Mutations in SHANK2 have been associated with ASD and reported to cause hyperconnectivity of human neurons (Zaslavsky et al., 2019). Leblond and coworkers reported two cases of patients with SHANK2 mutations who also presented a duplication of *CHRNA7*, the gene coding for the α7AChR protein (Leblond et al., 2012). A similar dysfunction of SHANK2 and α7AChR in a patient with intellectual disability and language impairment was reported (Chilian et al., 2013). These studies support a model of epistatic interactions of genes which upon disruption may lead to ASD. *SHANK3* mutant mice also exhibit an ASD-like phenotype. *SHANK3B-/-* mice have decreased thickness and length of the PSD in the striatum. They also present a reduction in spine density accompanied by reduced excitatory synaptic transmission (Peça et al., 2011). In addition, overexpression of *SHANK3* mutants (R12C and R300C) in cultured hippocampal neurons causes SHANK3 dysfunction, disrupting spine induction and maturation (Durand et al., 2012). It is interesting to note that one of the earliest clinical symptoms in SHANK3-associated ASD is neonatal skeletal muscle hypotonia (Lutz et al., 2020). SHANK3 deficiency in motoneurons alters the clustering of postsynaptic muscle-type nAChRs in myotubes (Lutz et al., 2020). Alterations in SHANK3 have been shown to provoke changes in synaptic function that requires trans-synaptic signaling through neurexin/neuroligin complexes (Arons et al., 2012). A recent study further discusses the disruption and loss of α7nAChR-GFP patches in *Drosophila* mushroom body calyces observed in $SHANK^{8k}$ mutants (Wu et al., 2017).

All in all, these findings suggest that cholinergic signaling is regulated by multiple interactions with different scaffolding molecules and that further research is required to unravel the diverse mechanisms linking cholinergic function with the key scaffolding protein(s) that are essential for synaptic function and plasticity.

Spine defects are also associated with other ASD-related monogenetic disorders such as fragile X syndrome (FXS) and Rett Syndrome. The former is a neurodevelopmental disorder that causes intellectual disability and behavioral deficits (Cheng et al., 2019). In FXS, neurons show immature postsynaptic dendritic spine protrusions as thin, actin-rich filopodia on the surface of dendrites (Cruz-Martín, Crespo, & Portera-Cailliau, 2010; Wijetunge, Angibaud, Frick, Kind, & Nägerl, 2014). Studies performed on FXS patients or mice demonstrated an overabundance of dendritic spines with immature morphologies in the cortical pyramidal neurons, cortex and hippocampus (Arroyo, Fiole, Mantri, Huang, & Portera-Cailliau, 2019; Grossman et al., 2010; McKinney, Grossman, Elisseou, & Greenough, 2005).

Fragile X mental retardation protein (FMRP) and the cytoplasmatic-interacting protein 1 (CYFIP1) have recently been associated with ASD and are believed to alter the balance between synaptic excitation and inhibition (Davenport et al., 2019). Disruption of CYFIP expression is associated with intellectual disability and ASD (Lo & Lai, 2020). Diminished FMRP production affects dendritic mRNA transport and translational regulation as well as dendritic spine maturation (Bagni, Tassone, Neri, & Hagerman, 2012). CYFIP is also part of the WAVE complex that promotes actin polymerization and likely contributes to the abnormal spine morphogenesis in FXS. Furthermore, the messenger RNA encoding bone morphogenetic protein type II receptor (BMPR2) is a target of FMRP. FMRP depletion causes increased expression of BMPR2, which activates the LIM domain kinase 1 (LIMK1) to increase phosphorylation and inhibition of the actin-depolymerization factor cofilin, hence altering actin dynamics and spine growth (Kashima et al., 2016).

Rett syndrome is an X-linked autism spectrum disorder that involves mutations of methyl-CpG-binding protein 2 (MeCP2), a gene encoding a methylated DNA-binding protein that regulates gene transcription, mRNA splicing, and chromatin structure in 80% of individuals with this neurodevelopmental disorder (Bird, 2008; Phillips & Pozzo-Miller, 2014). Brain abnormalities have been reported in Rett syndrome, including reduced neuronal size but increased cell density in several brain regions (Bauman, Kemper, & Arin, 1995a, 1995b). Reduced dendritic arborization (Armstrong, Dunn, Antalffy, & Trivedi, 1995) and spine dysgenesis has also been described in this disorder (Chapleau et al., 2009a; Kaufmann, Taylor, Hohmann, Sanwal, & Naidu, 1997). *Mecp2 k*nockout mice present impaired dendritic complexity (Nguyen et al., 2012) and lower dendritic spine density (Chapleau et al., 2012; Landi et al., 2011) as well as a lack of mushroom spines in cortical and hippocampal neurons (Chao, Zoghbi, & Rosenmund, 2007).

As previously discussed, BDNF has an important role in neuronal growth, synapse formation and activity-dependent plasticity through the activation of tropomyosin-related kinase B(TrkB) receptors (Chapleau, Larimore, Theibert, & Pozzo-Miller, 2009b; Ohira & Hayashi, 2009; Panja & Bramham, 2014). MeCP2 binds to the BDNF promoter and modulates BDNF expression in an activity-dependent manner (Chahrour & Zoghbi, 2007; Chen et al., 2003; Martinowich et al., 2003). Lower levels of BDNF have been reported in MeCP2 mutant mice while overexpression of BDNF rescues several cellular and behavioral deficits in these mice (Chang, Khare, Dani, Nelson, & Jaenisch, 2006; Pozzo-Miller, Pati, & Percy, 2015). Mutations in MeCP2 also lead to lower expression of mGluR (Zhong, Li, & Chang, 2012). Therefore, deficiencies in both signaling pathways may account for insufficient activity-dependent protein translation in dendrites.

7. Concluding remarks

Understanding the biochemistry and cell biology of synaptic function and plasticity is the first step towards deciphering brain disorders. Spine dynamics are cellular phenomena with a multiplicity of important implications for cognitive functions and memory, and impairment of such dynamics is strongly associated with various neurological and neuropsychiatric diseases. This is also the case with autistic spectrum disorder, a heterogeneous palette of syndromes clinically manifested in multiple variants, with a common denominator of intellectual disabilities, abnormal behaviors, and dysfunctional communication abilities.

Dendritic spines are anatomical and functional elements essential for the connectivity of the brain and are considered the major—if not key—anatomical substrates of various neuropsychiatric diseases. The focus on the altered morphology of dendritic spines in postsynaptic neurons, initially reported to be smaller in ASD, has now shifted to the remarkably accelerated dynamics of spine turnover (spine biogenesis and pruning), probably at the root of the accelerated compensatory rewiring of neuronal connectivity and feeble synaptic tenacity associated with the disorder. At the molecular level, the source of these alterations is increasingly being narrowed down to members of the spine proteome involved in synaptic recognition and adhesion, such as neuroligins, SHANK proteins, and other cell-adhesion proteins.

Underpinning the importance of nAChRs in excitatory transmission, many studies have demonstrated that nAChRs of the α7 subtype facilitate glutamate release from presynaptic terminals. In addition, due to their high Ca^{2+} permeability, α7 nAChRs have an important role as neuromodulators, contributing to neuronal maturation, synaptogenesis, regulation of neural circuits in the CNS, and through their involvement in LTP and LTD mechanisms, contribute to the neuroplasticity phenomena in brain. Advances in the field of the spine proteome should provide important clues about the etiopathogenic mechanisms involved in syndromic ASD as in RETT or fragile X syndromes.

References

Ahlgren, H., Bas-Orth, C., Freitag, H. E., Hellwig, A., Ottersen, O. P., & Bading, H. (2014). The nuclear calcium signaling target, activating transcription factor 3 (ATF3), protects against dendrotoxicity and facilitates the recovery of synaptic transmission after an excitotoxic insult. *The Journal of Biological Chemistry*, *289*, 9970–9982. https://doi.org/10.1074/jbc.M113.502914.

Akhondzadeh, S., Tajdar, H., Mohammadi, M. R., Mohammadi, M., Nouroozinejad, G. H., Shabstari, O. L., et al. (2008). A double-blind placebo controlled trial of piracetam added to risperidone in patients with autistic disorder. *Child Psychiatry and Human Development*, *39*, 237–245. https://doi.org/10.1007/s10578-007-0084-3.

Allen, J. A., Halverson-Tamboli, R. A., & Rasenick, M. M. (2007). Lipid raft microdomains and neurotransmitter signalling. *Nature Reviews. Neuroscience*, *8*, 128–140. https://doi.org/10.1038/nrn2059.

Al-Noori, S., & Swann, J. W. (2000). A role for sodium and chloride in kainic acid-induced beading of inhibitory interneuron dendrites. *Neuroscience*, *101*, 337–348. https://doi.org/10.1016/S0306-4522(00)00384-5.

Aramakis, V. B., & Metherate, R. (1998). Nicotine selectively enhances NMDA receptor-mediated synaptic transmission during postnatal development in sensory neocortex. *The Journal of Neuroscience*, *18*, 8485–8495. https://doi.org/10.1523/JNEUROSCI.18-20-08485.1998.

Arellano, J. I., Benavides-Piccione, R., DeFelipe, J., & Yuste, R. (2007). Ultrastructure of dendritic spines: Correlation between synaptic and spine morphologies. *Frontiers in Neuroscience*, *1*, 131–143. https://doi.org/10.3389/neuro.01.1.1.010.2007.

Armstrong, D., Dunn, J. K., Antalffy, B., & Trivedi, R. (1995). Selective dendritic alterations in the cortex of Rett syndrome. *Journal of Neuropathology and Experimental Neurology*, *54*, 195–201. https://doi.org/10.1097/00005072-199503000-00006.

Arons, M. H., Thynne, C. J., Grabrucker, A. M., Li, D., Schoen, M., Cheyne, J. E., et al. (2012). Autism-associated mutations in ProSAP2/Shank3 impair synaptic transmission and neurexin-neuroligin-mediated transsynaptic signaling. *The Journal of Neuroscience*, *32*, 14966–14978. https://doi.org/10.1523/JNEUROSCI.2215-12.2012.

Arroyo, A. I., Camoletto, P. G., Morando, L., Sassoe-Pognetto, M., Giustetto, M., Van Veldhoven, P. P., et al. (2014). Pharmacological reversion of sphingomyelin-induced dendritic spine anomalies in a Niemann Pick disease type A mouse model. *EMBO Molecular Medicine*, *6*, 398–413. https://doi.org/10.1002/emmm.201302649.

Arroyo, E. D., Fiole, D., Mantri, S. S., Huang, C., & Portera-Cailliau, C. (2019). Dendritic spines in early postnatal fragile X mice are insensitive to novel sensory experience. *The Journal of Neuroscience*, *39*, 412–419. https://doi.org/10.1523/JNEUROSCI.1734-18.2018.

Bagni, C., Tassone, F., Neri, G., & Hagerman, R. (2012). Fragile X syndrome: Causes, diagnosis, mechanisms, and therapeutics. *The Journal of Clinical Investigation*, *122*, 4314–4322. https://doi.org/10.1172/JCI63141.

Bailey, C. D. C., De Biasi, M., Fletcher, P. J., & Lambe, E. K. (2010). The nicotinic acetylcholine receptor alpha5 subunit plays a key role in attention circuitry and accuracy. *The Journal of Neuroscience*, *30*, 9241–9252. https://doi.org/10.1523/JNEUROSCI.2258-10.2010.

Bailey, C. D. C., Alves, N. C., Nashmi, R., De Biasi, M., & Lambe, E. K. (2012). Nicotinic α5 subunits drive developmental changes in the activation and morphology of prefrontal cortex layer VI neurons. *Biological Psychiatry*, *71*, 120–128. https://doi.org/10.1016/j.biopsych.2011.09.011.

Bailey, C. D. C., Tian, M. K., Kang, L., O'Reilly, R., & Lambe, E. K. (2014). Chrna5 genotype determines the long-lasting effects of developmental in vivo nicotine exposure on prefrontal attention circuitry. *Neuropharmacology*, *77*, 145–155. https://doi.org/10.1016/j.neuropharm.2013.09.003.

Ballesteros-Yáñez, I., Benavides-Piccione, R., Bourgeois, J. P., Changeux, J. P., & DeFelipe, J. (2010). Alterations of cortical pyramidal neurons in mice lacking high-affinity nicotinic receptors. *Proceedings of the National Academy of Sciences of the United States of America*, *107*, 11567–11572. https://doi.org/10.1073/pnas.1006269107.

Barbosa, A. C., Kim, M.-S., Ertunc, M., Adachi, M., Nelson, E. D., McAnally, J., et al. (2008). MEF2C, a transcription factor that facilitates learning and memory by negative regulation of synapse numbers and function. *Proceedings of the National Academy of Sciences of the United States of America*, *105*, 9391–9396. https://doi.org/10.1073/pnas.0802679105.

Bauman, M. L., Kemper, T. L., & Arin, D. M. (1995a). Microscopic observations of the brain in Rett syndrome. *Neuropediatrics*, *26*, 105–108. https://doi.org/10.1055/s-2007-979737.

Bauman, M. L., Kemper, T. L., & Arin, D. M. (1995b). Pervasive neuroanatomic abnormalities of the brain in three cases of Rett's syndrome. *Neurology, 45*, 1581–1586. https://doi.org/10.1212/wnl.45.8.1581.

Bemben, M. A., Shipman, S. L., Nicoll, R. A., & Roche, K. W. (2015). The cellular and molecular landscape of neuroligins. *Trends in Neurosciences, 38*, 496–505. https://doi.org/10.1016/j.tins.2015.06.004.

Benito, E., & Barco, A. (2015). The neuronal activity-driven transcriptome. *Molecular Neurobiology, 51*, 1071–1088. https://doi.org/10.1007/s12035-014-8772-z.

Bentley, P., Driver, J., & Dolan, R. J. (2011). Cholinergic modulation of cognition: Insights from human pharmacological functional neuroimaging. *Progress in Neurobiology, 94*, 360–388. https://doi.org/10.1016/j.pneurobio.2011.06.002.

Berg, D. K., & Conroy, W. G. (2002). Nicotinic alpha 7 receptors: Synaptic options and downstream signaling in neurons. *Journal of Neurobiology, 53*, 512–523. https://doi.org/10.1002/neu.10116.

Bertrand, D., Galzi, J. L., Devillers-Thiéry, A., Bertrand, S., & Changeux, J. P. (1993). Mutations at two distinct sites within the channel domain M2 alter calcium permeability of neuronal alpha 7 nicotinic receptor. *Proceedings of the National Academy of Sciences of the United States of America, 90*, 6971–6975. https://doi.org/10.1073/pnas.90.15.6971.

Bird, A. (2008). The methyl-CpG-binding protein MeCP2 and neurological disease. *Biochemical Society Transactions, 36*, 575–583. https://doi.org/10.1042/BST0360575.

Blundell, J., Blaiss, C. A., Etherton, M. R., Espinosa, F., Tabuchi, K., Walz, C., et al. (2010). Neuroligin-1 deletion results in impaired spatial memory and increased repetitive behavior. *Journal of Neuroscience, 30*(2115), LP–2129. https://doi.org/10.1523/JNEUROSCI.4517-09.2010.

Bonnet-Brilhault, F. (2017). Autism: An early neurodevelopmental disorder. *Archives de Pédiatrie, 24*, 384–390. https://doi.org/10.1016/j.arcped.2017.01.014.

Borovac, J., Bosch, M., & Okamoto, K. (2018). Regulation of actin dynamics during structural plasticity of dendritic spines: Signaling messengers and actin-binding proteins. *Molecular and Cellular Neurosciences, 91*, 122–130. https://doi.org/10.1016/j.mcn.2018.07.001.

Borroni, V., Kamerbeek, C., Pediconi, M. F., & Barrantes, F. J. (2020). Lovastatin differentially regulates α7 and α4 neuronal nicotinic acetylcholine receptor levels in rat hippocampal neurons. *Molecules, 25*, 4838. https://doi.org/10.3390/molecules25204838.

Bramham, C. R., & Messaoudi, E. (2005). BDNF function in adult synaptic plasticity: The synaptic consolidation hypothesis. *Progress in Neurobiology, 76*, 99–125. https://doi.org/10.1016/j.pneurobio.2005.06.003.

Brose, N., Betz, A., & Wegmeyer, H. (2004). Divergent and convergent signaling by the diacylglycerol second messenger pathway in mammals. *Current Opinion in Neurobiology, 14*, 328–340. https://doi.org/10.1016/j.conb.2004.05.006.

Brusés, J. L., Chauvet, N., & Rutishauser, U. (2001). Membrane lipid rafts are necessary for the maintenance of the (alpha)7 nicotinic acetylcholine receptor in somatic spines of ciliary neurons. *The Journal of Neuroscience, 21*, 504–512. https://doi.org/10.1523/JNEUROSCI.21-02-00504.2001.

Campbell, N. R., Fernandes, C. C., Halff, A. W., & Berg, D. K. (2010). Endogenous signaling through α7-containing nicotinic receptors promotes maturation and integration of adult-born neurons in the hippocampus. *The Journal of Neuroscience, 30*, 8734–8744. https://doi.org/10.1523/JNEUROSCI.0931-10.2010.

Carrasco, P., Sahún, I., McDonald, J., Ramírez, S., Jacas, J., Gratacós, E., et al. (2012). Ceramide levels regulated by carnitine palmitoyltransferase 1C control dendritic spine maturation and cognition. *The Journal of Biological Chemistry, 287*, 21224–21232. https://doi.org/10.1074/jbc.M111.337493.

Cartocci, V., Segatto, M., Di Tunno, I., Leone, S., Pfrieger, F. W., & Pallottini, V. (2016). Modulation of the isoprenoid/cholesterol biosynthetic pathway during neuronal differentiation in vitro. *Journal of Cellular Biochemistry*, *117*, 2036–2044. https://doi.org/10.1002/jcb.25500.

Cartocci, V., Catallo, M., Tempestilli, M., Segatto, M., Pfrieger, F. W., Bronzuoli, M. R., et al. (2018). Altered brain cholesterol/isoprenoid metabolism in a rat model of autism spectrum disorders. *Neuroscience*, *372*, 27–37. https://doi.org/10.1016/j.neuroscience.2017.12.053.

Chahrour, M., & Zoghbi, H. Y. (2007). The story of Rett syndrome: From clinic to neurobiology. *Neuron*, *56*, 422–437. https://doi.org/10.1016/j.neuron.2007.10.001.

Chang, Q., Khare, G., Dani, V., Nelson, S., & Jaenisch, R. (2006). The disease progression of Mecp2 mutant mice is affected by the level of BDNF expression. *Neuron*, *49*, 341–348. https://doi.org/10.1016/j.neuron.2005.12.027.

Chang, J.-Y., Parra-Bueno, P., Laviv, T., Szatmari, E. M., Lee, S.-J. R., & Yasuda, R. (2017). CaMKII autophosphorylation is necessary for optimal integration of Ca(2+) signals during LTP induction, but not maintenance. *Neuron*, *94*, 800–808.e4. https://doi.org/10.1016/j.neuron.2017.04.041.

Chao, H.-T., Zoghbi, H. Y., & Rosenmund, C. (2007). MeCP2 controls excitatory synaptic strength by regulating glutamatergic synapse number. *Neuron*, *56*, 58–65. https://doi.org/10.1016/j.neuron.2007.08.018.

Chapleau, C. A., Calfa, G. D., Lane, M. C., Albertson, A. J., Larimore, J. L., Kudo, S., et al. (2009a). Dendritic spine pathologies in hippocampal pyramidal neurons from Rett syndrome brain and after expression of Rett-associated MECP2 mutations. *Neurobiology of Disease*, *35*, 219–233. https://doi.org/10.1016/j.nbd.2009.05.001.

Chapleau, C. A., Larimore, J. L., Theibert, A., & Pozzo-Miller, L. (2009b). Modulation of dendritic spine development and plasticity by BDNF and vesicular trafficking: Fundamental roles in neurodevelopmental disorders associated with mental retardation and autism. *Journal of Neurodevelopmental Disorders*, *1*, 185–196. https://doi.org/10.1007/s11689-009-9027-6.

Chapleau, C. A., Boggio, E. M., Calfa, G., Percy, A. K., Giustetto, M., & Pozzo-Miller, L. (2012). Hippocampal CA1 pyramidal neurons of Mecp2 mutant mice show a dendritic spine phenotype only in the presymptomatic stage. *Neural Plasticity*, *2012*, 976164. https://doi.org/10.1155/2012/976164.

Charrin, S., Manié, S., Thiele, C., Billard, M., Gerlier, D., Boucheix, C., et al. (2003). A physical and functional link between cholesterol and tetraspanins. *European Journal of Immunology*, *33*, 2479–2489. https://doi.org/10.1002/eji.200323884.

Chaudhary, R., Agarwal, V., Kaushik, A. S., & Rehman, M. (2021). Involvement of myocyte enhancer factor 2c in the pathogenesis of autism spectrum disorder. *Heliyon*, 7, e06854. https://doi.org/10.1016/j.heliyon.2021.e06854.

Chen, W. G., Chang, Q., Lin, Y., Meissner, A., West, A. E., Griffith, E. C., et al. (2003). Derepression of BDNF transcription involves calcium-dependent phosphorylation of MeCP2. *Science*, *302*, 885–889. https://doi.org/10.1126/science.1086446.

Cheng, K., Chen, Y. S., Yue, C. X., Zhang, S. M., Pei, Y. P., Cheng, G. R., et al. (2019). Calsyntenin-1 negatively regulates ICAM5 accumulation in postsynaptic membrane and influences dendritic spine maturation in a mouse model of fragile X syndrome. *Frontiers in Neuroscience*, *13*. https://doi.org/10.3389/fnins.2019.01098.

Chilian, B., Abdollahpour, H., Bierhals, T., Haltrich, I., Fekete, G., Nagel, I., et al. (2013). Dysfunction of SHANK2 and CHRNA7 in a patient with intellectual disability and language impairment supports genetic epistasis of the two loci. *Clinical Genetics*, *84*, 560–565. https://doi.org/10.1111/cge.12105.

Cijsouw, T., Ramsey, A. M., Lam, T. K. T., Carbone, B. E., Blanpied, T. A., & Biederer, T. (2018). Mapping the proteome of the synaptic cleft through proximity labeling reveals new cleft proteins. *Proteome, 6*. https://doi.org/10.3390/proteomes 6040048.

Conroy, W. G., Nai, Q., Ross, B., Naughton, G., & Berg, D. K. (2007). Postsynaptic neuroligin enhances presynaptic inputs at neuronal nicotinic synapses. *Developmental Biology, 307*, 79–91. https://doi.org/10.1016/j.ydbio.2007.04.017.

Costa, J. F., Dines, M., & Lamprecht, R. (2020). The role of Rac GTPase in dendritic spine morphogenesis and memory. *Frontiers in Synaptic Neuroscience, 12*, 12. https://doi.org/10.3389/fnsyn.2020.00012.

Cruz-Martín, A., Crespo, M., & Portera-Cailliau, C. (2010). Delayed stabilization of dendritic spines in fragile X mice. *The Journal of Neuroscience, 30*, 7793–7803. https://doi.org/10.1523/JNEUROSCI.0577-10.2010.

Dani, J. A., & Bertrand, D. (2007). Nicotinic acetylcholine receptors and nicotinic cholinergic mechanisms of the central nervous system. *Annual Review of Pharmacology and Toxicology, 47*, 699–729. https://doi.org/10.1146/annurev.pharmtox.47.120505.105214.

Davenport, E. C., Szulc, B. R., Drew, J., Taylor, J., Morgan, T., Higgs, N. F., et al. (2019). Autism and schizophrenia-associated CYFIP1 regulates the balance of synaptic excitation and inhibition. *Cell Reports, 26*, 2037–2051.e6. https://doi.org/10.1016/j.celrep.2019.01.092.

Denker, A., Bethani, I., Kröhnert, K., Körber, C., Horstmann, H., Wilhelm, B. G., et al. (2011). A small pool of vesicles maintains synaptic activity in vivo. *Proceedings of the National Academy of Sciences of the United States of America, 108*, 17177–17182. https://doi.org/10.1073/pnas.1112688108.

Dieterich, D. C., & Kreutz, M. R. (2016). Proteomics of the synapse—A quantitative approach to neuronal plasticity. *Molecular & Cellular Proteomics, 15*, 368–381. https://doi.org/10.1074/MCP.R115.051482.

Dietschy, J. M., & Turley, S. D. (2004). Thematic review series: Brain lipids cholesterol metabolism in the central nervous system during early development and in the mature animal. *Journal of Lipid Research, 45*, 1375–1397. https://doi.org/10.1194/jlr.R400004-JLR200.

Dosemeci, A., Weinberg, R. J., Reese, T. S., & Tao-Cheng, J. H. (2016). The postsynaptic density: There is more than meets the eye. *Frontiers in Synaptic Neuroscience, 8*, 23. https://doi.org/10.3389/fnsyn.2016.00023.

Doyle, C. A., & McDougle, C. J. (2012). Pharmacologic treatments for the behavioral symptoms associated with autism spectrum disorders across the lifespan. *Dialogues in Clinical Neuroscience, 14*, 263–279. https://doi.org/10.31887/dcns.2012.14.3/cdoyle.

Dumitriu, D., Hao, J., Hara, Y., Kaufmann, J., Janssen, W. G. M., Lou, W., et al. (2010). Selective changes in thin spine density and morphology in monkey prefrontal cortex correlate with aging-related cognitive impairment. *The Journal of Neuroscience, 30*, 7507–7515. https://doi.org/10.1523/JNEUROSCI.6410-09.2010.

Dunaevsky, A., Tashiro, A., Majewska, A., Mason, C., & Yuste, R. (1999). Developmental regulation of spine motility in the mammalian central nervous system. *Proceedings of the National Academy of Sciences of the United States of America, 96*, 13438–13443. https://doi.org/10.1073/pnas.96.23.13438.

Durand, C. M., Perroy, J., Loll, F., Perrais, D., Fagni, L., Bourgeron, T., et al. (2012). SHANK3 mutations identified in autism lead to modification of dendritic spine morphology via an actin-dependent mechanism. *Molecular Psychiatry, 17*, 71–84. https://doi.org/10.1038/mp.2011.57.

Dziobek, I., Gold, S. M., Wolf, O. T., & Convit, A. (2007). Hypercholesterolemia in Asperger syndrome: Independence from lifestyle, obsessive-compulsive behavior, and social anxiety. *Psychiatry Research, 149*, 321–324. https://doi.org/10.1016/j.psychres.2006.02.003.

Echeverria, V., Yarkov, A., & Aliev, G. (2016). Positive modulators of the α7 nicotinic receptor against neuroinflammation and cognitive impairment in Alzheimer's disease. *Progress in Neurobiology*, *144*, 142–157. https://doi.org/10.1016/J.PNEUROBIO.2016.01.002.

Eissa, N., Al-Houqani, M., Sadeq, A., Ojha, S. K., Sasse, A., & Sadek, B. (2018). Current enlightenment about etiology and pharmacological treatment of autism spectrum disorder. *Frontiers in Neuroscience*, *12*, 304. https://doi.org/10.3389/fnins.2018.00304.

El-Husseini, A. E.-D., Schnell, E., Chetkovich, D. M., Nicoll, R. A., & Bredt, D. S. (2000). PSD-95 involvement in maturation of excitatory synapses. *Science*, *290*(80), 1364–1368. https://doi.org/10.1126/SCIENCE.290.5495.1364.

Eltokhi, A., Rappold, G., & Sprengel, R. (2018). Distinct phenotypes of shank2 mouse models reflect neuropsychiatric spectrum disorders of human patients with SHANK2 variants. *Frontiers in Molecular Neuroscience*, *11*, 1–15. https://doi.org/10.3389/fnmol.2018.00240.

Eltokhi, A., Santuy, A., Merchan-Perez, A., & Sprengel, R. (2021). Glutamatergic dysfunction and synaptic ultrastructural alterations in schizophrenia and autism spectrum disorder: Evidence from human and rodent studies. *International Journal of Molecular Sciences*, *22*, 1–26. https://doi.org/10.3390/ijms22010059.

Erickson, C. A., Posey, D. J., Stigler, K. A., Mullett, J., Katschke, A. R., & McDougle, C. J. (2007). A retrospective study of memantine in children and adolescents with pervasive developmental disorders. *Psychopharmacology*, *191*, 141–147. https://doi.org/10.1007/s00213-006-0518-9.

Erickson, C. A., Early, M., Stigler, K. A., Wink, L. K., Mullett, J. E., & McDougle, C. J. (2011). An open-label naturalistic pilot study of acamprosate in youth with autistic disorder. *Journal of Child and Adolescent Psychopharmacology*, *21*, 565–569. https://doi.org/10.1089/cap.2011.0034.

Erickson, C. A., Wink, L. K., Ray, B., Early, M. C., Stiegelmeyer, E., Mathieu-Frasier, L., et al. (2013). Impact of acamprosate on behavior and brain-derived neurotrophic factor: An open-label study in youth with fragile X syndrome. *Psychopharmacology*, *228*, 75–84. https://doi.org/10.1007/s00213-013-3022-z.

Fantini, J., & Barrantes, F. J. (2013). How cholesterol interacts with membrane proteins: An exploration of cholesterol-binding sites including CRAC, CARC, and tilted domains. *Frontiers in Physiology*, *4*, 31. https://doi.org/10.3389/fphys.2013.00031.

Fernandes, C. C., Berg, D. K., & Gómez-Varela, D. (2010). Lateral mobility of nicotinic acetylcholine receptors on neurons is determined by receptor composition, local domain, and cell type. *The Journal of Neuroscience*, *30*, 8841–8851. https://doi.org/10.1523/JNEUROSCI.6236-09.2010.

Fiala, J. C., Feinberg, M., Popov, V., & Harris, K. M. (1998). Synaptogenesis via dendritic filopodia in developing hippocampal area CA1. *The Journal of Neuroscience*, *18*, 8900–8911. https://doi.org/10.1523/jneurosci.18-21-08900.1998.

Flavell, S. W., Cowan, C. W., Kim, T.-K., Greer, P. L., Lin, Y., Paradis, S., et al. (2006). Activity-dependent regulation of MEF2 transcription factors suppresses excitatory synapse number. *Science*, *311*, 1008–1012. https://doi.org/10.1126/science.1122511.

Flavell, S. W., Kim, T.-K., Gray, J. M., Harmin, D. A., Hemberg, M., Hong, E. J., et al. (2008). Genome-wide analysis of MEF2 transcriptional program reveals synaptic target genes and neuronal activity-dependent polyadenylation site selection. *Neuron*, *60*, 1022–1038. https://doi.org/10.1016/j.neuron.2008.11.029.

Föger, N., Marhaba, R., & Zöller, M. (2001). Involvement of CD44 in cytoskeleton rearrangement and raft reorganization in T cells. *Journal of Cell Science*, *114*, 1169–1178. https://doi.org/10.1242/jcs.114.6.1169.

Forrest, M. P., Parnell, E., & Penzes, P. (2018). Dendritic structural plasticity and neuropsychiatric disease. *Nature Reviews. Neuroscience*, *19*, 215–234. https://doi.org/10.1038/nrn.2018.16.

Franco-Villanueva, A., Fernández-López, E., Gabandé-Rodríguez, E., Bañón-Rodríguez, I., Esteban, J. A., Antón, I. M., et al. (2014). WIP modulates dendritic spine actin cytoskeleton by transcriptional control of lipid metabolic enzymes. *Human Molecular Genetics*, *23*, 4383–4395. https://doi.org/10.1093/hmg/ddu155.

Frank, C., Rufini, S., Tancredi, V., Forcina, R., Grossi, D., & D'Arcangelo, G. (2008). Cholesterol depletion inhibits synaptic transmission and synaptic plasticity in rat hippocampus. *Experimental Neurology*, *212*, 407–414. https://doi.org/10.1016/j.expneurol.2008.04.019.

Fujii, H., Inoue, M., Okuno, H., Sano, Y., Takemoto-Kimura, S., Kitamura, K., et al. (2013). Nonlinear decoding and asymmetric representation of neuronal input information by CaMKIIα and calcineurin. *Cell Reports*, *3*, 978–987. https://doi.org/10.1016/j.celrep.2013.03.033.

Gillberg, C., Fernell, E., Kočovská, E., Minnis, H., Bourgeron, T., Thompson, L., et al. (2017). The role of cholesterol metabolism and various steroid abnormalities in autism spectrum disorders: A hypothesis paper. *Autism Research*, *10*, 1022–1044. https://doi.org/10.1002/aur.1777.

Glantz, L. A., & Lewis, D. A. (2000). Decreased dendritic spine density on prefrontal cortical pyramidal neurons in schizophrenia. *Archives of General Psychiatry*, *57*, 65–73. https://doi.org/10.1001/archpsyc.57.1.65.

Goodell, D. J., Zaegel, V., Coultrap, S. J., Hell, J. W., & Bayer, K. U. (2017). DAPK1 mediates LTD by making CaMKII/GluN2B binding LTP specific. *Cell Reports*, *19*, 2231–2243. https://doi.org/10.1016/j.celrep.2017.05.068.

Grizzell, J. A., Patel, S., Barreto, G. E., & Echeverria, V. (2017). Cotinine improves visual recognition memory and decreases cortical Tau phosphorylation in the Tg6799 mice. *Progress in Neuro-Psychopharmacology & Biological Psychiatry*, *78*, 75–81. https://doi.org/10.1016/j.pnpbp.2017.05.010.

Grossman, A. W., Aldridge, G. M., Lee, K. J., Zeman, M. K., Jun, C. S., Azam, H. S., et al. (2010). Developmental characteristics of dendritic spines in the dentate gyrus of Fmr1 knockout mice. *Brain Research*, *1355*, 221–227. https://doi.org/10.1016/j.brainres.2010.07.090.

Hagenston, A. M., & Bading, H. (2011). Calcium signaling in synapse-to-nucleus communication. *Cold Spring Harbor Perspectives in Biology*, *3*, a004564. https://doi.org/10.1101/cshperspect.a004564.

Halff, A. W., Gómez-Varela, D., John, D., & Berg, D. K. (2014). A novel mechanism for nicotinic potentiation of glutamatergic synapses. *The Journal of Neuroscience*, *34*, 2051–2064. https://doi.org/10.1523/JNEUROSCI.2795-13.2014.

Hammond, G. R. V., & Schiavo, G. (2007). Polyphosphoinositol lipids: Under-PPInning synaptic function in health and disease. *Developmental Neurobiology*, *67*, 1232–1247. https://doi.org/10.1002/dneu.20509.

Hardan, A. Y., & Handen, B. L. (2002). A retrospective open trial of adjunctive donepezil in children and adolescents with autistic disorder. *Journal of Child and Adolescent Psychopharmacology*, *12*, 237–241. https://doi.org/10.1089/104454602760386923.

Hering, H., & Sheng, M. (2001). Dentritic spines: structure, dynamics and regulation. *Nature Reviews. Neuroscience*, *2*, 880–888. https://doi.org/10.1038/35104061.

Hering, H., Lin, C. C., & Sheng, M. (2003). Lipid rafts in the maintenance of synapses, dendritic spines, and surface AMPA receptor stability. *The Journal of Neuroscience*, *23*, 3262–3271. https://doi.org/10.1523/jneurosci.23-08-03262.2003.

Hung, A. Y., Futai, K., Sala, C., Valtschanoff, J. G., Ryu, J., Woodworth, M. A., et al. (2008). Smaller dendritic spines, weaker synaptic transmission, but enhanced spatial learning in mice lacking Shank1. *The Journal of Neuroscience*, *28*, 1697–1708. https://doi.org/10.1523/JNEUROSCI.3032-07.2008.

Hutsler, J. J., & Zhang, H. (2010). Increased dendritic spine densities on cortical projection neurons in autism spectrum disorders. *Brain Research*, *1309*, 83–94. https://doi.org/10.1016/j.brainres.2009.09.120.

Itoh, K., Sakakibara, M., Yamasaki, S., Takeuchi, A., Arase, H., Miyazaki, M., et al. (2002). Cutting edge: Negative regulation of immune synapse formation by anchoring lipid raft to cytoskeleton through Cbp-EBP50-ERM assembly. *Journal of Immunology, 168*, 541–544. https://doi.org/10.4049/jimmunol.168.2.541.

Jamain, S., Quach, H., Betancur, C., Råstam, M., Colineaux, C., Gillberg, I. C., et al. (2003). Mutations of the X-linked genes encoding neuroligins NLGN3 and NLGN4 are associated with autism. *Nature Genetics, 34*, 27–29. https://doi.org/10.1038/ng1136.

Jones, A. K., & Sattelle, D. B. (2010). Diversity of insect nicotinic acetylcholine receptor subunits. *Advances in Experimental Medicine and Biology, 683*, 25–43. https://doi.org/10.1007/978-1-4419-6445-8_3.

Kang, L., Tian, M. K., Bailey, C. D. C., & Lambe, E. K. (2015). Dendritic spine density of prefrontal layer 6 pyramidal neurons in relation to apical dendrite sculpting by nicotinic acetylcholine receptors. *Frontiers in Cellular Neuroscience, 9*. https://doi.org/10.3389/fncel.2015.00398.

Kasai, H., Matsuzaki, M., Noguchi, J., Yasumatsu, N., & Nakahara, H. (2003). Structure-stability-function relationships of dendritic spines. *Trends in Neurosciences, 26*, 360–368. https://doi.org/10.1016/S0166-2236(03)00162-0.

Kashima, R., Roy, S., Ascano, M., Martinez-Cerdeno, V., Ariza-Torres, J., Kim, S., et al. (2016). Augmented noncanonical BMP type II receptor signaling mediates the synaptic abnormality of fragile X syndrome. *Science Signaling, 9*, ra58. https://doi.org/10.1126/scisignal.aaf6060.

Kaufmann, W. E., Taylor, C. V., Hohmann, C. F., Sanwal, I. B., & Naidu, S. (1997). Abnormalities in neuronal maturation in Rett syndrome neocortex: Preliminary molecular correlates. *European Child & Adolescent Psychiatry, 6*(Suppl. 1), 75–77.

Kennedy, M. B. (2000). Signal-processing machines at the postsynaptic density. *Science, 290*, 750–754. https://doi.org/10.1126/science.290.5492.750.

Kim, K., Lakhanpal, G., Lu, H. E., Khan, M., Suzuki, A., Hayashi, M. K., et al. (2015). A temporary gating of actin remodeling during synaptic plasticity consists of the interplay between the kinase and structural functions of CaMKII. *Neuron, 87*, 813–826. https://doi.org/10.1016/j.neuron.2015.07.023.

Kim, J.-W., Park, K., Kang, R. J., Gonzales, E. L. T., Kim, D. G., Oh, H. A., et al. (2019). Pharmacological modulation of AMPA receptor rescues social impairments in animal models of autism. *Neuropsychopharmacology: Official Publication of the American College of Neuropsychopharmacology, 44*, 314–323. https://doi.org/10.1038/s41386-018-0098-5.

Koudinov, A. R., & Koudinova, N. V. (2001). Essential role for cholesterol in synaptic plasticity and neuronal degeneration. *The FASEB Journal, 15*, 1858–1860. https://doi.org/10.1096/fj.00-0815fje.

Krönke, M. (1999). Biophysics of ceramide signaling: Interaction with proteins and phase transition of membranes. *Chemistry and Physics of Lipids, 101*, 109–121. https://doi.org/10.1016/s0009-3084(99)00059-6.

Kumar, V., Zhang, M. X., Swank, M. W., Kunz, J., & Wu, G. Y. (2005). Regulation of dendritic morphogenesis by Ras-PI3K-Akt-mTOR and Ras-MAPK signaling pathways. *The Journal of Neuroscience, 25*, 11288–11299. https://doi.org/10.1523/JNEUROSCI.2284-05.2005.

Lai, M.-C., Lombardo, M. V., & Baron-Cohen, S. (2014). Autism. *Lancet (London, England), 383*, 896–910. https://doi.org/10.1016/S0140-6736(13)61539-1.

Lamprecht, R. (2014). The actin cytoskeleton in memory formation. *Progress in Neurobiology, 117*, 1–19. https://doi.org/10.1016/j.pneurobio.2014.02.001.

Landi, S., Putignano, E., Boggio, E. M., Giustetto, M., Pizzorusso, T., & Ratto, G. M. (2011). The short-time structural plasticity of dendritic spines is altered in a model of Rett syndrome. *Scientific Reports, 1*, 45. https://doi.org/10.1038/srep00045.

Laumonnier, F., Bonnet-Brilhault, F., Gomot, M., Blanc, R., David, A., Moizard, M.-P., et al. (2004). X-linked mental retardation and autism are associated with a mutation in the NLGN4 gene, a member of the neuroligin family. *American Journal of Human Genetics*, *74*, 552–557. https://doi.org/10.1086/382137.

Leblond, C. S., Heinrich, J., Delorme, R., Proepper, C., Betancur, C., Huguet, G., et al. (2012). Genetic and functional analyses of SHANK2 mutations suggest a multiple hit model of autism spectrum disorders. *PLoS Genetics*, *8*. https://doi.org/10.1371/journal.pgen.1002521.

Linetti, A., Fratangeli, A., Taverna, E., Valnegri, P., Francolini, M., Cappello, V., et al. (2010). Cholesterol reduction impairs exocytosis of synaptic vesicles. *Journal of Cell Science*, *123*, 595–605. https://doi.org/10.1242/jcs.060681.

Lo, L. H.-Y., & Lai, K.-O. (2020). Dysregulation of protein synthesis and dendritic spine morphogenesis in ASD: Studies in human pluripotent stem cells. *Molecular Autism*, *11*, 40. https://doi.org/10.1186/s13229-020-00349-y.

Lord, C. (1995). Follow-up of two-year-olds referred for possible autism. *Journal of Child Psychology and Psychiatry*, *36*, 1365–1382. https://doi.org/10.1111/j.1469-7610.1995.tb01669.x.

Lord, C., Elsabbagh, M., Baird, G., & Veenstra-Vanderweele, J. (2018). Autism spectrum disorder. *Lancet (London, England)*, *392*, 508–520. https://doi.org/10.1016/S0140-6736(18)31129-2.

Lozada, A. F., Wang, X., Gounko, N. V., Massey, K. A., Duan, J., Liu, Z., et al. (2012a). Glutamatergic synapse formation is promoted by α7-containing nicotinic acetylcholine receptors. *The Journal of Neuroscience*, *32*, 7651–7661. https://doi.org/10.1523/JNEUROSCI.6246-11.2012.

Lozada, A. F., Wang, X., Gounko, N. V., Massey, K. A., Duan, J., Liu, Z., et al. (2012b). Induction of dendritic spines by β2-containing nicotinic receptors. *The Journal of Neuroscience*, *32*, 8391–8400. https://doi.org/10.1523/JNEUROSCI.6247-11.2012.

Lutz, A.-K., Pfaender, S., Incearap, B., Ioannidis, V., Ottonelli, I., Föhr, K., et al. (2020). Autism-associated SHANK3 mutations impair maturation of neuromuscular junctions and striated muscles. *Science Translational Medicine*, *12*, eaaz3267. https://doi.org/10.1126/scitranslmed.aaz3267.

Maggi, L., Le Magueresse, C., Changeux, J.-P., & Cherubini, E. (2003). Nicotine activates immature "silent" connections in the developing hippocampus. *Proceedings of the National Academy of Sciences of the United States of America*, *100*, 2059–2064. https://doi.org/10.1073/pnas.0437947100.

Marchand, S., Devillers-Thiéry, A., Pons, S., Changeux, J.-P., & Cartaud, J. (2002). Rapsyn escorts the nicotinic acetylcholine receptor along the exocytic pathway via association with lipid rafts. *The Journal of Neuroscience*, *22*, 8891–8901. https://doi.org/10.1523/JNEUROSCI.22-20-08891.2002.

Marchi, M., Risso, F., Viola, C., Cavazzani, P., & Raiteri, M. (2002). Direct evidence that release-stimulating alpha7* nicotinic cholinergic receptors are localized on human and rat brain glutamatergic axon terminals. *Journal of Neurochemistry*, *80*, 1071–1078. https://doi.org/10.1046/j.0022-3042.2002.00805.x.

Marotta, R., Risoleo, M. C., Messina, G., Parisi, L., Carotenuto, M., Vetri, L., et al. (2020). The Neurochemistry of Autism. *Brain Sciences*, *10*. https://doi.org/10.3390/brainsci10030163.

Martin, J., Cooper, M., Hamshere, M. L., Pocklington, A., Scherer, S. W., Kent, L., et al. (2014a). Biological overlap of attention-deficit/hyperactivity disorder and autism spectrum disorder: Evidence from copy number variants. *Journal of the American Academy of Child and Adolescent Psychiatry*, 53. https://doi.org/10.1016/j.jaac.2014.03.004. 761–70.e26.

Martin, M. G., Ahmed, T., Korovaichuk, A., Venero, C., Menchón, S. A., Salas, I., et al. (2014b). Constitutive hippocampal cholesterol loss underlies poor cognition in old rodents. *EMBO Molecular Medicine*, *6*, 902–917. https://doi.org/10.15252/emmm.201303711.

Martínez-Cerdeño, V. (2017). Dendrite and spine modifications in autism and related neurodevelopmental disorders in patients and animal models. *Developmental Neurobiology*, *77*, 393–404. https://doi.org/10.1002/dneu.22417.

Martinowich, K., Hattori, D., Wu, H., Fouse, S., He, F., Hu, Y., et al. (2003). DNA methylation-related chromatin remodeling in activity-dependent BDNF gene regulation. *Science*, *302*, 890–893. https://doi.org/10.1126/science.1090842.

Massey, K. A., Zago, W. M., & Berg, D. K. (2006). BDNF up-regulates α7 nicotinic acetylcholine receptor levels on subpopulations of hippocampal interneurons. *Molecular and Cellular Neurosciences*, *33*, 381–388. https://doi.org/10.1016/J.MCN.2006.08.011.

Matsuzaki, M., Ellis-Davies, G. C. R., Nemoto, T., Miyashita, Y., Iino, M., & Kasai, H. (2001). Dendritic spine geometry is critical for AMPA receptor expression in hippocampal CA1 pyramidal neurons. *Nature Neuroscience*, *4*, 1086–1092. https://doi.org/10.1038/nn736.

Maurer, S. V., & Williams, C. L. (2017). The cholinergic system modulates memory and hippocampal plasticity via its interactions with non-neuronal cells. *Frontiers in Immunology*, *8*, 1489. https://doi.org/10.3389/fimmu.2017.01489.

McKinney, R. A. (2010). Excitatory amino acid involvement in dendritic spine formation, maintenance and remodelling. *The Journal of Physiology*, *588*, 107–116. https://doi.org/10.1113/jphysiol.2009.178905.

McKinney, B. C., Grossman, A. W., Elisseou, N. M., & Greenough, W. T. (2005). Dendritic spine abnormalities in the occipital cortex of C57BL/6 Fmr1 knockout mice. *American Journal of Medical Genetics. Part B, Neuropsychiatric Genetics: The Official Publication of the International Society of Psychiatric Genetics*, *136B*, 98–102. https://doi.org/10.1002/ajmg.b.30183.

McQuiston, A. R. (2014). Acetylcholine release and inhibitory interneuron activity in hippocampal CA1. *Frontiers in Synaptic Neuroscience*, *6*, 20. https://doi.org/10.3389/fnsyn.2014.00020.

Mencel, M., Nash, M., & Jacobson, C. (2013). Neuregulin upregulates microglial α7 nicotinic acetylcholine receptor expression in immortalized cell lines: implications for regulating neuroinflammation. *PLoS One*, *8*, e70338. https://doi.org/10.1371/journal.pone.0070338.

Moretto, E., Longatti, A., Murru, L., Chamma, I., Sessa, A., Zapata, J., et al. (2019). TSPAN5 enriched microdomains provide a platform for dendritic spine maturation through neuroligin-1 clustering. *Cell Reports*, *29*, 1130–1146.e8. https://doi.org/10.1016/j.celrep.2019.09.051.

Murakoshi, H., Wang, H., & Yasuda, R. (2011). Local, persistent activation of Rho GTPases during plasticity of single dendritic spines. *Nature*, *472*, 100. https://doi.org/10.1038/NATURE09823.

Murru, L., Vezzoli, E., Longatti, A., Ponzoni, L., Falqui, A., Folci, A., et al. (2017). Pharmacological modulation of AMPAR rescues intellectual disability-like phenotype in Tm4sf2-/y mice. *Cerebral Cortex*, *27*, 5369–5384. https://doi.org/10.1093/cercor/bhx221.

Murru, L., Moretto, E., Martano, G., & Passafaro, M. (2018). Tetraspanins shape the synapse. *Molecular and Cellular Neurosciences*, *91*, 76–81. https://doi.org/10.1016/j.mcn.2018.04.001.

Mychasiuk, R., Muhammad, A., Gibb, R., & Kolb, B. (2013). Long-term alterations to dendritic morphology and spine density associated with prenatal exposure to nicotine. *Brain Research*, *1499*, 53–60. https://doi.org/10.1016/j.brainres.2012.12.021.

Nagappan, G., & Lu, B. (2005). Activity-dependent modulation of the BDNF receptor TrkB: Mechanisms and implications. *Trends in Neurosciences*, *28*, 464–471. https://doi.org/10.1016/j.tins.2005.07.003.

Nagy, J., Kobolák, J., Berzsenyi, S., Ábrahám, Z., Avci, H. X., Bock, I., et al. (2017). Altered neurite morphology and cholinergic function of induced pluripotent stem cell-derived neurons from a patient with Kleefstra syndrome and autism. *Translational Psychiatry*, *7*, e1179. https://doi.org/10.1038/tp.2017.144.

Naisbitt, S., Kim, E., Tu, J. C., Xiao, B., Sala, C., Valtschanoff, J., et al. (1999). Shank, a novel family of postsynaptic density proteins that binds to the NMDA receptor/PSD-95/GKAP complex and cortactin. *Neuron*, *23*, 569–582. https://doi.org/10.1016/s0896-6273(00)80809-0.

Nakahata, Y., & Yasuda, R. (2018). Plasticity of spine structure: Local signaling, translation and cytoskeletal reorganization. *Frontiers in Synaptic Neuroscience*, *10*, 29. https://doi.org/10.3389/FNSYN.2018.00029.

Nguyen, M. V. C., Du, F., Felice, C. A., Shan, X., Nigam, A., Mandel, G., et al. (2012). MeCP2 is critical for maintaining mature neuronal networks and global brain anatomy during late stages of postnatal brain development and in the mature adult brain. *The Journal of Neuroscience*, *32*, 10021–10034. https://doi.org/10.1523/JNEUROSCI.1316-12.2012.

Oda, A., Yamagata, K., Nakagomi, S., Uejima, H., Wiriyasermkul, P., Ohgaki, R., et al. (2014). Nicotine induces dendritic spine remodeling in cultured hippocampal neurons. *Journal of Neurochemistry*, *128*, 246–255. https://doi.org/10.1111/jnc.12470.

Oeseburg, B., Dijkstra, G. J., Groothoff, J. W., Reijneveld, S. A., & Jansen, D. E. M. C. (2011). Prevalence of chronic health conditions in children with intellectual disability: a systematic literature review. *Intellectual and Developmental Disabilities*, *49*, 59–85. https://doi.org/10.1352/1934-9556-49.2.59.

Ohira, K., & Hayashi, M. (2009). A new aspect of the TrkB signaling pathway in neural plasticity. *Current Neuropharmacology*, *7*, 276–285. https://doi.org/10.2174/157015909790031210.

Okamoto, K.-I., Nagai, T., Miyawaki, A., & Hayashi, Y. (2004). Rapid and persistent modulation of actin dynamics regulates postsynaptic reorganization underlying bidirectional plasticity. *Nature Neuroscience*, *7*, 1104–1112. https://doi.org/10.1038/nn1311.

Okamoto, K.-I., Narayanan, R., Lee, S. H., Murata, K., & Hayashi, Y. (2007). The role of CaMKII as an F-actin-bundling protein crucial for maintenance of dendritic spine structure. *Proceedings of the National Academy of Sciences of the United States of America*, *104*, 6418–6423. https://doi.org/10.1073/pnas.0701656104.

Okamura, K., Tanaka, H., Yagita, Y., Saeki, Y., Taguchi, A., Hiraoka, Y., et al. (2004). Cadherin activity is required for activity-induced spine remodeling. *The Journal of Cell Biology*, *167*, 961–972. https://doi.org/10.1083/jcb.200406030.

Orth, M., & Bellosta, S. (2012). Cholesterol: Its regulation and role in central nervous system disorders. *Cholesterol*, *2012*, 292598. https://doi.org/10.1155/2012/292598.

Oshikawa, J., Toya, Y., Fujita, T., Egawa, M., Kawabe, J., Umemura, S., et al. (2003). Nicotinic acetylcholine receptor alpha 7 regulates cAMP signal within lipid rafts. *American Journal of Physiology. Cell Physiology*, *285*, C567–C574. https://doi.org/10.1152/ajpcell.00422.2002.

Panja, D., & Bramham, C. R. (2014). BDNF mechanisms in late LTP formation: A synthesis and breakdown. *Neuropharmacology*, *76*(Pt C), 664–676. https://doi.org/10.1016/j.neuropharm.2013.06.024.

Parnass, Z., Tashiro, A., & Yuste, R. (2000). Analysis of spine morphological plasticity in developing hippocampal pyramidal neurons. *Hippocampus*, *10*, 561–568. https://doi.org/10.1002/1098-1063(2000)10:5<561::AID-HIPO6>3.0.CO;2-X.

Peça, J., Feliciano, C., Ting, J. T., Wang, W., Wells, M. F., Venkatraman, T. N., et al. (2011). Shank3 mutant mice display autistic-like behaviours and striatal dysfunction. *Nature, 472*, 437–442. https://doi.org/10.1038/nature09965.

Penzes, P., & Rafalovich, I. (2012). Regulation of the actin cytoskeleton in dendritic spines. *Adv. Exp. Med. Biol., 970*, 81–95. https://doi.org/10.1007/978-3-7091-0932-8_4.

Penzes, P., Cahill, M. E., Jones, K. A., VanLeeuwen, J.-E., & Woolfrey, K. M. (2011). Dendritic spine pathology in neuropsychiatric disorders. *Nature Neuroscience, 14*, 285–293. https://doi.org/10.1038/nn.2741.

Perez, A. S., & Bredt, D. S. (1998). The N-terminal PDZ-containing region of postsynatpic density-95 mediates association with caveolar-like lipid domains. *Neuroscience Letters, 258*, 121–123. https://doi.org/10.1016/S0304-3940(98)00846-5.

Petrov, A. M., Kasimov, M. R., & Zefirov, A. L. (2017). Cholesterol in the pathogenesis of Alzheimer's, Parkinson's diseases and autism: Link to synaptic dysfunction. *Acta Naturae, 9*, 26–37.

Phillips, M., & Pozzo-Miller, L. (2014). Dendritic spine dysgenesis in autism related disorders. *Neuroscience Letters, 601*, 30–40. https://doi.org/10.1016/j.neulet.2015.01.011.

Pike, L. J. (2006). Rafts defined: A report on the keystone symposium on lipid rafts and cell function. *Journal of Lipid Research, 47*, 1597–1598. https://doi.org/10.1194/jlr.E600002-JLR200.

Piton, A., Gauthier, J., Hamdan, F. F., Lafrenière, R. G., Yang, Y., Henrion, E., et al. (2011). Systematic resequencing of X-chromosome synaptic genes in autism spectrum disorder and schizophrenia. *Molecular Psychiatry, 16*, 867–880. https://doi.org/10.1038/mp.2010.54.

Pozzo-Miller, L., Pati, S., & Percy, A. K. (2015). Rett Syndrome: Reaching for clinical trials. *Neurotherapeutics, 12*, 631–640. https://doi.org/10.1007/s13311-015-0353-y.

Prado, V. F., Janickova, H., Al-Onaizi, M. A., & Prado, M. A. M. (2017). Cholinergic circuits in cognitive flexibility. *Neuroscience, 345*, 130–141. https://doi.org/10.1016/j.neuroscience.2016.09.013.

Rezaei, V., Mohammadi, M. R., Ghanizadeh, A., Sahraian, A., Tabrizi, M., Rezazadeh, S. A., et al. (2010). Double-blind, placebo-controlled trial of risperidone plus topiramate in children with autistic disorder. *Progress in Neuro-Psychopharmacology & Biological Psychiatry, 34*, 1269–1272. https://doi.org/10.1016/j.pnpbp.2010.07.005.

Richards, D. A., Mateos, J. M., Hugel, S., De Paola, V., Caroni, P., Gähwiler, B. H., et al. (2005). Glutamate induces the rapid formation of spine head protrusions in hippocampal slice cultures. *Proceedings of the National Academy of Sciences of the United States of America, 102*, 6166–6171. https://doi.org/10.1073/pnas.0501881102.

Richter, K. N., Wildhagen, H., Helm, M. S., Ußling, J.-E., Schikorski, T., & Rizzoli, S. O. (2018). Comparative synaptosome imaging: A semi-quantitative method to obtain copy numbers for synaptic and neuronal proteins. *Scientific Reports, 8*, 14838. https://doi.org/10.1038/s41598-018-33130-6.

Rinaldi, T., Kulangara, K., Antoniello, K., & Markram, H. (2007). Elevated NMDA receptor levels and enhanced postsynaptic long-term potentiation induced by prenatal exposure to valproic acid. *Proceedings of the National Academy of Sciences of the United States of America, 104*, 13501–13506. https://doi.org/10.1073/pnas.0704391104.

Rojas, D. C. (2014). The role of glutamate and its receptors in autism and the use of glutamate receptor antagonists in treatment. *Journal of Neural Transmission, 121*, 891–905. https://doi.org/10.1007/s00702-014-1216-0.

Ross, B. S., & Conroy, W. G. (2008). Capabilities of neurexins in the chick ciliary ganglion. *Developmental Neurobiology, 68*, 409–419. https://doi.org/10.1002/dneu.20598.

Runge, K., Cardoso, C., & de Chevigny, A. (2020). Dendritic spine plasticity: Function and mechanisms. *Frontiers in Synaptic Neuroscience, 12*, 36. https://doi.org/10.3389/FNSYN.2020.00036.

Sabatini, B. L., Oertner, T. G., & Svoboda, K. (2002). The life cycle of Ca(2+) ions in dendritic spines. *Neuron, 33*, 439–452. https://doi.org/10.1016/s0896-6273(02)00573-1.

Sala, C., Piëch, V., Wilson, N. R., Passafaro, M., Liu, G., & Sheng, M. (2001). Regulation of dendritic spine morphology and synaptic function by Shank and Homer. *Neuron, 31*, 115–130. https://doi.org/10.1016/s0896-6273(01)00339-7.

Sala, C., Vicidomini, C., Bigi, I., Mossa, A., & Verpelli, C. (2015). Shank synaptic scaffold proteins: Keys to understanding the pathogenesis of autism and other synaptic disorders. *Journal of Neurochemistry, 135*, 849–858. https://doi.org/10.1111/jnc.13232.

Sanders, S. J., He, X., Willsey, A. J., Ercan-Sencicek, A. G., Samocha, K. E., Cicek, A. E., et al. (2015). Insights into autism spectrum disorder genomic architecture and biology from 71 risk loci. *Neuron, 87*, 1215–1233. https://doi.org/10.1016/j.neuron.2015.09.016.

Sato, Y., & Okabe, S. (2019). Nano-scale analysis of synapse morphology in an autism mouse model with 15q11-13 copy number variation using focused ion beam milling and scanning electron microscopy. *Microscopy, 68*, 122–132. https://doi.org/10.1093/jmicro/dfy128.

Sato, D., Lionel, A. C., Leblond, C. S., Prasad, A., Pinto, D., Walker, S., et al. (2012). SHANK1 deletions in males with autism spectrum disorder. *American Journal of Human Genetics, 90*, 879–887. https://doi.org/10.1016/j.ajhg.2012.03.017.

Schengrund, C.-L., Ali-Rahmani, F., & Ramer, J. C. (2012). Cholesterol, GM1, and autism. *Neurochemical Research, 37*, 1201–1207. https://doi.org/10.1007/s11064-011-0697-6.

Schwartz, C. E., & Neri, G. (2012). Autism and intellectual disability: Two sides of the same coin. *American Journal of Medical Genetics. Part C, Seminars in Medical Genetics, 160C*, 89–90. https://doi.org/10.1002/ajmg.c.31329.

Segatto, M., Leboffe, L., Trapani, L., & Pallottini, V. (2014). Cholesterol homeostasis failure in the brain: Implications for synaptic dysfunction and cognitive decline. *Current Medicinal Chemistry, 21*, 2788–2802. https://doi.org/10.2174/0929867321666140303142902.

Segatto, M., Tonini, C., Pfrieger, F. W., Trezza, V., & Pallottini, V. (2019). Loss of mevalonate/cholesterol homeostasis in the brain: A focus on autism spectrum disorder and Rett syndrome. *International Journal of Molecular Sciences, 20*, 3317. https://doi.org/10.3390/ijms20133317.

Séguéla, P., Wadiche, J., Dineley-Miller, K., Dani, J. A., & Patrick, J. W. (1993). Molecular cloning, functional properties, and distribution of rat brain alpha 7: A nicotinic cation channel highly permeable to calcium. *The Journal of Neuroscience, 13*, 596–604. https://doi.org/10.1523/JNEUROSCI.13-02-00596.1993.

Selkoe, D. J. (2002). Alzheimer's disease is a synaptic failure. *Science, 298*(80), 789–791. https://doi.org/10.1126/science.1074069.

Sheng, M., & Pak, D. T. (2000). Ligand-gated ion channel interactions with cytoskeletal and signaling proteins. *Annual Review of Physiology, 62*, 755–778. https://doi.org/10.1146/annurev.physiol.62.1.755.

Sheng, M., & Sala, C. (2001). PDZ domains and the organization of supramolecular complexes. *Annual Review of Neuroscience, 24*, 1–29. https://doi.org/10.1146/annurev.neuro.24.1.1.

Simons, K., & Ikonen, E. (1997). Functional rafts in cell membranes. *Nature, 387*, 569–572. https://doi.org/10.1038/42408.

Sodero, A. O., Weissmann, C., Ledesma, M. D., & Dotti, C. G. (2011). Cellular stress from excitatory neurotransmission contributes to cholesterol loss in hippocampal neurons aging in vitro. *Neurobiology of Aging, 32*, 1043–1053. https://doi.org/10.1016/j.neurobiolaging.2010.06.001.

Spence, E. F., & Soderling, S. H. (2015). Actin out: Regulation of the synaptic cytoskeleton. *The Journal of Biological Chemistry, 290*, 28613–28622. https://doi.org/10.1074/jbc.R115.655118.

Stessman, H. A. F., Xiong, B., Coe, B. P., Wang, T., Hoekzema, K., Fenckova, M., et al. (2017). Targeted sequencing identifies 91 neurodevelopmental-disorder risk genes with autism and developmental-disability biases. *Nature Genetics*, *49*, 515–526. https://doi.org/10.1038/ng.3792.

Südhof, T. C. (2008). Neuroligins and neurexins link synaptic function to cognitive disease. *Nature*, *455*, 903–911. https://doi.org/10.1038/nature07456.

Suzuki, T. (2002). Lipid rafts at postsynaptic sites: Distribution, function and linkage to postsynaptic density. *Neuroscience Research*, *44*, 1–9. https://doi.org/10.1016/s0168-0102(02)00080-9.

Suzuki, T., Zhang, J., Miyazawa, S., Liu, Q., Farzan, M. R., & Yao, W.-D. (2011). Association of membrane rafts and postsynaptic density: proteomics, biochemical, and ultrastructural analyses. *Journal of Neurochemistry*, *119*, 64–77. https://doi.org/10.1111/j.1471-4159.2011.07404.x.

Sweet, R. A., Henteleff, R. A., Zhang, W., Sampson, A. R., & Lewis, D. A. (2009). Reduced dendritic spine density in auditory cortex of subjects with schizophrenia. *Neuropsychopharmacology*, *34*, 374–389. https://doi.org/10.1038/npp.2008.67.

Taber, K. H., Hurley, R. A., & Yudofsky, S. C. (2010). Diagnosis and treatment of neuropsychiatric disorders. *Annual Review of Medicine*, *61*, 121–133. https://doi.org/10.1146/annurev.med.051408.105018.

Tackenberg, C., Ghori, A., & Brandt, R. (2009). Thin, stubby or mushroom: Spine pathology in Alzheimers disease. *Current Alzheimer Research*, *6*, 261–268. https://doi.org/10.2174/156720509788486554.

Tada, T., & Sheng, M. (2006). Molecular mechanisms of dendritic spine morphogenesis. *Current Opinion in Neurobiology*, *16*, 95–101. https://doi.org/10.1016/j.conb.2005.12.001.

Taly, A., Corringer, P.-J., Guedin, D., Lestage, P., & Changeux, J.-P. (2009). Nicotinic receptors: Allosteric transitions and therapeutic targets in the nervous system. *Nature Reviews. Drug Discovery*, *8*, 733–750. https://doi.org/10.1038/nrd2927.

Tønnesen, J., Katona, G., Rózsa, B., & Nägerl, U. V. (2014). Spine neck plasticity regulates compartmentalization of synapses. *Nature Neuroscience*, *17*, 678–685. https://doi.org/10.1038/nn.3682.

Toro, R., Konyukh, M., Delorme, R., Leblond, C., Chaste, P., Fauchereau, F., et al. (2010). Key role for gene dosage and synaptic homeostasis in autism spectrum disorders. *Trends in Genetics*, *26*, 363–372. https://doi.org/10.1016/j.tig.2010.05.007.

Ueda, Y., & Hayashi, Y. (2013). PIP3 regulates spinule formation in dendritic spines during structural long-term potentiation. *The Journal of Neuroscience*, *33*, 11040–11047. https://doi.org/10.1523/JNEUROSCI.3122-12.2013.

Um, S. M., Ha, S., Lee, H., Kim, J., Kim, K., Shin, W., et al. (2018). NGL-2 deletion leads to autistic-like behaviors responsive to NMDAR modulation. *Cell Reports*, *23*, 3839–3851. https://doi.org/10.1016/j.celrep.2018.05.087.

Urbano, M., Okwara, L., Manser, P., Hartmann, K., & Deutsch, S. I. (2015). A trial of D-cycloserine to treat the social deficit in older adolescents and young adults with autism spectrum disorders. *The Journal of Neuropsychiatry and Clinical Neurosciences*, *27*, 133–138. https://doi.org/10.1176/appi.neuropsych.13070155.

Uzunova, G., Hollander, E., & Shepherd, J. (2014). The role of ionotropic glutamate receptors in childhood neurodevelopmental disorders: Autism spectrum disorders and fragile X syndrome. *Current Neuropharmacology*, *12*, 71–98. https://doi.org/10.2174/1570159x113116660046.

Vallés, A. S., & Barrantes, F. J. (2012). Chaperoning α7 neuronal nicotinic acetylcholine receptors. *Biochimica et Biophysica Acta, Biomembranes*, *1818*, 718–729. https://doi.org/10.1016/J.BBAMEM.2011.10.012.

van Woerden, G. M., Hoebeek, F. E., Gao, Z., Nagaraja, R. Y., Hoogenraad, C. C., Kushner, S. A., et al. (2009). betaCaMKII controls the direction of plasticity at parallel fiber-Purkinje cell synapses. *Nature Neuroscience, 12*, 823–825. https://doi.org/10.1038/nn.2329.

Vanderklish, P. W., & Edelman, G. M. (2002). Dendritic spines elongate after stimulation of group 1 metabotropic glutamate receptors in cultured hippocampal neurons. *Proceedings of the National Academy of Sciences of the United States of America, 99*, 1639–1644. https://doi.org/10.1073/pnas.032681099.

Wang, H. (2014). Lipid rafts: A signaling platform linking cholesterol metabolism to synaptic deficits in autism spectrum disorders. *Frontiers in Behavioral Neuroscience, 8*, 104. https://doi.org/10.3389/fnbeh.2014.00104.

Wang, L., Almeida, L. E. F., Spornick, N. A., Kenyon, N., Kamimura, S., Khaibullina, A., et al. (2015). Modulation of social deficits and repetitive behaviors in a mouse model of autism: The role of the nicotinic cholinergic system. *Psychopharmacology, 232*, 4303–4316. https://doi.org/10.1007/s00213-015-4058-z.

Wefelmeyer, W., Puhl, C. J., & Burrone, J. (2016). Homeostatic plasticity of subcellular neuronal structures: From inputs to outputs. *Trends in Neurosciences, 39*, 656–667. https://doi.org/10.1016/j.tins.2016.08.004.

WHO | World Health Statistics. (2017). *Monitoring health for the SDGs (2017)*. WHO. Available at: http://www.who.int/gho/publications/world_health_statistics/2017/en/. (Accessed 4 September 2018).

Wijetunge, L. S., Angibaud, J., Frick, A., Kind, P. C., & Nägerl, U. V. (2014). Stimulated emission depletion (STED) microscopy reveals nanoscale defects in the developmental trajectory of dendritic spine morphogenesis in a mouse model of fragile X syndrome. *The Journal of Neuroscience, 34*, 6405–6412. https://doi.org/10.1523/JNEUROSCI.5302-13.2014.

Wilhelm, B. G., Mandad, S., Truckenbrodt, S., Kröhnert, K., Schäfer, C., Rammner, B., et al. (2014). Composition of isolated synaptic boutons reveals the amounts of vesicle trafficking proteins. *Science, 344*, 1023–1028. https://doi.org/10.1126/science.1252884.

Wilkerson, J. R., Tsai, N.-P., Maksimova, M. A., Wu, H., Cabalo, N. P., Loerwald, K. W., et al. (2014). A role for dendritic mGluR5-mediated local translation of Arc/Arg3.1 in MEF2-dependent synapse elimination. *Cell Reports*, 7, 1589–1600. https://doi.org/10.1016/j.celrep.2014.04.035.

Wu, S., Gan, G., Zhang, Z., Sun, J., Wang, Q., Gao, Z., et al. (2017). A presynaptic function of shank protein in Drosophila. *The Journal of Neuroscience, 37*, 11592–11604. https://doi.org/10.1523/JNEUROSCI.0893-17.2017.

Yap, E.-L., & Greenberg, M. E. (2018). Activity-regulated transcription: Bridging the gap between neural activity and behavior. *Neuron, 100*, 330–348. https://doi.org/10.1016/j.neuron.2018.10.013.

Yin, H. L., & Janmey, P. A. (2003). Phosphoinositide regulation of the actin cytoskeleton. *Annual Review of Physiology, 65*, 761–789. https://doi.org/10.1146/annurev.physiol.65.092101.142517.

Zaslavsky, K., Zhang, W. B., McCready, F. P., Rodrigues, D. C., Deneault, E., Loo, C., et al. (2019). SHANK2 mutations associated with autism spectrum disorder cause hyperconnectivity of human neurons. *Nature Neuroscience, 22*, 556–564. https://doi.org/10.1038/s41593-019-0365-8.

Zhang, H., Webb, D. J., Asmussen, H., Niu, S., & Horwitz, A. F. (2005). A GIT1/PIX/Rac/PAK signaling module regulates spine morphogenesis and synapse formation through MLC. *The Journal of Neuroscience, 25*, 3379–3388. https://doi.org/10.1523/JNEUROSCI.3553-04.2005.

Zheng, Z., Zhu, T., Qu, Y., & Mu, D. (2016). Blood glutamate levels in autism spectrum disorder: A systematic review and meta-analysis. *PLoS One*, *11*, e0158688. https://doi.org/10.1371/journal.pone.0158688.
Zhong, X., Li, H., & Chang, Q. (2012). MeCP2 phosphorylation is required for modulating synaptic scaling through mGluR5. *The Journal of Neuroscience*, *32*, 12841–12847. https://doi.org/10.1523/JNEUROSCI.2784-12.2012.
Zhou, X., Nai, Q., Chen, M., Dittus, J. D., Howard, M. J., & Margiotta, J. F. (2004). Brain-derived neurotrophic factor and trkB signaling in parasympathetic neurons: relevance to regulating alpha7-containing nicotinic receptors and synaptic function. *The Journal of Neuroscience*, *24*, 4340–4350. https://doi.org/10.1523/JNEUROSCI.0055-04.2004.
Zhu, D., Xiong, W. C., & Mei, L. (2006). Lipid rafts serve as a signaling platform for nicotinic acetylcholine receptor clustering. *The Journal of Neuroscience*, *26*, 4841–4851. https://doi.org/10.1523/JNEUROSCI.2807-05.2006.
Zikopoulos, B., & Barbas, H. (2010). Changes in prefrontal axons may disrupt the network in autism. *The Journal of Neuroscience*, *30*, 14595–14609. https://doi.org/10.1523/JNEUROSCI.2257-10.2010.
Ziv, N. E., & Smith, S. J. (1996). Evidence for a role of dendritic filopodia in synaptogenesis and spine formation. *Neuron*, *17*, 91–102. https://doi.org/10.1016/S0896-6273(00)80283-4.

Kolibri 360 GmbH
Ferdinand-Jühlke-Straße 7
99095 Erfurt, Deutschland
produktsicherheit@kolibri360.de